OPTICAL PROCESSES
IN SEMICONDUCTORS

OPTICAL PROCESSES IN SEMICONDUCTORS

Jacques I. Pankove
David Sarnoff Research Center
RCA Laboratories

Dover Publications, Inc.
New York

Published in Canada by General Publishing Company, Ltd., 30 Lesmill Road, Don Mills, Toronto, Ontario.
Published in the United Kingdom by Constable and Company, Ltd., 10 Orange Street, London WC 2.

This Dover edition, first published in 1975, is an unabridged republication, with slight corrections, of the work originally published by Prentice-Hall, Inc., Englewood Cliffs, New Jersey, in 1971.

International Standard Book Number: 0-486-60275-3
Library of Congress Catalog Card Number: 75-16756

Manufactured in the United States of America
Dover Publications, Inc.
180 Varick Street
New York, N.Y. 10014

to Ethel
and to Martin and Simon

PREFACE

This text has been the basis for a series of lectures presented at the University of California in Berkeley during the 1968–69 academic year. It grew out of a realization that in order to become familiar with all the phenomena involving light in semiconductors, one must consult a great variety of sources, search many journals, and then one finds that the material is packaged for the specialist. I had already done this task to nourish many years of experimental research. Hence it seemed desirable to weave together all this information into a coherent form, adding new concepts collected from the current literature and from recent conferences. In the process of writing this book, new insights evolved which are published here for the first time.

This book deals with the interactions among photons, electrons and atoms in semiconductor crystals. These interactions comprise the absorption, transformation, modulation and generation of light. In the range of phenomenological complexity, one always associates the most intricate processes with biological effects, whereas the physics of solids appears relatively simple. In fact, what could be simpler than a perfect crystal at low temperature where everything is still? One soon discovers excitons, complexes of excitons, polaritons. . . . Then, imperfections and impurities, which are unavoidable, produce new states. Thus, the diversity of possible interactions among electrons in the various levels, photons and phonons increases. The application of external influences such as pressure, temperature, electric and magnetic fields creates further perturbations of an already complex microcosm.

Spectroscopic data reveals a wealth of information about physical processes involving radiation. The position of an emission or absorption peak indicates the energy separating levels between which a strong interaction occurs. The lowest photon energy at which a spectral structure begins marks the threshold for a class of transitions, while the shape of the spectrum is a measure of the transition probability or of the distribution of states. Absorption links all the states which are empty and thus covers a broad spectrum. Emission, on the other hand, results from a nonequilibrium situation and since carriers relax to the lowest available states, such as a band edge, the emission spectrum covers a narrow range. When traps are involved, slow temperature-dependent effects are observed. Nonradiative recombination also often exhibits a temperature dependence. Light can excite electrons to a sufficiently high energy for them to overcome barriers and to be emitted into vacuum. Electron emission allows a study of states far from the band edges. Reflectance modulation is another technique for probing the distribution of states and for assigning definitive values to critical points in the band structure. Light can stimulate the absorption or desorption of atoms on the surface of semiconductors and engender other photochemical reactions. With the availability of lasers, it is possible to explore the scattering of intense monochromatic radiation by internal oscillators, thus shifting the characteristic frequency of these oscillations to a more convenient spectral range and gaining additional information about selection rules in the process. The practical applications of optical processes in semiconductors are too numerous for a detailed treatment in this text; however, the underlying principles have been covered. Electroluminescence and lasing are practical sources of radiation. The absorption of light is used in a variety of photodetectors, photoconductors, photovoltaic cells and photoemitters.

My purpose is to give the student an insight into the relevant phenomena already uncovered and to present them in a form that couples readily to physical intuition. Many references have been included so that the reader, fascinated by some of these effects, can be guided to greater depths and to a more rigorous treatment of the subject. Problems have been devised to help the student firmly grasp the material by having an active interaction with the concepts. Although this book is primarily a text for graduate students, I hope it will be useful also to the seasoned researcher who is curious about optical phenomena in semiconductors.

I am grateful to the Department of Electrical Engineering and Computer Sciences of the University of California in Berkeley and to its Chairman, Professor Kuh, for inviting me to this, my first teaching experience. The management of the RCA Laboratories has kindly made this undertaking possible. I acknowledge that I have benefited from "Student Power" which has requested my teaching this course two terms instead of one. The students' attentive interest was a strong encouragement to complete my writing. Technical discussions with Professors L. Falicov, and S. Wang have been most helpful. I am also grateful to C. Bulucea, and P. Hoff for a critical reading of the manuscript, and to Mrs. I. Brown, Miss B. Bulivent, and Miss K. Shields for painstakingly transcribing from my manuscript.

Four of the chapters were completed after my return to the RCA Laboratories. For a thorough revision of the first draft, I drew upon the diverse expertise of many colleagues: W. Burke, N. Byer, B. Faughnan, A. Firester, W. Fonger, B. Goldstein, W. F. Kosonocky, H. Kressel, J. D. Levine, A. Miller, A. R. Moore, D. Redfield, R. E. Simon, R. W. Smith, and B. F. Williams. It is a pleasure to acknowledge their kind help. I am also thankful to A. Bahraman for assistance in designing some of the problems.

Jacques I. Pankove

Princeton, N. J.

CONTENTS

1 **ENERGY STATES IN SEMICONDUCTORS 1**

1-A Band Structure 1
1-A-1 Banding of Atomic Levels 1
1-A-2 Distribution in Momentum Space 3
1-A-3 Density-of-States Distribution 6
1-A-4 Carrier Concentration 7
1-B Impurity States 8
1-C Band Tailing 10
1-D Excitons 12
1-D-1 Free Excitons 12
1-D-2 Excitonic Complexes 14
1-D-3 Polaritons 16
1-E Donor–Acceptor Pairs 17
1-F States in Semiconducting Alloys 18

2 **PERTURBATION OF SEMICONDUCTORS BY EXTERNAL PARAMETERS 22**

2-A Pressure Effects 22
2-A-1 Hydrostatic Pressure 22
2-A-2 Uniaxial Strain 26
2-B Temperature Effects 27
2-C Electric-Field Effects 28
2-C-1 Stark Effects 28
2-C-2 Franz-Keldysh Effect 29
2-C-3 Ionization Effects 29
2-D Magnetic-Field Effects 30
2-D-1 Landau Splitting 30
2-D-2 Zeeman Effect 32

3 **ABSORPTION 34**

3-A Fundamental Absorption 34
3-A-1 Allowed Direct Transitions 35

3-A-2 Forbidden Direct Transitions 36
3-A-3 Indirect Transitions between Indirect Valleys 37
3-A-4 Indirect Transitions between Direct Valleys 42
3-A-5 Transitions between Band Tails 43
3-A-6 Fundamental Absorption in the Presence of a Strong Electric Field 46
3-B Higher-energy Transitions 52
3-C Exciton Absorption 57
3-C-1 Direct and Indirect Excitons 57
3-C-2 Exciton Absorption in the Presence of an Electric Field 60
3-D Absorption due to Isoelectronic Traps 61
3-E Transitions between a Band and an Impurity Level 62
3-F Acceptor-to-Donor Transitions 66
3-G Intraband Transitions 67
3-G-1 p-Type Semiconductors 67
3-G-2 n-Type Semiconductors 71
3-H Free-carrier Absorption 74
3-I Lattice Absorption 76
3-J Vibrational Absorption of Impurities 80
3-K Hot-Electron-Assisted Absorption 81

4 RELATIONSHIPS BETWEEN OPTICAL CONSTANTS 87

4-A Absorption Coefficient 87
4-B Index of Refraction 88
4-C The Kramers–Kronig Relations 89
4-D Reflection Coefficient 90
4-E Determination of Carrier Effective Mass 91
4-F Plasma Resonance 92
4-G Transmission 93
4-H Interference Effects 94

5 ABSORPTION SPECTROSCOPY 96

6 RADIATIVE TRANSITIONS 107

6-A The Van Roosbroeck–Shockley Relation 108
6-B Radiative Efficiency 111
6-C The Configuration Diagram 113
6-D Fundamental Transitions 114
6-D-1 Exciton Recombination 114
6-D-2 Conduction-Band-to-Valence-Band Transitions 124

6-E Transition between a Band and an Impurity Level 131
6-E-1 Shallow Transitions 131
6-E-2 Deep Transitions 132
6-E-3 Transitions to Deep Levels 138
6-F Donor–Acceptor Transitions 143
6-F-1 Spectral Structure 143
6-F-2 Transition Probability 147
6-F-3 Time Dependence of Donor-to-Acceptor Transitions 152
6-G Intraband Transition 154

7 NONRADIATIVE RECOMBINATION 160

7-A Auger Effect 161
7-B Surface Recombination 164
7-C Recombination through Defects or Inclusions 165
7-D Configuration Diagram 166
7-E Multiple-Phonon Emission 167

8 PROCESSES IN p-n JUNCTIONS 170

8-A Nature of the *p-n* Junction 170
8-A-1 The Depletion Layer 171
8-A-2 Junction Capacitance 173
8-A-3 Electric Field in the p-n Junction 174
8-B Forward-bias Processes 174
8-B-1 Band-to-Band Tunneling 175
8-B-2 Photon-Assisted Tunneling 177
8-B-3 Injection 180
8-B-4 Tunneling to Deep Levels 181
8-B-5 Donor-to-Acceptor Photon-Assisted Tunneling 183
8-B-6 Band Filling 189
8-B-7 Injection Luminescence in Lightly Doped Junctions 192
8-B-8 Optical Refrigeration 193
8-C Heterojunctions 197
8-D Reverse-Bias Processes 201
8-D-1 Saturation Current and Photoconductivity 201
8-D-2 Zener Breakdown 203
8-D-3 Avalanche Breakdown 203

9 STIMULATED EMISSION 213

9-A Relationship between Spontaneous and Stimulated Emission 213

9-B Criteria for Lasing in a Semiconductor 215

10 SEMICONDUCTOR LASERS 222

10-A Cavity and Modes 222
10-B Waveguiding Properties of the Active Region 225
10-C Far-Field Pattern 227
10-D Temperature Dependence 229
10-D-1 Effect of the Cavity 229
10-D-2 Temperature Dependence of Losses, of Efficiency and of Threshold Current Density 229
10-D-3 Power Dissipation 233
10-E Optimum Design for Injection Laser 234
10-F Influence of a Magnetic Field 238
10-G Pressure Effects 239

11 EXCITATION OF LUMINESCENCE AND LASING IN SEMICONDUCTORS 242

11-A Electroluminescence 242
11-A-1 Forward Biased p-n Junction 242
11-A-2 Forward-Biased Surface Barrier 245
11-A-3 Tunneling through an Insulating Layer 247
11-A-4 Bulk Excitation by Impact Ionization 247
11-B Optical Excitation 249
11-C Electron-Beam Excitation 252

12 PROCESSES INVOLVING COHERENT RADIATION 258

12-A Photon–Photon Interactions in Semiconductors 258
12-A-1 Quenching of a Laser by Another Laser 259
12-A-2 Amplification 261
12-A-3 Harmonic Generation 265
12-A-4 Two-photon Absorption 268
12-A-5 Frequency Mixing 270
12-B Photon–Phonon Interactions in Semiconductors 271
12-B-1 Raman Scattering 273
12-B-2 Brillouin Scattering 275
12-C Optical Properties of Acoustoelectric Domains 278
12-C-1 The Acoustoelectric Effect 279
12-C-2 Light Transmission at Acoustoelectric Domain 281
12-C-3 Light Emission by Acoustoelectric Domain 283
12-C-4 Brillouin Scattering Studies of Acoustoelectric Domains 285

13 PHOTOELECTRIC EMISSION 287

13-A Threshold for Emission 287
13-B Photoelectric Yield 289
13-C Effect of Surface Conditions 294
13-D Energy Distribution of Emitted Electrons 299

14 PHOTOVOLTAIC EFFECTS 302

14-A Photovoltaic Effect at p-n Junctions 302
14-A-1 Electrical Characteristics 302
14-A-2 Spectral Characteristics 305
14-A-3 The Solar Cell 307
14-B Photovoltaic Effects at Schottky Barriers 312
14-B-1 The Schottky Barrier 312
14-B-2 Photo-Effects 314
14-B-3 Particle Detectors 319
14-C Bulk Photovoltaic Effects 320
14-C-1 Dember Effect 320
14-C-2 Photomagnetoelectric Effect 321
14-D Anomalous Photovoltaic Effect 323
14-D-1 Characteristics of Anomalous Photovoltaic Cells 324
14-D-2 Conditions for Obtaining the Anomalous Photovoltaic Effect 325
14-D-3 Models for the Anomalous Photovoltaic Effect 327
14-D-4 Angular Dependence of Photovoltaic Effects 329
14-E Other Photovoltaic Effects 331
14-E-1 Lateral Photoeffect 331
14-E-2 Optically Induced Barriers 331
14-E-3 Photovoltaic Effect at a Graded Energy Gap 333

15 POLARIZATION EFFECTS 337

15-A Birefringence 337
15-A-1 Birefringence in Uniaxial Crystals 338
15-A-2 Elliptical Polarization 339
15-A-3 Birefringence in Biaxial Crystals 340
15-B Induced Optical Anisotropy 344
15-B-1 Electro-Optic Kerr Effect 344
15-B-2 Pockels Effect or Linear Electro-Optic Effect 345
15-B-3 Faraday Effect 346
15-B-4 Voigt Effect 347

15-B-5 Strain-Induced Birefringence 347
15-B-6 Deflection and Modulation of a Light Beam 348

16 PHOTOCHEMICAL EFFECTS 352

16-A Photochemistry with a Gaseous Ambient 353
16-A-1 Surface States 353
16-A-2 Adsorption and Desorption 354
16-A-3 Photocatalysis 358
16-A-4 Spectroscopic Analysis of Adsorbed Species 358
16-A-5 Epitaxial Growth 360
16-B Photochemistry with a Liquid Ambient 361
16-B-1 Chemical Etching 361
16-B-2 Electrolytic Etching 363
16-B-3 Plating 363
16-C Photochemical Reactions Inside the Crystal 365
16-C-1 Photo-Induced Annealing 365
16-C-2 Photochromism 366

17 EFFECT OF TRAPS ON LUMINESCENCE 370

17-A Growth and Decay of Luminescence 370
17-B Thermoluminescence 371
17-C Infrared Stimulated Luminescence 376
17-D Quenching of Luminescence 376
17-E Trapping Effects in Lasers 377
17-E-1 Time Delay in Lasers 377
17-E-2 Trap Storage Time 379
17-E-3 Traps as Saturable Absorbers 379
17-E-4 Temperature Dependence of Trapping in GaAs Injection Lasers 383
17-E-5 Double-Acceptor Model 385
17-E-6 Internal Q-Switching 386
17-F Triboluminescence 387
17-F-1 Strain-Excited Luminescence 387
17-F-2 Strain-Stimulated Luminescence 389
17-F-3 Fracture Luminescence 390

18 REFLECTANCE MODULATION 391

18-A Dependence of Reflectance on the Band Structure 392
18-B Reflectance-modulation Techniques 396
18-B-1 Electroreflectance 396
18-B-2 Optical Modulation of Reflectance 399
18-B-3 Cathodoreflectance Modulation 399

18-B-4 Piezoreflectance Modulation 400
18-B-5 Thermoreflectance Modulation 401
18-B-6 Wavelength Modulation 401
18-C Some Results 402

APPENDICES 409

I Table of Symbols 410
II Properties of Semiconductors 412
III Nomograph of the Temperature Dependence of the Fermi Level in a Degenerate Parabolic Band 414
IV Physical Constants 416

INDEX 417

OPTICAL PROCESSES
IN SEMICONDUCTORS

ENERGY STATES IN SEMICONDUCTORS

1

In this chapter we shall sketch how the assemblage of similar atoms into an array leads to the formation of bands of allowed states separated by an energy gap. Then we shall show that the energy gap can be filled with a great variety of allowed states, some localized due to impurities and others permeating the crystal (excitons). We shall also describe how the various particles can interact to form complexes.

The relevance of these levels to the optical properties of the semiconductors lies in the fact that optical effects deal with transitions between various states and, therefore, it is well to review these first and to see how they come about.

1-A Band Structure

1-A-1 BANDING OF ATOMIC LEVELS

To understand the nature of semiconductors one must consider what happens when similar atoms are brought together to form a solid such as a crystal. As two similar atoms approach each other the wave functions of their electrons begin to overlap. To satisfy Pauli's exclusion principle, the states of all spin-paired electrons acquire energies which are slightly different from their values in the isolated atom. Thus if N atoms are packed within a range of interaction, $2N$ electrons of the same orbital can occupy $2N$ different states, forming a band of states instead of a discrete level as in the isolated atom.

The energy distribution of the states depends strongly on the interatomic

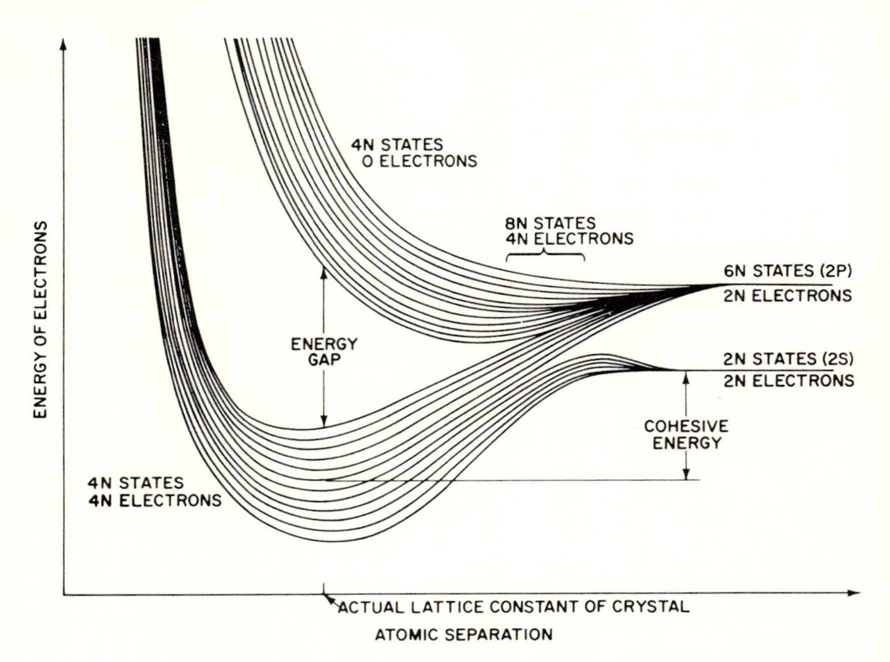

Fig. 1-1 Energy banding of allowed levels in diamond as a function of spacing between atoms.[1]

distance. This is illustrated in Fig. 1-1 for an assemblage of carbon atoms. The lowest-energy states are depressed to a minimum value when the diamond crystal is formed. The average amount by which the potential energy has dropped is related to the cohesive energy of the crystal. Notice that some of the higher-energy states (2P) merge with the band of 2S states. As a result of this mixing of states, the lower band contains as many states as electrons. This band is called the valence band and is characterized by the fact that it is completely filled with electrons. Such a filled band cannot carry a current. The upper band of states, which contains no electron, is called the conduction band. If an electron were placed in this band, it could acquire a net drift under the influence of an electric field.

Clearly, since in the energy gap there are no allowed states, one would not expect to find an electron within that range of energies.

It is the extent of the energy gap and the relative availability of electrons that determine whether a solid is a metal, a semiconductor, or an insulator. In a semiconductor the energy gap usually extends over less than about three electron-volts and the density of electrons in the upper band (or of holes in

[1] G. E. Kimball, *J. Chem. Phys.* **3**, 560 (1935).

the lower band) is usually less than 10^{20} cm^{-3}. By contrast, in a metal the upper band is populated with electrons far above the energy gap and the electron concentration is of the order of 10^{23} cm^{-3}. Insulators, on the other hand, have a large energy gap—usually greater than 3 eV—and have a negligible electron concentration in the upper band (and practically no holes in the lower band).

Since the interatomic distance in a crystal is not isotropic but rather varies with the crystallographic direction, one would expect this directional variation to affect the banding of states. Thus, although the energy gap which characterizes a semiconductor has the same minimum value in each unit cell, its topography within each unit cell can be extremely complex.

1-A-2 DISTRIBUTION IN MOMENTUM SPACE

We have just seen that allowed states have definite energy assignments. Now we must consider how the allowed states are distributed in momentum space. The importance of this consideration will be evident later when we find that in optical transitions we must conserve both energy and momentum.

The kinetic energy of an electron is related to its momentum p by the classical relation:

$$E = \frac{p^2}{2m^*} \tag{1-1}$$

where m^* is the electron effective mass (which may be different from the value in vacuum). From quantum mechanics we have the following expression:

$$p = k\hbar \tag{1-2}$$

where $\hbar$ is Dirac's constant $= h/2\pi$, h being Planck's constant; and k is the wave vector. Because of the relation (1-2), and to better couple to classical intuition, we shall call k the "momentum vector." If we conceive of the crystal as a square well potential with an infinite barrier and a bottom of width L, we shall find that k can have the discrete values $k = n(\pi/L)$, where n is any nonzero integer. Note that L is an integral number N of unit lattice cells having a periodicity, a. Therefore, a is the smallest potential well one could construct. Hence, when $n = N$, $k = \pi/a$ is the maximum significant value of k. This maximum value occurs at the edge of the Brillouin zone. A Brillouin zone is the volume of k-space containing all the values of k up to π/a, where a varies with direction. Larger values of the momentum vector k' just move the system in to the next Brillouin zone, which is identical to the first zone and, therefore, the system can be treated as having a momentum-vector $k = k' - \pi/a$. The kinetic energy of the electron can be expressed as

$$E = \frac{k^2\hbar^2}{2m^*} \tag{1-3}$$

If the whole crystal, a cube whose sides have a length L, is the potential well, the allowed energies are

$$E = \frac{h^2}{2m^*L^2}(n_x^2 + n_y^2 + n_z^2) \tag{1-4}$$

Although E varies in discrete steps, since the quantum numbers n are integers, the steps are so small ($\sim 10^{-18}$ eV for a 1-cm^3 crystal) that E appears as a quasi-continuum.

Let us first consider how the energy varies with momentum along one direction of momentum space. Figure 1-2 illustrates the parabolic dependence

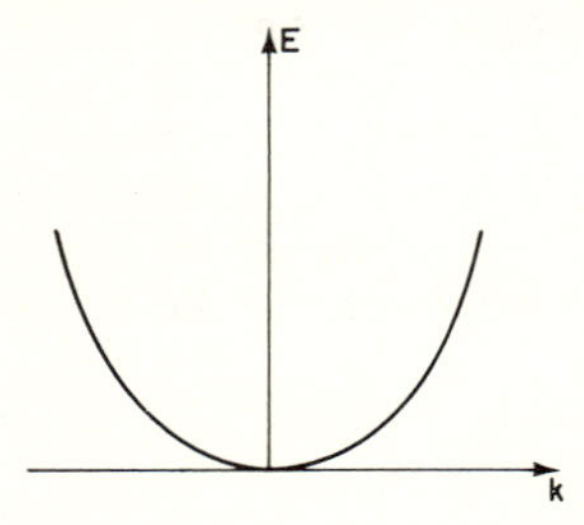

Fig. 1-2 Parabolic dependence of energy vs. momentum.

of E on k. Hence such a distribution of states is called a parabolic valley—the pictorial impact is even more pronounced in a three-dimensional representation of E vs. k_x and k_y. In a three-dimensional momentum space, a constant-energy surface forms a closed shell and, with every increment in momentum, the energy of successive shells increases quadratically. One often takes the top of the valence bands as the reference level. Then the bottom of the conduction band is located at a higher potential corresponding to the energy gap (Fig. 1-3). The significance of the downward curvature of the

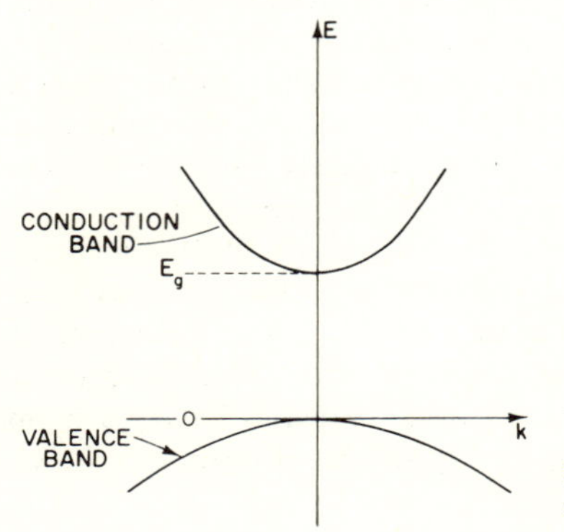

Fig. 1-3 Energy vs. momentum in a direct-gap two-band system.

valence band is that if electrons could have a net motion in the valence band (if it were not completely filled), the electrons would be accelerated in the direction opposite to that in which they would move if they were in the conduction band, as if they had negative mass.

The separation between nearest atoms varies in different directions. Therefore, the shape of a constant-energy surface must deviate from that of a perfect sphere.[2,3] Furthermore, because of the cumulative interactions from nearest neighbors, next-nearest neighbors, and all the higher-order neighbors, the minimum of the valley may occur not at $k_x = k_y = k_z = 0$ but at some point defining a specific crystallographic direction such as [111] (Fig. 1-4). Because of crystal symmetry, the same distribution $E(k_x, k_y, k_z)$

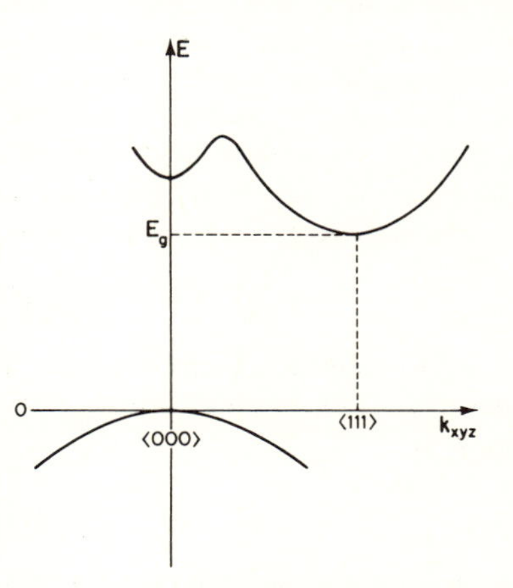

Fig. 1-4 Energy vs. momentum diagram for a semiconductor with conduction band valleys at $k = \langle 000 \rangle$ and $k = \langle 111 \rangle$.

must be repeated at all equivalent directions. Thus there can be four or eight $\langle 111 \rangle$ valleys and three or six $\langle 100 \rangle$ valleys. The lower number is obtained

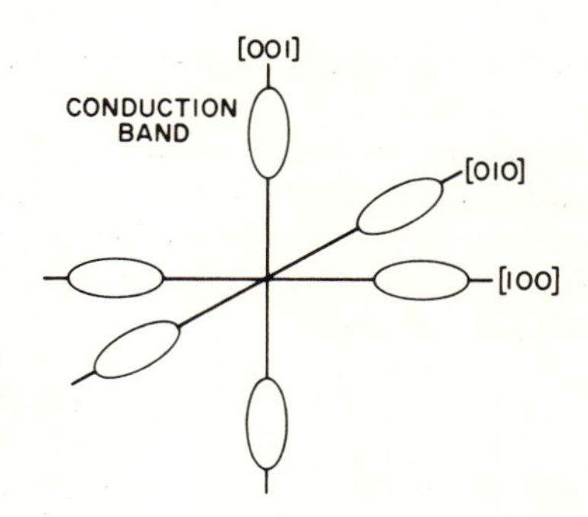

Fig. 1-5 Constant-energy contours of valleys in the conduction band edge of silicon, forming six ellipsoidal valleys in the [100] directions.

[2] F. Herman, *Proc. IRE* **43,** 1703 (1955).
[3] R. N. Dexter, H. Zeiger, and B. Lax, *Phys. Rev.* **104** , 637 (1956).

when the valleys occur at the edges of the Brillouin zone (at $k = \pi/a$, a is the lattice constant) and are shared by adjacent zones. For example, in germanium the four $\langle 111 \rangle$ valleys consist of cigar-shaped ellipsoids (the parabolicity is not isotropic), and their longitudinal axis is oriented along the [111] directions. The higher number of valleys is obtained when the valleys are inside the Brillouin zone (for example, the $\langle 100 \rangle$ valleys of silicon, as in Fig. 1-5).

1-A-3 DENSITY-OF-STATES DISTRIBUTION

In momentum space the density of allowed points is uniform. The surfaces of constant energy are, to first approximation, spherical (isotropic-medium); then the volume of k-space between spheres of energy E and $E + dE$ is $4\pi k^2\, dk$. Here E is measured with respect to the edge of the parabolic band. Since a single state occupies in momentum space a volume $8\pi^3/V$ (V is the actual volume of the crystal) and there are two states per level, one finds that the number of energy states in the interval E and $E + dE$ is

$$N(E)\, dE = \frac{1}{2\pi^2 \hbar^3}(2m^*)^{3/2} E^{1/2}\, dE \tag{1-5}$$

where m^* is the electron effective mass. For convenience, V is taken as a unit volume (e.g., 1 cm^3 in cgs units). The total density of states up to some energy E is

$$N = \frac{1}{3\pi^2}\frac{1}{\hbar^3}(2m^* E)^{3/2} \tag{1-6}$$

Since, in general, the valleys are rotational ellipsoids instead of spherical surfaces, the effective mass is not isotropic; then an average density of state effective mass is used:

$$m^* = (m_l^* m_{t_1}^* m_{t_2}^*)^{1/3} \tag{1-7}$$

where m_l^* is the longitudinal effective mass and $m_{t_1}^*$ and $m_{t_2}^*$ are the two transverse masses.

Each valley contributes its own set of states; therefore, each energy level may consist of states from several valleys. Hence to find the density of states one must add the contributions of all the valleys. Thus in a multivalley semiconductor the number of states between, say, the bottom of the conduction band and some energy E is

$$N = \frac{1}{3\pi^2 \hbar^3} \sum_j g_j (2m_j^*)^{3/2} (E - E_j)^{3/2} \tag{1-8}$$

where g_j is the number of valleys of type j, m_j^* is the average effective mass of a j-valley, and E_j is the energy at the bottom of the j-valley. A similar treatment applies to states in the valence band.

We have stated above that the effective mass is usually not perfectly isotropic. We shall now blur this picture somewhat further by pointing out that the valleys are parabolic in energy–momentum space only over a limited range near the bottom of the valley. This limitation could be expected, since the quasi-continuum of states makes a gradual connection between all the valleys (Fig. 1-4). Furthermore, spin–orbit interactions induce a perturbation which results in subbands and in deviations from parabolicity at potentials away from the edge of the valley. However, the assumption of a parabolic band is usually a good first approximation, and we shall see later that in practice the subtleties of the theoretical model are obscured by the imperfections of nature.

1-A-4 CARRIER CONCENTRATION

So far we have dealt with band states and their distribution in energy and momentum space. Now we must consider the occupancy of these states. When photons are interacting with electrons, the intensity of the interaction will depend on the number of electrons involved. The density of electrons is simply the product of the density of states and the Fermi–Dirac function (see Fig. 1-6):

$$f(E) = \frac{1}{\exp \frac{(E - E_F)}{kT} + 1} \tag{1-9}$$

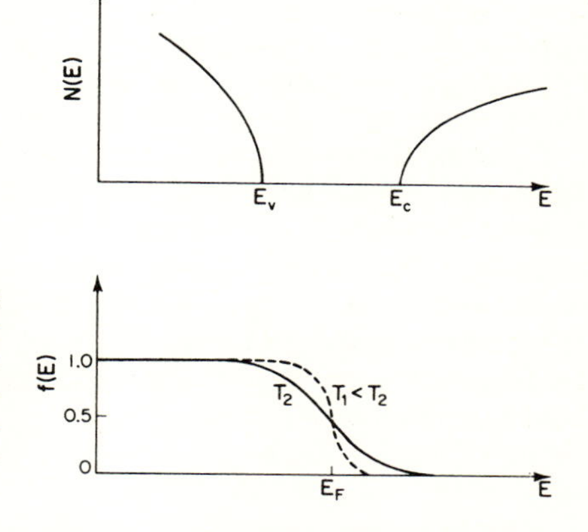

Fig. 1-6 The upper diagram shows the variation of the density of states near the energy gap; the lower diagram shows the Fermi–Dirac function at two temperatures. The product of the two ordinates determines the electron concentration at various energies.

where k is Boltzmann's constant and T is the absolute temperature; E_F, the Fermi level, is the energy at which the expectation of finding a state occupied by an electron is $\frac{1}{2}$. Pauli's principle allows each state to be occupied by at most two electrons; however, a given energy level may consist of more than one state (in this case it is said to be degenerate). The density of holes is the product of the density of states and the probability of the state being empty. This probability is given by

$$1 - f(E) = \frac{1}{\exp\left(\frac{E_F - E}{kT}\right) + 1} \tag{1-10}$$

1-B Impurity States

When an impurity atom is introduced in a lattice, it produces several types of interactions. If the impurity atom replaces one of the constituent atoms of the crystal and provides the crystal with one or more additional electrons than the atom it replaced, the impurity is a donor. Thus As on a Ge-site in a germanium crystal is a donor, and Te on an As-site in GaAs is a donor, as is Si on a Ga-site in GaAs. If the impurity atom provides less electrons than the atom it replaces, it forms an acceptor (e.g., Zn on a Ga-site in GaAs, or Si on an As-site in GaAs).

Instead of replacing an atom of the host crystal, the impurity may lodge itself in an interstitial position. Then its outer-shell electrons are available for conduction and the interstitial impurity is a donor.

A missing atom results in a vacancy and deprives the crystal of one electron per broken bond. This makes the vacancy an acceptor. Vacancies and interstitial impurities often combine to form a molecular impurity which may be either a donor or an acceptor.

In compound semiconductors a deviation from stoichiometry generates donors or acceptors depending on whether it is the cation or the anion which is in excess. However, it has been shown that in PbTe it is not the excess ion but rather the vacancy which determines whether the material is *n*-type or *p*-type.[4] Accordingly, a Pb-vacancy in the Te-rich PbTe transfers two states from the valence band to the conduction band; since 4 electrons are associated with each Pb atom, the Pb-vacancy leaves two holes in the valence band, which makes the semiconductor *p*-type. On the other hand, a Te-vacancy in Pb-rich material transfers 8 levels from the valence to the conduction band and removes the 6 Te-electrons. Hence the two electrons which no longer can be accommodated in the valence band occupy the lowest two states of the conduction band, making the Pb-rich PbTe *n*-type.

The extra electron of the donor is attracted most strongly to the positive charge of the impurity nucleus. Thus it acts as the electron of a hydrogen

[4] N. J. Parada and G. W. Pratt, Jr., *Phys. Rev. Letters* **22**, 180 (1969).

atom immersed in the high dielectric constant ϵ of the crystal. This enables us to calculate the energy binding the electron to the impurity, i.e., its ionization energy:

$$E_i = \frac{m^* q^4}{2h^2\epsilon^2 n^2} = \frac{m^*}{m\epsilon^2 n^2} 13.6 \text{ eV} \tag{1-11}$$

where q is the electron charge, m is the mass of the electron in vacuum, and n is a quantum number ≥ 1. Ionization from successively higher quantum states requires rapidly decreasing increments in energy. The ionization energy from the ground state to the conduction band is obtained by making $n = 1$ in Eq. (1-11). Since ϵ is of the order of 10 and the effective-mass ratio is less than 1, the ionization energy is usually less than 0.1 eV. When the electron of the donor is in the conduction band, it is essentially free; therefore, its ground state, the donor level, is one ionization energy below the conduction band. Similarly, the acceptor level is one binding or ionization energy above the valence band. Note that since the effective masses of electrons and holes are usually different, the donor and acceptor binding energies can be different. It should be pointed out that the hydrogenic model is a very crude approximation because the effective mass varies considerably around the impurity atom.

Along with the reduced binding energy of a hydrogen-like impurity imbedded in a high dielectric medium, one finds that the electron orbit around the impurity atom becomes very large. If one makes the analogy with a hydrogen-like atom, the radius of the first Bohr obit is

$$a = \frac{\hbar^2 \epsilon}{q^2 m^*} = \frac{\epsilon}{m^*/m} a_0 \tag{1-12}$$

where a_0 is the radius of the first Bohr orbit of hydrogen (equal to 0.53×10^{-8} cm). Hence the electron bound to the donor is not localized at the donor but rather travels through many lattice sites in the neighborhood of the impurity.

As the impurity concentration is increased, the electron wave functions at the impurity level begin to overlap. This obviously happens at a concentration of the order of $1/a^3$ (i.e., 10^{20} cm^{-3}). In practice, the wave-function overlap occurs already at impurity concentrations as low as 10^{16} cm^{-3}. An overlap of wave functions is an interaction which changes slightly the potential of each level, resulting in the formation of a band of states in the region of overlap.[5] As the impurity concentration is increased further, the impurity band broadens and eventually merges with the nearest intrinsic band.

When the impurity atom can contribute more than one extra carrier (electron or hole), it is called a multiple donor or a multiple acceptor. The

[5]F. Stern and R. M. Talley, *Phys. Rev.* **100**, 1638 (1955).

multiple impurity has a state for each carrier it can contribute. Obviously, when the multiple donor has released one electron it is singly ionized. When it is doubly ionized, the donor is doubly charged and, therefore, the corresponding binding energy is much greater than for the singly ionized state. Hence as the degree of ionization increases, the various donor levels go deeper below the conduction-band edges.

Some impurities do not agree, by far, with the simple hydrogen model, and form levels which may lie deep in the energy gap. All the transition elements seem to form deep levels. The reasons for which certain impurities form a deep level are not yet completely understood.

1-C Band Tailing

While impurity-band formation is an obvious consequence of increased impurity concentration, another important effect occurs: a perturbation of the bands by the formation of tails of states extending the bands into the energy gap. The problem of band tailing has received much theoretical attention.[6-13] An ionized donor exerts an attractive force on the conduction electrons and a repulsive force on the valence holes (acceptors act conversely). Since impurities are distributed randomly in the host crystal, the local interaction will be more or less strong depending on the local crowding of impurities (Fig. 1-7). It should be noted that the local energy gap—the separation between the top of the valence band and the bottom of the conduction band—is everywhere maintained constant. But the density-of-states distribution which integrates the number of states at each energy inside the whole volume shows that there are conduction-band states at relatively low potentials and valence-band states in high-potential regions. It must be remembered that, in this model, the states of each tail are spatially separated, as is evident on the left side of Fig. 1-7.

Deep impurity states move up and down with the potential of the associated band edge (e.g., acceptors move with the valence band edge). Hence at high concentrations, the impurity states form a band whose distribution tails into the energy gap like the associated band edge.[14]

[6] W. Baltensperger, *Phil. Mag.* **44**, 1355 (1953).

[7] P. Aigrain, *Physica* **20**, 978 (1954).

[8] R. H. Parmenter, *Phys. Rev.* **97**, 587 (1955).

[9] V. L. Bonch-Bruevich, *Proc. 1962, Int. Conf. Phys. Semiconductors*, Exeter, p. 216; *Soviet Phys. Solid-State* **5**, 1353 (1964); *The Electronic Theory of Heavily Doped Semiconductors*, American Elsevier (1966).

[10] P. A. Wolff, *Proc. 1962, Int. Conf. Phys. Semiconductors*, Exeter, p. 220.

[11] E. M. Conwell and B. W. Levinger *Proc. 1962, Int. Conf. Phys. Semiconductors*, Exeter, p. 227.

[12] E. O. Kane, *Phys. Rev.* **131**, 79 (1963).

[13] B. I. Halperin and M. Lax, *Phys. Rev.*, **148**, 772 (1966).

[14] T. N. Morgan, *Phys. Rev.* **139**, A343 (1965).

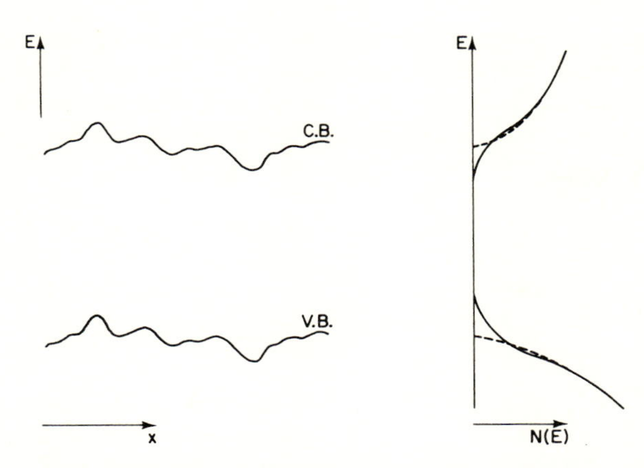

Fig. 1-7 The left diagram shows the perturbation of the band edges by Coulomb interaction with inhomogeneously distributed impurities. This leads to the formation of tails of states shown on the right side. The dashed lines show the distribution of states in the unperturbed case.

There is still another type of interaction between impurities and the surrounding crystal: the deformation potential.[15] Since the impurity is usually either larger or smaller than an atom of the host lattice, a local mechanical strain is obtained (a compression as in Fig. 1-8, or a dilation). As is evident

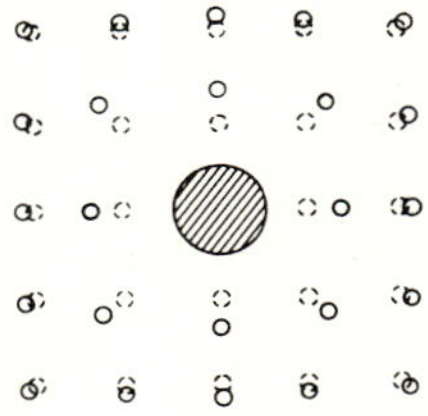

Fig. 1-8 Compressional strain induced by the incorporation of a large impurity.

from Fig. 1-1, in some materials compression will increase the energy gap and dilation will reduce it. This type of interaction will, therefore, also perturb the band edges. An interstitial atom evidently induces a deformation potential corresponding to compressional strain, whereas a vacancy will have the opposite effect, since it produces dilational strain. Usually, both interstititals and vacancies are present in addition to substitutional impurities.

Dislocations are also usually present in crystals. They occur at the edge of an extra plane of atoms. The misfit of such an extra plane results in compressional and dilational strains, with the consequent onset of both lowering

[15] W. Shockley and J. Bardeen, *Phys. Rev.* **77**, 407 (1950).

and raising of the potentials in the neighborhood of the dislocation (Fig. 1-9).

Hence we can say that impurities will induce tails in the density states by perturbing the band edge via deformation potential, via coulomb interaction, and by forming a band of impurity states.

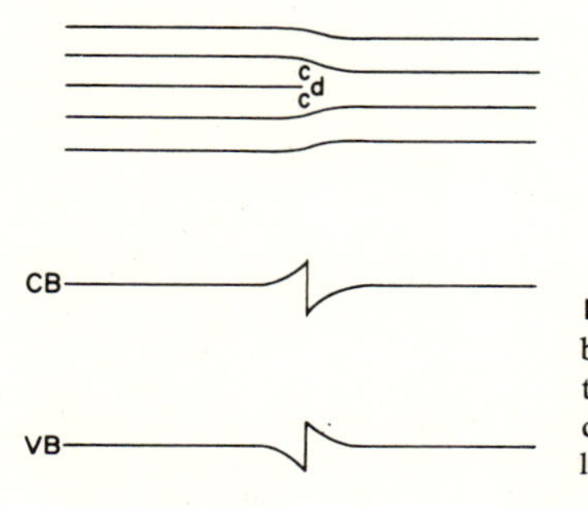

Fig. 1-9 An edge dislocation produces both compressional strains (c) and dilational strain (d) which result in the deformation potential shown in the lower diagram.

1-D Excitons[16]

1-D-1 FREE EXCITONS

A free hole and a free electron as a pair of opposite charges experience a coulomb attraction. Hence the electron can orbit about the hole as if this were a hydrogen-like atom. The ionization energy for such a system is then

$$E_x = \frac{-m_r^* q^4}{2h^2\epsilon^2}\frac{1}{n^2}$$

where n is an integer ≥ 1 indicating the various exciton states and m_r^* is the reduced mass:

$$\frac{1}{m_r^*} = \frac{1}{m_e^*} + \frac{1}{m_h^*}$$

m_e^* and m_h^* being the electron and hole effective masses, respectively. In an impurity atom—say, a donor—or in the hydrogen atom, the effective mass of the nucleus is very large and, therefore, the reduced mass is equal to that of the electron. But in an exciton the reduced mass is lower than the effective mass of the electron because m_e^* and m_h^* are more nearly of the same order of magnitude. Hence we should expect the exciton binding energy to be lower than either the donor or the acceptor binding energies.

The exciton can wander through the crystal (the electron and the hole

[16] A detailed discussion of excitons can be found in R. S. Knox, "Theory of Excitons," *Solid State Physics*, ed. F. Seitz and D. Turnbull, Academic Press (1963), Supplement 5.

are now only relatively free because they are associated as a mobile pair). Because of this mobility, the exciton is not a set of spatially localized states. Furthermore, the exciton states do not have a well-defined potential in the semiconductor's energy diagram. However, it is customary to use the conduction-band edge as a reference level and to make this edge the continuum state ($n = \infty$). Then the various states of the exciton are represented as shown in Fig. 1-10.

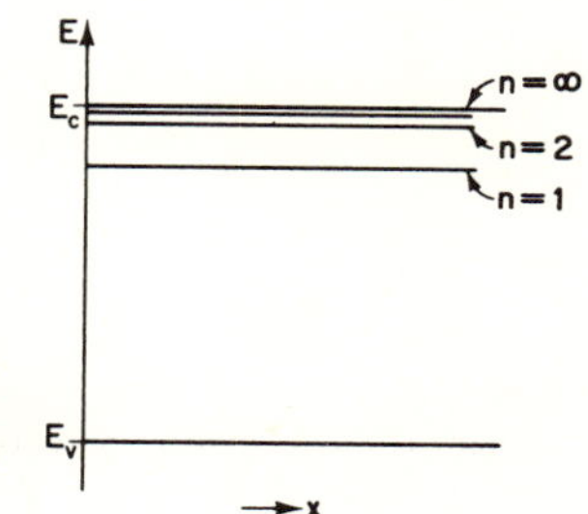

Fig. 1-10 Energy level diagram for the exciton and its excited states, exciton energy being referred to the edge of the conduction band.

Note that when a free electron and a free hole have the same momentum k, in general they move with different velocities: $\hbar(dE_c/dk)$ for the electron and $\hbar(dE_v/dk)$ for the hole (E_c referring to the conduction band and E_v to the valence band). Since the electron and the hole of an exciton must move together through the crystal, their translational velocities must be identical. This condition places a restriction on the regions in $(E - k)$-space where excitons can be found, namely at the "critical points":

$$\left[\frac{dE}{dk}\right]_{\text{electron}} = \left[\frac{dE}{dk}\right]_{\text{hole}}$$

Since the effective mass of the hole is many orders of magnitude smaller than that of the proton, the analogy to the hydrogen atom must be modified: the center of gravity of the exciton may be located many lattice spaces away from the hole. The moving exciton has a kinetic energy

$$\frac{h^2 K^2}{2(m_e^* + m_h^*)}$$

where K is the momentum vector associated with the motion of the center of gravity. The addition of the kinetic energy means that exciton levels are slightly broadened into bands.

At high electron and hole concentrations, electron–electron and hole–hole coulombic repulsion tend to reduce the range over which the attractive coulomb interaction can occur (screened coulomb interaction), but pairing still can occur.[17] However, at high doping, the potential fluctuations of the band edges generate internal fields. Local fields in the semiconductor exert

[17]G. D. Mahan, *Phys. Rev.* **153**, 882 (1967).

a force on the electron and the hole separately. These forces can act in opposite directions for the electron and the hole as shown in Fig. 1-11(a). When the intensity of the local field exceeds the coulomb field inside the exciton, the exciton dissociates.

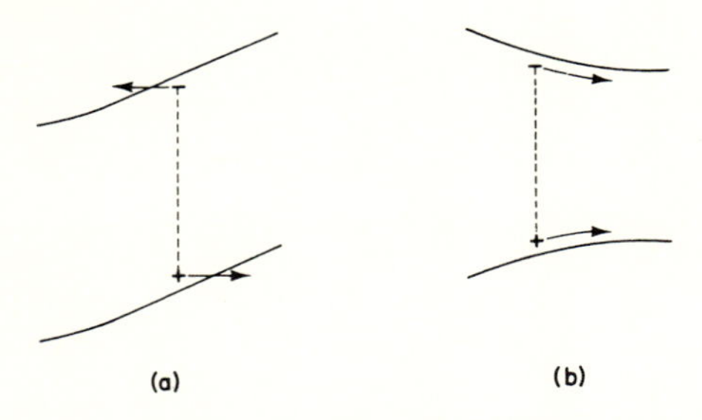

Fig. 1-11 Exciton in a region of perturbed band potentials: (a) strong local field; (b) deformation potential.

However, when the local field is due to a deformation potential [Fig. 1-11(b)], the forces act on the electron and the hole in the same direction. These forces cause the exciton to drift to the region of minimum-energy gap without breaking up.

Note that when the exciton dissociates, it creates a free electron and a free hole. When the lifetime of an exciton is very short, the energies of the exciton states are broadened via the uncertainty principle.

1-D-2 EXCITONIC COMPLEXES

It is conceivable that three or more particles combine to form ion-like or molecule-like complexes.[18] The simplest set of possible complexes is reproduced below:

$\oplus$	$+$	$-$		H_2^+	excitonic ions
$\oplus$	$-$	$-$		H^-	
$+$	$+$	$-$		H_2^+	
$+$	$+$	$-$	$-$	H_2	excitonic molecules
$\oplus$	$+$	$-$	$-$	H_2	

where $\oplus$ represents a donor, $+$ represents a hole, and $-$ represents an electron. A similar table could be made with an acceptor $\ominus$. Additional distinctions can be made in all the above models, depending on the ratio of the effective masses m_h^*/m_e^*. We shall only allude to the extreme complexity that might be expected: each electronic level has a fine structure corresponding

[18] M. A. Lampert, *Phys. Rev. Letters* **1**, 450 (1958).

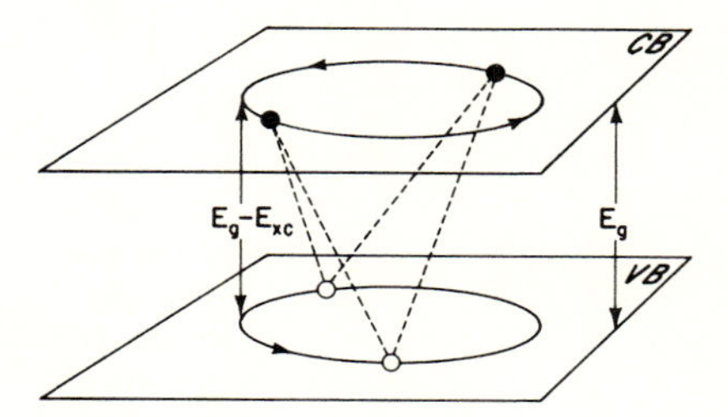

Fig. 1-12 Excitonic complex consisting of two electrons associated with two holes.

to rotational and vibrational modes. Further complication is expected from the fact that the effective mass of the carriers is usually not isotropic. However, all these complications should be small effects which mostly broaden the dominant energy levels. Two free holes and two free electrons (++−−) can combine to form a positronium-like molecule (Fig. 1-12). Such a complex would have a lower energy than two free excitons (Fig. 1-13), since each car-

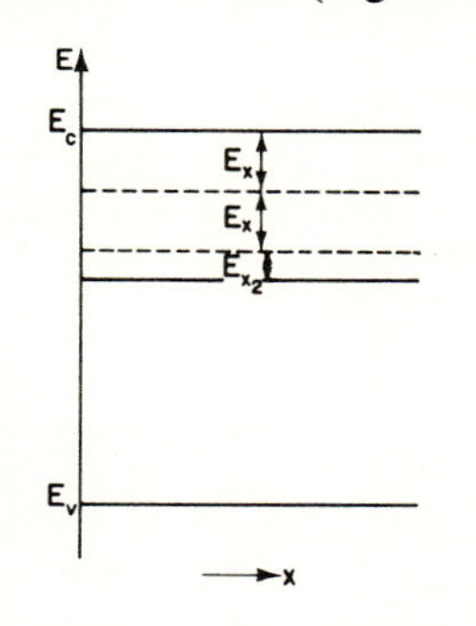

Fig. 1-13 Energy level diagram for the excitonic complex of Fig. 1-12. E_x is the binding energy of a free exciton (hole-electron pair); E_{x_2} is the binding energy of two free excitons.

rier sees the coulombic attraction of not one but two opposite charges. Such a complex has been discovered in silicon[19] and in a number of other semiconductors.

A free hole can combine with a neutral donor to form a positively charged excitonic ion. In this case, the electron bound to the donor still travels in a wide orbit about the donor. The associated hole which moves in the electrostatic field of the "fixed" dipole, determined by the instantaneous position of the electron, then also travels about this donor (Fig. 1-14); for this reason, this complex is called a "bound exciton." An electron associated with a neutral acceptor is also a bound exciton. Both of these types of bound excitons were also first observed in silicon.[20] It has been found experimentally that the

[19] J. R. Haynes, *Phys. Rev. Letters* **17**, 860 (1966).
[20] J. R. Haynes, *Phys. Rev. Letters* **4**, 361 (1960).

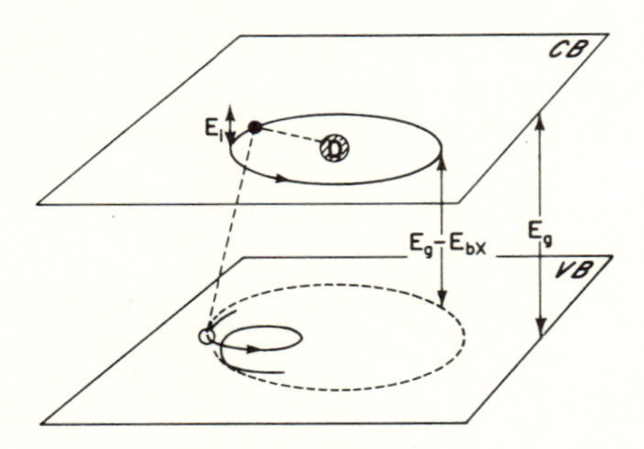

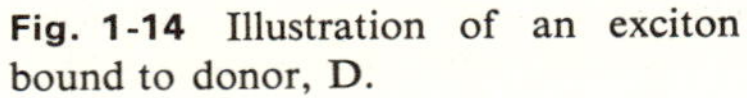
Fig. 1-14 Illustration of an exciton bound to donor, D.

binding energy of the excitonic complex in silicon is about one-tenth of the binding energy of the impurity (donor or acceptor).[20] These values agree with a theoretical estimate[21] that the binding energy E_{x_2} of the excitonic complex should be within the limits $0.055E_i < E_{x_2} < 0.35E_i$, the lower limit corresponding to the removal of an electron from the negative hydrogen-like ion, the upper limit corresponding to the dissociation of the hydrogen-like molecule.

All the possible excitonic complexes, free or bound to one neutral impurity are illustrated in Fig. 1-15.

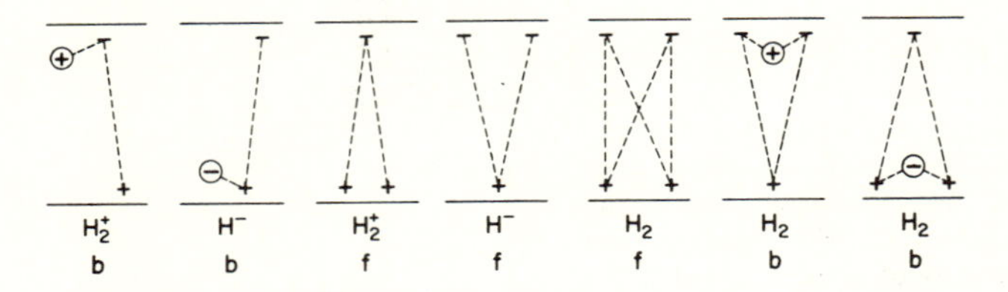

Fig. 1-15 Diagram of excitonic complexes. Bound excitons are labelled "b", free excitons are labelled "f". The symbols are: – electron, + hole, ⊕ donor, ⊖ acceptor.

1-D-3 POLARITONS[22,23]

A polariton is the complex resulting from the polarizing interaction between an electromagnetic wave and an oscillator resonant at the same frequency. The oscillator can be one or more atoms, electrons, or holes or their combination. Although polaritons initially have designated the interaction

[20] J. R. Haynes, *Phys. Rev. Letters* **4,** 361 (1960).

[21] W. Kohn, footnote 7 in Haynes (above); see also J. J. Hopfield, "The Quantum Chemistry of Bound Exciton Complexes," *Proc. 7th Int. Conf. Phys. Semiconductors*, Paris (1964) Dunod, p. 725.

[22] J. J. Hopfield "Radiative Recombination at Shallow Centers," *II–VI Semiconducting Compounds*, ed. D. G. Thomas, Benjamin (1967) p. 786.

[23] W. C. Tait, D. A. Cambell, J. R. Packard, and R. L. Weiher, "Luminescence from Inelastic Scattering of Polaritons by Longitudinal Optical Phonons," *II–VI Semiconducting Compounds*, ed. D. G. Thomas, Benjamin (1967), p. 370.

between excitons and photons, they can also represent the interaction between photons and optical phonons and between photons and plasmons. Phonons are collective vibrational modes of the atoms forming the crystal; plasmons are collective oscillations of the free carriers.

Let us consider the case of the interaction between excitons and photons. The dispersion curve for a free exciton is the parabola of Fig. 1-16, whereas that of the photon is the straight line. In the vicinity of the intersection of these two curves, the polariton-dispersion curve exhibits a maximum interaction between the photon and the exciton. Only the lowest branch of the dispersion curve will be of interest to us. Above the knee of the polariton curve the particle will behave as a free exciton; below the knee it will behave as a photon.

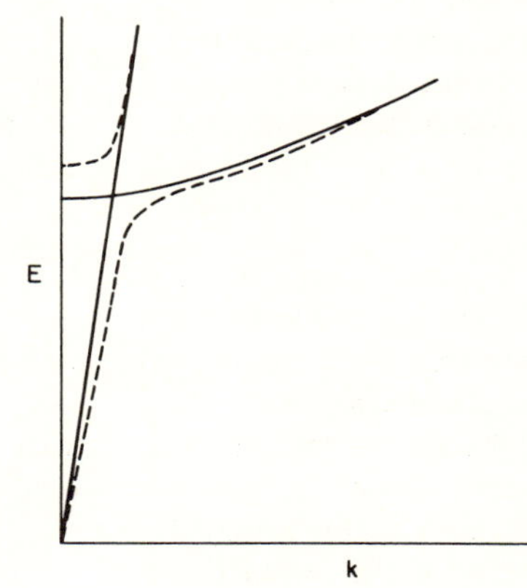

Fig. 1-16 Dispersion curves for polaritons (dotted line) and free excitons and photons (solid lines).

The polariton is not to be confused with the polaron[24] of ionic crystals, which results from an interaction between the electron and the lattice and, therefore, consists of a free electron (or hole) and associated phonons. In ionic crystals the atoms nearest the electron are displaced by coulomb interaction with the electron. Because of this polarization effect, the energy and the effective mass of the polaron are slightly different than their values for the free carrier.

1-E Donor–Acceptor Pairs

Donors and acceptors can form pairs and act as stationary molecules imbedded in the host crystal. The coulomb interaction between a donor and an acceptor results in a lowering of their binding energies. This can be viewed in the following simple argument. As the neutral donor and the neutral ac-

[24] D. M. Eagles, *Phys. Rev.* **145**, 645 (1966) and references therein.

ceptor are brought closer together, the donor's electron becomes increasingly shared by the acceptor. In other words, the donor and the acceptor become increasingly more ionized. In the fully ionized state, the binding energy is zero and the corresponding level lies at the band edge. The amount by which the impurity levels are shifted due to this pairing interaction is simply the coulomb interaction inside a medium of dielectric constant ϵ:

$$\Delta E = \frac{q^2}{\epsilon r}$$

where r is the donor–acceptor pair separation. Since the electron is shared by the donor–acceptor pair, it is irrelevant to say what fraction of ΔE modifies the ground state of either of the two impurities. This is similar to the exciton case, where we could divide the exciton binding energy between an electron state and a hole state and refer those binding energies to the appropriate band edges. In the donor–acceptor pair case it is convenient to consider only the separation between the donor and the acceptor level:

$$E_{\text{pair}} = E_g - E_D - E_A + \frac{q^2}{\epsilon r} \tag{1-13}$$

where E_D and E_A are the respective ionization energies of the donor and the acceptor as isolated impurities.

Note that since the impurities are located at discrete sites in the lattice (e.g., substitutional sites), the distance r varies by finite increments. For nearest neighbors, r is smallest; for more distant pairs, r increases in ever-smaller increments. Thus the pair interaction provides a possible range of states: from E_D and E_A for a very distant pair (a negligible pairing) to states which may lie inside the conduction and valence bands for near neighbors such that $(q^2/\epsilon r) > E_D + E_A$.[25]

In compound semiconductors a distinction can be made depending on the lattice site occupied by the impurities.[26] The anions and cations form similar but separate sublattices. Substitutional impurities fit into one or the other sublattice. If the donor and the acceptor occupy the same sublattice, they form a type-I donor–acceptor pair (e.g., Si and Te on P-sites in GaP); if, on the other hand, they occupy opposite sublattices, they form a type-II donor–acceptor pair (e.g., Zn on a Ga-site and S on a P-site in GaP).

1-F States in Semiconducting Alloys

When an alloy is made of two semiconductors, it is expected that the energy gap of the alloy will assume a value intermediate between the gaps of the two pure semiconductors and that the gap will vary in proportion to the

[25] H. Reiss, C. S. Fuller, and F. J. Morin, *Bell Syst. Tech. J.* **35**, 535 (1956).
[26] J. J. Hopfield, D. G. Thomas, and M. Gershenzon, *Phys. Rev. Letters* **10**, 162 (1963).

composition. However, the rate of change of the energy gap with composition depends on the nature of the lowest conduction-band valley. Thus Ge and Si form a solid solution, $Ge_{1-x}Si_x$, continuously miscible over the whole compositional range $0 < x < 1$.[27] Relative to the top of the valence band, the $\langle 111 \rangle$ valleys move to higher energies with increasing Si-concentration faster than the $\langle 100 \rangle$ valleys (Fig. 1-17). The two sets of valleys have the same potential energy when $x \approx 0.15$. Note that the energy gap of the alloy does not vary linearly with the composition. The deviation from linearity may be accounted for by the formation of band tails due to the random perturbation of the lattice by the minority atoms.

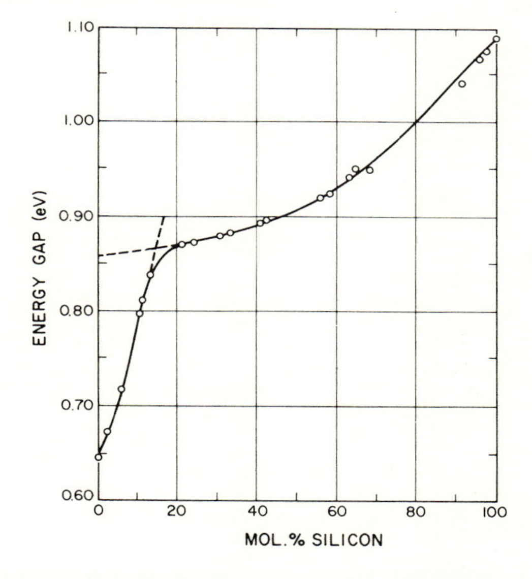

Fig. 1-17 Composition dependence of energy gap in Ge–Si alloys at 296°K.[27]

It is interesting that Si and C form a compound with a well-defined sublattice for each element, whereas Si and Ge do not.

In some ternary alloys the energy gap seems to vary linearly with the composition. Here again, the rate of change of the energy gap with the composition depends on the nature of the lowest valley. Thus in $GaAs_{1-x}P_x$ the direct valley increases at a higher rate with the GaP-concentration x than does the indirect valley. The direct and the indirect valleys have the same potential when the composition is $x = 0.44$. Hence the energy gap of $GaAs_{1-x}P_x$ is direct over the range $0 < x < 0.44$, and indirect over the range $0.44 < x < 1$.

[27] R. Braunstein, A. R. Moore, and F. Herman, *Phys. Rev.* **109**, 695 (1958).

The variation of $E_g(x)$ is approximately linear over both ranges of composition as shown in Fig. 1-18. Ternary alloys between II–VI compounds preserve a direct gap over the whole compositional range. In $ZnS_{1-x}Se_x$ the energy gap varies linearly with composition; whereas $ZnS_{1-x}Te_x$ and $ZnSe_{1-x}Te_x$ have an anomalously high nonlinear dependence, which still lacks an explanation.[28]

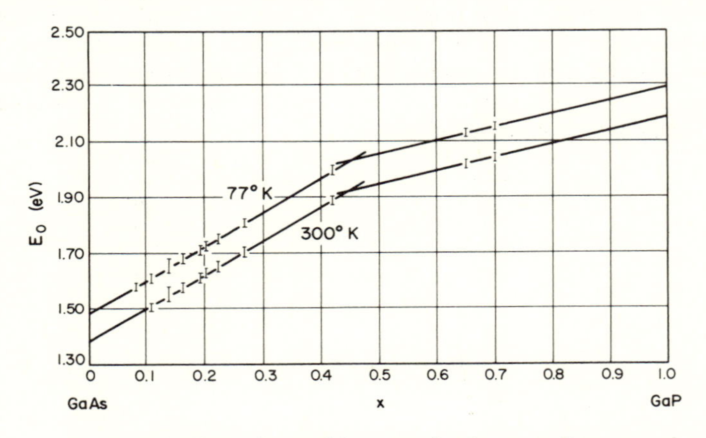

Fig. 1-18 Dependence of the energy band gap on alloy composition in the system, $GaAs_{1-x}P_x$.[29]

In the system $Hg_{1-x}Cd_xTe$ the energy gap varies linearly between 1.6 eV for CdTe and −0.14 eV for HgTe.[30] The negative sign of the energy gap signifies that the conduction and valence bands overlap, transforming the semiconductor into a semimetal. The ability to tailor the energy gap to such small values has made this alloy very important for infrared detection.

Quarternary alloys between III–V and II–VI semiconducting compounds have been synthesized over a large range of compositions.[31-34] However, their optical properties have shown a highly non-linear dependence of the energy gap on composition (Fig. 1-19).

Problem 1. The conduction band of germanium consists of ⟨111⟩ valleys at the extreme of the Brillouin zone, a "direct" ⟨000⟩ valley 0.15 eV above the bottom

[28] S. Larach, R. E. Shrader, and C. F. Stocker, *Phys. Rev.* **108**, 587 (1957).

[29] J. J. Tietjen and J. A. Amick, *J. Electrochem. Soc.* **113**, 724 (1966).

[30] T. C. Harman, W. H. Kleiner, A. J. Strauss, G. B. Wright, J. G. Mavroides, J. M. Honig, D. H. Dickey, Solid State Comm. **2**, 305 (1964); also D. Long and J. L. Schmit, "$Hg_{1-x}Cd_xTe$ and Closely Related Alloys," *Semiconductors and Semimetals*, ed. R. K. Willardson and A. C. Beer, Academic Press (to be published).

[31] I. Bertoti, M. Farkas-Jahnke, M. Harsy, T. Nemeth, and K. Richter, *Proc. International Conf. on Luminescence*, p. 1261, Hungarian Academy of Science, Budapest (1968).

[32] W. M. Yim, *J. Appl. Phys.* **40**, 2617 (1969).

[33] S. M. Ku and L. J. Bodi, *J. Phys. Chem. Solids* **29**, 2077 (1968).

[34] E. F. Hockings and J. W. Robinson, private communication.

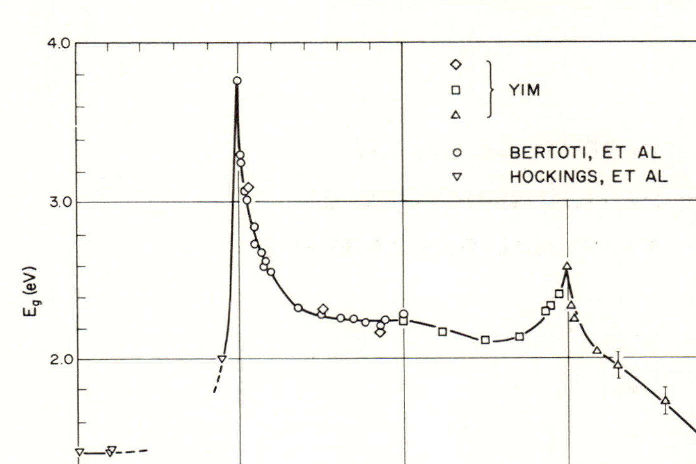

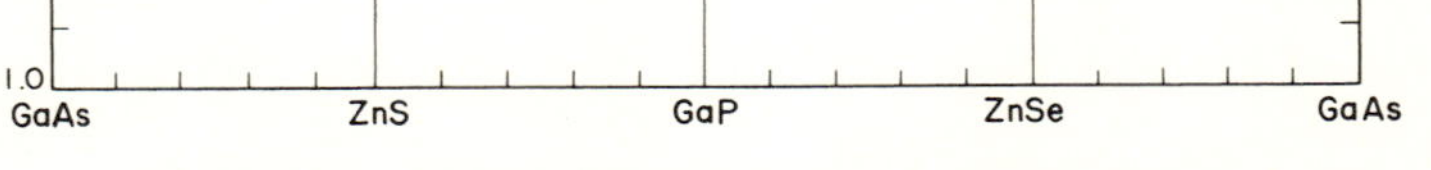

Fig. 1-19 Bandgap energy vs. alloy composition in four systems of the type $(\text{II–VI})_{1-x}$–$(\text{III–V})_x$.[35]

of the conduction band, and six $\langle 100 \rangle$ valleys 0.18 eV above the bottom of the conduction band.

The longitudinal and transverse effective masses are

$$\langle 111 \rangle \qquad m_l^* = 1.58m, \qquad m_t^* = 0.082m$$

$$\langle 000 \rangle \qquad m_l^* = m_t^* = 0.036m$$

$$\langle 100 \rangle \qquad m_l^* = 0.19m, \qquad m_t^* = 0.97m$$

where m is the mass of the free electron $= 9 \times 10^{-28}$ gr.

Find the position of the Fermi level above the bottom of the conduction band at 0°K when the electron concentration is 2×10^{18} cm^{-3}.

Problem 2. Write down the expression for the energy of a conduction-band electron in a semiconductor with ellipsoidal constant-energy surface. Assume that the coordinate axes in the $\bar{k}$-space coincide with the principal directions of the ellipsoidal constant-energy surface. Derive the expression for the density of states per unit energy per unit volume.

[35] S. Bloom, private communication.

PERTURBATION OF SEMICONDUCTORS BY EXTERNAL PARAMETERS

2

Now we shall see how externally applied perturbations affect the various energy levels of semiconductors. The parameters which can be controlled externally are: pressure, temperature, and electric and magnetic fields. The pressure may be hydrostatic or uniaxial; it affects the potential energy of all the levels. Temperature affects the potential of the levels via the internal-pressure changes it generates; it also affects the population of the various states.

Electric fields polarize the lattice and, when intense, induce a splitting of levels (Stark effect). But most important is the Franz–Keldysh effect, which effectively broadens all the states.

Magnetic fields cause a Landau splitting of all the states in the conduction and valence bands and a Zeeman splitting of impurity and exciton states.

A knowledge of all these effects permits us to differentiate between the various levels which might be involved in an optical process where the detail of the mechanism may be ambiguous.

2-A Pressure Effects

2-A-1 HYDROSTATIC PRESSURE

Hydrostatic pressure brings all the atoms closer together. As illustrated in Fig. 1-1, a smaller atomic separation will result in a larger energy gap. (However, in some materials, e.g., Te and PbSe, the energy gap decreases with decreasing interatomic spacing—i.e., with increasing pressure.) For a small change Δa in lattice constant, the energy E of a level may be assumed to vary linearly with Δa:

$$E = E_0 + E_1 \Delta a$$

where E_0 is the energy of the level considered at zero pressure and E_1 is a coefficient; E_1 will in general be different for different levels. Therefore, the pressure dependence of the energy gap will be due to

$$\Delta E_g = (E_{1_c} + E_{1_v}) \Delta a$$

where E_{1_c} and E_{1_v} are the pressure coefficients for the edges of the conduction and valence bands, respectively. Since the various valleys will have different coefficients (some positive, some negative), at sufficiently high deformations the semiconductor may switch from being a direct-gap material to becoming an indirect-gap material (or vice versa). The pressure dependence of the energy gap in germanium is shown in Fig. 2-1. Values of the pressure dependences of the separation between the top of the valence-band and the bottoms of various conduction-band valleys in a number of semiconductors are shown in Table 2-1.

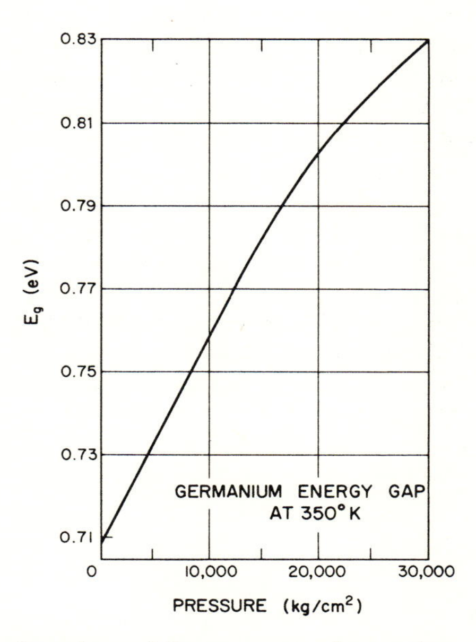

Fig. 2-1 Pressure dependence of the energy gap of germanium. The nonlinearity at the higher pressures is to be noted.[1]

[1]W. Paul and H. Brooks, *Phys. Rev.* **94**, 1128 (1954).

Table 2-1[3]

Coefficients for change of energy gap with pressure†

Material	$(\partial E_g/\partial P)_T \times 10^6$(eV-cm²/kg) *for conduction-band minima*‡ *at*			Reference
	Γ§	*X*	*L*	
C	——	<1.0	——	4
Si	——	−1.5	5.0	4, 4
Ge	12.0	0 to −2.0	5.0	4, 4, 5
Sn	——	——	5.0	6
AlSb	——	−1.6	——	7
GaP	10.7	−1.1, −1.7, −1.8	——	8, 8, 9, 4
GaAs	9.4, 12.0	−8.7	——	9, 4, 9
InP	4.6, 8.4	−10.0	——	7, 10, 7
GaSb	12.0, 16.0	<0	5.0, 7.3	7, 11, 7, 4, 7
InAs	4.8, 8.5, 5.5, 10.0	——	3.2	7, 11, 12, 10, 7
InSb	14.2, 15.5	——	——	13, 14

†Some coefficients have been deduced from electrical measurements as well as optical measurements.

‡Confirmed or speculated.

§Γ occurs at $k = 000$, X at $k = 100$, and L at $k = 111$.

Note that the pressure dependences of the direct energy gaps **Γ** in many III–V compounds and in germanium are similar. The different rates and signs of the pressure dependences of each valley can cause the energy gap to go through a maximum or a minimum at sufficiently high pressures. Thus in germanium the energy gap goes through a maximum at 50 kilobars.[2] This is readily understood on the basis that while the set of ⟨111⟩ valleys moves to higher energies, the set of ⟨100⟩ valleys moves to lower energies. At 50 kbars the bottoms of both sets of valleys would be at the same potential.

In GaAs the direct valley of the conduction band moves to higher potentials faster than the indirect valley. Hence as the pressure increases, the material eventually becomes an indirect-gap semiconductor. When the two valleys

[2] T. E. Slykhouse and H. G. Drickamer, *J. Phys. Chem. Solids* **7**, 210 (1958).

[3] E. J. Johnson, *Semiconductors and Semimetals*, ed. R. K. Willardson and A. C. Beer, Academic Press, Vol. 3, p. 200.

[4] W. Paul, *J. Appl. Phys. Suppl.* **32**, 2082 (1961).

[5] O. Madelung, *Physics of* III–V *Compounds*, Wiley (1964).

[6] S. H. Groves and W. Paul, *Phys Rev. Letters* **11**, 194 (1963).

[7] A. L. Edwards and H. G. Drickamer, *Phys. Rev.* **122**, 1149 (1961).

[8] R. Zallen and W. Paul, *Phys. Rev.* **134**, A1628 (1964).

[9] A. L. Edwards, T. E. Slykhouse, and H. G. Drickamer, *J. Phys. Chem. Solids* **11**, 140 (1959).

[10] R. Zallen, Ph. D. thesis, Harvard University (1964).

[11] J. H. Taylor, *Bull. Am. Phys. Soc.* **3**, 121 (1958).

[12] J. H. Taylor, *Phys. Rev.* **100**, 1593 (1958).

[13] D. Long, *Phys. Rev.* **99**, 388 (1955).

[14] R. W. Keyes, *Phys. Rev.* **99**, 490 (1955).

cross, the electron mobility suddenly drops and the electron–hole pair radiative-recombination process becomes less efficient.

The effect of hydrostatic pressure on the binding energy of free excitons and excitonic molecules should be small and depend only on the strain dependence of the dielectric constant. However, when pressure causes a crossing of the valleys, the effective mass in the lowest conduction-band states suddenly changes and the exciton binding energy varies accordingly.

Pressure affects only slightly the ionization energy of impurities. For example, the pressure coefficients for the ionization energies of impurities in germanium and silicon are shown in Table 2-2. Because the ionization energy of impurities does not change appreciably with pressure, the impurity level follows the potential of the band edge with which it is associated.

Table 2-2[15]

Pressure coefficients of ionization energies of impurities in silicon and germanium

Element	*System*	*Ionization energy* (electron volts)	*Pressure coefficient* (eV-cm²/kg)
silicon	arsenic donor	0.05	$\sim -5 \times 10^{-8}$
	aluminum acceptor	0.06	$\sim +1 \times 10^{-8}$
	indium acceptor	0.16	$\sim +5 \times 10^{-8}$
	gold acceptor‡	0.54†	-1.2×10^{-6}
	gold acceptor‡	0.62§	-0.3×10^{-6}
	gold donor	0.35§	$\leq \lvert 5 \times 10^{-8} \rvert$
germanium	gold donor	0.04§	$+0.11 \pm 0.02 \times 10^{-8}$
	gold acceptor	0.15§	$+0.55 + 0.06 - 0.02 \times 10^{-6}$
	gold acceptor	0.19§	$+2.9 \pm 0.1 \times 10^{-6}$
	gold acceptor	0.04†	$+2.1 \pm 0.1 \times 10^{-6}$

†Measured with respect to the conduction band.
‡These are the same level.
§Measured with respect to the valence band.

The largest effect of pressure on the impurities is the change in their population at temperatures such that $kT \approx E_i$, but this effect is limited to a narrow temperature range.

A possibly large effect of pressure on impurities may result from the crossing of valleys. The ionization energy of the impurity varies with the effective mass of the lowest valley. When the valleys cross, the effective mass changes and so does the ionization energy. Hence at a given temperature the impurity level may be either empty or full depending on whether E_i is larger or smaller than kT.

[15]W. Paul and D. M. Warschauer, *Solids Under Pressure*, McGraw-Hill (1963), p. 222.

A pressure-induced change in energy gap may affect the population of the conduction and valence bands because the concentration of intrinsic carriers has an exponential dependence on the energy gap:

$$n_i^2 = N_c N_v \exp \frac{-E_g(P)}{kT} \tag{2-1}$$

2-A-2 UNIAXIAL STRAIN

Strains shift the distribution of the density of states by some amount which is different in various crystallographic directions. The changes in energy are given by the elastic coefficients of a strain tensor. Thus compression of a silicon crystal along the [100] axis decreases the energy gap in the direction of the compression; but because of a simultaneous dilation in the transverse directions, the conduction-band valleys lying in the transverse plane move to a higher energy (Fig. 2-2). The redistribution of the states causes a rearrangement of the electrons which spill into the lower-energy states made available by the strain. This reduces the free energy of the system and affects the elastic constants. The electronic contribution to the elastic constants of semiconductors has been extensively studied by Keyes.[16]

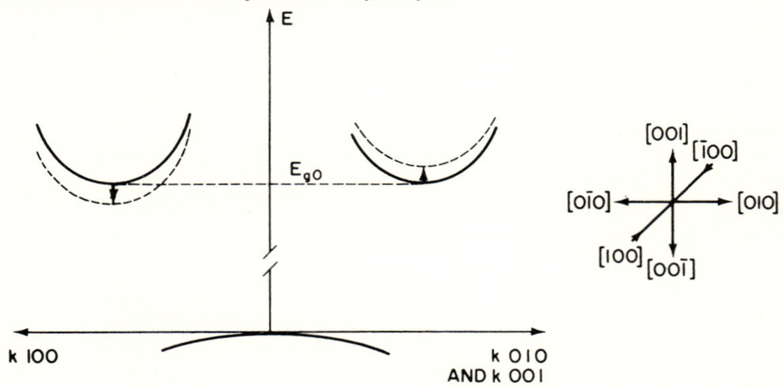

Fig. 2-2 Longitudinal and transverse shifts of conduction band valleys in silicon due to uniaxial pressure along the [100] direction.

Although the impurity concentration is not an externally controllable parameter, it is appropriate to mention here the effect of impurities. Increasing the doping of a semiconductor introduces a deformation potential, and causes a redistribution of electrons and a decrease of the elastic constants.

In compound semiconductors, because of the polar nature of the interatomic bonding, strong local fields already exist in the vicinity of each lattice site. A uniaxial strain will enhance these local fields. In III–V compounds,

[16] R. W. Keyes, *Solid State Physics* **20**, 37 (1967), ed. F. Seitz, D. Turnbull, H. Ehrenreich, Academic Press.

cubic symmetry allows one longitudinal (hydrostatic) deformation and two transverse (shear) deformations. But in II–VI compounds the effects are more complex because the hexagonal symmetry allows two longitudinal and four transverse deformations.

2-B Temperature Effects

As the temperature of a semiconductor increases, the lattice expands and the oscillations of the atoms about their equilibrium lattice points increase. We have already seen how dilation leads to a change (increase or decrease) in the energy gap. Besides this shift of the band edges, the increased motion of the atoms broadens the energy levels.

Just as in the case of pressure-induced deformation, temperature will affect the potential assigned to an impurity level via the band edge with which it is associated, keeping the ionization energy constant. There is also an electron–lattice interaction which depends strongly on temperature. At temperatures much lower than the Debye temperature, the energy gap varies proportionately to the square of the temperature, whereas much above the Debye temperature the energy gap varies linearly with the temperature. The temperature dependence of the gap for many semiconductors has been fitted by the following empirical relation:[17]

$$E_g(T) = E_g(0) - \frac{\alpha T^2}{T + \beta} \tag{2-2}$$

where $E_g(0)$ is the value of the energy gap at 0°K and α and β are constants. Table 2-3 lists these values for several semiconductors.

Table 2-3[17]

Values of the parameters in Eq. (2-2)

Substance	*Type of gap*	E_g (0) (eV)	α ($\times 10^{-4}$)	β
diamond	E_{gi}	5.4125	−1.979	−1437
Si	E_{gi}	1.1557	7.021	1108
Ge	E_{gi}	0.7412	4.561	210
Ge	E_{gd}	0.8893	6.842	398
6H SiC	E_{gi}	3.024	−0.3055	−311
GaAs	E_{gd}	1.5216	8.871	572
InP	E_{gd}	1.4206	4.906	327
InAs	E_{gd}	0.426	3.158	93

E_{gi} = indirect gap E_{gd} = direct gap

[17] Y. P. Varshni, *Physica* **34**, 149 (1967).

The dominant effect of temperature is to vary exponentially the population of the states, as evident from Eq. (2-1). Note that the electron population of the conduction band and the hole population of the valence band can be changed also by optical excitation or by electron bombardment, which we shall discuss in Secs. 11-B and C. When a large concentration of free carriers is introduced in a semiconductor, the resulting screened coulomb interaction between carriers perturbs the band edges and induces a tailing of the band states into the energy gap, an effect similar to the tailing discussed in Sec. 1-C. This free-carrier–induced gap shrinkage has been observed in germanium[18] and in gallium arsenide.[19]

Excitons are thermally dissociated when the thermal energy exceeds the exciton binding energy ($kT > E_x$). Hence when excitons are formed at temperatures $T \geq E_x/k$, their lifetime is reduced and their energy broadens via the uncertainty principle: $\Delta E = \hbar/\Delta t$ where Δt is the exciton lifetime.

2-C Electric-Field Effects

2-C-1 STARK EFFECTS

An electric field tends to orient the elliptical orbits of electrons so that the center of gravity of the ellipse and the focus of the ellipse (the nucleus) are aligned with the electric field $\mathscr{E}$ (Fig. 2-3). Of course, circular orbits, like the ground state of a hydrogenic impurity, are not affected by the electric field. But the excited state of an exciton and all the P-states in the conduction band correspond to elongated orbits.

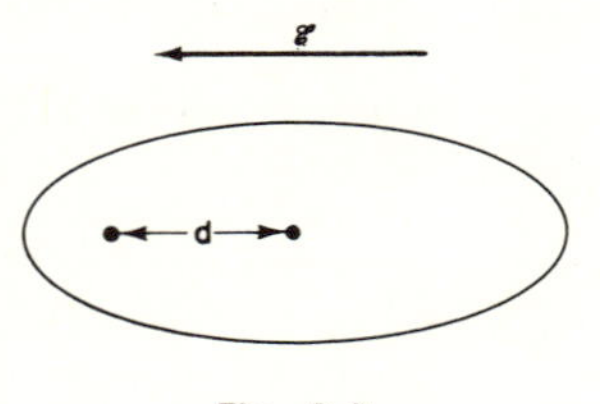

Fig. 2-3

When the electron precesses slowly, its orbit lines up with the weakest field; therefore, the energy shift of that state is

$$\Delta E = qd\,\mathscr{E} \tag{2-3}$$

[18] V. M. Asnin and A. A. Rogatchev, *Soviet Phys.-Solid State* **5**, 1730 (1963).
[19] N. Holonyak Jr., M. R. Johnson and J. A. Rossi, *Applied Phys. Letters* **12**, 151 (1968).

where d is the eccentricity of the orbit and $\mathscr{E}$ is the electric field. This is the first-order Stark effect.

When the orbit precesses rapidly, the average position of the electron is centered on the nucleus. Turning on the electric field will gradually displace the average position of the electron in the direction of the field proportionately to the field and induce a dipole. Then the energy shift of this state will vary as the square of the field—this is the second-order Stark shift.

2-C-2 FRANZ–KELDYSH EFFECT[20]

In the presence of an electric field, $\mathscr{E}$, the band edges are tilted. An electron moving inside a band a distance x away from the edge and maintaining a constant total energy acquires a kinetic energy $q\mathscr{E}x$. Its momentum is real, corresponding to the running wavefunctions similar to those describing the band. If it were possible for an electron to move away from the band edge into the gap, maintaining a constant total energy, its kinetic energy would become negative and the corresponding momentum would be imaginary. The wave function of the electron in the gap, e^{ikx}, (here k is imaginary) represents a damped wave. Hence the probability of finding an electron in the energy gap away from the band edge decreases according to:

$$\exp - \frac{|E - E_e|}{\Delta E}$$

where E is the energy in the gap at position x, E_e is the energy of the band edge at x and ΔE is a parameter which depends on the field. Therefore the states at the band edges are effectively broadened exponentially into the gap by an average deviation ΔE which depends on the electric field. It can be shown that this dependence is of the form:[21]

$$\Delta E = \tfrac{3}{2}(m^*)^{-1/3}(q\hbar\mathscr{E})^{2/3} \qquad (2\text{-}4)$$

Later on, we shall see the tremendous importance of this effect in optical and transport properties of semiconductors (and insulators).

2-C-3 IONIZATION EFFECTS

An electric field can exert a stronger force on the electron than the local forces binding the electron to an impurity or to an excitonic state. In this case, the center is ionized and the carrier is free to move in the appropriate band. If donors are ionized, free electrons appear in the conduction band. If acceptors are ionized by the field, free holes appear in the valence band. The ionization of excitons produce both free holes and free electrons.

[20] L. V. Keldysh, *Soviet Physics—JETP* **7**, 788 (1958); W. Franz, *Z. Naturforsch.*, **13a**, 484 (1958).

[21] V. S. Vavilov, *Soviet Phys.—Uspekhi* **4**, 761 (1962).

The free carriers are subsequently accelerated by the applied field. The resulting high kinetic energies allow the "hot" carriers to interact with the lattice and with each other in a variety of ways, some of which we shall see in Secs. 3-K, 6-G, 8-D-3, and 11-A-4.

2-D Magnetic-Field Effects

2-D-1 LANDAU SPLITTING

When the semiconductor is placed in a magnetic field H_z, the electron motion in the z-direction is not affected, but motion components in the transverse direction result in a periodic circular motion with an angular frequency:

$$\omega_c = \frac{qH_z}{m^*c}$$

where ω_c is called the cyclotron frequency. These orbits are quantized and the allowed transverse energies become

$$E_{xy} = \frac{q\hbar H_z}{m^*c}\left(n + \frac{1}{2}\right)$$

where n is an integer ≥ 0. In the longitudinal direction the energy of the electron in a parabolic band is independent of the magnetic field and is still given by

$$E_z = \frac{\hbar^2 k^2}{2m^*}$$

But an accounting[22,23] of the number of states at each energy shows that now, instead of the parabolic density of states, the magnetic field coalesces the states into discrete subbands, as shown in Fig. 2-4. The density of states is given as[23]

$$\frac{dN}{dE} = \left(\frac{2m^*}{\hbar^2}\right)^{1/2} \frac{1}{(2\pi l)^2} \sum_{n,\pm} \left[E - \left(n + \frac{1}{2} \pm \frac{\nu}{2}\right)\hbar\omega_c\right]^{-1/2}$$

where l is the radius of the lowest cyclotron orbit, given by

$$\frac{1}{l^2} = \frac{m^*}{\hbar}\omega_c$$

and ν is given by $\nu = m^*g/2m$, where g is the gyromagnetic ratio. (These values for various III–V and IV–VI compounds are tabulated in Tables 2-4 and 2-5.)

Note that the magnetic field raises the bottom of the conduction band by $(\frac{1}{2}q\hbar H/m_c^*c)$. The top of the valence band is shifted to lower energies by

[22] F. Seitz, *The Modern Theory of Solids*, McGraw-Hill (1940), p. 583.

[23] L. M. Roth and P. N. Argyres, *Semiconductors and Semimetals*, ed. R. K. Willardson and A. C. Beer, Vol. 1, Academic Press (1966), p. 159.

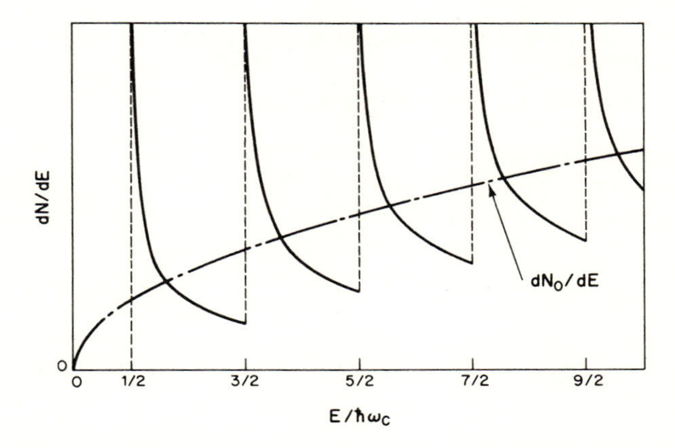

Fig. 2-4 Density of states in the presence of a magnetic field. The zero field case is also shown.[23]

Table 2-4[24]

Effective masses and g-factors for electrons at $k = 0$ in III–V compounds

Material	m/m^*	g	$\lvert \nu \rvert = \lvert gm^*/2m \rvert$
InSb	66	−44	0.33
InAs	36	−12	0.17
InP	14	0.60	0.021
GaSb	22	−6.1	0.15
GaAs	12	0.32	0.013
GaP	7.7	1.76	0.11
AlSb	9.1	0.4	0.022

Table 2-5[25]

Effective masses and g-factors for electrons at $k = 111$ in lead salts

Material	m/m^*	g
PbTe	4.55	$\lvert 29 \rvert$
PbSe	25.8	$\lvert 19 \rvert$
PbS	20	$\lvert 11.5 \rvert$

[24]Based on calculation of M. Cardona, *J. Phys. Chem. Solids* **24**, 1543 (1963).

[25]J. F. Butler and A. R. Calawa, *Proc. Physics of Quantum Electronics Conference*, San Juan, Puerto Rico, McGraw-Hill (1966), p. 458.

$\frac{1}{2}(q\hbar H/m_v^* c)$. The net shift of the energy gap is then

$$\Delta E_G = \frac{1}{2}\left(\frac{1}{m_c^*} + \frac{1}{m_v^*}\right)\frac{q\hbar H}{c}$$

It should be noted also that the conduction and valence bands are made up of several valleys, each of which has its own ladder of discrete energy levels. Furthermore, since the effective mass varies with direction, the levels may be broadened (impurities and thermal vibrations cause a further broadening). The directional dependence of the cyclotron orbit has been used to determine the effective mass of the carriers in various crystallographic directions and thus to define the shape of the valleys in momentum space. A detailed treatment and survey of cyclotron resonance has been made by Lax and Mavroides.[26]

2-D-2 ZEEMAN EFFECT[27]

The interaction of a magnetic field with an orbiting electron results in a splitting of the energy level formerly occupied by the electron (when H was nil). This splitting, corresponding to the Larmor precession of the orbit,

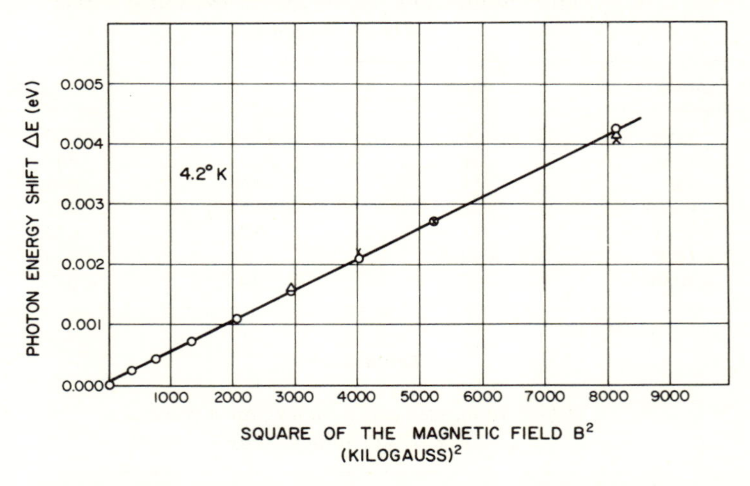

Fig. 2-5 Quadratic shift of emission energy in GaAs with magnetic field. This shift has been attributed to the magnetic field dependence of a hydrogenic donor.[28]

[26] B. Lax and J. G. Mavroides, *Solid State Physics*, Vol. 11, ed. Seitz and Turnbull, Academic Press (1960), p. 261.

[27] A treatment of linear and quadratic Zeeman effects can be found in L. I. Schiff, *Quantum Mechanics*, McGraw-Hill (1949).

[28] F. L. Galeener, G. B. Wright, W. E. Krag, T. M. Quist, and H. J. Zeiger, *Phys. Rev. Letters* **10**, 472 (1963).

amounts, at low fields, to an energy change of $\pm(qH\hbar/2m^*c)$, i.e., one-half of the Landau splitting. At high magnetic fields and for large orbits the splitting becomes quadratic in H. This nonlinear behavior occurs when H is greater than the internal field due to the combined action of the spin and the orbit. Large orbits occur in the case of excitons and in the case of the ground state of impurities. Therefore, impurities and excitons can exhibit a quadratic Zeeman effect (Fig. 2-5). By contrast, band states exhibit larger but linear shifts.

Problem 1. In conjunction with the Franz–Keldysh effect, the probability $P(E)$ of finding an electron in the gap at an energy E below the conduction-band edge E_c is proportional to

$$\exp\left(\frac{E - E_c}{\Delta E}\right)^{3/2}$$

where ΔE is given by Eq. (2.4).

Draw the band edge in the presence of an electric field and express the above exponential in terms of field $\mathscr{E}$, and position x. Resolve the apparent contradictions between $P(x, \mathscr{E})$ and $P(E)$, the latter showing the that probability of finding an electron below the conduction-band edge increases with the field.

ABSORPTION

3

The most direct and perhaps the simplest method for probing the band structure of semiconductors is to measure the absorption spectrum. In the absorption process, a photon of a known energy excites an electron from a lower- to a higher-energy state. Thus by inserting a slab of semiconductor at the output of a monochromator and studying the changes in the transmitted radiation, one can discover all the possible transitions an electron can make and learn much about the distribution of states.

Therefore, in this chapter we shall review all the possible transitions: band-to-band, excitons, between subbands, between impurities and bands, transitions by free carriers within a band, and also the resonances due to vibrational states of the lattice and of the impurities. In the following two chapters we shall describe how various optical constants can be obtained, concluding with an illustration of the general experimental procedure for measuring absorption spectra. Absorption is expressed in terms of a coefficient $\alpha(h\nu)$ which is defined as the relative rate of decrease in light intensity $L\ (h\nu)$ along its propagation path:

$$\alpha = \frac{1}{L(h\nu)} \frac{d[L(h\nu)]}{dx}$$

3-A Fundamental Absorption

The fundamental absorption refers to band-to-band or to exciton transitions, i.e., to the excitation of an electron from the valence band to the conduction band. The fundamental absorption, which manifests itself by a rapid rise in absorption, can be used to determine the energy gap of the semiconductor. However, because the transitions are subject to certain selection rules, the estimation of the energy gap from the "absorption edge" is not a straightforward process—even if competing absorption processes can be accounted for.

Because the momentum of a photon, h/λ, (λ is the wavelength of light, thousands of angstroms), is very small compared to the crystal momentum h/a (a is the lattice constant, a few angstroms), the photon-absorption process should conserve the momentum of the electron. The absorption coefficient $\alpha(h\nu)$ for a given photon energy $h\nu$ is proportional to the probability P_{if} for the transition from the initial state to the final state and to the density of electrons in the initial state, n_i, and also to the density of available (empty) final states, n_f, and this process must be summed for all possible transitions between states separated by an energy difference equal to $h\nu$:

$$\alpha(h\nu) = A \sum P_{if} n_i n_f \tag{3-1}$$

In what follows, for simplicity we shall assume that all the lower states are filled and that all the upper states are empty, a condition which is true for undoped semiconductors at 0°K.

3-A-1 ALLOWED DIRECT TRANSITIONS

Let us consider absorption transitions between two direct valleys where all the momentum-conserving transitions are allowed (Fig. 3-1), i.e., the transition probability P_{if} is independent of photon energy. Every initial state at E_i is associated with a final state at E_f such that

$$E_f = h\nu - |E_i|$$

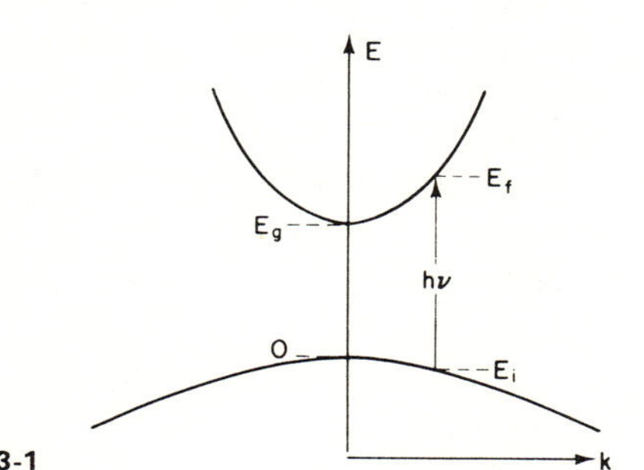

Fig. 3-1

But in parabolic bands,

$$E_f - E_g = \frac{\hbar^2 k^2}{2m_e^*}$$

and

$$E_i = \frac{\hbar^2 k^2}{2m_h^*}$$

Therefore,

$$h\nu - E_g = \frac{\hbar^2 k^2}{2}\left(\frac{1}{m_e^*} + \frac{1}{m_h^*}\right)$$

The density of directly associated states can then be found as was done in Sec. 1-A-3:

$$N(h\nu)\,d(h\nu) = \frac{8\pi k^2\,dk}{(2\pi)^3}$$
$$= \frac{(2m_r)^{3/2}}{2\pi^2\hbar^3}(h\nu - E_g)^{1/2}\,d(h\nu)$$

where m_r is the reduced mass given by $1/m_r = 1/m_e^* + 1/m_h^*$. Hence the absorption coefficient is

$$\alpha(h\nu) = A^*(h\nu - E_g)^{1/2} \tag{3-2}$$

where A^* is given by[1]:

$$A^* \approx \frac{q^2\left(2\dfrac{m_h^* m_e^*}{m_h^* + m_e^*}\right)^{3/2}}{nch^2 m_e^*} \tag{3-3}$$

For an index of refraction $n = 4$ and assuming the hole and electron effective masses equal the free electron mass, one gets

$$\alpha(h\nu) \approx 2 \times 10^4 (h\nu - E_g)^{1/2}\ \text{cm}^{-1}$$

$h\nu$ and E_g being expressed in eV.

3-A-2 FORBIDDEN DIRECT TRANSITIONS

In some materials, quantum selection rules forbid direct transitions at $k = 0$ but allow them at $k \neq 0$, the transition probability increasing with k^2. In the model of Fig. 3-1 this means that the transition probability increases proportionately to $(h\nu - E_g)$. Since the density of states linked in direct transitions is proportional to $(h\nu - E_g)^{1/2}$, the absorption coefficient has the following spectral dependence:

$$\alpha(h\nu) = A'(h\nu - E_g)^{3/2} \tag{3-4}$$

where A' is given by[1]

$$A' = \frac{4}{3}\,\frac{q^2\left(\dfrac{m_h^* m_e^*}{m_h^* + m_e^*}\right)^{5/2}}{nch^2 m_e^* m_h^* h\nu} \tag{3-5}$$

Again, for $n = 4$ and $m_h^* = m_e^* = m$:

$$\alpha(h\nu) = 1.3 \times 10^4 \frac{(h\nu - E_g)^{3/2}}{h\nu}\ \text{cm}^{-1}$$

Note that the $h\nu$ in the denominator varies slowly compared to $(h\nu - E_g)^{3/2}$.

[1]J. Bardeen, F. J. Blatt, and L. H. Hall, *Proc. of Atlantic City Photoconductivity Conference* (1954), J. Wiley and Chapman and Hall (1956), p. 146.

3-A-3 INDIRECT TRANSITIONS BETWEEN INDIRECT VALLEYS

When a transition requires a change in both energy and momentum, a double, or two-step, process is required because the photon cannot provide a change in momentum. Momentum is conserved via a phonon interaction as illustrated in Fig. 3-2. A phonon is a quantum of lattice vibration. Although a broad spectrum of phonons is available, only those with the required momentum change are usable. These are usually the longitudinal- and the transverse-acoustic phonons. Each of these phonons has a characteristic energy E_p. Hence to complete the transition E_i to E_f, a phonon is either emitted or absorbed. These two processes are given respectively by

$$\left.\begin{aligned} h\nu_e &= E_f - E_i + E_p \\ h\nu_a &= E_f - E_i - E_p \end{aligned}\right\} \tag{3-6}$$

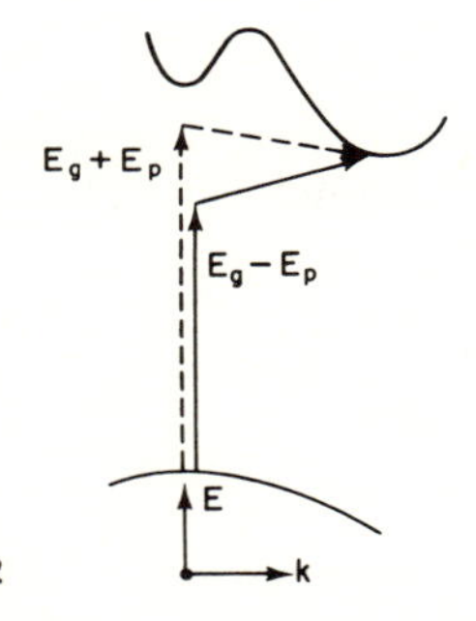

Fig. 3-2

In indirect transitions, all the occupied states of the valence band can connect to all the empty states of the conduction band. The density of initial states at an energy E_i is

$$N(E_i) = \frac{1}{2\pi^2\hbar^3}(2m_h^*)^{3/2}\,|E_i|^{1/2} \tag{3-7}$$

The density of states at E_f is

$$N(E_f) = \frac{1}{2\pi^2\hbar^3}(2m_e^*)^{3/2}(E_f - E_g)^{1/2}$$

Substituting Eq. (3-6),

$$N(E_f) = \frac{1}{2\pi^2\hbar^3}(2m_e^*)^{3/2}(h\nu - E_g \mp E_p + E_i)^{1/2} \tag{3-8}$$

The absorption coefficient is proportional to the product of the densities of initial states given by Eq. (3-7) and final states given by Eq. (3-8) integrated over all possible combinations of states separated by $h\nu \pm E_p$; α is also proportional to the probability of interacting with phonons, which is itself a function $f(N_p)$ of the number N_p of phonon of energy E_p. The number of phonons is given by Bose–Einstein statistics[2]:

$$N_p = \frac{1}{\exp \frac{E_p}{kT} - 1} \tag{3-9}$$

Hence

$$\alpha(h\nu) = Af(N_p) \int_0^{-(h\nu - E_g \mp E_p)} |E_i|^{1/2}(h\nu - E_g \mp E_p + E_i)^{1/2}\, dE_i \tag{3-10}$$

After integration and substituting Eq. (3-9) into (3-10), the absorption coefficient for a transition with phonon absorption is

$$\alpha_a(h\nu) = \frac{A(h\nu - E_g + E_p)^2}{\exp \frac{E_p}{kT} - 1} \tag{3-11}$$

for $h\nu > E_g - E_p$. The probability of phonon emission is proportional to $N_p + 1$; hence the absorption coefficient for a transition with phonon emission is

$$\alpha_e(h\nu) = \frac{A(h\nu - E_g - E_p)^2}{1 - \exp\left(-\frac{E_p}{kT}\right)} \tag{3-12}$$

for $h\nu > E_g + E_p$.

Since both phonon emission and phonon absorption are possible when $h\nu > E_g + E_p$, the absorption coefficient is then

$$\alpha(h\nu) = \alpha_a(h\nu) + \alpha_e(h\nu) \tag{3-13}$$

for $h\nu > E_g + E_p$.

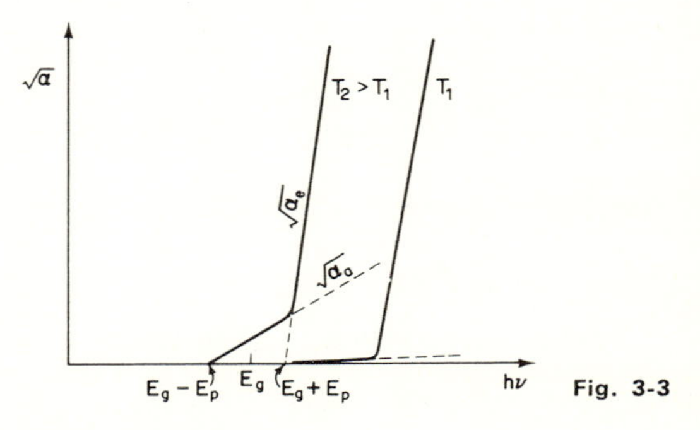

Fig. 3-3

[2]S. Wang, *Solid State Electronics*, McGraw-Hill (1966), p. 46.

At very low temperatures, the phonon density is very small [large denominator in Eq. (3-11)]; therefore, α_a is also small. The temperature dependences of α_a and α_e are illustrated in Fig. 3-3, where the square root of α is plotted to yield a linear dependence on $h\nu$. Such a plot, by extrapolation to $\alpha = 0$, gives the values of $E_g - E_p$ and $E_g + E_p$. Note that E_g has been shifted with temperature to reflect the temperature dependence of the energy gap.

As mentioned earlier, there are several types of phonons, one longitudinal-acoustic and two transverse-acoustic, which can participate in the transition process. In fact, they all participate, but with different probabilities.[3]

If the semiconductor is heavily doped, the Fermi level is inside the band (the conduction band in an n-type material) by a quantity ξ_n (Fig. 3-4). Since

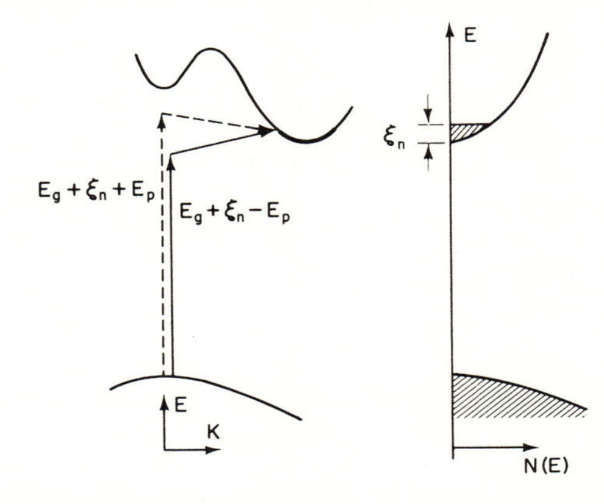

Fig. 3-4 Energy–momentum diagram for degenerate n-type germanium in the [111] direction. Two phonon-assisted transitions are shown to illustrate the usual photon absorption mechanism.

the states below ξ_n are already filled, fundamental transitions to states below $E_g + \xi_n$ are forbidden; hence the absorption edge should shift to higher energies by about ξ_n. The shift of the absorption edge due to band filling is sometimes called the Burstein–Moss shift.[4,5] A calculation of the absorption coefficient was made for heavily doped n-type germanium[6]; the results are reproduced in Fig. 3-5. At 0°K, only the phonon-emission process is possible; $\sqrt{\alpha_e}$ for pure germanium intercepts the abscissa at $E_g + E_p$. The calculated intercept shifts by ξ_n, as expected. The drop of absorption at a given $h\nu > E_g + E_p + \xi_n$ with increasing doping is due to the decrease in the number of available final states.

[3]G. G. MacFarlane, T. P. McLean, J. E. Quarrington, and V. Roberts, *Phys. Rev.* **108**, 1137 (1957) and **111**, 1245 (1958).
[4]E. Burstein, *Phys. Rev.* **93**, 632 (1954).
[5]T. S. Moss, *Proc. Phys. Soc.* (London) **B76**, 775 (1954).
[6]J. I. Pankove and P. Aigrain, *Phys. Rev.* **126**, 956 (1962).

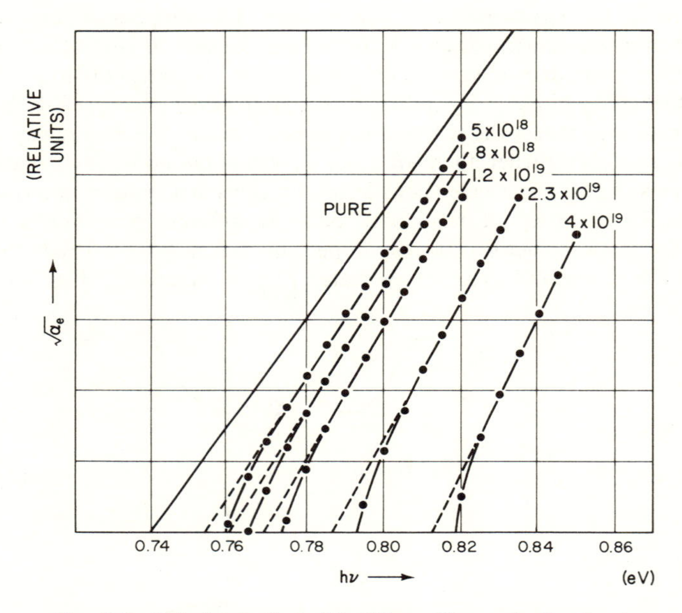

Fig. 3-5 Calculated effect of the filling of the conduction band on optical absorption in germanium at 4.2°K.[6]

In heavily doped indirect-gap semiconductors it is possible to conserve momentum by a scattering process† such as electron–electron scattering[7,8] or by impurity scattering.[6] In these cases the scattering probability is proportional to the number N of scatterers, and phonon assistance is not needed. Then the absorption coefficient becomes

$$\alpha(h\nu) = A\, N(h\nu - E_g - \xi_n)^2 \tag{3-14}$$

where A is a constant. The data for As-doped germanium is shown in Fig. 3-6. Note that the slope of the absorption edge increases with doping. In fact, the functional relationship of Eq. (3-14), which predicts that $d\sqrt{\alpha}/d(h\nu)$ is proportional to $N^{1/2}$, is verified in Fig. 3-7. The study of the absorption edge of germanium coupled with many other types of measurements has shown that heavy doping leads to an effective shrinkage of the energy gap,

[6]J. I. Pankove and P. Aigrain, *Phys. Rev.* **126**, 956 (1962).

†For a quantum-mechanical treatment of scattering processes the reader is referred to E. J. Johnson, *Semiconductors and Semimetals*, ed. R. K. Willardson and A. C. Beer, Academic Press (1967), Vol. 3, p. 183.

[7]C. Hass, *Phys. Rev.*, **125**, 1965 (1962).

[8]S. M. Ryvkin, *Physica Status Solidi*, **11**, 285 (1965).

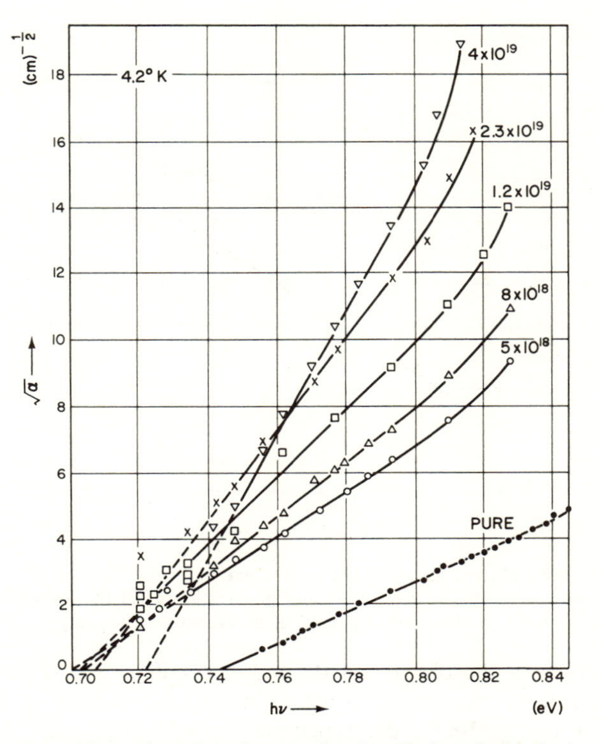

Fig. 3-6 Absorption edge of germanium heavily doped with As.[6]

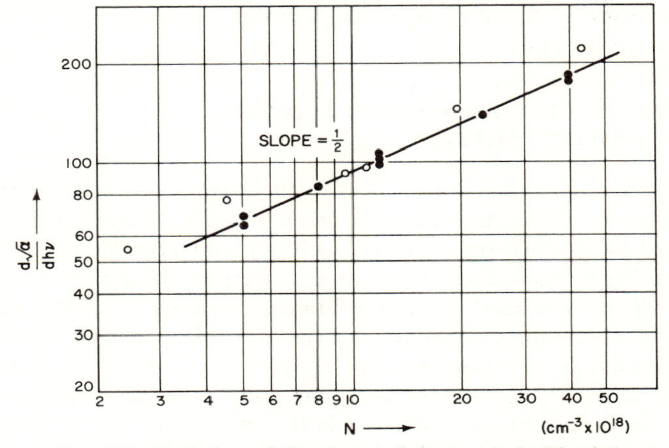

Fig. 3-7 Variation of the slopes of the curves in Fig. 3-6 as a function of carrier concentration.[6] The open circles have been calculated from the 80°K data of Haas.[7]

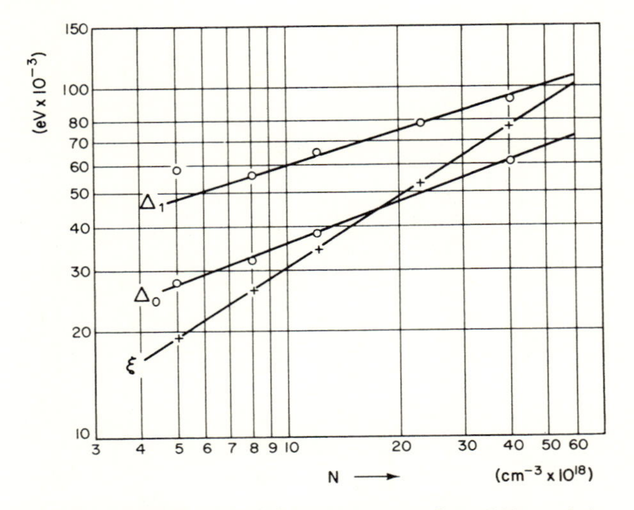

Fig. 3-8 Shrinkages Δ_0 of the energy gap at $k = \langle 000 \rangle$ and Δ_1 of the energy gap at $k = \langle 111 \rangle$ and penetration ξ_n of the Fermi level into the conduction band at 4.2°K as a function of doping.[9]

which affects both the direct and the indirect valleys[9] (see Fig. 3-8). The shrinkage Δ_1 of the indirect gap is determined by subtracting the zero intercepts of $\sqrt{\alpha}$ in Fig. 3-6 for the doped material $(E'_g + \xi_n)$ from that for the pure material $(E_g + E_p)$ and then adding the calculated ξ_n and subtracting E_p. The shrinkage Δ_0 of the direct gap is evaluated from the absorption edge at higher values of α.

3-A-4 INDIRECT TRANSITIONS BETWEEN DIRECT VALLEYS

Indirect transitions between direct valleys (Fig. 3-9) is very similar to our previous case of transitions between indirect valleys. Momentum is conserved by a second-order process such as phonon emission or absorption or scattering by impurities or by carriers. Here again, any occupied initial state in the valence band is connected to all the empty states of the conduction band. Hence the absorption coefficient for this case is given by the formula in Eqs. (3-11) through (3-13) if phonons are involved, and by the formula in Eq. (3-14) if phonons are not used to conserve momentum. In either case, the absorption coefficient in the absorption edge increases as the square of the value by which the photon energy exceeds some threshold. Such indirect transitions, being two-step processes, have a lower probability than

[9] J. I. Pankove, *Progress in Semiconductors* **9**, 48 (1965), ed. A. F. Gibson and R. E. Burgess, Heywood and Company.

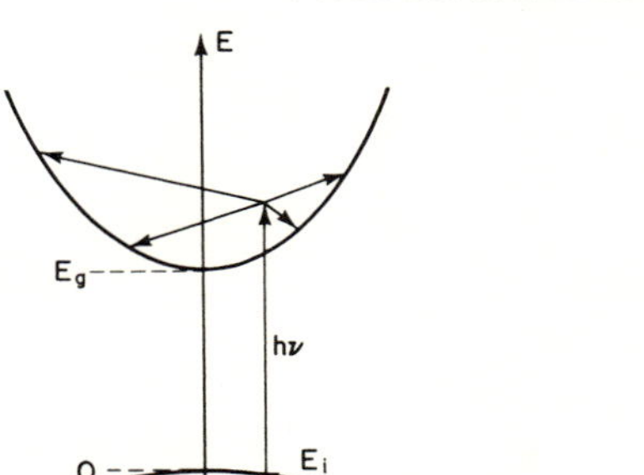

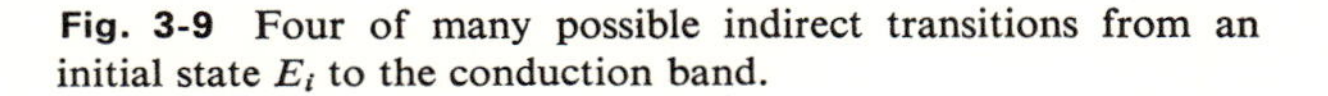

Fig. 3-9 Four of many possible indirect transitions from an initial state E_i to the conduction band.

the direct transition which would occur concurrently. The actual absorption coefficient would be the sum of the two contributions.

3-A-5 TRANSITIONS BETWEEN BAND TAILS

We have seen that momentum-conserving transitions between parabolic bands should result in an absorption edge which obeys the Eq. (3-2), i.e., where the absorption constant increases with the square root of the photon energy in excess of the gap energy. One usually plots the absorption edge semilogarithmically, as shown in Fig. 3-10. For direct transitions one expects no absorption below the energy gap and, therefore, a steeply rising absorption edge. But in practice one usually finds an exponentially increasing absorption edge.[10-20] In a number of materials, it is found that $d(\ln \alpha)/d(h\nu) = 1/kT$— this is known as Urbach's rule.[21] The exponential absorption edge of GaAs was found to correlate quite well with transitions involving band tails, which can be controlled by doping.[20]

Let us examine how an exponential distribution of states would affect the absorption coefficient. Consider the case of a degenerate p-type material.

[10] T. S. Moss and T. D. F. Hawkins, *Infrared Phys.* **1**, 111 (1961).
[11] T. S. Moss, *J. Appl. Phys.* **32**, 2136 (1961).
[12] M. D. Sturge, *Phys. Rev.* **127**, 768 (1962).
[13] W. J. Turner and W. E. Reese, *J. Applied Phys.* **35**, 350 (1964).
[14] D. E. Hill, *Phys. Rev.* **133**, A866 (1964).
[15] C. M. Chang, Stanford Electronics Laboratories Technical Report No. 5064–2 (unpublished).
[16] G. Lucovsky, *Appl. Phys. Letters* **5**, 37 (1964).
[17] I. Kudman and L. J. Vieland, *J. Phys. Chem. Solids* **24**, 967 (1963).
[18] W. G. Spitzer and J. M. Whelan, *Phys. Rev.* **114**, 59 (1959).
[19] D. Redfield, *Phys. Rev.* **140**, A2056 (1965).
[20] J. I. Pankove, *Phys. Rev.* **140**, A2059 (1965).
[21] F. Urbach, *Phys. Rev.* **92**, 1324 (1953).

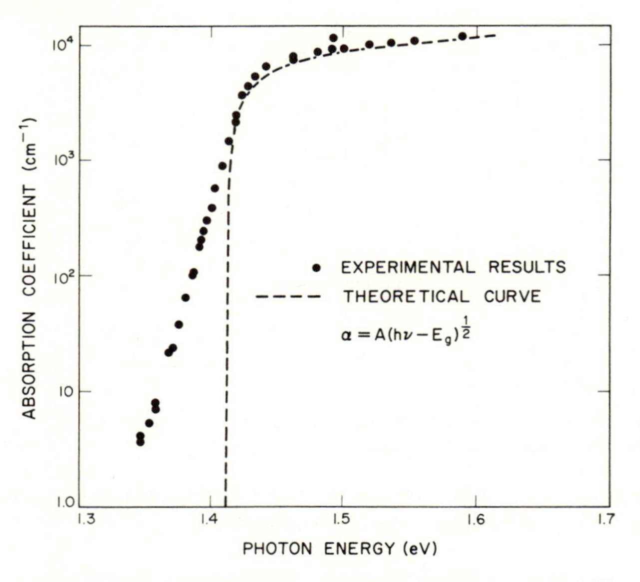

Fig. 3-10 Absorption edge of GaAs at room temperature.[10]

The Fermi level is taken to be in the parabolic portion of the valence band, so that the perturbed part of the valence band lies above the Fermi level. Then the density of initial states, N_i, is proportional to $|E_v|^{1/2}$, where E_v is the

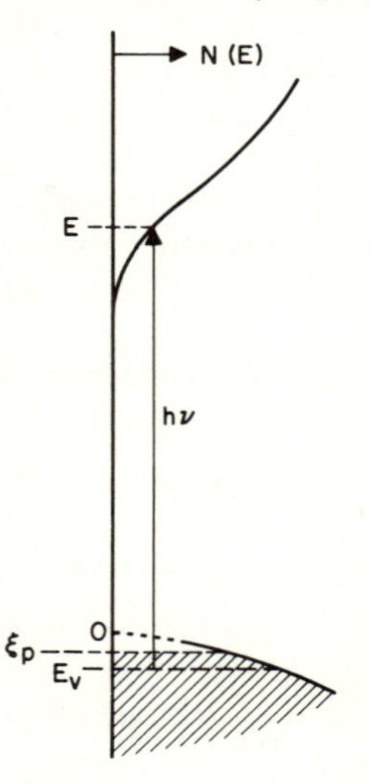

Fig. 3-11 Energy diagram illustrating how absorption probes the conduction band tail of states in a *p*-type semiconductor. The tail of the valence band has been omitted since, being empty, it does not participate in the absorption process.

[10] T. S. Moss and T. D. F. Hawkins, *Infrared Phys.* **1**, 111 (1961).

energy of the state with respect to what would be the edge of the parabolic valence band (Fig. 3-11). The final states form an exponential tail to the conduction band and their density at some energy E is given by

$$N_f = N_0 e^{E/E_0}$$

where E_0 is an empirical parameter having the dimensions of energy; E_0 describes the distribution of states but not their energy assignment. Let us assume that momentum conservation presents no problem in our optical transitions and that the matrix element for the transitions is constant, i.e., independent of the photon energy $h\nu = E - E_v$. As we have seen in Sec. 3-A-3, this is a reasonable assumption for heavily doped semiconductors.

The absorption coefficient is proportional to the product of the densities of initial and final states integrated over all the possible transitions for a given $h\nu$:

$$\alpha(h\nu) = A \int_{\xi_p}^{h\nu - \xi_p} |E_v|^{1/2} \exp \frac{E}{E_0} \, dE \tag{3-15}$$

where A is a constant. Let us substitute $E - h\nu$ for E_v and make the following change of variable:

$$x = \frac{h\nu - E}{E_0}$$

Equation 3-15 can be written

$$\alpha(h\nu) = - A e^{h\nu/E_0} (E_0)^{3/2} \int_{(h\nu + \xi_p)/E_0}^{\xi_p/E_0} x^{1/2} e^{-x} \, dx \tag{3-16}$$

The lower limit is set to ∞ because $h\nu \gg E_0$; this makes the integral independent of $h\nu$ and leads to the solution

$$\alpha(h\nu) = A(E_0)^{3/2} e^{h\nu/E_0} \left[\tfrac{1}{2}(\pi)^{1/2} - \int_0^{\xi_p/E_0} x^{1/2} e^{-x} \, dx \right]$$

The slope of the absorption edge on a semilogarithmic plot then gives

$$E_0 = \left[\frac{d(\ln \alpha)}{d(h\nu)} \right]^{-1} \tag{3-17}$$

As mentioned earlier, the experimentally determined absorption edge fits an exponential dependence of $h\nu$. Hence a quantity E_0 can be obtained and correlated with the impurity concentration as in Fig. 3-12.

We have assumed in our model that doping perturbs both the conduction and the valence bands, and that the Fermi level is inside the parabolic portion of the appropriate band. The transitions will link a parabolic band with the tail of the opposite band. Hence in n-type material, it is the tail of the valence band which is measured, while in p-type material it is the conduction-band tail which affects the measurement.

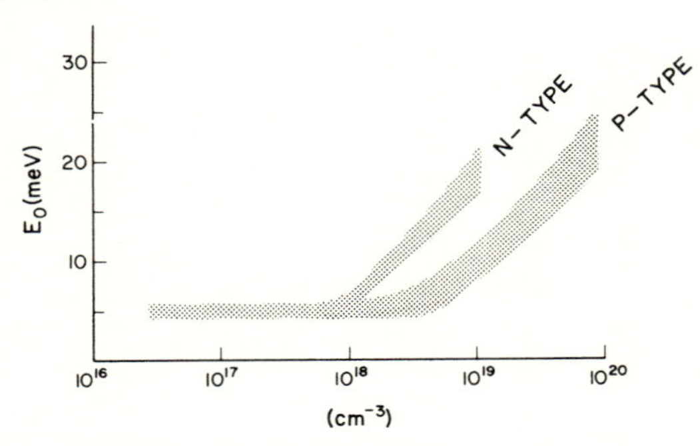

Fig. 3-12 Variation of the parameter E_0 with carrier concentration.[20]

The model of a tailing density of states has been supported indirectly by many observations other than the absorption spectrum. Perhaps the most direct observation of the tailing of states is provided by tunneling spectroscopy.[22] This technique is beyond our scope and, therefore, will not be discussed here, but it might become such an important new tool for finding the distribution of states that all future solid-state physicists should be aware of it.

We have seen in Chapter 1 how tails of states can result from the perturbation of the band edges by charged impurities. The perturbed band edges remain parallel at every point in space and, therefore, the energy gap is everywhere constant. But such a perturbation produces a local electric field which may be quite large, of the order of 10^5 V/cm. In Chapter 2, we indicated that the Franz–Keldysh effect smears the band edges. The connection between the local internal field and the Franz–Keldysh effect was made by Redfield, who proposed that the fluctuations of the internal field were responsible for the exponential absorption edge.[23] Before discussing the absorption edge from the point of view of variable local fields, it is appropriate to describe how a uniform field affects the absorption edge.

3-A-6 FUNDAMENTAL ABSORPTION IN THE PRESENCE OF A STRONG ELECTRIC FIELD

A simple way of expressing the Franz–Keldysh effect is to say that at a given energy there is a greater probability of finding the electron (or the hole) inside the energy gap (Fig. 3-13), or that the tunneling probability is increased when an electric field $\mathscr{E}$ is present.

[20] J. I. Pankove, *Phys. Rev.* **140**, A2059 (1965).
[22] G. D. Mahan and J. W. Conley, *Appl. Phys. Letters* **11**, 29 (1967).
[23] D. Redfield, *Phys. Rev.*, **130**, 916 (1963).

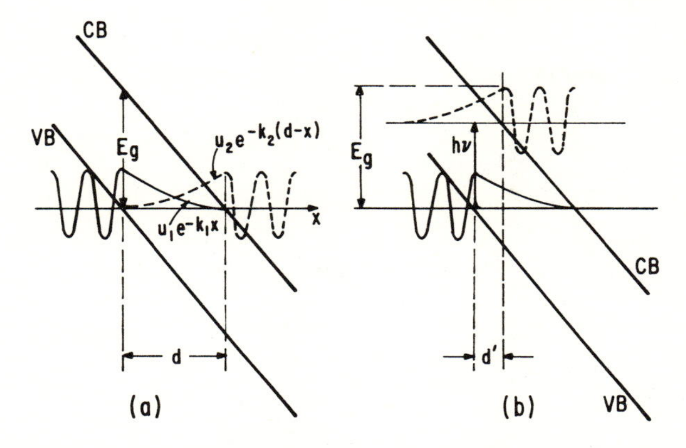

Fig. 3-13 Electron tunneling (a) without change in energy, (b) with photon absorption.

In the presence of an electric field, the probability of finding an electron in the energy gap is described by an exponentially decaying function ue^{ikx} where k is imaginary[24] (Fig. 3-13). The valence electron must tunnel through a triangular barrier to appear in the conduction band. The height of this barrier is E_g and its thickness has a value d such that

$$d = E_g/q\mathscr{E} \tag{3-18}$$

where $\mathscr{E}$ is the electric field. As the field is increased, the tunneling distance is reduced, and the overlap of the wave functions describing the probability of finding an electron in the gap is increased.

As shown in Fig. 3-13(b), the assistance of a photon $h\nu$ is equivalent to reducing the barrier thickness to a value

$$d' = (E_g - h\nu)/q\mathscr{E} \tag{3-19}$$

then the overlap of the wave functions is increased making the tunneling transfer more probable.

This tunneling involves, of course, only the longitudinal component of momentum. The imaginary values of k occur in the longitudinal direction, i.e., in the direction of tunneling. Only the transverse momentum is conserved. The longitudinal momentum, parallel to the electric field, goes to zero at the band edge (the turning point). To take these important details into account requires a higher level of treatment, which can be found in Kane[25] who treats the case of tunneling at constant total energy; in Morgan,[26]

[24]See for example S. Wang, *Solid State Electronics*, McGraw-Hill (1966), p. 369.
[25]E. O. Kane, *J. Phys. Chem. Solids* **12**, 181 (1959).
[26]T. N. Morgan, *Phys. Rev.* **148**, 890 (1966).

who treats the case of tunneling assisted by photon emission; and in Keldysh,[27] who treats the case of tunneling assisted by photon absorption.

The Franz–Keldysh effect in fundamental absorption—i.e., photon-assisted tunneling—was first observed in CdS.† It appeared as a uniform shift of the absorption edge to a lower energy. Similar observations have been made in Si,[28] GaAs[29-31] (Fig. 3-14) and in Ge.[32] Note that tunneling-assisted absorption is made more probable by the presence of an electric

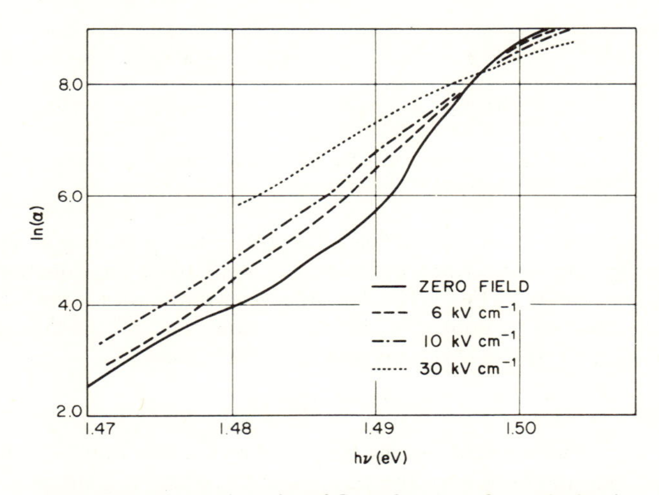

Fig. 3-14 Absorption edge of GaAs from set of crystals showing good agreement in regions of overlapping data. Zero-field, 6-, 10-, and 30-kV cm^{-1} curves are derived from approximately 200, 100, 100, and 50 points, respectively. Estimated standard error in ln (α) for each point is typically 0.03. Optical resolution 500μ eV.[30]

field. Therefore, for a discrete level or a narrow band of states (for example, an exciton), it is a broadening effect and not a shift that one should observe. However, when only the low-energy edge is observed, as in the present case, the Franz–Keldysh broadening of this edge appears as a shift to lower energies.

[27]L. V. Keldysh, *J. Exptl. Theoret. Phys.* (USSR) **47**, 1945 (1964), transl. *Sov. Phys. JETP* **20**, 1307 (1965).

[28]V. S. Vavilov and K. I. Britsyn, *Fizika Tverdogo Tela* (USSR) **2**, 1936 (1960), transl. *Soviet Physics—Solid State* **2**, 1746 (1961)); *Fizika Tverdogo Tela* **3**, 2497 (1961), transl. *Soviet Physics—Solid State* **3**, 1816 (1962).

[29]T. S. Moss, *J. Appl. Phys. Supplement* **32**, 2136 (1961).

[30]E. G. S. Paige and H. D. Rees, *Phys. Rev. Letters* **16**, 444 (1966).

[31]C. M. Penchina, A. Frova, and P. Handler, *Bull. A. P. S.*, Series II, **9**, 714 (1964).

[32]A. Frova and P. Handler, *Phys. Rev.* **137**, A1857 (1965).

†R. Williams, *Phys. Rev.* **117**, 1487 (1960).

While we are on the subject of the effect of a high electric field, let us consider briefly what has been called a "Stark splitting" of the band levels. Actually, the dominant effect is that for photon energies larger than the gap energy, the transition probability assumes an oscillatory dependence on energy.[33] The theory predicts that at very high fields the absorption spectrum should form a damped oscillation. A periodicity of this type has been observed in GaAs, as shown in Fig. 3-15.[30]

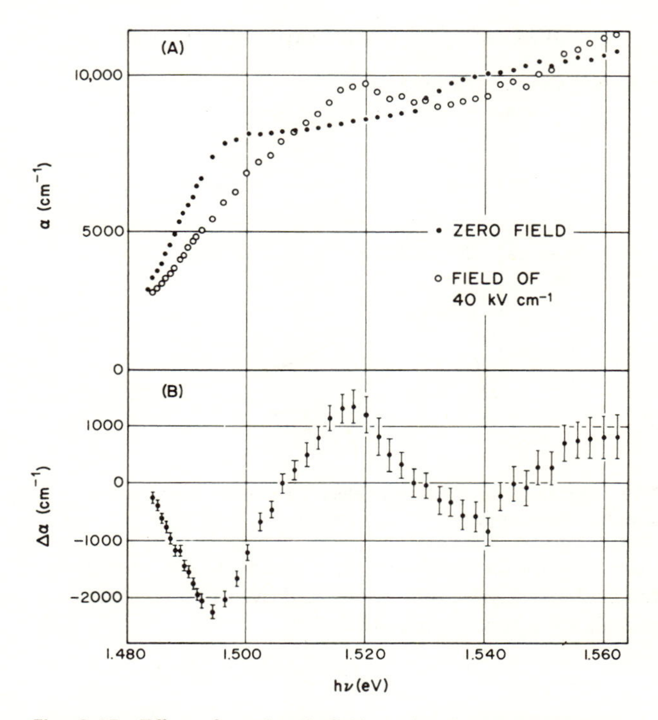

Fig. 3-15 Effect of an electric field on the absorption spectrum above the fundamental edge: (a) absorption spectra of GaAs at zero field and 40 kv cm^{-1}; (b) difference between spectra. Optical resolution 1 meV.[30]

Let us return now to the influence of local internal fields on the absorption edge. These vary from point to point in the crystal. To find an average value of the local field, one assumes that the impurities or defects are uniformly distributed in the crystal. If their concentration is N per cm^3, their mean spacing r_0 is given by the volume reserved to each impurity:

[33] J. Callaway, *Phys. Rev.* **134**, A998 (1964).

$$\tfrac{4}{3}\pi(r_o)^3 = \frac{1}{N} \tag{3-20}$$

If the impurity is singly charged, the "normal" field is defined by[34]

$$\mathscr{E}_o = \frac{q}{k(r_o)^2} = 2.6\frac{q}{k}N^{2/3} \tag{3-21}$$

The probability distribution of electric fields of various intensities $\mathscr{E}$ is described by the expression

$$W(\mathscr{E})\,d\mathscr{E} = \frac{3}{2\mathscr{E}}\left(\frac{\mathscr{E}_o}{\mathscr{E}}\right)^{3/2} \exp\left[-\left(\frac{\mathscr{E}_o}{\mathscr{E}}\right)^{3/2}\right] d\mathscr{E} \tag{3-22}$$

This function is shown in Fig. 3-16.

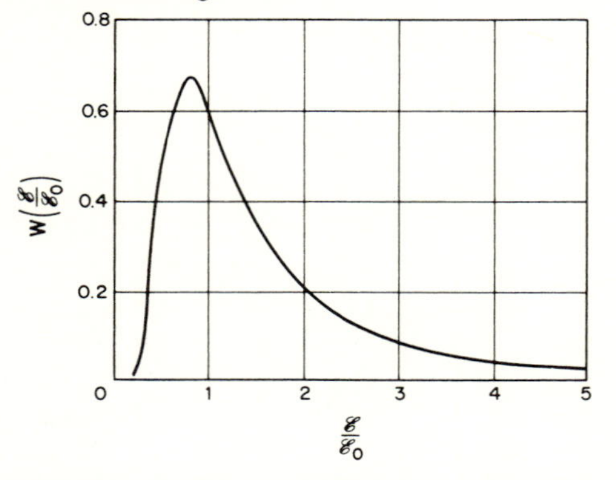

Fig. 3-16 Probability distribution of electric fields due to charged impurities in a semiconductor.[34]

To find the absorption coefficient at a given photon energy $h\nu$, the local-absorption coefficient $A(h\nu, \mathscr{E})$ is integrated over all the values of $\mathscr{E}$ weighted by the probability of occurrence, $W(\mathscr{E})$[35]:

$$\alpha(h\nu) = \int_0^\infty A(h\nu, \mathscr{E})\, W(\mathscr{E})\, d\mathscr{E} \tag{3-23}$$

Such a calculation was made for the case where the field is generated by a surface potential.[36] The field decreased from a maximum at the surface [an upper value for the integration of Eq. (3-23)]. The resulting absorption edge is exponential over at least four orders of magnitude. The use of a surface potential as a source of internal field is attractive from an experimental point of view because it can be operated upon in a variety of ways (chemical, electrical, optical) and because the surface internal fields can be separated from bulk internal fields by changing the thickness of the specimen.

[34] D. Redfield, *Phys. Rev.* **130**, 914 (1963).
[35] D. Redfield, *Phys. Rev.* **130**, 916 (1963).
[36] D. Redfield, *Phys. Rev.* **140**, A2056 (1965).

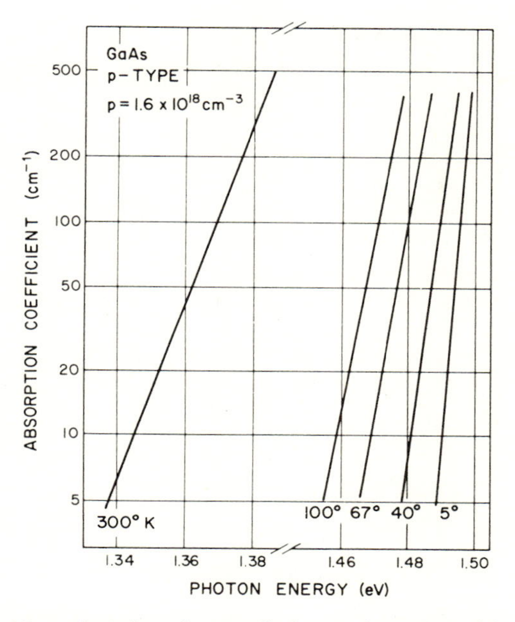

Fig. 3-17 Absorption edge of *p*-type GaAs over its exponential range corrected for free carrier absorption.[37]

The most striking experimental demonstration that the exponential absorption edge results from photon-assisted tunneling between tails of states induced by coulomb interaction with charged impurities is as follows.[37] The slope of the exponential absorption edge of *p*-type GaAs varies with temperature as shown in Fig. 3-17. This was interpreted as resulting from the temperature-dependent net charge at the acceptors. As the temperature increases, more acceptors become ionized (negatively charged) and the perturbation of the band edges (extent of tails) becomes greater. This interpretation was tested by using slightly overcompensated GaAs with about the same impurity concentration. Now all the acceptors have received an electron from the donors; therefore, they are fully ionized. An extensive exponential absorption edge is obtained (Fig. 3-18). Raising the temperature cannot increase the ionization of the already fully charged acceptors; therefore, the perturbation of the band edges does not change and the slope of the absorption edge remains constant. The displacement of the absorption edge to lower energy as the temperature increases is due to the temperature dependence of the energy gap.

[37]D. Redfield and M. A. Afromowitz, *Appl. Phys. Letters* **11**, 138 (1967).

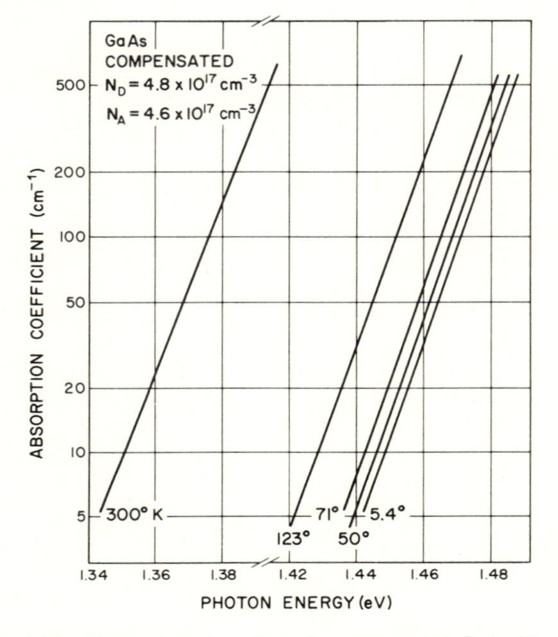

Fig. 3-18 Absorption edge of nearly compensated GaAs.[37]

3-B Higher-energy Transitions

We have seen in the energy–momentum diagrams that higher energies separate the conduction band from the valence band as one goes away from the lowest minimum in the conduction band. Direct transitions from valence band to conduction band can occur at almost all points in momentum space (except those forbidden by selection rules). These transitions have energies greater than the energy gap. Furthermore, as shown in Fig. 3-19, there are subbands in both conduction and valence bands which allow still higher transitions. The valence band in most type-IV and type-III–V semiconductors is split by spin–orbit interaction. This interaction lowers one of the three subbands to appreciably lower energies. Thus in the absorption spectrum a direct transition from the top of the valence band can be accompanied by a small peak or step at a slightly higher energy corresponding to a transition from the lower subband (Fig. 3-20).

The study of higher-energy transitions by transmission spectroscopy requires that the specimen be extremely thin, since the absorption coefficient α is very high (the radiation is attenuated by a factor $e^{-\alpha x}$ over a distance x).

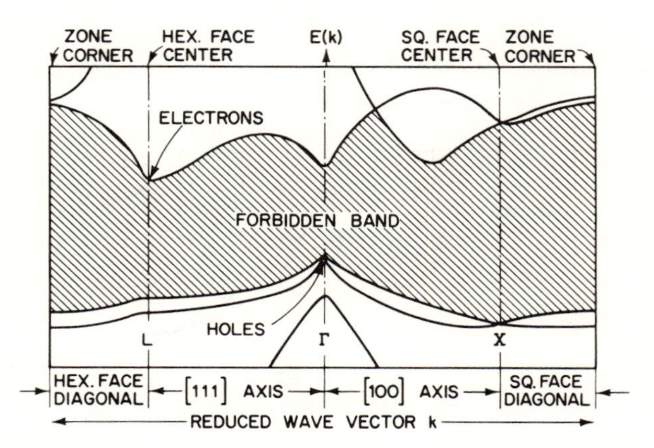

Fig. 3-19 Schematic diagram of the energy band structure of germanium. The conduction band edge occurs at the hexagonal face centers, while the valence band edge occurs at the central zone point. The width of the forbidden band is 0.65 eV at room temperature. The spin-orbit splitting at the Γ valence band edge is approximately 0.28 eV. The lowest conduction band has three types of minima: four $\langle 111 \rangle$ minima (at the band edge); one $\langle 000 \rangle$ minimum; and six $\langle 100 \rangle$ minima. The $\langle 000 \rangle$ minimum lies 0.15 eV above the $\langle 111 \rangle$ minima; the $\langle 100 \rangle$ minima lie 0.18 eV above the $\langle 111 \rangle$ minima.[38]

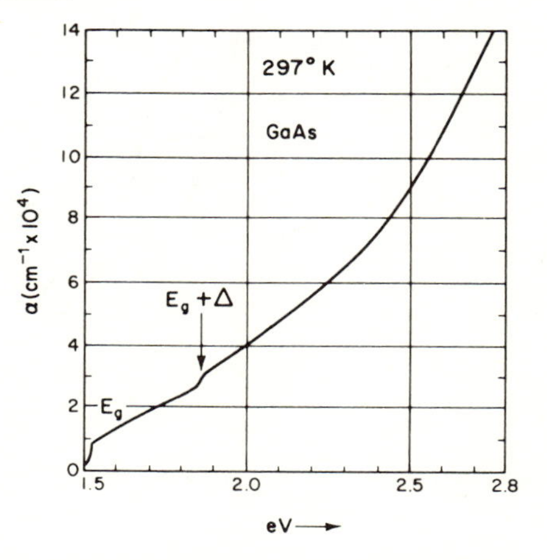

Fig. 3-20 Absorption of a thin sample of GaAs showing the spin-orbit splitting $E_g + \Delta$ of the fundamental absorption edge E_g.[12]

[38] F. Herman, *Proc. IRE*, 1716 (1955).
[12] M. D. Sturge, *Phys. Rev.* **127**, 768 (1962).

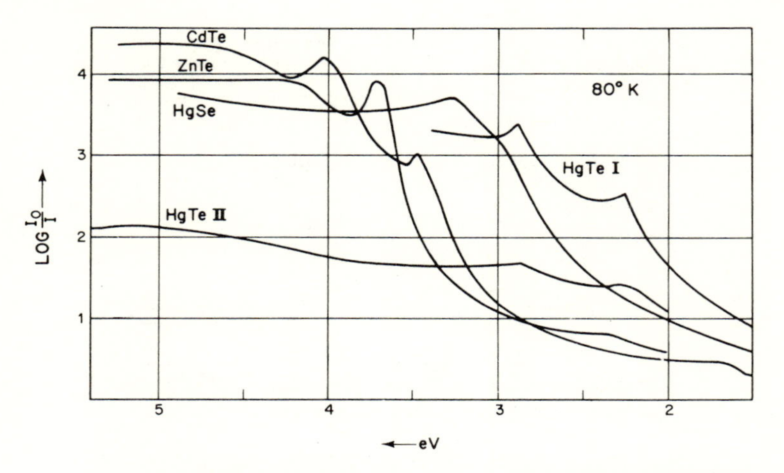

Fig. 3-21 Absorption of films of some II–VI compounds at 80°K as a function of photon energy.[40] Thickness 0.25μ (CdTe), 0.09μ (ZnTe), 0.15μ (HgTe I), ~0.05μ (HgTe II), and 0.06μ (HgSe).

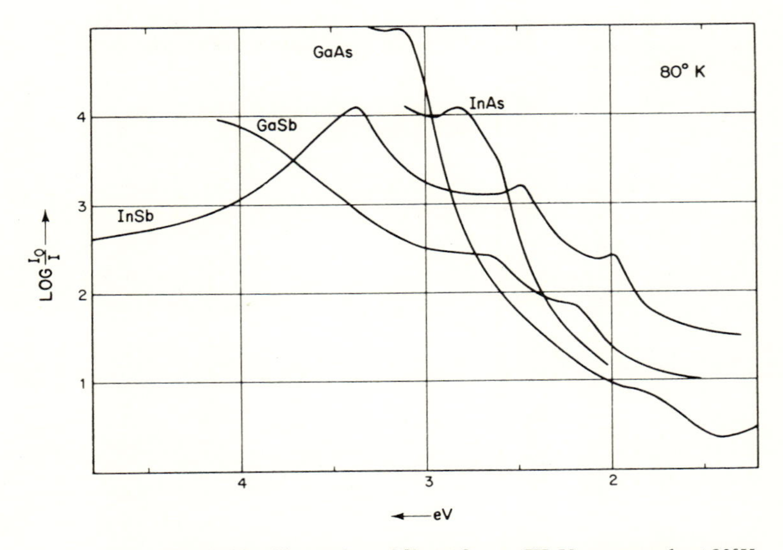

Fig. 3-22 Absorption of films of some III–V compounds at 80°K as a function of photon energy.[40] Thickness 0.25μ (InSb), 0.08μ (GaSb), 0.18μ (InAs), and 0.24μ (GaAs).

[40] M. Cardona and G. Harbeke, *J. Appl. Phys.* **34**, 813 (1963).

Table 3-1[40]

Energies of absorption singularities of thin films and of the corresponding reflectivity peaks of bulk material (all in eV) and their temperature coefficients $10^{-4} eV/°K$

	Material	Ge	InSb	InAs	GaSb	GaAs	HgTe	HgSe	CdTe	ZnTe	ZnSe
	E_1 (295°K)	2.15	1.89	2.50	2.07	2.97	2.14	2.87	3.38	3.52	4.80
	$E_1 + \Delta_1$ (295°K)	2.35	2.44	2.75	2.56	3.17	2.78	3.20	3.96	4.19	5.10
Absorption	$-(dE_1/dT)$, $T < 295°K$	4.15±0.2	4.0±0.2		4.0±0.4	4.2±0.5	4.5±0.2	4.6±0.2	5.0±0.3	5.2±0.3	5.2±0.3
	$-(dE_1 + \Delta_1)/dT$, $T < 295°K$		3.6±0.2		3.6±0.5		3.8±0.3	3.8±0.3	3.8±0.3	4.1±0.3	4.1±0.5
	E_1 (295°K)	2.01 2.015	1.82 1.835	2.52 2.48	2.02 2.015	2.94 2.895	2.08	2.82	3.29	3.57	4.75
Reflection	$E_1 + \Delta_1$ (295°K)	2.28 2.30	2.38 2.35	2.81 2.745	2.48 2.46	3.20 3.13	2.77	3.13	3.84	4.13	5.10
	$-(dE_1/dT)$ $T < 295°K$	4.2±0.4			4.2±0.6			4.3±0.6	5.0±0.6		
	$T > 295°K$	4.2±0.3	5.3±0.3	5.4±0.3	4.6±0.3	4.6±0.3					
	$-(dE_1 + \Delta_1)/dT$, $T > 295°K$	4.4±0.3	4.9±0.3	5.5±0.3	5.4±0.3	6.2±0.3					
	$T < 295°K$	4.2±0.4					6.4		6.0	6.4	

[40] M. Cardona and G. Harbeke, *J. Appl. Phys.* **34**, 813 (1963).

Thin samples are difficult to prepare and are always strained. Hence they are rarely used. Extremely thin films can be prepared by evaporation; these also are usually strained and polycrystalline. However, absorption spectra of such films reveal higher-energy absorption edges and peaks which can be interpreted with the help of theoretical predictions[39] from band-structure calculations, and these in turn provide numerical fixes for higher-order calculations. Examples of structure in the absorption spectra of II–VI and III–V compounds due to higher-energy transitions are shown in Figs. 3-21 and 3-22. In Table 3-1 this data is interpreted in terms of the L-transition across the gap (near $k = 111$) and in terms of the L-transition from the split-off valence subband. In this same table the absorption data is compared with data obtained by reflectance measurements.

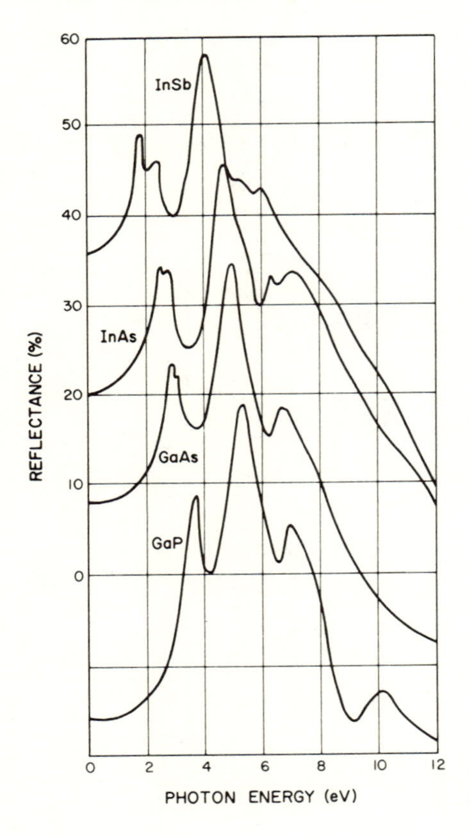

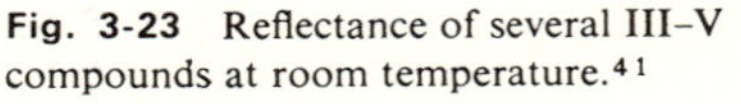
Fig. 3-23 Reflectance of several III–V compounds at room temperature.[41]

[39] M. Cardona, *J. Phys. Chem. Solids* **24**, 1543 (1963).
[41] H. Ehrenreich, H. R. Philipp, and J. C. Phillips, *Phys. Rev. Letters* **8**, 59 (1962).

The visible and UV reflectance spectrum from polished, cleaved, or etched surfaces of bulk semiconductors shows a series of peaks (Fig. 3-23) which correspond to direct transitions at higher energies. The intensity and sharpness of these transitions results from the high densities of initial and final states involved. Thus the peaks are quite intense for transitions where the two levels have the same energy difference over a large range of position in momentum space, i.e., where the valence and conduction subbands have parallel branches in E–K space. Up to about 10 eV, the structure characterizes transitions from the valence band to higher levels in the conduction band. From about 10 to 16 eV, a decrease in reflectance is obtained; this decrease is attributed to the excitation of lossy collective plasma oscillations of the valence electrons. Above about 16 eV, the reflectance spectrum exhibits a broad maximum, which is attributed to transition from a filled band below the valence band to empty states in the conduction band.[42]

3-C Exciton Absorption

3-C-1 DIRECT AND INDIRECT EXCITONS

The formation of excitons usually appears as narrow peaks in the absorption edge of direct-gap semiconductors, or as steps in the absorption edge of indirect-gap semiconductors. The theory of exciton absorption has been worked out by Elliott.[43]

In direct-gap materials the free exciton occurs when the photon energy is $h\nu = E_g - E_x$ (E_x is the binding energy of the exciton). At $k = 0$, this is a very pronounced transition, which broadens with temperature as shown in Fig. 3-24. Since excitons can be created with some kinetic energy, it is evident that they can also be created by higher-energy photons, thus contributing a component to the absorption coefficient in the region of band-to-band transitions.

In indirect-gap materials phonon participation is needed to conserve momentum. Therefore, an increase in absorption coefficient is obtained at

$$h\nu = E_g - E_p - E_x$$

for the transition with phonon absorption, and

$$h\nu = E_g + E_p - E_x$$

for the transition with phonon emission.

There are two transverse (T) and one longitudinal (L) phonons in each of the acoustic (A) and optical (O) branches of the phonon spectrum. More than

[42] H. R. Philipp and H. Ehrenreich, *Phys. Rev. Letters* **8**, 1 (1962).
[43] R. J. Elliott, *Phys. Rev.* **108**, 1384 (1957).

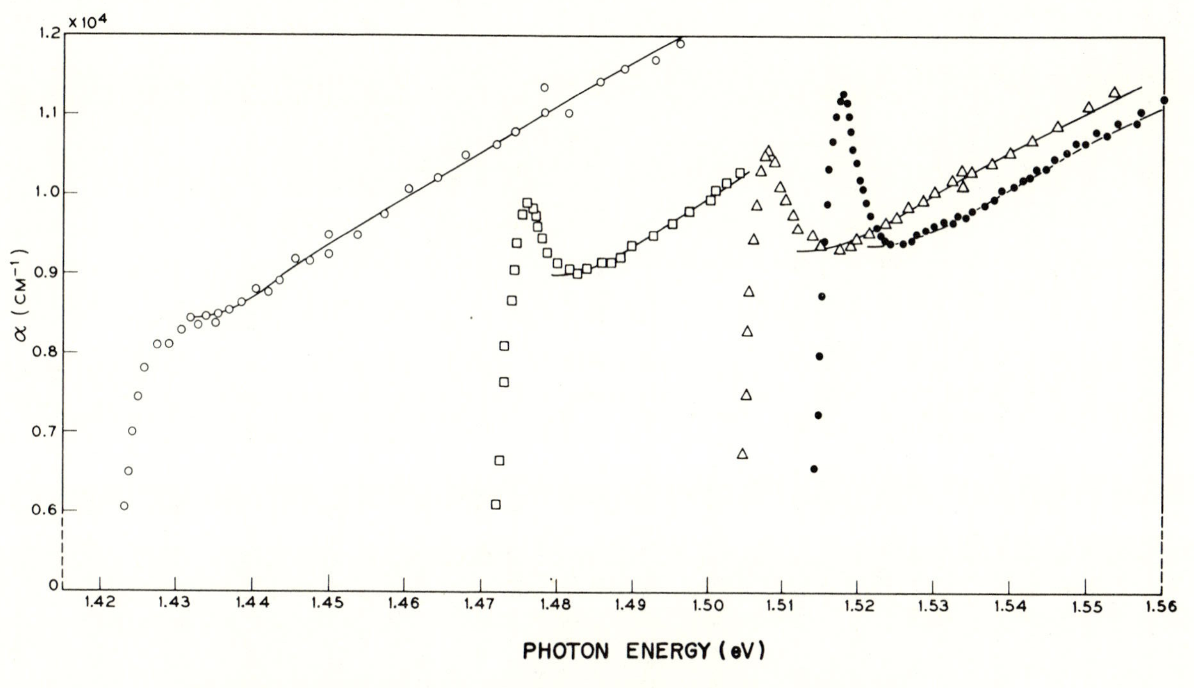

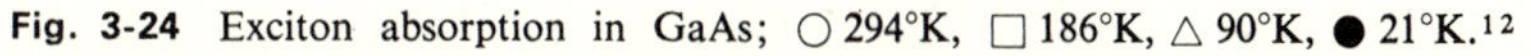

Fig. 3-24 Exciton absorption in GaAs; ○ 294°K, □ 186°K, △ 90°K, ● 21°K.[12]

[12]M. D. Sturge, *Phys. Rev.* **127**, 768, (1962).

one phonon can participate in the transitions, and they can be absorbed or emitted in various combinations. Hence a great number of steps in the absorption edge can be obtained—as, for example, in Fig. 3-25.

In indirect transitions, one observes steps in the absorption edge rather than peaks, because the phonons allow connecting states with the same velocity dE/dk at photon energies greater than those connecting the band-edge excitons (where $dE/dk = 0$).

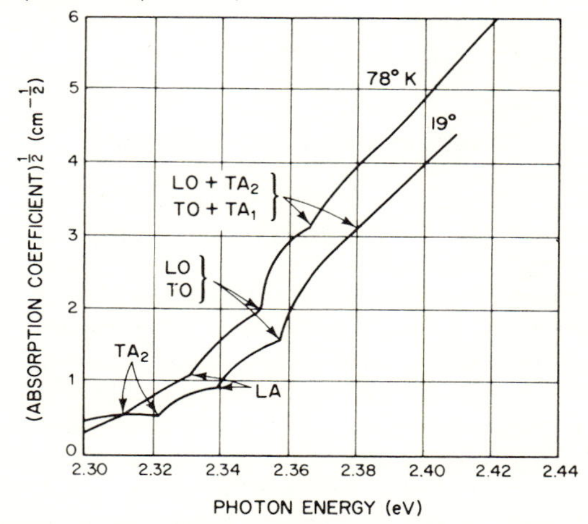

Fig. 3-25 Threshold energies for the formation of excitons with phonon emission in the absorption edge of GaP.[44]

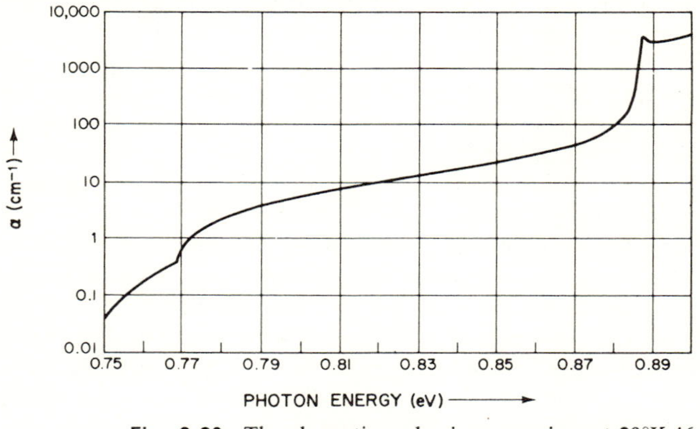

Fig. 3-26 The absorption edge in germanium at 20°K.[46]

[44] M. Gershenzon, D. G. Thomas, and R. E. Dietz, *Proc. Int. Conference on Semiconductor Physics*, Exeter, Inst. of Phys. and Phys. Soc., London (1962), p. 752.

[46] T. P. McLean, "The Absorption Edge Spectrum of Semiconductors," *Progress in Semiconductors*, ed. A. F. Gibson et al., Heywood and Company (1960), Vol. 5, p. 87.

Even in indirect-gap semiconductors it is possible to create direct excitons provided there is a set of direct critical points. In germanium one such set occurs at $k = 0$ (the direct gap). Figure 3-26 shows the absorption edge of germanium over a large range of absorption coefficients: one distinguishes the phonon-absorption branch at low energies (this scale does not reveal the indirect-exciton structure[45]); then a step at about 0.77 eV corresponds to the onset of the phonon-emission branch; and, at the highest energies, the absorption coefficient increases rapidly where direct transitions are possible. Here the direct-exciton peak is clearly resolved.

3-C-2 EXCITON ABSORPTION IN THE PRESENCE OF AN ELECTRIC FIELD[47]

In nearly pure semiconductors at low temperatures the few impurities contained in the crystal are neutral. However, a small electric field (5 to 30 V/cm) is able to ionize the impurities.[48] The ionization of impurities results in two cooperative effects which can be seen in absorption: (1) the field-ionized impurities perturb the bands by a coulomb interaction which induces strong local fields and tails of states and thus changes the slope of the absorption edge; (2) the ionized carriers screen the coulomb interaction between electrons and holes, thereby reducing the probability for exciton formation and eliminating the exciton peak. Figure 3-27(a) shows the 4.2°K absorption spectrum of antimony-doped germanium. Curve 1 was obtained without applied electric field; curve 2 was obtained with a field of 100 V/cm. Note the disappearance of the exciton peak and the more gradual slope of the direct absorption edge. Note also that the slope of the direct absorption edge when the donors are emptied by the electric field is the same as when the donors are emptied thermally (curve 3). The change in absorption coefficient due to the electric field is shown in Fig. 3-27(b) for three differently doped specimens. A marked decrease in absorption occurs at the exciton peak, while the absorption increases on either side of the peak, as if the exciton line had been greatly broadened by the electric field.

The ionization can be readily determined by monitoring the $I(V)$ characteristic: it is found that the absorption modulation sets in at the same field as that at which impact ionization occurs. The modulation of the absorbence saturates with increasing applied field, corresponding to the complete ionization of all the impurities. Hence the modulation is associated with the appearance of free carriers. Note that the fields used in these experiments are much lower than those needed to ionize excitons ($\sim 10^3$ V/cm). Hence the

[45] G. C. MacFarlane, T. P. McLean, J. E. Quarrington, and V. Roberts, *Phys. Rev.* **108**, 1377 (1957).

[47] V. M. Asnin, G. L. Eristavi, and A. A. Rogachev, *Phys. Stat. Sol.* **29**, 443 (1968).

[48] E. I. Zavaritskaya, *Soviet Phys. Solid State* **7**, 1983 (1966).

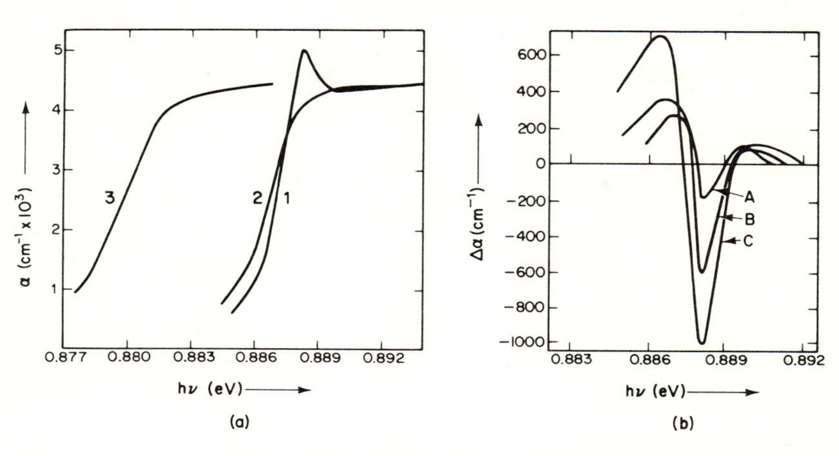

Fig. 3-27 (a) Absorption spectra of antimony-doped germanium with $N_d = 8 \times 10^{15}$ cm^{-3}: (1) $T = 4.2$°K, without field; (2) $T = 4.2$°K, with field $E = 100$ V/cm; (3) $T = 77$°K. (b) Spectra of germanium absorption coefficient modulation at 4.2°K in electric fields corresponding to total ionization of impurity centers: (A) $N_d = 6 \times 10^{14}$ cm^{-3}; (B) $N_d = 2 \times 10^{15}$ cm^{-3}; (C) $N_d = 8 \times 10^{15}$ cm^{-3}.

present effect is well accounted for by the screening of electrons and holes by the free carriers released from the impurities.

3-D Absorption due to Isoelectronic Traps

Isoelectronic centers are formed by replacing substitutionally one atom of the crystal by another atom of the same valence. This is the case of nitrogen substituting for phosphorus in GaP (or for bismuth on the Ga-sites of GaP). The isoelectronic trap can bind an exciton. The details of the trapping process can be visualized as follows: The nitrogen isoelectronic trap is a very localized potential well which can trap an electron, thus becoming charged; the resulting coulomb field attracts a hole; now the two trapped carriers form an exciton bound to the isoelectronic trap. When these centers are close together, they may interact with each other in pairs and form a new set of energy levels for their own bound excitons.[49] The binding energy of an exciton bound to isoelectronic centers depends on the distance between the two atoms of a pair. The strongest binding energy occurs for the nearest pairs. The absorption spectrum due to such excitons is shown in Fig. 3-28(a). The lowest-energy absorption peak (a dip in transmission) corresponds to the

[49] D. G. Thomas, J. J. Hopfield, and C. J. Frosch, *Phys. Rev. Letters* **15**, 857 (1965).

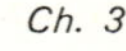

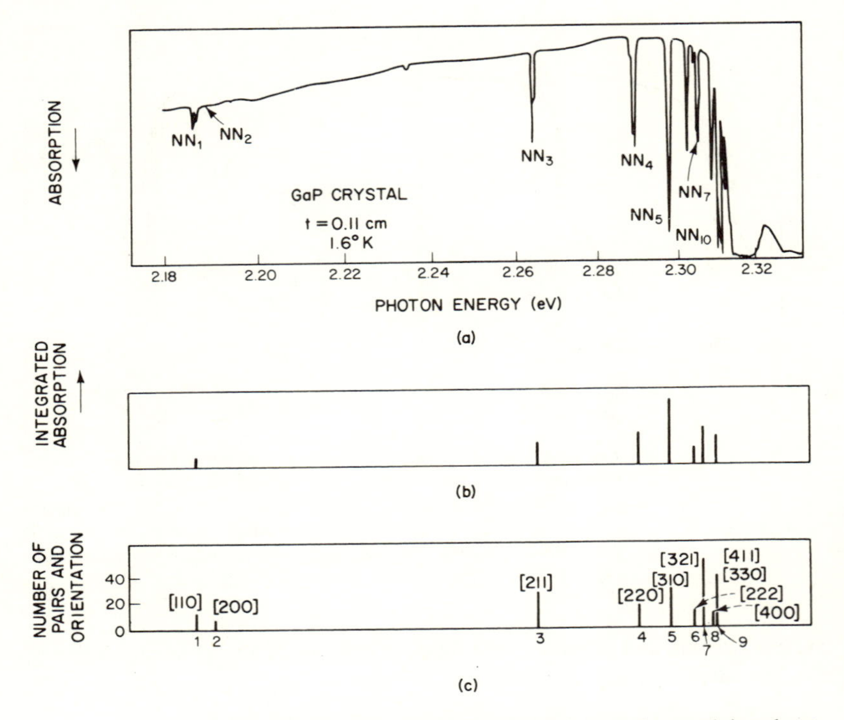

Fig. 3-28 *NN* lines (a) in absorption in crystals containing about 10^{19} *N* atoms/cc. (b) shows the approximate relative strengths of the lines in absorption. (c) shows the expected relative intensities of the lines if it is assumed that the nitrogen atoms are arranged randomly and that relative numbers alone control the transition intensities. The orientations of the pairs are also indicated.[49]

most strongly bound exciton of a nearest neighbor *N–N* pair; other absorption peaks correspond to more distant pairings. The intensity of the absorption peak depends on the number of pairings which can form about one nitrogen atom in a given crystallographic direction. This relationship is compared in Fig. 3-28(b) and (c).

3-E Transitions between a Band and an Impurity Level

The transition between a neutral donor and the conduction band or between the valence band and the neutral acceptor can occur by the absorption

[49]D. G. Thomas, J. J. Hopfield, and C. J. Frosch, *Phys. Rev. Letters* **15**, 857 (1965).

of a low-energy photon (Fig. 3-29, transitions *a* and *b*). For this absorption process the energy of the photon must be at least equal to the ionization energy E_i of the impurity. Typically, this energy corresponds to the far infrared region of the spectrum. A good example of such an absorption spectrum is that of Fig. 3-30. No absorption is obtained until the photon energy equals the first-excitation energy of the donor [quantum number $n = 1$ in Eq. (1-11)]. Absorption peaks are obtained for excitation to states $n = 1$, 2, and 3. The higher modes merge into a band corresponding to the complete ionization of the donor. Note that although the density of final states (the conduction band) increases with energy, the absorption coefficient for complete ionization of the impurity (transition to the conduction band) decreases with energy. This decrease of the absorption coefficient with energy beyond the broad peak of Fig. 3-30 is due to a rapid decrease of the transition probability away

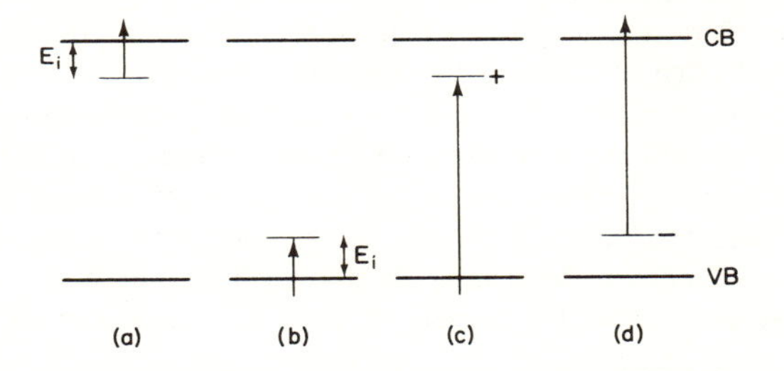

Fig. 3-29 Absorbing transitions between impurities and bands: (a) donor to conduction band; (b) valence band to acceptor; (c) valence band to donor; (d) acceptor to conduction band.

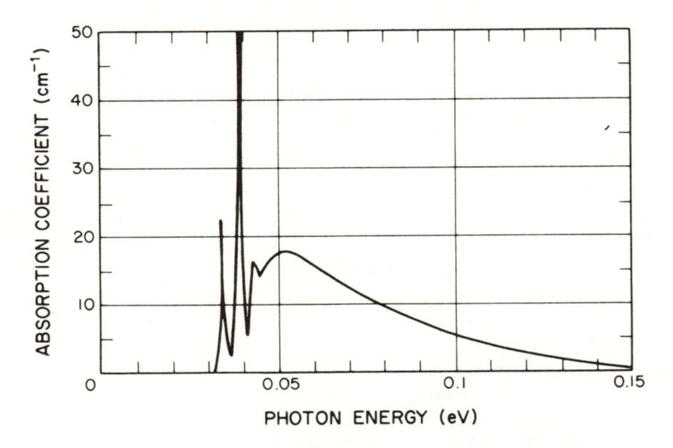

Fig. 3-30 Absorption coefficient of a boron-doped silicon sample as a function of photon energy.[50]

[50]E. Burstein, G. S. Picus, and N. Sclar, *Proc. Photoconductivity Conference, Atlantic City*, Wiley (1956), p. 353.

from the bottom of the conduction band. At $k \neq k_o$, where k_o is the crystal momentum at the bottom of the conduction band, the wave function of the impurity decreases approximately as $1/[1 + (k - k_o)^2]^2$.[51] Hence the probability of finding an electron in the lower state for a direct transition to the conduction band at larger $(k - k_o)$, i.e., at higher energies, decreases. The limited extent of the donor in momentum space is then responsible for the declining absorption at photon energies greater than the binding energy of the donor.

The transition between the valence band and an ionized donor (it must be empty to allow the transition) or between an ionized acceptor and the conduction band occurs at photon energies which are given by

$$h\nu > E_g - E_i$$

[See Fig. 3-29, transitions (c) and (d).]

Unlike the exciton absorption which occurs between a discrete level and the well-defined edge of a band, the transition between an impurity and a band involves the whole band of levels. Hence transitions between an impurity and a band should manifest themselves by a shoulder in the absorption edge at a threshold lower than the energy gap by an amount E_i, as illustrated in Fig. 3-31. The absorption coefficient for transitions involving the impurity level covers a much smaller range than transitions between valence and conduction bands because the density of impurity states is much lower than

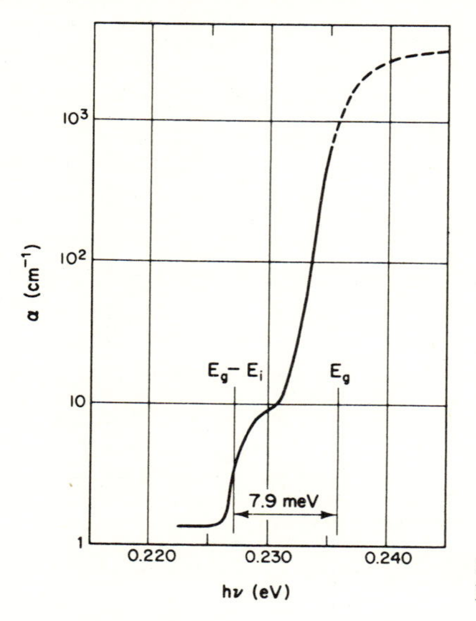

Fig. 3-31 Impurity absorption in InSb, $T \sim 10°K$.[52]

[51] W. Kohn, *Solid State Physics*, Academic Press (1957), Vol. 5, p. 257.
[52] E. J. Johnson and H. Y. Fan, *Phys. Rev.* **139**, A1991 (1965).

the density of states in the bands. In practice, the shallow impurities are seldom resolved from the background of absorption due to transitions involving tails of states. In the case of deep levels, when E_i is large compared to the width of the absorption edge, the impurity can contribute a definite step in the absorption spectrum.

In the low-energy transition between the impurity level and the nearest band edge there is no problem with momentum conservation since the band edge is an excited state of the impurity. However, in the higher-energy transitions [cases (c) and (d) of Fig. 3-29] the additional step of phonon emission or absorption may be needed to complete the transition if the transition is indirect and if other scattering processes (discussed in Sec. 3-A-3) are not available. If phonons are involved, the absorption threshold in Fig. 3-31 would be shifted by the amount of phonon energy supplied toward lower energies for phonon absorption or toward higher energies for phonon emission.

In compound semiconductors the transition probability should depend on the sublattice in which the impurity finds a substitutional site (or the sublattice with which the interstitial impurity interacts most.[53] Thus in the case of GaP, donors which substitute on P-sites (S, Se, or Te) couple to the conduction branch at X_1 (Fig. 3-32) and X_1 forms the $\langle 100 \rangle$ minimum. On the other hand, donors which substitute on Ga-sites (Si) couple to the conduction branch at X_3, making X_3 the $\langle 100 \rangle$ minimum. The transition probability between the impurity and the top of the valence band depends on the extent of the impurity wave function at $k = 0$, i.e., on the probability of finding the donor's electron at $k = 0$. This probability is lower for the donor associated

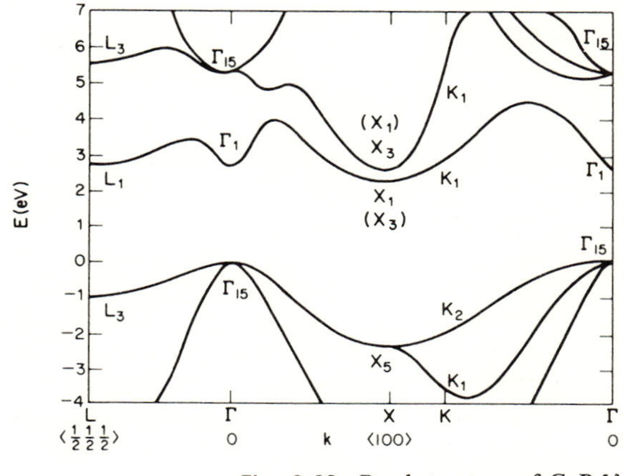

Fig. 3-32 Band structure of GaP.[53]

[53] T. N. Morgan, *Phys. Rev. Letters* **21**, 819 (1968).

to the X_3–Γ_{15} subband than for the donor associated with the lower X_1–Γ_1 subband. Thus transitions involving donors on P-sites are more probable than those involving donors on Ga-sites in GaP.

3-F Acceptor-to-Donor Transitions

When both donors and acceptors are simultaneously present in the crystal, the material is partly compensated, fully compensated, or overcompensated depending on the ratio of donors to acceptors being <1, 1, or >1, respectively. In either case, the acceptors' states are at least partly occupied and the donors' states are at least partly empty. Then it is possible to absorb a photon by promoting an electron from an acceptor state to a donor state. As we have seen in Sec. 1-E, due to the spacing-dependent coulomb interaction between donor and acceptor, a large set of transitions is possible, allowing absorption at the following photon energies:

$$h\nu = E_g - E_D - E_A + \frac{q^2}{\epsilon r}$$

Note that the absorption structure due to acceptor-to-donor transitions should be quite different from that due to isoelectronic traps. In Sec. 3-D, we saw that the nearest isoelectronic impurities provided the highest binding energies and, therefore, the lowest energy peak in the absorption spectrum, while the continuum occurred at the higher energies, where it merged with the fundamental edge. In the present case of transitions from acceptors to donors, however, the lowest-energy transitions should occur for the more distant pairs—i.e., for the continuum—while the discrete structure should occur at the higher energies near the fundamental edge. Because the discrete structure appears so near the fundamental edge, it is difficult to find it in absorption experiments. However, we shall see later that emission experiments make this structure evident.

Surface states comprise both donors and acceptors which are separated by an energy lower than the gap energy. It is possible to excite an electron from the acceptor to the donor to measure this energy difference. However, since the surface states are located in an infinitesimally thin layer, their absorbance is very difficult to detect. The arrangement in the inset of Fig. 3-33 has been used to detect the absorption due to surface states in Ge.[54] To probe the surface many times, a technique of internal-reflection spectroscopy[55] is employed. The sensitivity is further enhanced by a change in the ambient (oxidation), which prevents the optical transitions and thus permits a differential measurement specific to the surface states. After oxidation, the

[54]G. Chiarotti, G. Del Signore, and S. Nannarone, *Phys. Rev. Letters* **21**, 1170 (1968).
[55]N. J. Harrick, *Internal Reflection Spectroscopy*, Wiley (1967).

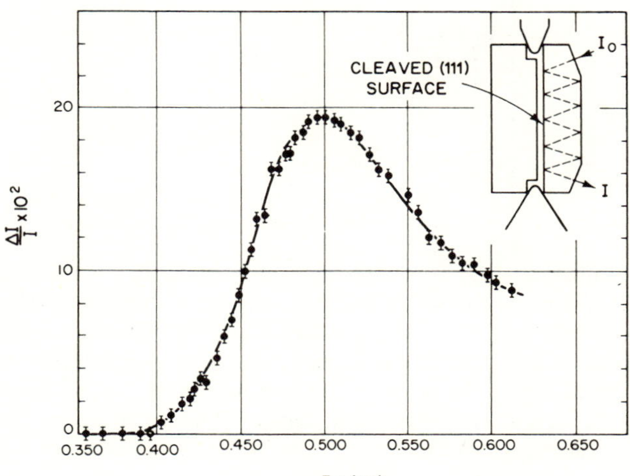

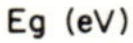

Fig. 3-33 Variation of the intensity of light totally reflected at a cleaved surface of intrinsic Ge at room temperature after the oxidation of the surface, as a function of the energy of the photons: $\Delta I/I = (I_{\text{oxidized}} - I_{\text{clean}})/I_{\text{oxidized}}$. This spectrum which lies below the energy gap E_g is attributed to transitions from acceptors to donors. The path of the light is shown in the inset.[54]

semiconductor is so strongly *p*-type at the surface that the acceptors are empty and, therefore, can no longer participate in the absorption process. The data of Fig. 3-33 has been attributed to a transition between a set of surface acceptor states practically coincident with the top of the valence band and a set of donor surface states about 0.16 eV below the bottom of the conduction band.[54]

3-G Intraband Transitions

3-G-1 *p*-TYPE SEMICONDUCTORS

The valence band of most semiconductors consists of three subbands[56] which are separated by spin–orbit interaction:[57] The resulting band structure is shown in Fig. 3-34. In *p*-type semiconductors, when the top of the valence band is populated with holes, it is possible to make three types of photon-

[56]F. Herman and J. Callaway, *Phys. Rev.* **89**, 518 (1953).
[57]G. Dresselhaus, A. F. Kip, and C. Kittel, *Phys. Rev.* **95**, 568 (1954).

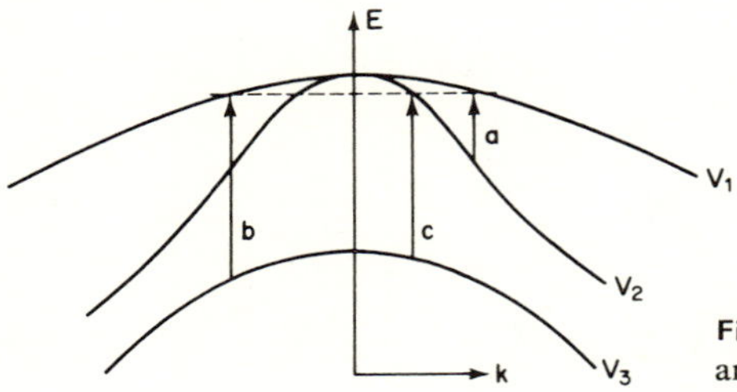

Fig. 3-34 Valence subband structure and intraband transitions.

absorbing transitions: (a) from the light-hole band V_2 to the heavy-hole band V_1; (b) from the split-off band V_3 to the heavy-hole band V_1; and (c) from the split-off band V_3 to the light-hole band V_2. These transitions have been observed in a number of semiconductors, and their interpretation can be verified by changing the position of the Fermi level, i.e., by doping. This absorption is proportional to the hole density,[58] and it disappears when the material is made *n*-type. A detailed calculation of the valence-band structure in germanium, using data from a different type of experiment (effective masses from cyclotron resonance), leads to a good fit between the theory of intra-valence-band absorption and the experimental results.[59] In the data of Fig. 3-35 for germanium, the 0.4-eV peak is attributed to ($V_3 \longrightarrow V_1$)-transition; the 0.3-eV peak is attributed to ($V_3 \longrightarrow V_2$)-transitions—these two peaks coincide at low temperatures where the Fermi level moves closer to the top of the valence band: and the broad peak at lower photon energies is attributed to ($V_2 \longrightarrow V_1$)-transitions.[60]

Similar interpretations have been made in *p*-type GaAs (Fig. 3-36).[61] The peaks at 0.42 eV and at 0.31 eV and the bump at 0.15 eV are due to $V_3 \longrightarrow V_1$, $V_3 \longrightarrow V_2$, and $V_2 \longrightarrow V_1$, respectively.

Note that as the doping or the temperature is changed, the intensities and positions of the peaks should change. As the Fermi level moves deeper into the valence band, the peak of $V_3 \longrightarrow V_1$ moves to higher energies, the peak of $V_3 \longrightarrow V_2$ to lower energies, and the low-energy edge of the ($V_2 \longrightarrow V_1$)-transition moves to higher energies. The effect of doping is especially well pronounced in *p*-type InSb.[62]

In semiconductors whose energy gap is smaller than the spin–orbit splitting energy, the ($V_3 \longrightarrow V_1$)- and ($V_3 \longrightarrow V_2$)-transitions may be masked by the fundamental absorption.

Semiconductors having warped $E(k)$ valence-band surfaces will exhibit a strongly changing assymmetry of the absorption peak corresponding to the ($V_2 \longrightarrow V_1$)-transitions.[62]

[58] H. B. Briggs and R. C. Fletcher, *Phys. Rev.* **87**, 1130 (1952) and **91**, 1342 (1952).
[59] E. O. Kane, *J. Phys. Chem. Solids* **1**, 82 (1956).
[60] W. Kaiser, R. J. Collins, and H. Y. Fan, *Phys. Rev.* **91**, 1380 (1953).
[61] R. Braunstein and E. O. Kane, *J. Phys. Chem. Solids* **23**, 1423 (1962).
[62] G. W. Gobeli and H. Y. Fan, *Phys. Rev.* **119**, 613 (1960).

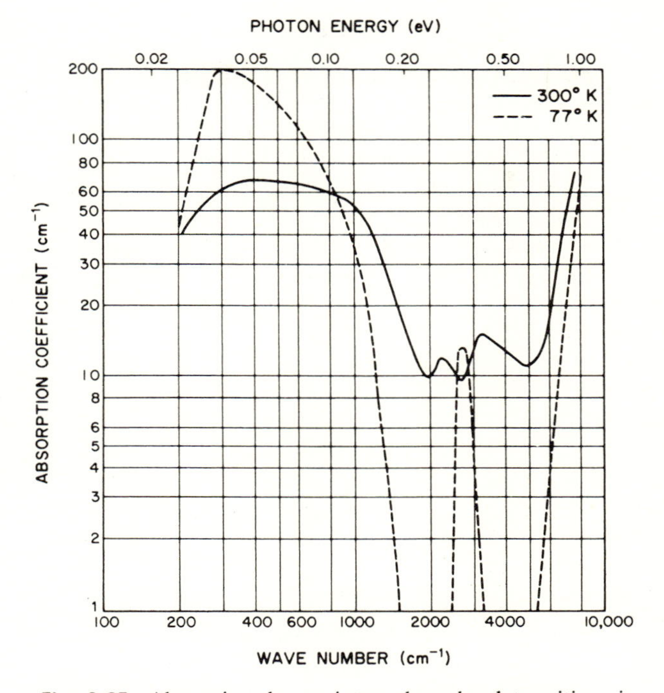

Fig. 3-35 Absorption due to intra-valence-band transitions in p-type germanium.[60]

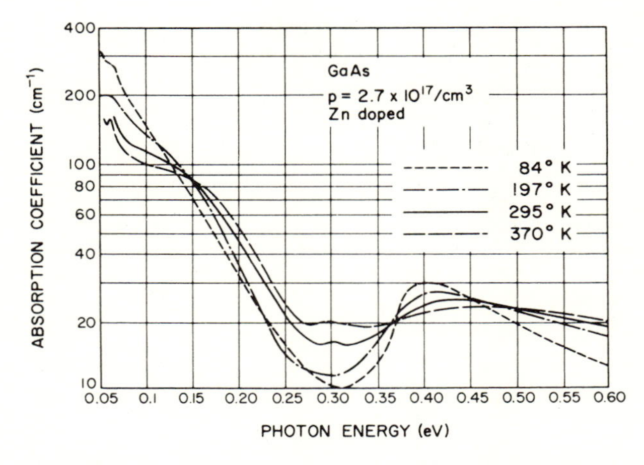

Fig. 3-36 Absorption due to intra-valence-band transitions in p-type GaAs.[61]

Table 3-2[61]

Spin–orbit splittings of the III–V compounds: E_g is the optical energy gap; Δ_0 (calc) is the estimated spin–orbit splitting at the center of the Brillouin zone; Δ_0 (exp) is the experimentally observed spin-splitting; Δ_0 is the spin–orbit splitting at the edge of the Brillouin zone

Material	E_g (eV at 0°K)	Δ_0(calc)	Δ_0(exp)	Δ_0
AlP	3.0	0.051	——	——
AlAs	2.2	0.29	——	——
AlSb	1.6	0.76	0.75	0.60
GaP	2.4	0.10	——	<0.15
GaAs	1.53	0.33	0.33	0.39
GaSb	0.80	0.81	——	0.72
InP	1.34	0.18	——	0.24
InAs	0.45	0.41	0.43	0.44
InSb	0.25	0.89	0.98	0.84

Table 3-3[61]

Effective masses of the III–V compounds: m_c(calc) is the estimated electron effective mass of the ($k = 0$)-minima; m_c (exp) is the measured electron effective mass; m_{v_2} (calc) and m_{v_3} (calc) are the estimated effective masses of the light-hole band and the spin–orbit split-off band, respectively; m_{v_2} (exp) and m_{v_3}(exp) are the corresponding observed masses; m_{hd} is the density-of-states hole effective masses determined from transport measurement; m_{v_1} (exp) is the heavy-hole mass determined from optical measurements

Material	m_c (calc)	m_c (exp)	m_{v_2} (calc)	m_{v_2} (exp)	m_{v_3} (calc)	m_{v_3} (exp)	m_{hd}	m_{v_1} (exp)
AlP	0.131	——	0.392	——	0.850	——	——	——
AlAs	0.110	——	0.220	——	0.680	——	——	——
AlSb	0.081	——	0.137	——	0.490	——	0.9	——
GaP	0.102	——	0.202	——	0.545	——	——	——
GaAs	0.075	0.074	0.130	0.12	0.388	0.20	0.5	0.68
GaSb	0.046	0.047	0.065	——	0.298	——	0.39	0.23
InP	0.066	0.073	0.112	——	0.310	——	0.40	——
InAs	0.026	0.021	0.035	0.025	0.154	0.083	0.33	0.41
InSb	0.0153	0.0143	0.015	0.012	0.190	——	0.20	——

[61] R. Braunstein and E. O. Kane, *J. Phys. Chem. Solids* **23**, 1423 (1962).

The values of the spin–orbit splitting of III–V compounds are shown in Table 3-2, and the values of the effectives masses are reported in Table 3-3.

It is interesting to point out that because of selection rules, the transition probabilities between the valence subbands vanish at $k = 0$ but increase with k^2.[61] This is an example of forbidden direct transitions treated in Sec. 3-A-2.

3-G-2 *n*-TYPE SEMICONDUCTORS

In an *n*-type semiconductor intraband transitions between the set of conduction subbands is conceivable. An absorption peak with a low-energy threshold at 0.27 eV has been observed in *n*-type GaP (Fig. 3-37).[63] This peak increases with the electron concentration. This absorption has been attributed to a direct intraband transition at the ⟨100⟩ minimum between the points X_1 and X_3 of Fig. 3-32.[64] In agreement with this model, measurements on $GaAs_{1-x}P_x$ alloys as a function of composition show that this absorption peak vanishes as soon as the Γ_1-valley becomes lowest and collects all the electrons.[65]

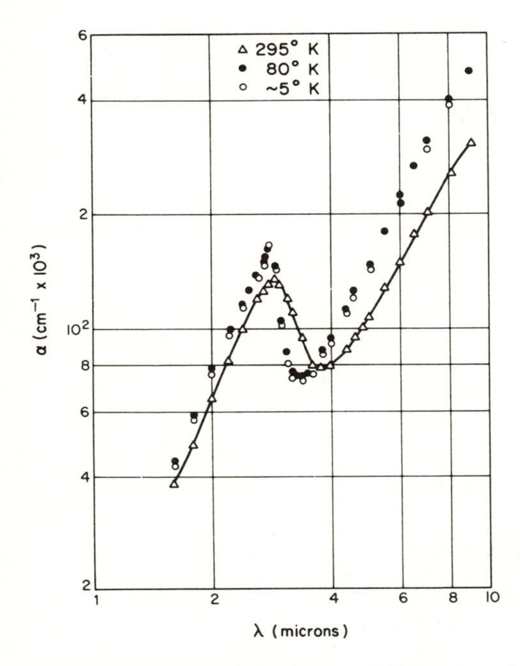

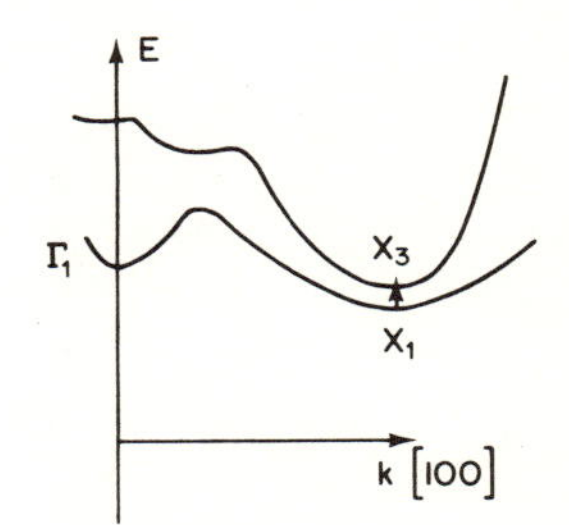

Fig. 3-37 Absorption spectrum of *n*-type GaP (N = 1 × 10^{18} cm^{-3}).[63] The rapid rise of α at long wavelengths is due to free carrier absorption to be discussed in the next section.

[63] W. G. Spitzer, M. Gershenzon, C. J. Frosch, and D. F. Gibbs, *J. Phys. Chem. Solids* **11**, 339 (1959).
[64] W. Paul, *J. Appl. Phys.* **32**, 2082 (1961).
[65] J. W. Allen and J. W. Hodby, *Proc. Phys. Soc.* (London) **82**, 315 (1962).

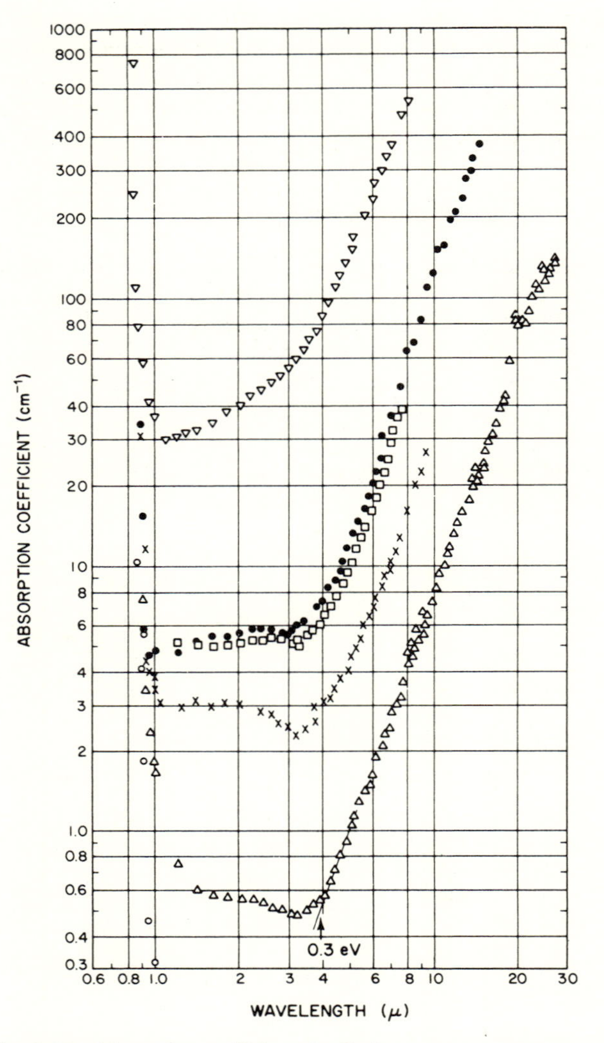

Fig. 3-38 Absorption coefficient of n-GaAs at room temperature for six samples of different doping, the electron concentration increasing from 1.3×10^{17} cm^{-3} for the lowest complete curve to 5.4×10^{18} cm^{-3} for the upper curve.[67] The rapid rise of α at short wavelengths is the fundamental edge. The rise of α at long wavelengths is due to free-carrier absorption to be discussed next.

[67] W. G. Spitzer and J. M. Whelan, *Phys. Rev.* **114**, 59 (1959).

A peak of 0.29 eV has also been observed[66] in n-type AlSb and also interpreted[64] as due to intraband transitions in the region of the $\langle 100 \rangle$ valley of the conduction band.

Indirect transitions between minima at different k of the same conduction subband have been proposed to explain a bump in the low-energy absorption edge of other n-type materials. This is shown in Fig. 3-38.[67] The compound GaAs exhibits a bump with a low-energy threshold at about 0.3 eV, which might be due to a transition form the $\langle 000 \rangle$ valley to the $\langle 100 \rangle$ valley. The intensity of this absorption increases with the electron concentration. This interpretation seems to agree with measurements on n-type $GaAs_{1-x}P_x$ alloys, where the separation between the two valleys can be varied.[68] Furthermore, this indirect process has an absorption cross-section α/N about two orders of magnitude lower than the direct process described above for the case of GaP.

A similar transition at 0.25 eV has been reported in n-type GaSb and tentatively attributed to a transition from the $\langle 000 \rangle$ valley to a $\langle 100 \rangle$ valley.[69] In n-type InP also, the transition from $\langle 000 \rangle$ to $\langle 100 \rangle$ appears as a bump below the fundamental absorption, as shown in Fig. 3-39.[70] In this case,

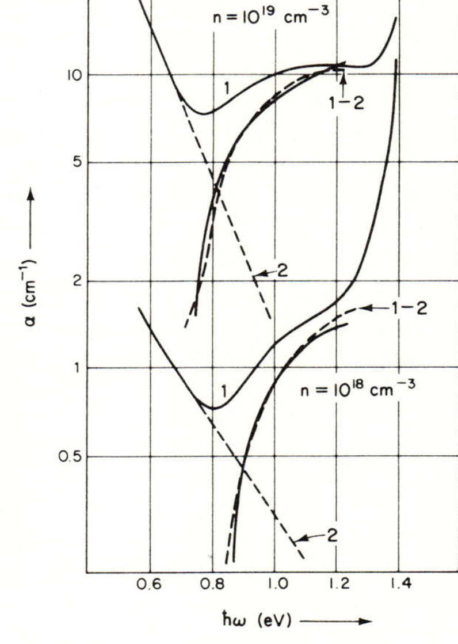

Fig. 3-39 Optical absorption α vs. photon energy $\hbar\omega$ in two samples of InP at 77°K. The total absorption curves 1 are broken up into the free-carrier absorption curves 2, and the intersubband absorption dotted curves 1–2. A reasonable fit to the intersubband absorption is provided by curves of the form $\alpha \approx (\hbar\omega - E_0)^{1/2}$ with $E_0 = 0.74$ for $n = 10^{19}$ cm^{-3} and $E_0 = 0.86$ for $n = 10^{18}$ cm^{-3}.[70]

[66] W. J. Turner and W. E. Reese, *Phys. Rev.* **117**, 1003 (1960).

[68] G. D. Clark and N. Holonyak, Jr., *Phys. Rev.* **156**, 913 (1967).

[69] W. M. Becker, A. K. Ramdas, and H. Y. Fan, *J. Appl. Phys.* **32**, 2094 (1961).

[70] M. R. Lorenz, W. Reuter, W. P. Dumke, R. J. Chicotka, G. D. Pettit, and J. M. Woodall, *Appl. Phys. Letters* **13**, 421 (1968).

when the free-carrier absorption (extrapolated over the spectral range of interest) is subtracted from the measured absorption, the remaining absorption, curve 1–2, follows approximately the law $(h\nu - E_o)^{1/2}$ expected for an indirect transition from a narrow range of filled states in the $\langle 000 \rangle$ valley to a parabolic $\langle 100 \rangle$ valley. The threshold E_o, which occurs at about 0.8 eV, includes the energy ΔE separating the two valleys of the conduction band, the phonon energy E_p involved in the indirect transition, and the position ξ_n of the Fermi level above the bottom of the lowest valley: $E_o = \Delta E + E_p - \xi_n$. For the two differently doped materials of Fig. 3-39, the lowest value of E_o appears in the most heavily doped crystal.

3-H Free-carrier Absorption

By "free carrier" we mean here a carrier free to move inside a band, i.e., a carrier which can interact with its ambient. Free-carrier absorption is characterized by a monotonic, often structureless, spectrum which grows as λ^p, where p can range from 1.5 to 3.5 and $\lambda = c/\nu$ is the photon wavelength. An example of free-carrier absorption spectrum can be found in the right-hand portion of Fig. 3-38 and as curves 2 in the left-hand portion of Fig. 3-39.

To absorb a photon, the electron must make a transition to a higher energy state within the same valley (Fig. 3-40). Such a transition requires an additional interaction to conserve momentum. The change in momentum can be provided by interaction with the lattice by way of phonons or by scattering from ionized impurities.

The Drude theory for the oscillation of an electron driven by a periodic electric field in a metal leads to a damping (attenuation) which increases as λ^2. The collision with the semiconductor lattice resulting in scattering by acoustic phonons leads[71] to an absorption increasing as $\lambda^{1.5}$. But scattering by optical phonons gives[72] a dependence in $\lambda^{2.5}$; while scattering by ionized

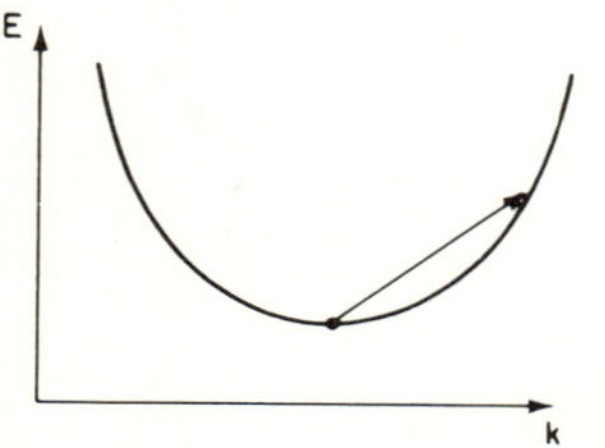

Fig. 3-40 Transition of a free electron in the conduction band.

[71] H. Y. Fan and M. Becker, *Semiconducting Materials*, Butterworth Scientific Publications Ltd. (1951), p. 132.
[72] S. Visvanathan, *Phy. Rev.* **120**, 376 (1960).

impurities gives a dependence in λ^3 or $\lambda^{3.5}$, depending on the approximations used in the theory.[73]

In general, all three modes of scattering will occur and the resultant free-carrier absorption α_f will be a weighted sum of the three processes:

$$\alpha_f = A\lambda^{1.5} + B\lambda^{2.5} + C\lambda^{3.5} \tag{3-24}$$

where A, B, and C are constants. The dominant mode of scattering will depend on the impurity concentration. The exponent p in the dependence λ^p should increase with doping or with compensation.

Table 3-4 shows the best fitting p and the capture cross-section (α_f/N) for various materials.

Table 3-4[74]

Free-carrier absorption in n-type compounds

	Carrier concentration (10^{17} cm^{-3})	α_f/N† (10^{-17} cm^{-2})	p
GaAs	1–5	3	3
InAs	0.3–8	4.7	3
GaSb	0.5	6	3.5
InSb	1–3	2.3	2
InP	0.4–4	4	2.5
GaP	10	(32)	(1.8)
AlSb	0.4–4	15	2
Ge	0.5–5	~4	~2

†The ratio α_f/N of absorption coefficient to carrier concentration is given for the wavelength of 9 μ. The parameter p expresses the wavelength dependence of absorption in the approximation $\alpha_f \propto \lambda^p$.

The classical formula for the free-carrier absorption coefficient α_f is

$$\alpha_f = \frac{Nq^2\lambda^2}{m^* 8\pi^2 \boldsymbol{n} c^3 \tau} \tag{3-25}$$

where N is the carrier concentration, $\boldsymbol{n}$ is the index of refraction, and τ is the relaxation time. Note that τ reflects the influence of the scatterers. Hence for ionized-impurity scattering, one would expect that the scattering probability would depend on the impurity. This dependence of the absorption coefficient on the chemical nature of the impurity has been verified in n-type germanium[75] where, at a given wavelength,

$$\alpha_f(\text{As}) > \alpha_f(\text{P}) > \alpha_f(\text{Sb})$$

[73]H. Y. Fan, W. G. Spitzer, and R. J. Collins, *Phys. Rev.* **101**, 566 (1956).

[74]H. Y. Fan, *Semiconductors and Semimetals*, ed. R. K. Willardson and A. C. Beer, Academic Press (1967), Vol. 3, p. 409.

[75]W. G. Spitzer, F. A. Trumbore, and R. C. Logan, *J. Appl. Phys.* **32**, 1822 (1961).

and in GaAs[76] where, also at a given wavelength,

$$\alpha_f(\mathrm{S}) > \alpha_f(\mathrm{Se}) > \alpha_f(\mathrm{Te})$$

Furthermore, the relaxation time depends on the concentration of scatterers. Therefore, at high doping one should find that α_f is not simply proportional to N as suggested by Eq. (3-25). Figure 3-41 shows that in arsenic-doped germanium, α_f is proportional to $N^{3/2}$. Since the effective mass is constant over this concentration range,[75] Eq. (3-25) implies that τ is proportional to $N^{-1/2}$.

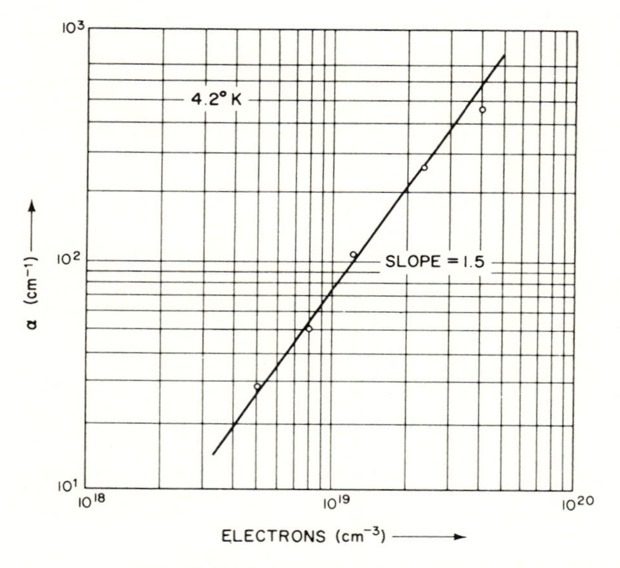

Fig. 3-41 Free-carrier absorption at 2.4μ in Ge.[9]

3-I Lattice Absorption[77]

In compound semiconductors, the bonding between atoms of different species forms a set of electric dipoles. These dipoles can absorb energy from an electromagnetic field, achieving a maximum coupling to the radiation when the frequency of the radiation equals a vibrational mode of the dipole. This

[76] E. P. Rashevskaya and V. I. Fistul, *Sov. Phys. Solid State* **9**, 1443 (1967).

[9] J. I. Pankove, *Progress in Semiconductors* **9**, 48 (1965).

[77] For an excellent review paper on this subject see W. G. Spitzer, *Semiconductors and Semimetals*, ed. R. K. Willardson and A. C. Beer, Academic Press (1957), Vol. 3, p. 17.

occurs in the far infrared region of the spectrum. Usually the vibrational mode is complex, consisting of several types of fundamental vibrations (multiphonon emission). The photon has a negligible momentum h/λ, whereas phonons can have a momentum as large as h/a (a is the lattice constant). Therefore, two or more phonons must be emitted to satisfy momentum conservation. Semiconductors have two transverse optical modes (TO), two transverse acoustic modes (TA), one longitudinal optical mode (LO), and a longitudinal acoustic mode (LA). Sometimes two transverse modes have similar dispersion characteristics $E_p(k)$. Furthermore, selection rules forbid certain combinations of phonons. Nevertheless, the multiplicity of possible combinations of all these modes is tremendous and explains the complex structure usually observed. Figure 3-42 shows a portion of the lattice-vibration spectrum of n-type GaAs.[78] The identification of these peaks and of peaks in spectral regions adjacent to the spectrum of Fig. 3-42 is shown in Table 3-5.

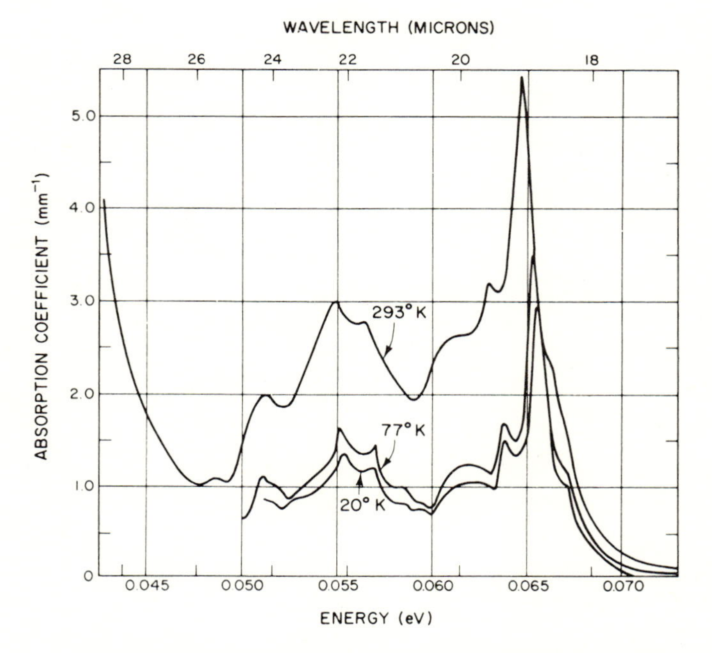

Fig. 3-42 Lattice absorption coefficient of high-resistivity n-type GaAs vs. wavelength from 18 to 28 microns at 20°, 77°, and 293°K.[78]

[78] W. Cochran, S. J. Fray, F. A. Johnson, J. E. Quarrington, and N. Williams, *J. Appl. Phys.* **32**, 2102 (1961).

In homopolar semiconductors there is no bonding dipole. Yet lattice-vibrational spectra are observed. Apparently a second-order process occurs:[79] the radiation induces a dipole which in turn has a stronger coupling to the

Table 3-5[78]

Characteristic phonon energies and assignment for GaAs

Position (eV)	*Assignment*
0.0955	$TO_1 + TO_2 + TO_2$ $0.0324 + 0.0316 + 0.0316$
0.0885	$TO_1 + TO_1 + LA$ $0.0324 + 0.0324 + 0.0237$
0.0860	$TO_2 + TO_2 + LA$ $0.0316 + 0.0316 + 0.0228$
0.0735	$TO_1 + TO_1 + TA$ $0.0324 + 0.0324 + 0.0087$
0.0716(?)	$TO_2 + TO_2 + TA$ $0.0316 + 0.0316 + 0.0084$
0.0648	$TO_1 + TO_1$ $0.0324 + 0.0324$
0.0631	$TO_2 + TO_2$ $0.0316 + 0.0316$
0.0612	$TO_1 + LO$ $0.0324 + 0.0288$ or $TO_2 + LO$ $0.0316 + 0.0296$
0.058	$LO + LO$ $0.029 + 0.029$
0.0565	$TO_1 + LA$ $0.0324 + 0.0241$
0.0548	$TO_2 + LA$ $0.0316 + 0.0232$
0.0510	$LO + LA$ $0.0288 + 0.0222$
0.048(?)	$LA + LA$ $0.024 + 0.024$
0.0413	$TO_1 + TA$ $0.0324 + 0.0089$
0.0398	$TO_2 + TA$ $0.0316 + 0.0082$
0.038	$LO + TA$ $0.029 + 0.009$

[79]M. Lax and E. Burstein, *Phys. Rev.* **97**, 39 (1955).

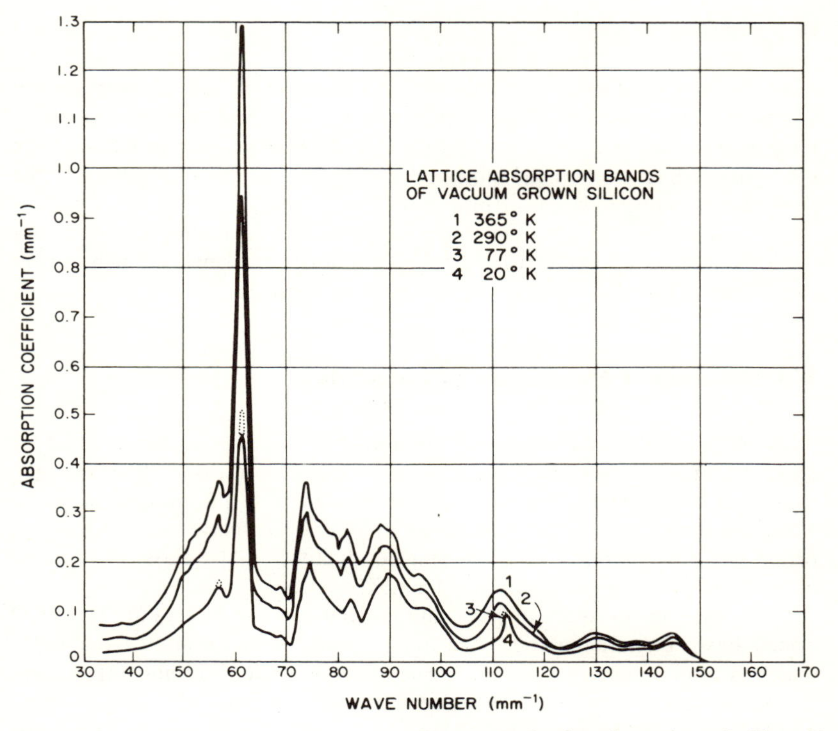

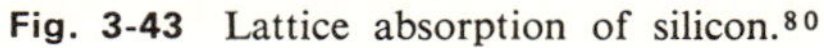
Fig. 3-43 Lattice absorption of silicon.[80]

Table 3-6[77,80]

Phonon assignments in silicon

Wave number (mm^{-1})	*Peak energy* (eV)	*Phonon assignment*†
144.8	0.1795	3TO
137.8	0.1708	2TO + LO
130.2	0.1614	2TO + LO
	——	2TO + LA
96.4	0.1195	2TO
89.6	0.1111	TO + LO
81.9	0.1015	TO + LA
74.0	0.0917	LO + LA
68.9	0.0756	TO + TA
61.0	0.0702	LO + TA

†TO = 0.0598 eV, LO = 0.0513 eV, LA = 0.0414 eV, TA = 0.0158 eV.

[80]F. A. Johnson, *Proc. Phys. Soc.* (London) **73**, 265 (1959).

radiation and produces more phonons. An example of lattice absorption in the homopolar material Si is shown in Fig. 3-43, and the corresponding phonon assignment is listed in Table 3-6. Even-higher-order processes have been invoked to explain a higher-energy absorption corresponding to the simultaneous emission of four phonons.[81]

It is interesting to note that introducing defects in homopolar semiconductors (such as by neutron irradiation) produces local fields which permit the normally forbidden single-phonon participation.[82]

3-J Vibrational Absorption of Impurities

Some impurities are so strongly bound that they cannot be detected by optical ionization (which requires photon energies greater than the gap energy, a range where band-to-band transition is the dominant process). Yet their presence may be manifested by abnormal scattering in transport properties (low mobility or poor heat conductivity). The classical example of such an impurity is oxygen in silicon.[83]

Oxygen is believed to form an SiO molecule in the Si crystal. The Si–O bonds have a characteristic vibrational spectrum at about 9μ (0.14 eV), which is reminiscent of a similar structure found in quartz and other compounds containing Si and O. Furthermore, the intensity of this absorption band depends on the exposure of the silicon to O_2 during the growth of the crystal or during subsequent heat treatment. The heat treatment may cause a rearrangement of the oxygen configuration into an SiO_2 phase which is less absorbing at 9μ but which can be detected by Rayleigh scattering from the SiO_2 aggregates.

3-K Hot-Electron-Assisted Absorption[84]

We have seen, in conjunction with indirect transitions, that it is possible to absorb photons of energy lower than the gap energy, the additional increment in energy needed to satisfy energy conservation coming from a simultaneously absorbed phonon. In this section we shall see that the energy increment can be obtained from a free carrier.

Consider first transitions between direct valleys as illustrated in Fig. 3-44(a) and (b). A photon with energy $h\nu = E_g - \Delta E$ might not be absorbed

[81] C. A. Klein and R. I. Rudko, *Appl. Phys. Letters* **13**, 129 (1968).
[82] M. Balkanski and W. Mazerewics, *J. Phys. Chem. Solid* **23**, 573 (1962).
[83] H. J. Hrostowski, "Infrared Absorption of Semiconductors," *Semiconductors*, ed. M. B. Hannay, Reinhold (1959), p. 474.
[84] S. M. Ryvkin, *Phys. Stat. Sol.* **11**, 285 (1965).

because of its energy "deficit" ΔE. But in the present process, if the energy deficit is small, the photon promotes an electron from the valence band to a virtual state a where it receives an additional boost of energy $\Delta E'$ and of momentum $\Delta k'$ to complete the transition to the conduction band. This additional boost could come from the phonon emitted by another "hot" electron in the conduction band. In this case this is a three-body interaction involving two electrons and a photon; one of the electrons transfers its excess energy to the other electron by a collision in which a phonon is exchanged. This is equivalent to an Auger effect with phonons. (We shall discuss the Auger effect with photons in Sec. 7-A.) Of course, the cooperating electron which emits the phonon must be sufficiently hot to supply the energy increment which will balance the deficit. Hot electrons can be obtained either by raising the temperature of the whole system or by applying a small electric field which will raise only the effective temperature T_e of the electrons.

In the case of indirect valleys [Fig. 3-44(c)], the present process is inadequate for an indirect transition with such a large momentum change. The average momentum that an electron with energy kT_e can contribute is only

$$\frac{(2kT_e m_e^*)^{1/2}}{\hbar}$$

which is much smaller than that required by momentum conservation in indirect-gap materials.

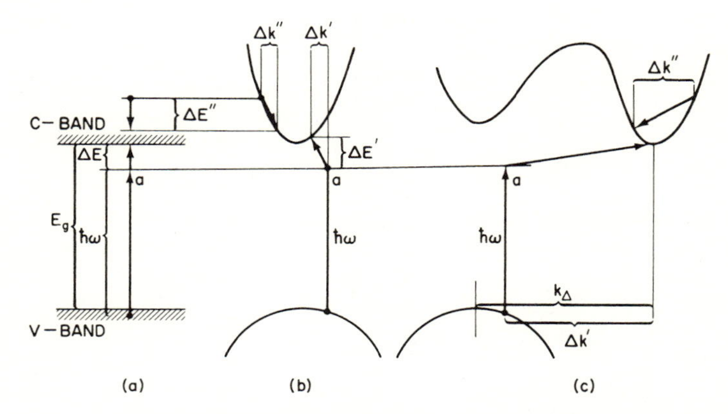

Fig. 3-44 Illustration of the conservation laws for the indirect absorption of photons having an energy deficit. (a) Conservation of energy. (b) Case of extrema located at the same point of k-space. An electron is transferred to the virtual state a (due to the photon energy $\hbar\omega$), and then to the C-band (due to the absorption of E'' from another electron). (c) Case of widely separated extrema; the absorption of a photon with an energy deficit is impossible because the momentum conservation law cannot be satisfied if $k_\Delta \gg \Delta k''$.[84]

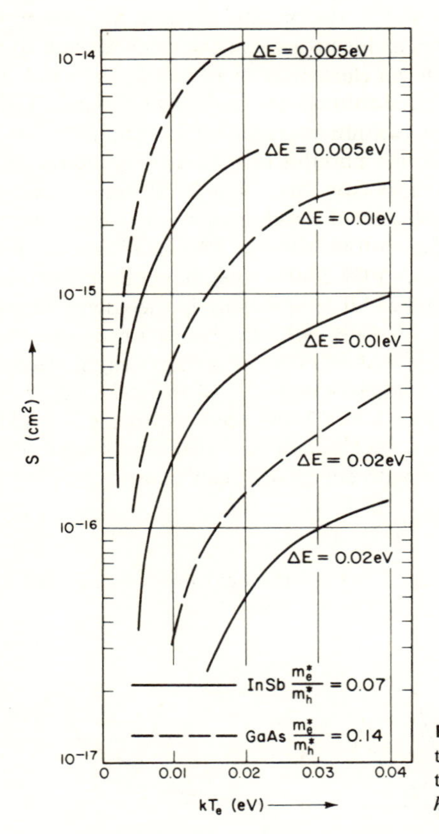

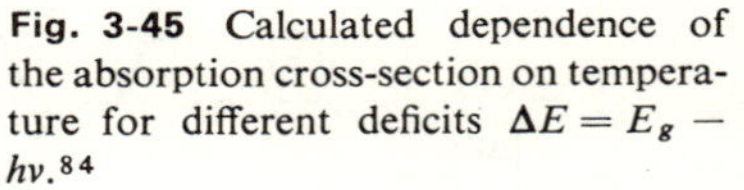
Fig. 3-45 Calculated dependence of the absorption cross-section on temperature for different deficits $\Delta E = E_g - h\nu$.[84]

In the direct-gap semiconductor GaAs, a calculation of the transition probability shows that hot-electron-assisted absorption is not a negligible effect. Thus for $kT_e = 0.01$ eV, $n = 10^{16}$ cm^{-3}, and $\Delta E = 0.01$ eV, the absorption coefficient turns out to be $\alpha \approx 6$ cm^{-1}.

The capture cross-section s (defined as $s = \alpha/N$, where N is the carrier concentration) varies with the electron temperature as shown in Fig. 3-45. At low temperatures $kT_e \ll \Delta E$, the number of electrons having an energy ΔE is given by Boltzmann's distribution; hence s is proportional to $\exp(-\Delta E/kT_e)$; when $kT_e \gg \Delta E$, the temperature dependence is weak.

The contribution of this process to the absorption spectrum is shown in Fig. 3-46. Hence it seems that it might be difficult to separate this contribution from other weak processes affecting the fundamental absorption (Sec. 3-A-4), transitions to impurity levels (Sec. 3-E), or intraband transitions (Sec. 3-G). However, the present process is the only one to exhibit a dependence on elec-

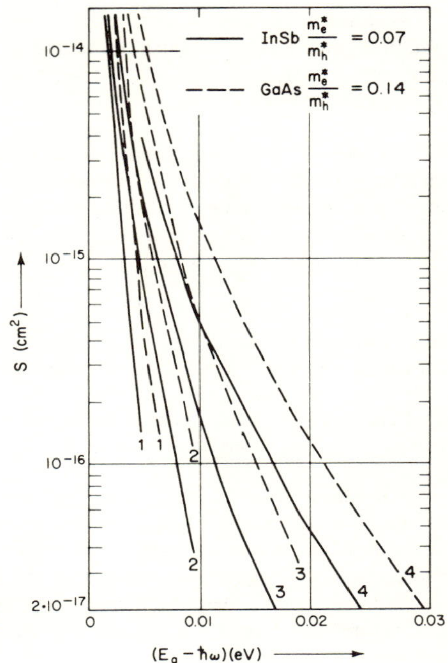

Fig. 3-46 Calculated spectral dependence of the absorption cross-section for different electron temperatures (increasing from 1 to 4).[84]

tron temperature (as distinguished from lattice temperature). The electron temperature can be varied by an electric field. As a further distinction, let us mention that the phonon-absorption mechanism, which depends on lattice temperature, cuts off at an energy deficit equal to the most energetic phonon mode, whereas the hot electron can contribute higher increments of energy.

Finally, let us mention the possibility of an "absorption avalanche" setting in. In the presence of a small electric field, each photon absorbed contributes an electron which becomes heated by the field and can assist further absorption of lower-than-gap photons. Thus the probability for the absorption increases with the number of electrons and the absorbance builds up rapidly in time. When the lifetime τ of the electrons and the light intensity I are large enough to give $\tau sI \gg 1$, the increment of carrier concentration builds up as

$$\Delta n = n_o[\exp(sIt) - 1]$$

where n_o is the equilibrium carrier concentration. The avalanching build-up increases until some mechanism, such as electron–electron scattering, sets in to reduce τ to a value such that $\tau sI < 1$.

Obviously, the hot-electron-assisted absorption can be observed only in relatively pure direct-gap semiconductors.

Problem 1. Assume that a heavily doped n-type crystal of GaAs retains its parabolic distribution of states (no tailing of states). Let $N_e = 2 \times 10^{18}$ electrons/cm^3, $m_e^* = 0.07$, and $m_h^* = 0.5$. Find the threshold photon energy for direct, allowed absorbing transitions.

Problem. 2 After subtracting the contribution of free-carrier absorption, the data of Fig. 3-39 shows an absorption process of the form $\alpha = (h_\nu - E_o)^{1/2}$. This process is believed to be a transition from the Fermi level in the $k = \langle 000 \rangle$ valley to the set of indirect valleys at $k = \langle 100 \rangle$. The following values of E_o are obtained:

$$E_o = 0.74 \text{ eV} \quad \text{for} \quad n = 10^{19} \text{ electrons/cm}^3$$

$$E_o = 0.86 \text{ eV} \quad \text{for} \quad n = 10^{18} \text{ electrons/cm}^3$$

Assuming that all the valleys are parabolic, find the density of state effective mass for the direct valley of the conduction band.

Problem 3. Discuss the suitability of germanium and silicon as a window material for CO_2 lasers.

Problem 4. Tensile stress is applied along the [100] axis of an n-type single crystal of silicon in liquid hydrogen. The Fermi level is located at 10 meV above the bottom of the conduction band at zero stress. Assume that the magnitudes of the transverse and longitudinal changes in the energy gap are equal and that they amount to 20 meV. Describe qualitatively the effect of this uniaxial stress on the absorption edge. Sketch the absorption edge with and without stress.

Problem 5. Consider a perfect crystal of GaAs at 0°K. (Perfect crystals are grown in textbooks.) The band structure of GaAs is as shown in Fig. 3P-5. Imagine that you are measuring its absorption spectrum in an apparatus where the sample can be subjected to electron bombardment which generates an abundance of electron–

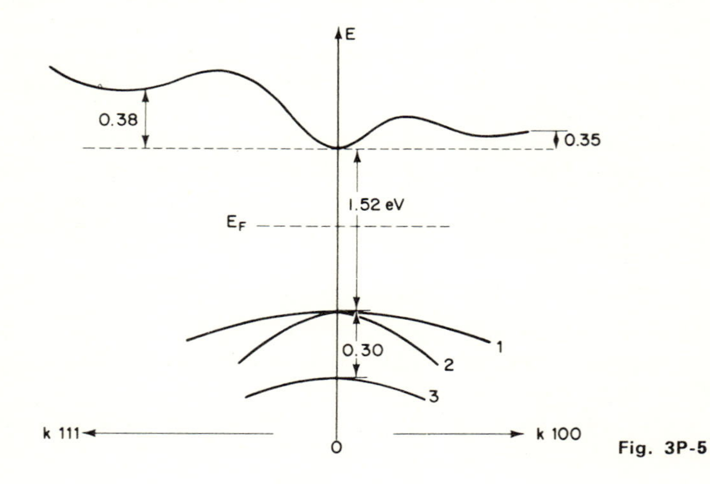

Fig. 3P-5

hole pairs. Find how the absorption spectrum evolves as the electron-beam excitation is increased. Assume that the carriers thermalize rapidly in the bands and that their recombination is not radiative (so as not to confuse the spectrum).

Problem 6. Prove that for a given energy deficit ΔE and a photon energy $h\nu$, a transition from the valence to the conduction band is possible by the process of Sec. 3-K if the hot electron has an energy of at least

$$E_{\min} = \frac{m_v + 2m_c}{m_v + m_c} \Delta E$$

Problem 7. Read the paper by E. O. Kane, *J. Phys. Chem. Solids* **1**, 83 (1956). Assume isotropic valence subbands. Describe qualitatively (in relative units) the absorption spectrum due to transitions between valence subbands in germanium at 0°K when the material contains 8×10^{17} and 2.5×10^{19} holes/cm³. The values of the effective masses of valence subbands in Ge at $k = 0$ are:

$$m_{V_1} = 0.3m \qquad m_{V_2} = 0.04m \qquad m_{V_3} \approx 0.3m$$

Note: Problems 4, 5, 6 of Chapter 6 (dealing with selection rules) can be treated as absorption problems. These problems and the four problems below are suitable for students who have had a course in quantum mechanics.

Problem 8. The information given here will be useful in the next few problems. For band-to-impurity transitions, it can be shown that the transition probability per initial and per final state is proportional to $|A(k)|^2$. The quantity $A(k)$ is a constant given by

$$A(k) = \frac{8\sqrt{\pi}(a_i)^{3/2}}{[1 + k^2(a_i)^2]^2\sqrt{V}}$$

where V is the volume of the crystal and a_i is the effective electron Bohr orbit associated with the impurity (donor or acceptor); $a_i = \hbar^2\epsilon/m^*e^2$. The factor $A(\bar{k})$ arises from writing the impurity state as $\phi_i(\bar{r}) = \sum_{\bar{k}} A(\bar{k})\psi_{\bar{k}}(\bar{r})$, where $\psi_{\bar{k}}$ is a Bloch state (see article by Kohn in *Solid State Phys.*, Vol. 5). This is the result of the effective-mass theory for shallow-impurity states, and the ionization energy measured from the band edge is given by $E_i = m^*q^4/2\hbar^2\epsilon^2$. The transition rate is obtained by multiplying the transition probability r by the optical density of states and their occupation probability.

Compare impurity absorption at room temperature and at low temperature for the following cases:

(a) nondegenerate n-type GaAs;
(b) degenerate n-type GaAs;
(c) nondegenerate p-type GaAs;
(d) degenerate p-type GaAs.

Illustrate this comparison by drawing the absorption coefficient vs. energy. Assume $E_i = 5$ meV for donors and 30 meV for acceptors. Also, $m_c/m_v = 1/7$, and $m_v = 0.5m$, where m is the free electron mass.

Problem 9. Use the information presented in the statement of Prob. 8. Compare the shape of the absorption edge in a direct-gap semiconductor (e.g., GaAs) for impurity absorption (valence-band–to–donors) and band-to-band absorption. How can you distinguish between the two? What can you say about the behavior of the two as a function of temperature? Assume degenerate p-type material but neglect band-perturbation effects. Discuss also the shape of the absorption edge as doping changes the semiconductor from a degenerate to a nondegenerate material.

Problem 10. Now consider compensated nondegenerate n-type GaAs with $N_o/N_a \approx 10$, and discuss the kind of absorption edge you expect at room temperature and at low temperature. Do you expect to find a structure with maxima in the absorption edge? If so, make quantitative comparisons at these points using the information in Prob. 8. Sketch the dependence ln α vs. $E = \hbar\omega$.

Problem 11. Read the statement of Prob. 10, but solve it for the case of degenerate compensated material. Assume the Fermi level is above the conduction-band edge by an amount equal to the acceptor ionization energy E_A (both at room and low temperatures). Take $m_v/m_c = 7$. Assume that the results of the effective-mass theory of shallow-impurity states given in Prob. 8 are applicable here.

Problem 12. Discuss the absorption edge in compensated degenerate p-type material with the Fermi level an amount ξ_p in the valence band. Compare the relative magnitude of each process in this case with those of Prob. 11. Consider room temperature and low temperature.

RELATIONSHIPS BETWEEN OPTICAL CONSTANTS

4

4-A Absorption Coefficient

Let the radiation be a plane wave of frequency ν propagating in the x-direction with a velocity v:

$$\mathscr{E} = \mathscr{E}_o \exp\{i2\pi\nu[t - (x/v)]\} \tag{4-1}$$

The velocity of propagation through a semiconductor having a complex index of refraction

$$\boldsymbol{n}_c = \boldsymbol{n} - i\boldsymbol{k} \tag{4-2}$$

is related to the velocity of propagation in vacuum, c, by

$$v = \frac{c}{\boldsymbol{n}_c} \tag{4-3}$$

Therefore,

$$\frac{1}{v} = \frac{\boldsymbol{n}}{c} - \frac{i\boldsymbol{k}}{c}$$

Substituting the latter into Eq. (4-1) gives

$$\mathscr{E} = \mathscr{E}_o \exp(i2\pi\nu t)\exp(-i2\pi x\boldsymbol{n}/c)\exp(-2\pi\nu\boldsymbol{k}x/c) \tag{4-4}$$

Note that the last term in Eq. (4-4) is a damping factor. The fraction of the incident power available after propagating a distance x through a material with a conductivity σ is

$$\frac{P(x)}{P(0)} = \frac{\sigma\mathscr{E}^2(x)}{\sigma\mathscr{E}^2(0)} = \exp(-4\pi\nu\boldsymbol{k}x/c) \tag{4-5}$$

In terms of the absorption coefficient α,

$$\frac{P(x)}{P(0)} = \exp(-\alpha x) \tag{4-6}$$

Therefore,

$$\alpha = \frac{4\pi\nu \boldsymbol{k}}{c} \tag{4-7}$$

where $\boldsymbol{k}$, the imaginary part of $\boldsymbol{n}_c$, is called the "extinction coefficient."

4-B Index of Refraction

Radiation propagating through an uncharged homogeneous semiconductor having a permeability μ, a dielectric constant ϵ, and an electrical conductivity σ obeys Maxwell's equations:

$$\nabla \times \mathscr{E} = -\frac{\mu}{c}\frac{dH}{dt} \tag{4-8}$$

$$\nabla \times H = \frac{4\pi\sigma}{c}\mathscr{E} + \frac{\epsilon}{c}\frac{d\mathscr{E}}{dt} \tag{4-9}$$

$$\nabla \cdot H = 0 \tag{4-10}$$

$$\nabla \cdot \mathscr{E} = 0 \tag{4-11}$$

We can combine Eqs. (4-8) and (4-9):

$$\begin{aligned}\nabla \times \nabla \times \mathscr{E} &= -\frac{\mu}{c}\frac{d}{dt}(\nabla \times H)\\ &= -\frac{\mu}{c^2}4\pi\sigma\frac{d\mathscr{E}}{dt} - \frac{\mu\epsilon}{c^2}\frac{d^2\mathscr{E}}{dt^2}\end{aligned}$$

But since $\nabla \times \nabla \times \mathscr{E} = \nabla(\nabla \cdot \mathscr{E}) - \nabla^2\mathscr{E}$, and considering Eq. (4-11), we can write

$$\frac{d^2\mathscr{E}}{dx^2} = \frac{\mu}{c^2}4\pi\sigma\frac{d\mathscr{E}}{dt} + \frac{\mu\epsilon}{c^2}\frac{d^2\mathscr{E}}{dt^2} \tag{4-12}$$

Now, we let Eq. (4-12) operate on Eq. (4-1) to obtain

$$-\frac{(2\pi\nu)^2}{v^2} = i2\pi\nu\frac{\mu}{c^2}4\pi\sigma - \frac{\mu\epsilon}{c^2}(2\pi\nu)^2$$

or

$$\frac{1}{v^2} = \frac{\mu\epsilon}{c^2} - i\frac{\mu 4\pi\sigma}{2\pi\nu c^2} \tag{4-13}$$

In all the semiconductors we shall consider, $\mu = 1$; thus Eq. (4-13) can be rewritten

$$\frac{1}{v^2} = \frac{\epsilon}{c^2} - i\frac{2\sigma}{\nu c^2} \tag{4-14}$$

Note that Eq. (4-3), gives

$$\frac{1}{v^2} = \frac{n_c^2}{c^2} = \frac{n^2}{c^2} - \frac{i2nk}{c^2} - \frac{k^2}{c^2} \tag{4-15}$$

Equating the real and imaginary terms of Eqs. (4-14) and (4-15) gives

$$n^2 - k^2 = \epsilon \tag{4-16}$$

$$nk = \frac{\sigma}{\nu} \tag{4-17}$$

Now we can solve for n and for k:

$$n^2 - k^2 = (n + k)(n - k)$$

$$\epsilon^2 = (n^2 + k^2 + 2nk)(n^2 + k^2 - 2nk)$$

$$= (n^2 + k^2)^2 - (2nk)^2$$

$$= (n^2 + k^2)^2 - \left(\frac{2\sigma}{\nu}\right)^2$$

$$n^2 + k^2 = \left[\epsilon^2 + \left(\frac{2\sigma}{\nu}\right)^2\right]^{1/2} \tag{4-18}$$

Combining Eqs. (4-16) and (4-18) gives

$$n^2 = \tfrac{1}{2}\epsilon\left\{\left[1 + \left(\frac{2\sigma}{\nu\epsilon}\right)^2\right]^{1/2} + 1\right\} \tag{4-19}$$

$$k^2 = \tfrac{1}{2}\epsilon\left\{\left[1 + \left(\frac{2\sigma}{\nu\epsilon}\right)^2\right]^{1/2} - 1\right\} \tag{4-20}$$

When σ tends to zero, as in insulating materials, n tends to $\sqrt{\epsilon}$ and the extinction coefficient k tends to zero. Hence the material becomes transparent. The values of the index of refraction in semiconductors are shown in Appendix II.

There seems to be an empirical relation between the index of refraction and the energy gap of a semiconductor:

$$n^4 E_g = 77$$

which is called the Moss rule and is obeyed by semiconductors whose value of n^4 is in the range of 30 to 440.[1]

4-C The Kramers–Kronig Relations[2]

Let us define a complex dielectric constant $\epsilon_c = n_c^2$ which depends on the frequency ω of the electromagnetic wave. The real and imaginary parts ϵ_1

[1]T. S. Moss, *Optical Properties of Semiconductors*, Butterworth (1959), p. 48.

[2]For a more detailed treatment see F. Stern, "Elementary Theory of the Optical Properties of Solids," *Solid State Physics*, ed. F. Seitz and D. Turnbull, Academic Press (1963), Vol. 15, p. 299; also T. S. Moss, *Optical Properties of Solids*, Butterworth (1959) p. 22, and A. Abragam, *Principles of Nuclear Magnetism*, Clarendon (1961), p. 93.

and ϵ_2, respectively, are interdependent according to the Kramers–Kronig relations:

$$\epsilon_1(\omega) = 1 + \frac{2}{\pi} P \int_0^\infty \frac{\omega' \epsilon_2(\omega')\, d\omega'}{[\omega']^2 - \omega^2} \tag{4-21}$$

$$\epsilon_2(\omega) = -\frac{2\omega}{\pi} P \int_0^\infty \frac{\epsilon_1(\omega')}{[\omega']^2 - \omega^2}\, d\omega' \tag{4-22}$$

where P is the Cauchy principal value of the integral:

$$P\int_0^\infty \equiv \lim_{a \to 0} \left(\int_0^{\omega - a} + \int_{\omega + a}^\infty \right)$$

Since optical-absorption spectra determine $\alpha(E)$, where $E = h\nu$ is the energy of the photon, we shall express the dispersion relation for the refractive index $\boldsymbol{n}(E)$ in terms of $\alpha(E)$. Thus a relation equivalent to Eq. (4-21) is obtained for the complex index of refraction ($\boldsymbol{n}_c = \boldsymbol{n} - i\boldsymbol{k}$):

$$\boldsymbol{n}(E) - 1 = \frac{2}{\pi} P \int_0^\infty \frac{E' \boldsymbol{k}(E')}{(E')^2 - E^2}\, dE' \tag{4-23}$$

Since $\boldsymbol{k}(E') = hc\alpha(E')/4\pi E'$ [as obtainable from Eq. (4-7)] the relation in Eq. (4-23) becomes

$$\boldsymbol{n}(E) - 1 = \frac{ch}{2\pi^2} P \int_0^\infty \frac{\alpha(E')}{(E')^2 - E^2}\, dE' \tag{4-24}$$

which permits a calculation of $\boldsymbol{n}(E)$ when the absorption spectrum $\alpha(E)$ is completely known. In practice, $\alpha(E)$ is determined only over a limited range, and an educated guess is made about the value of the integral beyond this range.

4-D Reflection Coefficient

For normal incidence, the reflection coefficient affecting the intensity of the radiation is given by

$$R = \frac{(\boldsymbol{n} - 1)^2 + \boldsymbol{k}^2}{(\boldsymbol{n} + 1)^2 + \boldsymbol{k}^2} \tag{4-25}$$

When $\boldsymbol{k} = 0$, i.e., in the transparent range, then

$$R = \frac{(\boldsymbol{n} - 1)^2}{(\boldsymbol{n} + 1)^2} \tag{4-26}$$

If $\boldsymbol{n} = 0$, $R = 1$ and the semiconductor is totally reflecting.

In both cases, when either $\boldsymbol{n}$ or $\boldsymbol{k}$ is nil, Eq. (4-17) requires that $\sigma = 0$, i.e., that the medium not be lossy. If σ is not zero, the material is neither perfectly transparent nor perfectly reflecting, and the radiation experiences losses. The losses manifest themselves through the absorption coefficient α as we saw in Eq. (4-7). Hence

$$k = \frac{c\alpha}{4\pi\nu} \tag{4-27}$$

or, substituting the value of k from Eq. (4-17),

$$\alpha = \frac{4\pi\sigma}{nc}$$

When σ is large, Eqs. (4-19) and (4-20) show that n and k can both become large and nearly equal. In that case also, the reflectance approaches unity.

4-E Determination of Carrier Effective Mass

The dielectric constant ϵ can be expressed in terms of the polarizability χ of the medium:

$$\epsilon = 1 + 4\pi\chi \tag{4-28}$$

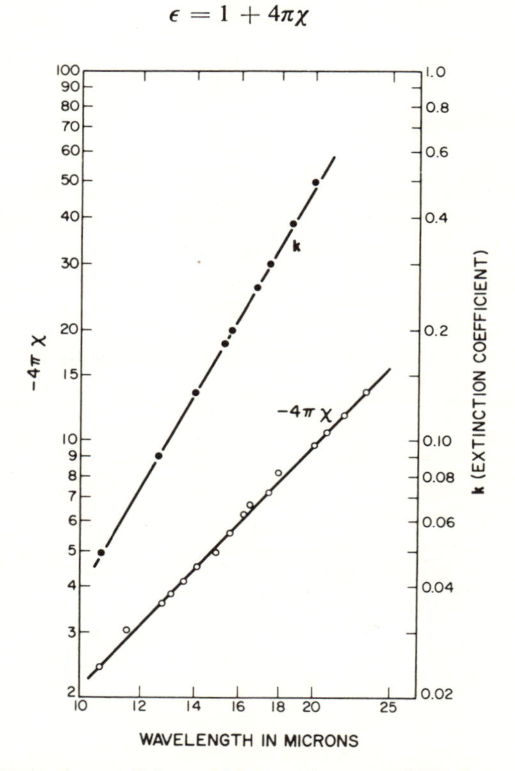

Fig. 4-1 Extinction coefficient and free carrier susceptibility for an n-type germanium sample[3] ($n = 3.9 \times 10^{18}/\text{cm}^3$).

[3]W. G. Spitzer and H. Y. Fan, *Phys. Rev.* **106**, 883 (1957).

When χ is contributed by $N\,(\mathrm{cm}^{-3})$ free electrons, its value is[3]

$$\chi = \frac{Nq^2}{(2\pi\nu)^2\langle m^*\rangle} \tag{4-29}$$

where $\langle m^*\rangle$ is an averaged effective mass. The averaging process gives more importance to the faster electrons. For spherical energy surfaces, one uses m^*. Hence if one makes the approximation of a spherical energy surface, it is possible to obtain the effective mass from Eq. (4-29) via Eqs. (4-28) and (4-16); $\boldsymbol{k}$ is derived from Eq. (4-27) and $\boldsymbol{n}$ from Eq. (4-17) or from Eq. (4-26), where it is a reasonable approximation; $\alpha(\nu)$ is the absorption spectrum; and N is obtained from a Hall-effect measurement. An example of the determination of $\boldsymbol{k}$ and χ for electrons in germanium using free-carrier absorption data is shown in Fig. 4-1.

Note that a knowledge of the effective mass coupled to conductivity data permits a calculation of the carrier relaxation time τ from the relation

$$\sigma = \frac{Nq^2}{m^*}\tau \tag{4-30}$$

4-F Plasma Resonance

In the infrared region the reflectivity R of a semiconductor undergoes an anomalous dispersion and tends to unity when the incident energy approaches the plasma frequency. The plasma frequency in a solid is given by

$$\omega_p = \left(\frac{4\pi Nq^2}{m^*\epsilon}\right)^{1/2} \tag{4-31}$$

where N is the carrier concentration. Since the plasma frequency increases

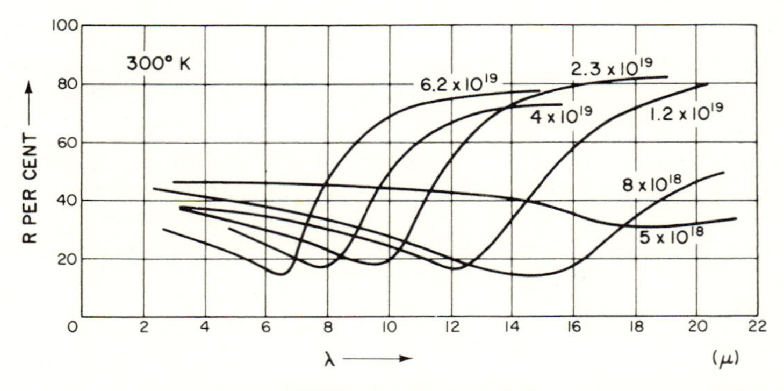

Fig. 4-2 Reflectivity of arsenic-doped germanium at 300°K.[4]

[4]J. I. Pankove, "Properties of Heavily Doped Germanium," *Progress in Semiconductors*, ed. A. F. Gibson and R. E. Burgess, Heywood (1965), Vol. 9, p. 67.

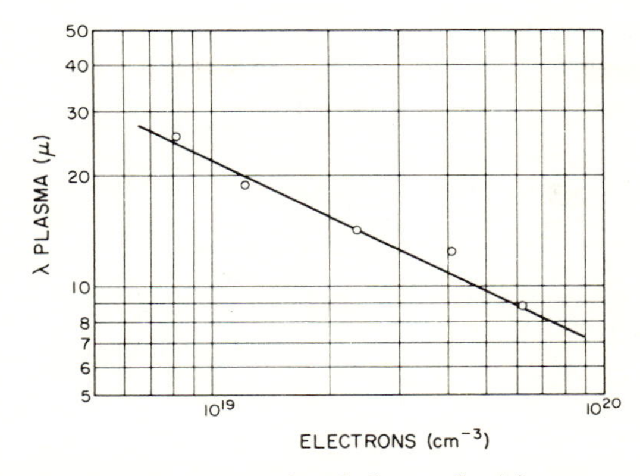

Fig. 4-3 Plasma resonance wavelength from reflectivity curves vs. carrier concentration.[4]

with the carrier density, higher reflectance occurs in impure materials at a shorter wavelength than in the purer materials. Figure 4-2 shows Ge-reflectivity curves obtained at room temperature for different doping concentrations, the impurity being arsenic. If the curves are extrapolated to unity reflectivity, one finds roughly the plasma wavelength λ_p. A plot of λ_p versus carrier concentration is shown in Fig. 4-3, which gives the expected square-root dependence of Eq. (4-31). Note that the plasma-frequency data can also be used to determine the carrier effective mass.

4-G Transmission

The transmission coefficient is defined as the ratio of transmitted to incident power, I/I_o. If the specimen has a thickness x, an absorption coefficient α, and a reflectivity R, the radiation traversing the first interface is $(1 - R)I_o$, the radiation reaching the second interface is $(1 - R)I_o \exp(-\alpha x)$, and only a fraction $(1 - R)(1 - R)I_o \exp(-\alpha x)$ emerges. The portion internally reflected eventually comes out, but considerably attenuated. These multiple internal reflections are illustrated in Fig. 4-4. The end-result is that the overall transmission is given by

$$T = \frac{(1 - R)^2 \exp(-\alpha x)}{1 - R^2 \exp(-2\alpha x)} \tag{4-32}$$

When the product αx is large, one can neglect the second term in the denominator; then

$$T \approx (1 - R)^2 e^{-\alpha x} \tag{4-33}$$

If R and x are known, the relation in Eq. (4-32) can be solved for α. If R is

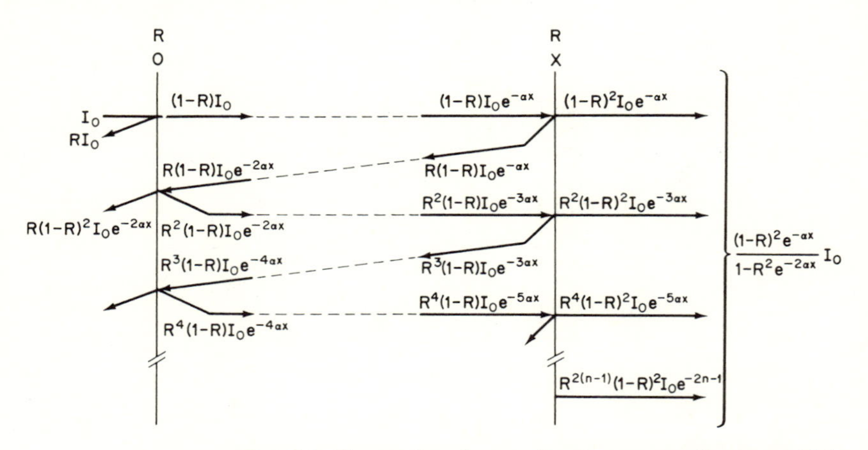

Fig. 4-4 Accounting of energy flow in a system allowing multiple internal reflections.

not known, one can measure the transmittance of two samples having different thicknesses x_1 and x_2. Then α is obtained from

$$\frac{T_1}{T_2} \approx e^{\alpha(x_2-x_1)} \tag{4-34}$$

Note that since $T_1 = I_1/I_o$ and $T_2 = I_2/I_o$, it is not necessary to know I_o to use Eq. (3-34), where one can replace T_1/T_2 by I_1/I_2.

4-H Interference Effects

The phase of the propagating electromagnetic wave varies with position according to Eq. (4-1), so that if the sample is a plane parallel wafer of thickness x, the phase of the propagating plane wave changes by

$$\theta = \frac{n}{c}2\pi\nu x = 2\pi\frac{n}{\lambda}x$$

upon reaching the exit facet. The internally reflected wave will return to the exit facet with a phase change of 3θ, then 5θ, etc. Constructive interference will result if

$$x = \frac{(2m+1)\lambda}{2n} \tag{4-35}$$

where m is an integer. And destructive interference will obtain when

$$x = \frac{(2m+1)\lambda}{4n} \tag{4-36}$$

Note that λ/n is the wavelength of the radiation inside the semiconductor. Hence the transmitted radiation is modulated by the phase factor $\cos\theta$. The transmitted radiation will then go through a maximum whenever Eq. (4-35) is satisfied—i.e., at all wavelengths such that an odd number of half-wavelengths can fit in the sample thickness.

Hence it is possible to find the index of refraction n by measuring the wavelengths at which two adjacent maxima occur. Then

$$n = \frac{1}{\left(\frac{1}{\lambda_2} - \frac{1}{\lambda_1}\right)x} \tag{4-37}$$

Let us mention in passing that the interference technique in conjunction with the high reflectivity of heavily doped semiconductors at long wavelengths has played a crucial role in the development of epitaxial growth. The thickness of a lightly doped epitaxial layer grown on a heavily doped substrate can be accurately determined from the interference pattern obtained in a reflection measurement.[5] The substrate exhibits a high reflectance at wavelengths greater than those corresponding to plasma resonance.

Problem 1. The conductivity introduced in Eqs. (4-19) and (4-20) describes the response of charged carriers to an applied electric field. Write out the equation of motion for the electrons in a degenerate semiconductor in the presence of electromagnetic radiation, and from this equation derive an expression for the frequency dependence of the conductivity. From this result and from Eq. (4-9), comment on the frequency dependence of the losses. Also give physical arguments as to why this result gives the right answer.

Problem 2. Write down Maxwell's equations for a loss-free medium of dielectric constant ϵ. By analogy to these equations, show that in the case of metals or semiconductors one can equivalently describe the response of the system to electromagnetic radiation by a complex dielectric constant ϵ_c, without introducing the conductivity σ. Derive the expressions relating ϵ_c to the complex index of refraction and to the conductivity.

Problem 3. Consider the response of a direct-gap semiconductor to electromagnetic radiation of frequency $\hbar\omega \geq E_g$. Discuss if the set of Eqs. (4-8)–(4-11), (4-19) and (4-20) can be applied in this case and, if not, what modifications are necessary. Also comment on the applicability of the results of Prob. 1. For $\hbar\omega \geq E_g$, give the relation between the absorption coefficient and the complex index of refraction. How can one relate the index of refraction to the absorption-coefficient data?

Problem 4. Equations (4-35) and (4-36) give the conditions for constructive and destructive interference for light passing through a film of thickness x. In the former case the transmitted intensity is maximum and in the latter case it is minimum. Find the ratio of the maximum intensity to the minimum intensity in terms of the reflectivity, thickness x, and the absorption coefficient of the medium. (The ratio is called "contrast" or "visibility.")

[5] W. G. Spitzer and M. Tannenbaum, *J. Appl. Phys.* **32**, 744 (1961).

ABSORPTION SPECTROSCOPY

5

To measure the absorption spectrum of a semiconductor, one uses a monochromator, which selects a narrow band $\Delta\lambda$ of wavelengths from a source of radiation. This spectral band is centered at a wavelength λ which can be varied. Hence the monochromator is a tunable filter having a band pass $\Delta\lambda$ or $\Delta\nu$ and a resolution $\Delta\lambda/\lambda = \Delta\nu/\nu$.

Note that monochromators and spectrometers are identical instruments; the name is determined only by how it is used. If the instrument is placed between the specimen and a detector, it is a spectrometer. If it is placed between a source of radiation and the specimen, the instrument is called a monochromator. Thus an emission spectrum is measured with a spectrometer, whereas for a transmission- or reflectance-spectrum measurement where one needs a monochromatic radiation, the machine is used as a monochromator. In a monochromator (or spectrometer) the desired wavelength is selected by the position of the exit slit with respect to the dispersed spectrum. In a spectrograph the exit slit is replaced by a broad-spectrum photographic plate which registers the intensity at each wavelength as a series of more or less opaque bands or lines. Later, the spectrogram thus obtained is scanned by a light-spot-and-detector head which records the density of the bands in the spectrogram as a function of wavelength. Note that the photographic-emulsion detector limits the spectrograph to the visible and the ultraviolet. However, if the radiation available is very intense (as that of a CO_2 laser), a spectrograph can be used in the infrared with a thermally sensitive detector, such as certain phosphors whose luminescence is quenched by infrared.

For absorption and reflectance measurements it is best to place the sample at the exit slit of the instrument in order to avoid the simultaneous excitation of competing processes which result from events occurring at wavelengths other than those of interest. For example, with broad spectrum illumination hole–electron pairs can be formed at the higher photon energies and recombine emitting photons at some lower energy where they may mask the absorption process.

Depending on the resolution desired and the spectral region to be covered, one uses a prism, or a grating, or two prisms in series, or two gratings in series, or a combination of prism and grating. The spectral range of greatest interest determines the prism material to be used or the blazing of the grating.

Figure 5-1 shows an example of a prism instrument. Here, G is the source of radiation, C is the specimen, and T is the detector. Mirror optics are used to render the optical system achromatic. Whenever lenses are employed to collect and focus the light, great care must be taken in limiting their use to the achromatic range, otherwise the light-gathering efficiency is greatly impaired. To minimize distortions, off-axis parabolic mirrors are often used. In the instrument of Fig. 5-1, the radiation goes through the prism P twice. The spectral selection—i.e., the choice of which wavelength goes through the exit slit S_2—is controlled by tilting the "Littrow" mirror M_4. In this instrument, prisms made of various materials can be easily interchanged, extending the usefulness of the instrument from about 2000 Å to 50 μ. The spectral ranges over which various materials are transparent are shown in Fig. 5-2. In addition to transparency, the index of refraction (Fig. 5-3) is an important characteristic of prisms. The data is presented in a more useful form in Fig. 5-4, which shows dispersion $dn/d\lambda$ vs. λ. In the near UV region, quartz is the favored prism material. Dense flint glass gives the best dispersion through the visible range and up to 2 μ; LiF extends the range to about 5 μ, CaF_2 to 8 μ, and NaCl to 15 μ; KBr permits measurements to about 35 μ and CsI to 50 μ. Above 50 μ, gratings are used; spectral measurements have been extended beyond 300 μ.

The source of radiation must also be carefully selected depending on the spectral range to be studied. For the infrared range, an incandescent "globar"

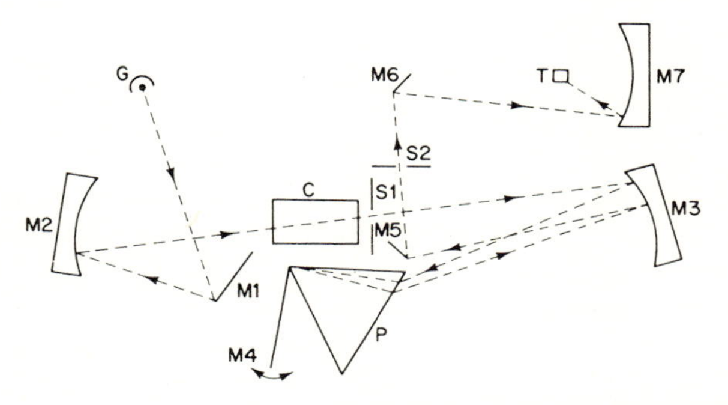

Fig. 5-1 The Perkin-Elmer spectrometer.

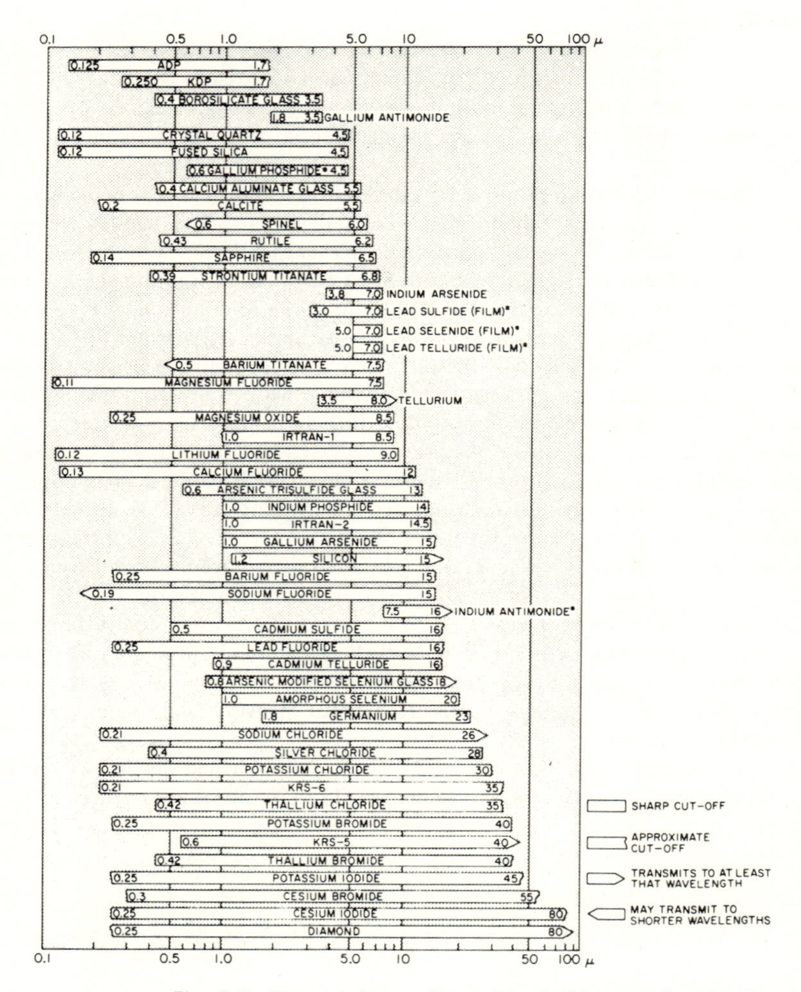

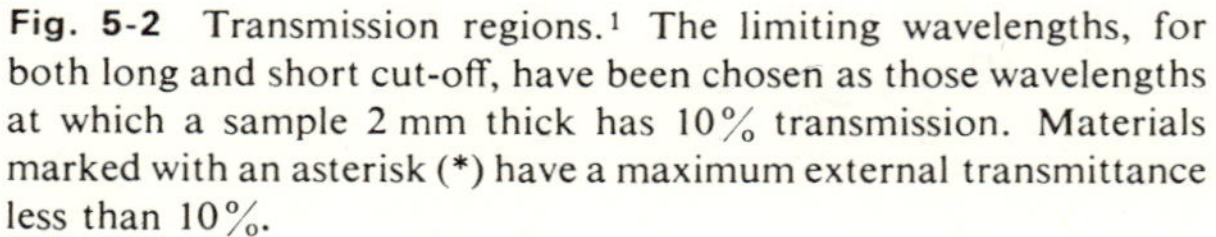

Fig. 5-2 Transmission regions.[1] The limiting wavelengths, for both long and short cut-off, have been chosen as those wavelengths at which a sample 2 mm thick has 10% transmission. Materials marked with an asterisk (*) have a maximum external transmittance less than 10%.

[1]S. S. Ballard, K. A. McCarthy, and W. L. Wolfe, *American Institute of Physics Handbook*, 2nd Ed., McGraw-Hill, New York (1963), Sec. 6, p. 11.

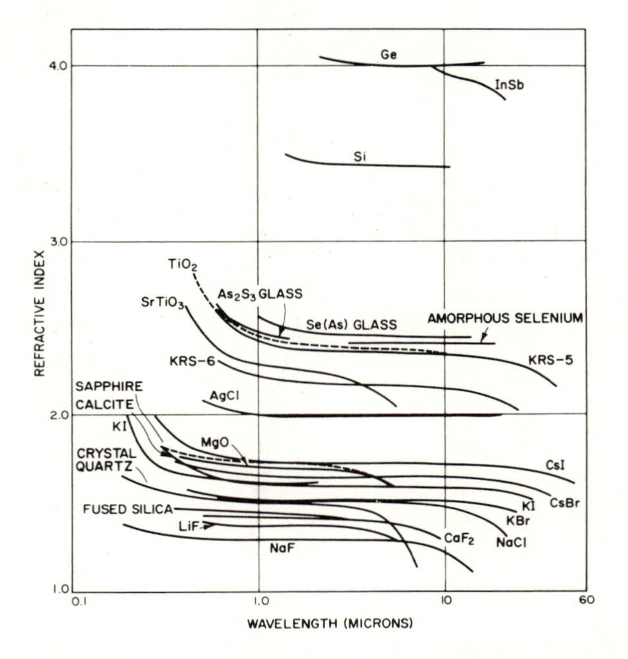

Fig. 5-3 Refractive index vs. wavelength for several optical materials.[1]

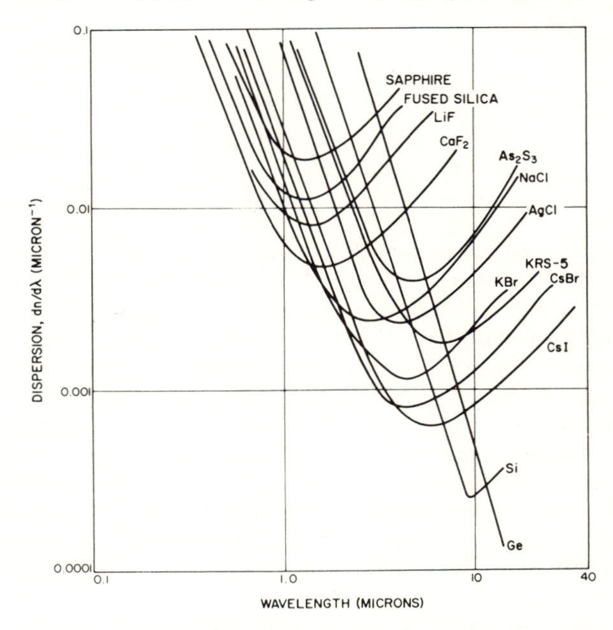

Fig. 5-4 Dispersion vs. wavelength for several optical materials.[1]

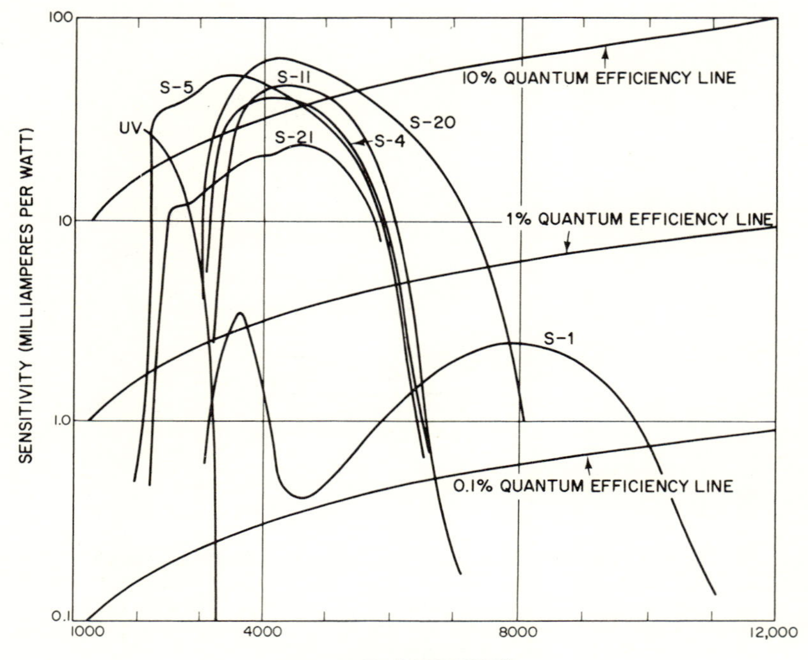

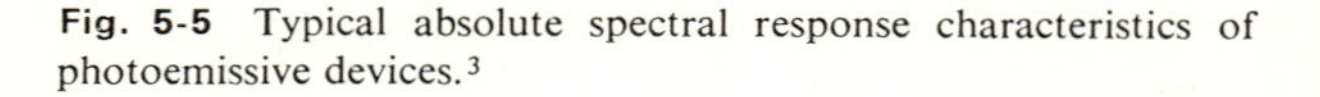

Fig. 5-5 Typical absolute spectral response characteristics of photoemissive devices.[3]

made of sintered silicon carbide is used. For the visible or near infrared (up to about 2.5 μ), a tungsten-ribbon lamp is usually employed. The UV range is generated by a gas-discharge lamp such as a hydrogen lamp. The emission spectrum of a gas-discharge lamp consists of many narrow lines riding a broad continuous spectrum. The wavelength of these lines is often known to at least five significant figures and is tabulated in various handbooks.[2] Hence the gas-emission, or arc, lines are often used to calibrate the instrument.

The detector, also, must be chosen for the proper spectral range. The sensitivity of various detectors is shown in Figs. 5-5, 5-6, and 5-7. In the far infrared region, thermocouples and bolometers are used. Sensitivity is sometimes sacrificed for a gain in speed of response. With extreme care, sensitivities of 10^{-12} W can be obtained.

[2]For example, the *American Institute of Physics Handbook*.
[3]"RCA Photo and Image tubes," Booklet 1Ce-269.

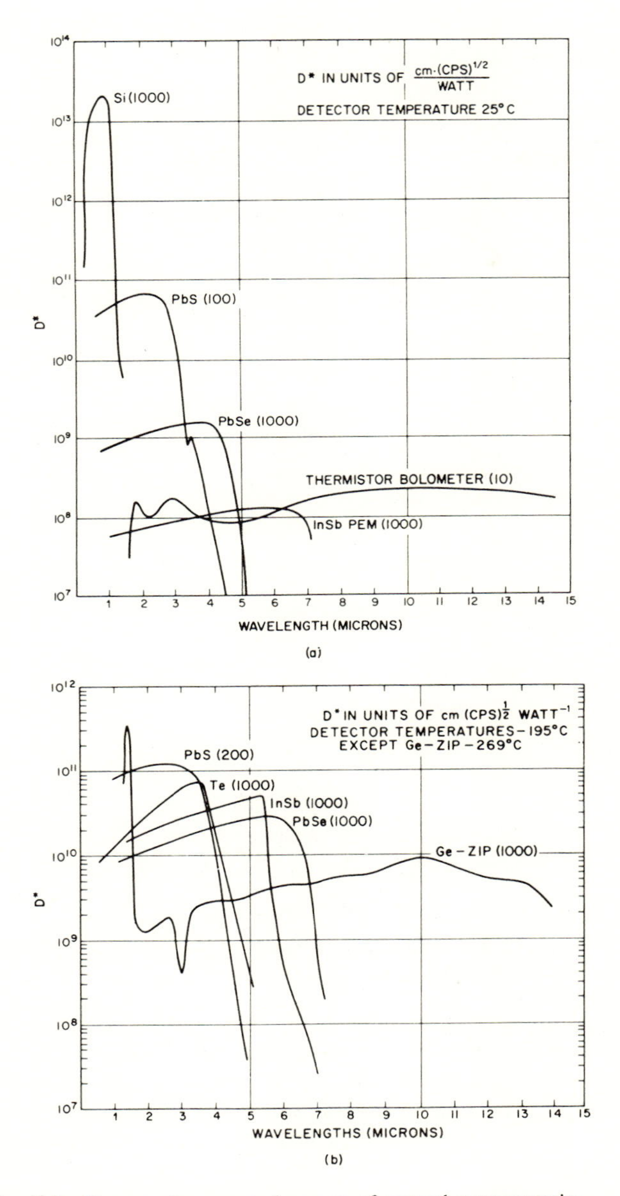

Fig. 5-6 Comparative spectral curves of several representative photodetectors at room temperature (a), and at low temperature (b). The data was obtained in a 1-Hz bandwidth at the frequency indicated in parentheses in Hz.[4]

[4]R. Clark Jones, *Proc. IRE* **47**, 1495 (1959).

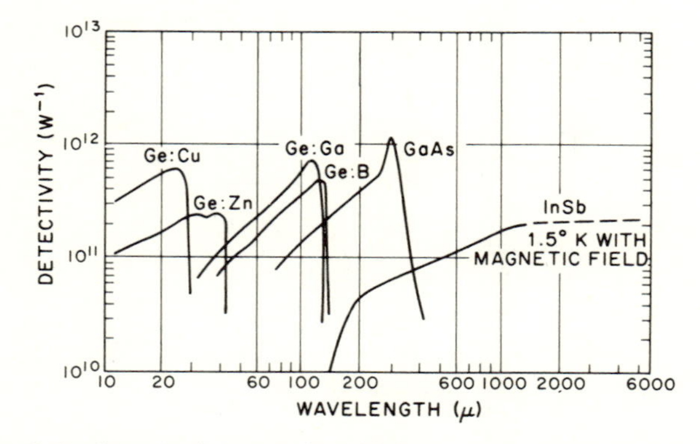

Fig. 5-7 Comparative spectral sensitivity of cooled semiconductor detectors for the far infrared.[5]

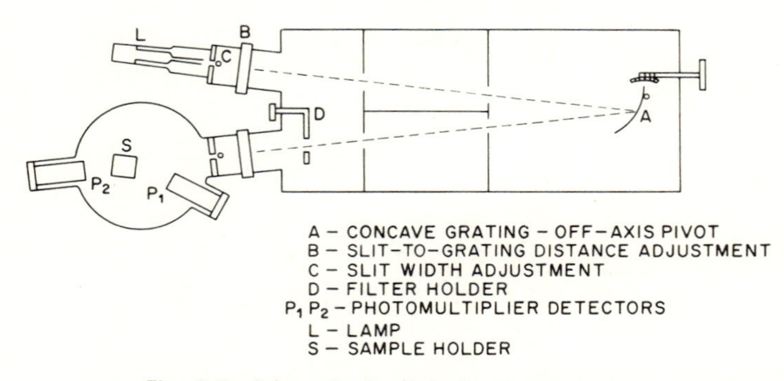
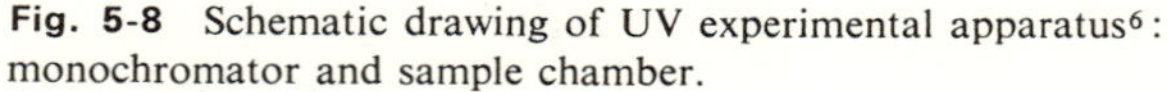

Fig. 5-8 Schematic drawing of UV experimental apparatus[6]: monochromator and sample chamber.

To operate far in the ultraviolet range (above 6 eV), the whole system—source, monochromator, sample, and detector—must be enclosed in a vacuum to prevent absorption of UV by air. A vacuum monochromator is illustrated in Fig. 5-8. This instrument was designed for reflectance studies. The sample S is withdrawn to measure the incident radiation by means of P_2. With the sample in position, detector P_1 measures the reflected radiation.

We shall now show an example of an absorption measurement, an example drawn from the author's own experience with GaAs.[7] The polished

[5]Kindly supplied by G. E. Stillman.

[6]H. R. Philipp and H. Ehrenreich, "Ultraviolet Optical Properties" *Semiconductors and Semimetals*, ed. R. K. Willardson and A. C. Beer, Academic Press (1967), Vol. 3, p. 96.

[7]J. I. Pankove, *Phys. Rev.* **140**, A-2059 (1965).

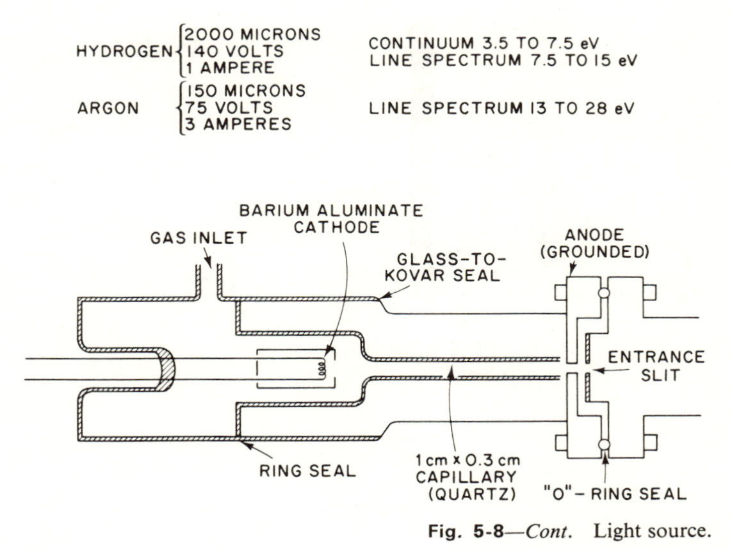

Fig. 5-8—*Cont.* Light source.

sample of GaAs is mounted on a copper slide provided with two identical apertures, the specimen being located over one of these apertures (Fig. 5-9). The copper slide is supported by a cold finger and surrounded by a thermal shield at the same temperature. The slide can be positioned magnetically to measure either the incident radiation I_o (through the unmasked aperture) or the radiation I transmitted by the sample.

The temperature of the cold finger is measured by either a carbon resistor or a GaAs thermometric diode. The carbon resistor is reliable below 50°K, while the GaAs diode is sensitive above 20°K. Thus a range of overlap is available for cross-checking.

The transmission of a sample of thickness t and reflectivity R is given by

$$I = \frac{I_o(1 - R)^2 e^{-\alpha t}}{1 - R^2 e^{-2\alpha t}}$$

which takes into account multiple internal reflections. In our case, with $R = 0.30$ the denominator can be made 1 when $\alpha t \geq 1.2$ (this makes an error of less than 1%); then

$$\alpha t = \ln I_o - \ln I + 2 \ln (1 - R)$$

The equipment can be arranged to solve this relation directly and to plot αt versus $h\nu$. This is done as illustrated in Fig. 5-10. The logarithm of the intensity of the incident radiation $I_o(h\nu)$ is stored on a cam driven by the mo-

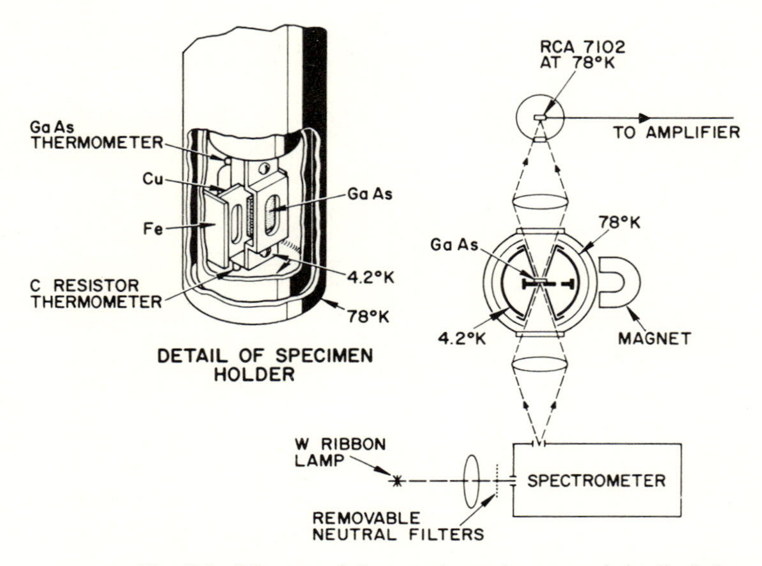

Fig. 5-9 Diagram of the experimental setup and detail of the specimen holder.

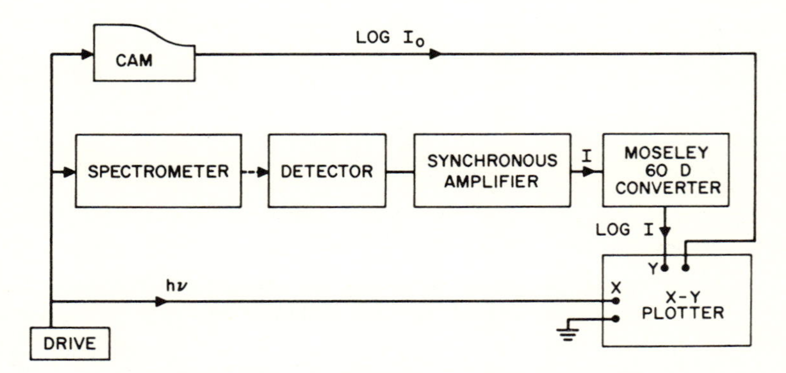

Fig. 5-10 Diagram of the instrumentation for plotting directly αt vs. $h\nu$.

nochromator. The cam itself drives a potentiometer giving an output proportional to $\ln I_o$. The signal due to the transmitted radiation $I(h\nu)$ is passed through a logarithmic converter and the resulting output $\ln I$ is subtracted from $\ln I_o$ at the input of an X–Y plotter, which is also driven by the spectrometer. A typical set of curves is reproduced in Fig. 5-11. The three calibration lines at $\alpha t = 1.23$, 3.18, and 5.12 are obtained by inserting the corresponding neutral attenuators while the light from the monochromator passes through the unmasked aperture of the copper slide. The ability to check easily the calibration lines makes it possible to verify the system performance at any time.

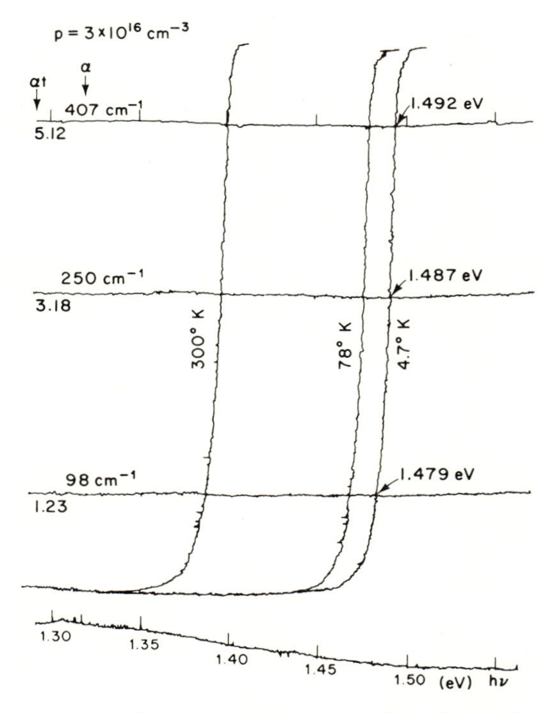

Fig. 5-11 A typical recording of αt vs. $h\nu$.

Problem 1. Specimens in the shape of a Weierstrass sphere were used in several early spectroscopic studies to improve the signal-to-noise ratio by increasing the light-gathering power of the optics. A Weierstrass sphere consists of a sphere of radius r from which a segment has been removed such that the exposed plane, the "Weierstrass plane," is located at a distance $r/\boldsymbol{n}$ from the center of the sphere, as shown in Fig. 5P-1 ($\boldsymbol{n}$ is the index of refraction). A key property of the Weier-

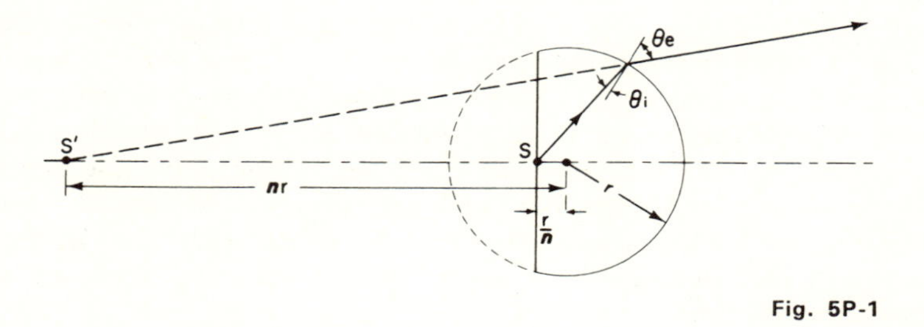

Fig. 5P-1

strass sphere is that rays originating at the center S of the Weierstrass plane are refracted at the spherical surface and appear to come from a virtual source S', located at a distance nr from the center of the sphere, as shown in Fig. 5P-1.

(a) Verify the above property by ray tracing. Use theorems about similar triangles and Snell's law ($\sin \theta_e = n \sin \theta_i$).

(b) For Ge, where $n = 4$, calculate the improvement in light collection by optics on the right-hand side of the specimen for a punctual isotropic light source at the "Weierstrass point" S (center of the Weierstrass plane) compared to an identical point light source on the left-hand side of a plane parallel slab of germanium, parallel to the Weierstrass plane.

(c) Mention one or two drawbacks associated with the use of the Weierstrass sphere.

RADIATIVE TRANSITIONS

6

Emission of radiation is the inverse of the absorption process. An electron occupying a higher energy state than it would under equilibrium conditions makes a transition to an empty lower-energy state and all or most of the energy difference between the two states can be emitted as electromagnetic radiation. The radiation rate is determined by the product of the density n_u of carriers in the upper state, the density n_l of empty lower states, and the probability P_{ul} for 1 carrier/cm^3 in the upper state to make a radiative transition to 1 vacancy/cm^3 in the lower state:

$$R = n_u n_l P_{ul} \tag{6-1}$$

This is somewhat similar to the expression [Eq. (3-1)] that we wrote for the absorption process; however, there absorption was described in terms of the mean free path for photon decay, whereas emission is expressed as a rate of photon generation per unit volume.

Most of the transitions that we have considered as examples of absorption mechanisms can occur in the opposite direction and produce a characteristic emission. However, there is an important difference between the information one can obtain by absorption and by emission in a semiconductor: the absorption process can involve all the states in the semiconductor—i.e., states on either side of the Fermi level, resulting in a broad spectrum—whereas the emission process couples a narrow band of states containing the thermalized electrons with a narrow band of empty states containing thermalized holes and, therefore, produces a narrow spectrum.

The main requirement for emission is that the system not be at equilibrium. This deviation from equilibrium requires some form of excitation. The light-emission process is generally called luminescence. Excitation by an electric current (injection or breakdown) results in electroluminescence. Optical excitation (via absorption) produces photoluminescence, and excitation with an electron beam causes cathodoluminescence. Triboluminescence is due to mechanical excitation. But thermoluminescence is not a simple thermal excitation—which is called incandescence; rather, thermoluminescence requires an excitation at low temperature in order to freeze the carriers in trapping states from which they can be subsequently released thermally. In chemiluminescence, light is emitted during a chemical reaction—but none has been reported for semiconductors.

Fluorescence is a luminescence which occurs only during excitation. Phosphorescence is a luminescence which continues for some time after the excitation is terminated.

We shall now discuss the relationship between emission and absorption.

6-A The van Roosbroeck–Shockley Relation[1]

The simple statement of the van Roosbroeck–Shockley relation is that at equilibrium, the rate of optical generation of electron–hole pairs is equal to their rate of radiative recombination. If one makes a detailed balance of these processes at the various photon frequencies ν, one can write that the rate of emission at frequency ν in an interval $d\nu$ is given by

$$R(\nu)\, d\nu = P(\nu)\, \rho(\nu)\, d\nu \tag{6-2}$$

where $P(\nu)$ is the probability per unit time of absorbing a photon of energy $h\nu$, and $\rho(\nu)\, d\nu$ is the density of photons of frequency ν in an interval $d\nu$. The latter is obtained from Planck's radiation law (assuming for simplicity that the index of refraction $\boldsymbol{n}$ is independent of ν):

$$\rho(\nu)\, d\nu = \frac{8\pi\nu^2\boldsymbol{n}^3}{c^3}\frac{1}{\exp\left(\frac{h\nu}{kT}\right) - 1}\, d\nu \tag{6-3}$$

The absorption probability is related to the mean lifetime $\tau(\nu)$ of the photon in the semiconductor:

$$P(\nu) = \frac{1}{\tau(\nu)}$$

The mean lifetime can be calculated from the mean free path $1/\alpha(\nu)$ of a photon traveling at a velocity $v = c/\boldsymbol{n}$ (again assuming a constant index of refraction):

$$\tau(\nu) = \frac{1}{\alpha(\nu)v}$$

Therefore,

$$P(\nu) = \alpha(\nu)\, v = \alpha(\nu)\frac{c}{\boldsymbol{n}} \tag{6-4}$$

[1] W. van Roosbroeck and W. Shockley, *Phys. Rev.* **94**, 1558 (1954).

Substituting Eqs. (6-3) and (6-4) into Eq. (6-2),

$$R(\nu)\,d\nu = \frac{\alpha(\nu)\,8\pi\nu^2\boldsymbol{n}^2}{c^2[\exp(h\nu/kT) - 1]}\,d\nu \tag{6-5}$$

Equation (6-5) is the fundamental relation between the expected emission spectrum and the observed absorption spectrum. It can be expressed in terms of the extinction coefficient $\boldsymbol{k}(\nu)$. We recall from Eq. (4-7) that

$$\alpha(\nu) = \frac{4\pi\nu\boldsymbol{k}(\nu)}{c} \tag{6-6}$$

Then Eq. (6-5) can be rewritten as

$$R(\nu)\,d\nu = \frac{32\pi^2\boldsymbol{k}(\nu)\boldsymbol{n}^2\nu^3}{c^3[\exp(h\nu/kT) - 1]}\,d\nu \tag{6-7}$$

The total number R of recombinations per unit volume per second is obtained by integrating over all photon frequencies. For this it is convenient to make a change of variable $u = h\nu/kT$. Note that, from Eq. (6-6),

$$u = \alpha(\nu)\frac{c}{4\pi\boldsymbol{k}(\nu)}\frac{h}{kT}$$

Then,

$$R = \frac{8\pi\boldsymbol{n}^2(kT)^3}{c^2h^3}\int_0^\infty \frac{\alpha(\nu)\,u^2}{e^u - 1}\,du \tag{6-8}$$

Note that in terms of u, Eq. (6-5) could be written

$$R(\nu)\,d\nu = \frac{8\pi}{c^2}\left(\frac{kT}{h}\right)^3\boldsymbol{n}^2\,\alpha(\nu)\frac{u^2}{e^u - 1}\,du$$

which consists of two factors which characterize the semiconductor, $\boldsymbol{n}^2$ and $\alpha(\nu)$, and a factor which is independent of the material:

$$U = \frac{8\pi}{c^2}\left(\frac{kT}{h}\right)^3\frac{u^2}{e^u - 1}$$

or if $\boldsymbol{k}$ is used[2] instead of α:

$$U' = \frac{32\pi^2}{c^3}\left(\frac{kT}{h}\right)^4\frac{u^3}{e^u - 1} = 1.785 \times 10^{22}\left(\frac{T}{300}\right)^4\frac{u^3}{e^u - 1}$$

Therefore, in principle one can transform the absorption spectrum of a semiconductor of known index of refraction $\boldsymbol{n}$ into the expected emission spectrum—for example, as shown in Fig. 6-1.

Equations (6-5) and (6-8) state a fundamental relationship between the recombination rate and the absorption coefficient. Although the above formulation was originally derived for band-to-band transitions, it is valid for transitions between any sets of states. These expressions were formulated for the case of thermal equilibrium; now let us examine what happens when deviations from thermal equilibrium occur.

[2] Y. P. Varshni, *Phys. Stat. Sol.* **19**, 459 (1967), and **20**, 9 (1967).

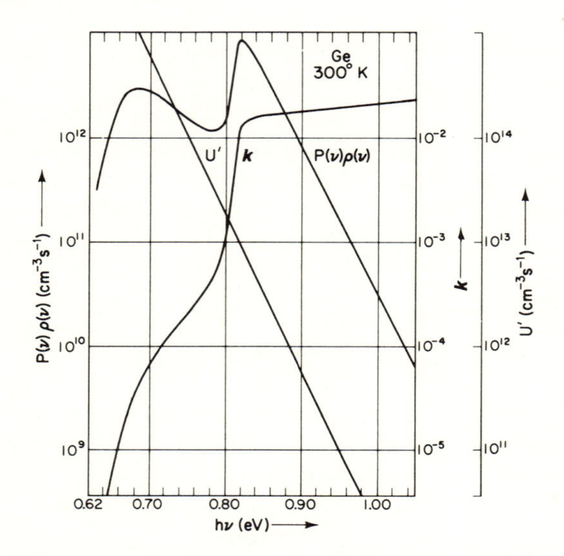

Fig. 6-1 Plot of $\boldsymbol{k} = c\alpha/4\pi\nu$, U' and $P(\nu)\rho(\nu)$ for germanium at 300°K.[2]

We shall retain the notation n and p for electron and hole concentrations, respectively, but it will be understood that p refers to the concentration of any kind of empty state below the Fermi level; i.e., they could be holes in the valence band or ionized donors or neutral acceptors or the hole of an exciton.

Let n_i be the intrinsic carrier concentration. The total radiative-recombination rate R_c is proportional to the concentrations of electrons n and empty states p:

$$R_c = \frac{np}{(n_i)^2} R \tag{6-9}$$

Equation 6-9 implies that the carrier lifetime decreases with increasing carrier concentration, and also that the total radiative-recombination rate becomes R when the product np approaches the equilibrium intrinsic value n_i^2.

Let us express the carrier concentration as a deviation from an equilibrium value n_0 (and p_0); then $n = n_0 + \Delta n$ and $p = p_0 + \Delta p$, and the recombination rate becomes

$$R + \Delta R = \frac{(n_0 + \Delta n)(p_0 + \Delta p)}{n_0 p_0} R = \frac{n_0 p_0 + p_0 \Delta n + n_0 \Delta p + \Delta n \Delta p}{n_0 p_0} R$$

or, neglecting the very small term $\Delta n \, \Delta p$,

$$\frac{\Delta R}{R} = \frac{\Delta n}{n_0} + \frac{\Delta p}{p_0} \tag{6-10}$$

Now we can find the radiative lifetime of excess carriers, assuming $\Delta n = \Delta p$:

$$\tau = \frac{\Delta n}{\Delta R} = \frac{1}{R} \frac{n_0 p_0}{n_0 + p_0} \tag{6-11}$$

for an intrinsic material, $n_0 = p_0 = n_i$; therefore,

$$\tau = \frac{n_i}{2R} \tag{6-12}$$

If we call B the probability for radiative recombination [this is equivalent to the P_{ul} of Eq. (6-1), where only discrete states were considered rather than a set of states], we get

$$B = \frac{R}{(n_i)^2} \tag{6-13}$$

Calculated values of τ and B for conduction-band–to–valence-band radiative transitions are shown in Table 6-1.

Table 6-1[3]

Radiative recombination at 300° K

Material	E_g	n_i	B	τ (intrinsic)	τ (for 10^{17} majority carriers/cm^3)
	eV	cm^{-3} × 10^{14}	cm^3/sec × 10^{-12}		μs
Si	1.08	0.00015	0.002	4.6 h	2500
Ge	0.66	0.24	0.034	0.61 sec	150
GaSb	0.71	0.043	13	0.009 sec	0.37
InAs	0.31	16	21	15 μs	0.24
InSb	0.18	200	40	0.62 μs	0.12
PbS	0.41	7.1	48	15 μs	0.21
PbTe	0.32	40	52	2.4 μs	0.19
PbSe	0.29	62	40	2.0 μs	0.25
GaP[4]	2.25	——	0.003	——	3000

6-B Radiative Efficiency

The transition from an upper- to a lower-energy state may also proceed via one or more intermediate states. The intermediate steps may or may not be radiative. Yet they have a profound effect on the actual efficiency of the radiative transition from the upper to the lower state.

[3] R. N. Hall, *Proc. Institution of Electrical Engineering* **106B**, Suppl. No. 17, 923 (1959).

[4] M. Gershenzon, "Radiative Recombination in the III–V Compounds," *Semiconductors and Semimetals*, ed. R. K. Willardson and A. C. Beers, Academic Press (1967), p. 305.

Let us consider the case of one intermediate state i (Fig. 6-2). The van Roosbroeck–Shockley treatment of the probability for radiative transition $u \longrightarrow l$ is based on the measured absorption coefficient for the reverse process $l \longrightarrow u$ involving the same photon energy $h\nu$. Clearly, in the discrete-state model of Fig. 6-2 the intermediate state does not participate in the absorption process at $h\nu$. But in the recombination process two paths are possible and, therefore, two competing processes occur: the radiative transition emitting a photon $h\nu$ and the transition $u \longrightarrow i \longrightarrow l$, which if radiative will result in photons of energy lower than $h\nu$. Often, the transition $u \longrightarrow i \longrightarrow l$ dominates and makes the radiative process at $h\nu$ appear inefficient. However, the radiative transition $u \longrightarrow l$ still obeys Eq. (6-5); but the number of carriers participating in this process is reduced by virtue of their decay via the competing process, which is characterized by a recombination time τ'. Hence the effective recombination time is given by

$$\frac{1}{\tau_{\text{eff}}} = \frac{1}{\tau} + \frac{1}{\tau'} \tag{6-14}$$

Putting Eq. (6-14) into (6-12) and the latter into (6-9), the total recombination rate is

$$R_T = \frac{1}{\tau_{\text{eff}}} \frac{np}{2n_i}$$

while the radiative-recombination rate is

$$R = \frac{1}{\tau} \frac{np}{2n_i}$$

Therefore, the radiative efficiency is

$$\eta = \frac{R}{R_T} = \frac{1}{1 + \tau/\tau'} \tag{6-15}$$

The statistics of recombination via an intermediate state have been treated as a function of minority- and majority-carrier concentrations by Shockley–Read[5] and Hall.[3]

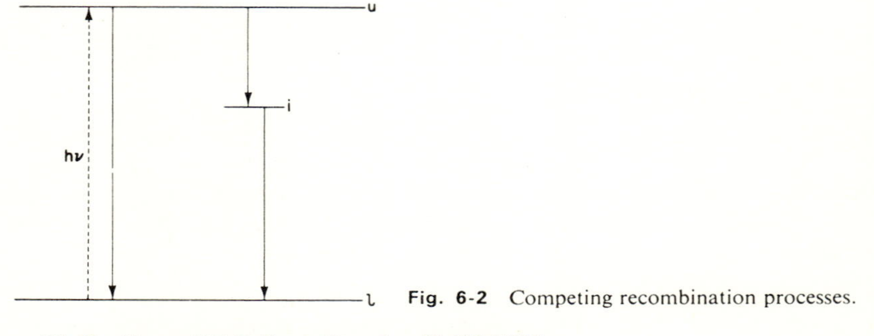

Fig. 6-2 Competing recombination processes.

[5] W. Shockley and W. T. Read, *Phys. Rev.* **87**, 835 (1952).

6-C The Configuration Diagram

The concept of the configuration diagram is introduced at this point because it is useful in explaining the phonon-emission process that can accompany many radiative transitions.

The configuration diagram (Fig. 6-3) presents the energy of the ground state and of the first excited state of an atom as a function of its position, this atom being either an impurity atom or a host lattice atom. Note that, at its equilibrium position, the ground state is at a minimum potential. This minimum a is the equilibrium potential of an electron in the ground state when the atom is stationary at its equilibrium position in a stationary crystal (which is a reasonable situation at a very low temperature). The excited state, however, may have a minimum potential for a slightly different position of the atom. Hence when the electron is excited from the ground state at a, the whole system partly "relaxes" to condition c. This means a displacement Δx of the atom. The excited electron loses some energy in the process and the energy thus lost is dissipated in the form of atomic displacement, i.e., as a phonon.

When the electron returns to the ground state at d, a new atomic displacement is necessary for the system to relax to its lowest energy at a. This displacement $d \longrightarrow a$ also takes the form of a phonon emission.

Although all or any of the six phonons may take part in this process, the LO-phonon is generally favored because it is the one that produces the strongest polarization field—i.e., the strongest change is potential per unit displacement.

The energy emitted in the transition $c \longrightarrow d$ is lower than the energy absorbed in the transition $a \longrightarrow b$. The difference between these two energies, is called the "Franck–Condon shift." In general, the degradation of optical energy is called a "Stokes shift." Therefore, the Franck–Condon shift is a Stokes shift due to the displacement of atoms following optical excitation.

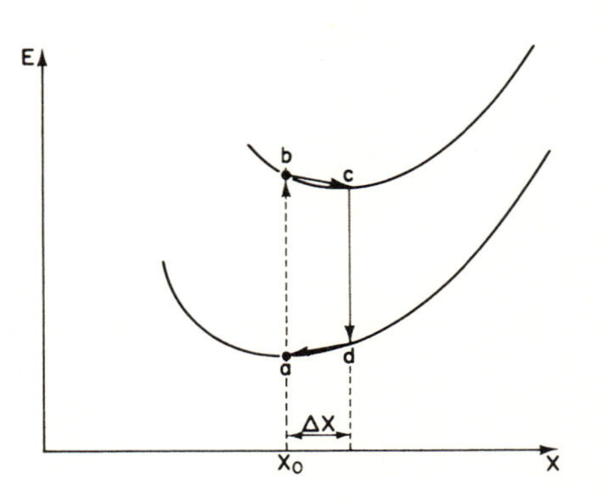

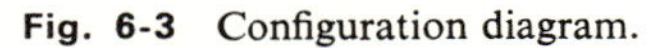

Fig. 6-3 Configuration diagram.

6-D Fundamental Transitions

Since emission requires that the system be in a nonequilibrium condition, let us assume that some means of excitation is acting on the semiconductor to produce hole–electron pairs. Let us first consider the fundamental transitions, those occurring at or near the band edges.

6-D-1 EXCITON RECOMBINATION

6-D-1-a Free Excitons

If the material is sufficiently pure, the electrons and holes pair off into excitons which then recombine, emitting a narrow spectral line. In a direct-gap semiconductor, where momentum is conserved in a simple radiative transition [Fig. 6-4(a)], the energy of the emitted photon is simply

$$h\nu = E_g - E_x \tag{6-16}$$

In Sec. 1-D-1 we saw that the exciton can have a series of excited states whose ionization energy is lower by a factor $1/n^2$ than that of the ground state E_{x_1} which corresponds to $(n = 1)$. Hence the free-exciton emission might consist of a series of narrow lines starting at $E_g - E_{x_1}$ and occurring at $E_g - (1/n^2)E_{x_1}$. However, the intensity of the higher-order peaks decreases rapidly[6] (as n^{-3}) and is difficult to observe in the presence of a background of other radiative processes. In GaAs the $(n = 1)$ and $(n = 2)$ free excitons have been identified.[7]

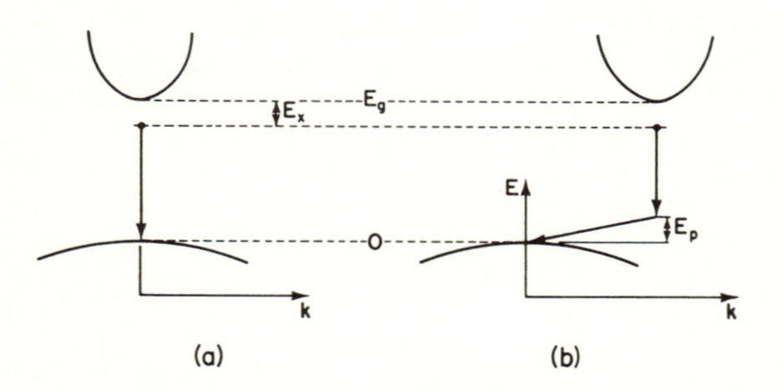

Fig. 6-4 Exciton recombination: (a) direct; (b) indirect.

In an indirect-gap semiconductor, momentum conservation requires that a phonon be emitted to complete the transition [Fig. 6-4(b)]. Then the energy

[6] R. J. Elliott, *Phys. Rev.* **108**, 1384 (1957).
[7] M. A. Gilleo, P. T. Bailey, and D. E. Hill, *Phys. Rev.* **174**, 898 (1968).

of the emitted photon is

$$h\nu = E_g - E_x - E_p$$

where E_p is the energy of the phonon involved.

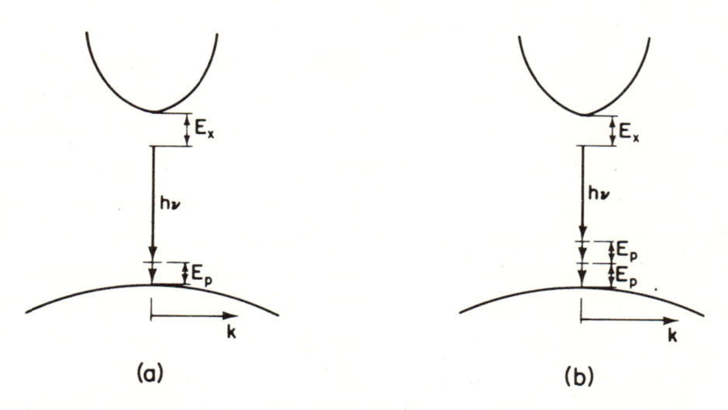

Fig. 6-5 Direct exciton recombination with (a) one optical phonon emission, (b) two optical phonon emission.

Note that, at the cost of a lower transition probability, a direct transition can also occur with the emission of one or more optical phonons (Fig. 6-5). In fact, our discussion of the configuration diagram (Sec. 6-C) showed that in general even direct transitions can lead to phonon emission, the favored mode being an optical phonon (optical phonon being possible at $k = 0$). Hence the narrow emission spectrum of the exciton can be replicated at several lower photon energies such that

$$h\nu = E_g - E_x - \boldsymbol{m}E_p$$

where $\boldsymbol{m}$ is the number of optical phonons emitted per transition. Obviously, the larger $\boldsymbol{m}$ is, the lower the transition probability and the weaker the corresponding emission line.

It should be noted that although the transition with phonon emission is less probable than the direct recombination, the resulting photon has a greater chance of escape because it occurs at a lower photon energy than the direct recombination, i.e., in a region of the spectrum where the semiconductor is more transparent. However, the reabsorption of exciton radiation is not necessarily a loss of energy because, in the process, a new exciton is formed which will provide another opportunity to emit.

6-D-1-b Exciton Transport and Polariton Recombination

Since free excitons can propagate through the crystal, it should be possible to produce them in one part of a crystal and to detect them in another region of the semiconductor.[8] Excitons can have a short lifetime (10^{-8} sec in CdS), at the end of which they become a photon. But this photon, having propagated a short distance through the crystal, can generate another exciton. This resonant interaction could transport energy over a considerable distance. The migration of excitons could be detected by the light emitted at a distant site where the exciton recombines radiatively, or by the photoconductivity produced when a strong local field breaks up the exciton into a pair of independent carriers. Some evidence for such an effect has been presented[8] but has led to much controversy, since it is difficult to separate exciton migration from internal light scattering when the photon energy is lower than the gap energy. The possibility of self-regeneration via the resonant recombination–generation process might cause a slight increase in the apparent lifetime of excitons. In CdS a decay time of 9 microseconds for the exciton emission was observed after flash excitation.[9] This is several orders of magnitude longer than the 10^{-8} sec expected for this process. However, trapping effects may be invoked to account for this slow luminescence decay.

In Sec. 1-D-3, we saw that polaritons represent the interaction between excitons and photons resulting in the dispersion curve of Fig. 1-16. When free excitons are formed, they can thermalize along the parabolic portion of the dispersion curve toward the knee of the curve, emitting acoustical phonons; they can also emit optical phonons, in which case the exciton transforms into a photon (below the knee of the curve) and can either propagate through the crystal or be emitted. Hence luminescence is obtained when a polariton is scattered by an optical phonon. Luminescence can also be obtained when the polariton collides with the surface.[10] The difference between the emission spectra from free-exciton recombination and polariton recombination is shown in Fig. 6-6. The exciton spectrum cuts off abruptly at low energy corresponding to a stationary exciton at $k = 0$, and if the recombination process occurs via an LO-phonon emission, the LO-phonon spectrum replica also cuts-off abruptly at the lower-energy edge. The polariton-emission spectrum, on the other hand, exhibits a low-energy tail continuing below the "free-exciton edge" corresponding to all the states below the knee of Fig. 1-16 within one LO-phonon energy.

6-D-1-c Bound Excitons

In the presence of impurities, bound excitons may be obtained. When these recombine, their emission is characterized by a narrow spectral width at a lower photon energy than that of the free exciton. A line width of the order of 0.1 meV has been found for excitons bound to shallow impurities in

[8] M. Balkanski and J. Broser, *Zeit. Electrochemie* **61**, 715 (1957).
[9] M. Balkanski and R. D. Waldron, *Phys. Rev.* **112**, 123 (1958).
[10] J. J. Hopfield, *Journ. Phys. Soc.* (Japan) **21**, supplement, 77 (1966).

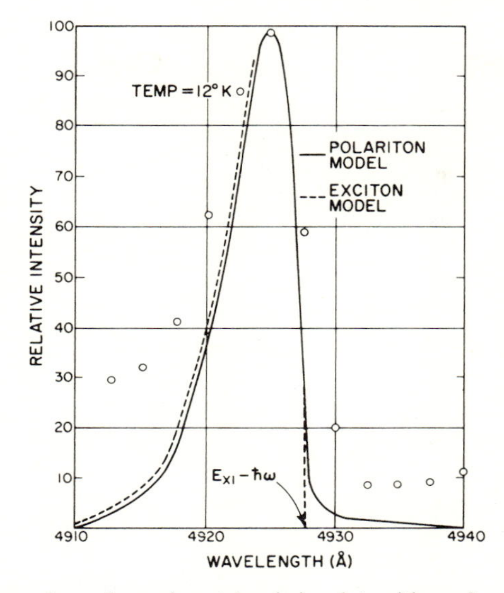

Fig. 6-6 Comparison of experimental emission data with results predicted by polariton (solid curve) and exciton (dashed curve) models for CdS at 12°K. The agreement between the experimental data and the theoretical curves would be improved if the contribution of adjacent emission peaks (not shown) were subtracted.[11]

GaAs.[7] Often, both free and bound excitons occur simultaneously in the same material, in which case each can be identified by its energy and line width (e.g., $\Delta h\nu = 0.1$ meV for the bound exciton compared to 1 meV for the free exciton at 1.4°K in high-quality GaAs).[7]

Examples of exciton emission are shown in Figs. 6-7 through 6-11.

Figure 6-7 presents the emission spectrum of fairly pure InP[12]. The emission peak labeled 1 is believed to be due to free-exciton recombination, while the lines 2, 3, 4, and 5 are attributed to the recombination of a bound exciton with 0-, 1-, 2-, and 3-phonon emission, respectively. The spacing between the lines 2, 3, 4, and 5 is 43 meV, which corresponds to the known value for the LO-phonon. The peaks I and II are due to other processes not involving excitons.

The radiative recombination of a bound exciton in CdS is shown in Fig. 6-8. The peak I_1 is the zero-phonon emission. The lower-energy peaks are associated with various phonon emissions as labeled.

[11]W. C. Tait, D. A. Campbell, J. R. Packard, and R. L. Weither, "Luminescence from Inelastic Scattering of Polaritons by Longitudinal Optical Phonons," II–VI Semiconducting Compounds, ed. D. G. Thomas, Benjamin (1967), p. 370.

[12]W. J. Turner and G. D. Pettit, *Appl. Phys. Letters* **3**, 102 (1963).

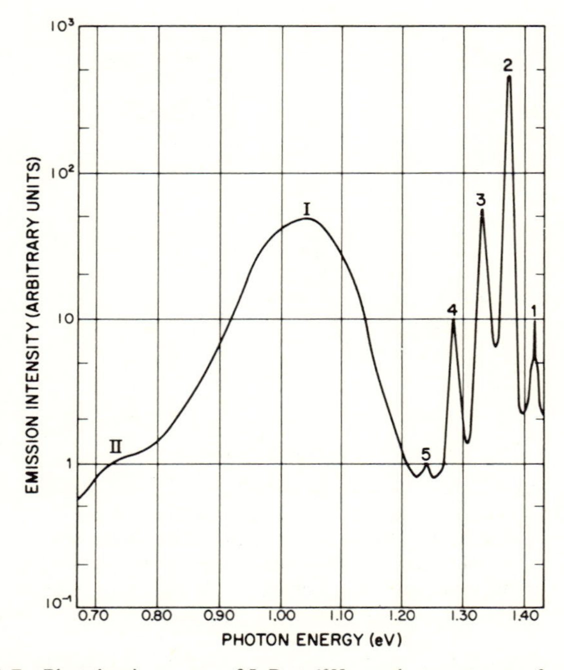

Fig. 6-7 Photoluminescence of InP at 6°K vs. photon energy of emitted light.[12]

Figure 6-9 presents the emission spectrum of high-quality GaAs at a low level of excitation to minimize carrier interactions. The free-exciton peak is identified as $n = 1$, and the exciton bound to the shallow Se-donor is labeled accordingly.

Excitons bound to oxygen impurities have been found in GaAs.[13] They produce three equally spaced emission peaks corresponding to 0-, 1-, and 2-phonon emissions. The identification of these peaks is deduced from the fact that their intensities vary linearly with the oxygen concentration.

Figure 6-10 shows, for GaP, the emission spectrum due to three different bound-exciton recombinations. The same overall transitions as that of line *A* but with emission of one and two phonons have been identified, and the different phonons involved are labeled on the spectrum. This complex identification was possible only after a comparison with the spectra of many other differently doped specimens. The identification also required knowing the

[13]M. I. Nathan and G. Burns, *Phys. Rev.* **129**, 125 (1963).

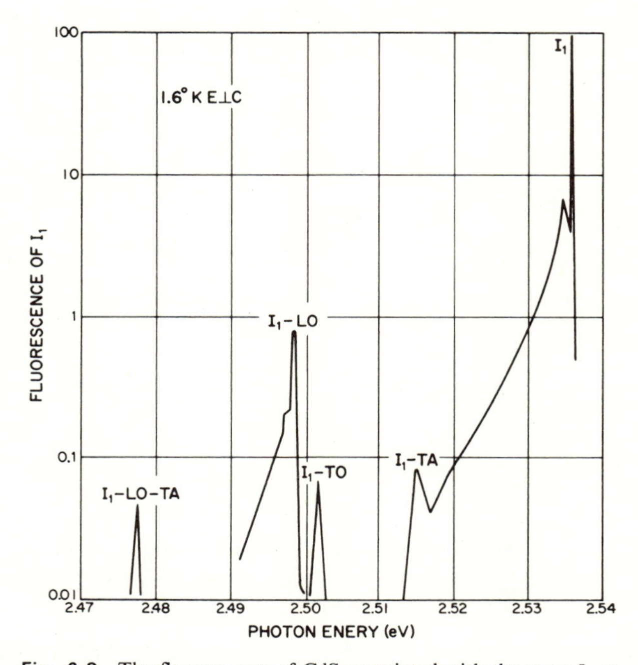

Fig. 6-8 The fluorescence of CdS associated with the state I_1 at 1.6°K. It is very similar at 4.2°K. The strong line at high energy is the "no-phonon" line, and the other fluorescence corresponds to I_1 decaying with the emission of various phonons. LO represents the longitudinal optical phonon at $k = 0$, the energy of which is well known (0.038 eV). The assignments of the transverse acoustic and transverse optical phonons with energies at $k = 0$ of 0.021 and 0.034 eV, respectively, are tentative.[15]

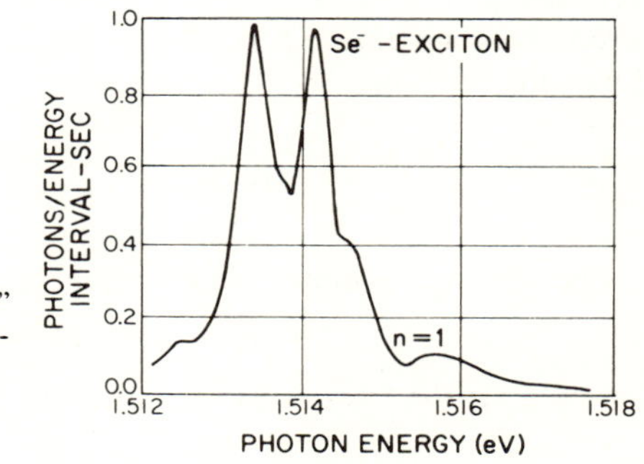

Fig. 6-9 Photoluminescence of "pure" GaAs due to Se-bound exciton, free exciton and free carrier recombination, spectrum at 2.12°K with 0.09-meV resolution.[7]

[7]M. A. Gilleo, P. T. Bailey, and D. E. Hill, *Phys. Rev.* **174**, 898 (1968).
[15]D. G. Thomas and J. J. Hopfield, *J. Appl. Phys.* **33**, 3243 (1962).

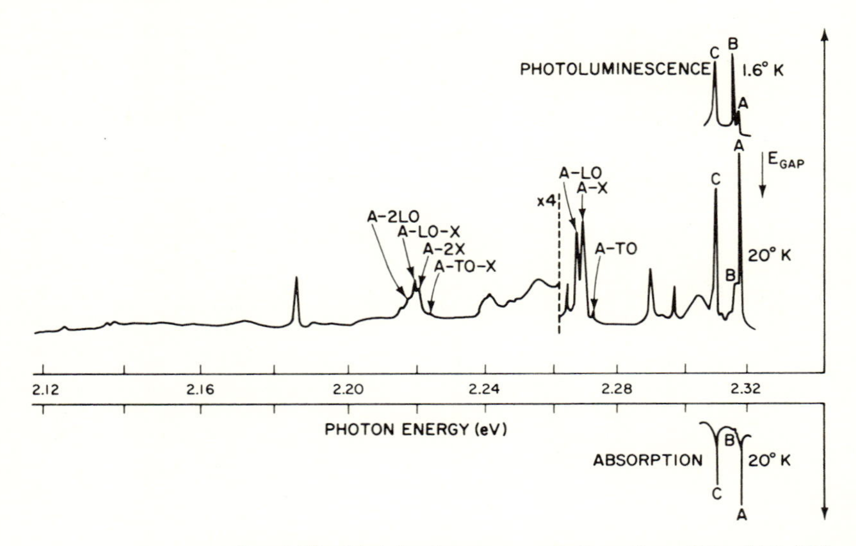

Fig. 6-10 Photoluminescence and absorption of the "A", "B", and "C" bound exciton transitions in GaP and the phonon (LO, TO, X) replicas of the "A" line.[14]

phonon energies available in GaP. A careful study of the Zeeman splitting of the emission spectrum established more definitively that exciton states were involved.[14]

A study of the intensity of the line-A exciton emission in GaP as a function of temperature shows the disappearance of excitons due to thermal ionization.[16] A similar study at 4.2°K as a function of an applied electric field shows the disappearance of excitons due to their field-induced ionization.[13] As the concentration of these bound excitons decreases with increasing temperature or field, the intensity of line A and of its phonon replica decreases. From the temperature dependence of the emission intensity which obeys the following relation—

$$\frac{L(T)}{L(0)} = \frac{1}{1 + CT^{3/2}\exp\left(-\frac{E_i}{kT}\right)}$$

—one obtains the ionization energy E_i for the bound exciton:

$$E_i = E_x + E_b = 0.021 \text{ eV}$$

where E_x is the free-exciton binding energy and E_b is the additional energy

[14] D. G. Thomas, M. Gershenzon, and J. J. Hopfield, *Phys. Rev.* **131**, 2397 (1963).

[16] B. M. Ashkinadze, I. P. Kretsu, S. L. Lyshkin, S. M. Ryvkin, and I. D. Yaroshetskii, *Fizika Tverdovo Tela* (USSR) **10**, 3681 (1968).

binding the free exciton to the center. Supporting evidence that the excitons are broken up into free carriers by thermal or field ionization is manifested by increased electrical conductivity.

Figure 6-11 shows the emission spectrum of excitons in Si. Lines A, B, D and E are due to free excitons. The line at E is produced by the emission of the transverse acoustical (TA) phonon, the principal line at D is produced by the emission of the transverse optical phonon, and the lines at B and A are produced by the emission of sums of two phonons. Line C is extremely sharp and depends on the phosphorus concentration; it has been attributed to a complex consisting of a hole bound by two electrons to a phosphorus ion.[17]

Let us just mention that bound-exciton emission has also been found in germanium,[18] in SiC,[19] and in several other semiconductors.

Exciton recombination is of particular importance in lasers, where, as we shall see in a later section, the narrowness of the emission spectrum plays an important role.

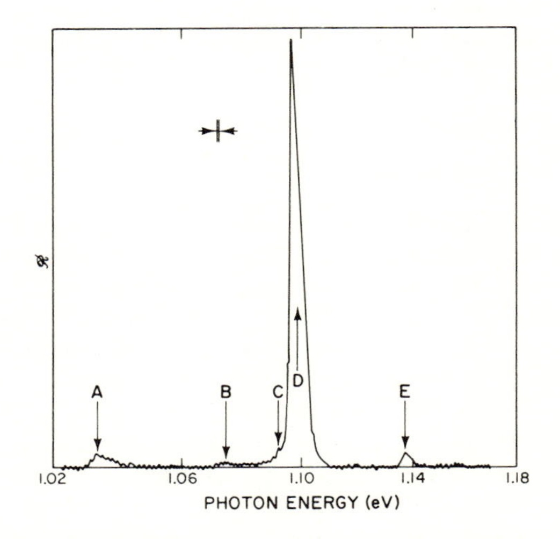

Fig. 6-11 Spectrometer response, $\mathscr{R}$ (nearly proportional to the number of photons/unit energy interval) as a function of the energy of the photons for a silicon crystal containing 2×10^{14} phosphorus atoms per cm^3 at 18°K.[17]

[17] J. R. Haynes, M. Lax, and W. F. Flood, *Intern. Conf. on Physics of Semiconductors, Prague*, 1960, Acad. of Sciences, Prague (1961), p. 423.

[18] C. Benoit-à-la-Guillaume and O. Parodi, *Intern. Conf. on Physics of Semiconductors, Prague, 1960*, Acad. of Sciences, Prague (1961), p. 426.

[19] W. J. Choyke and L. Patrick, *Phys. Rev.* **127**, 1868 (1962).

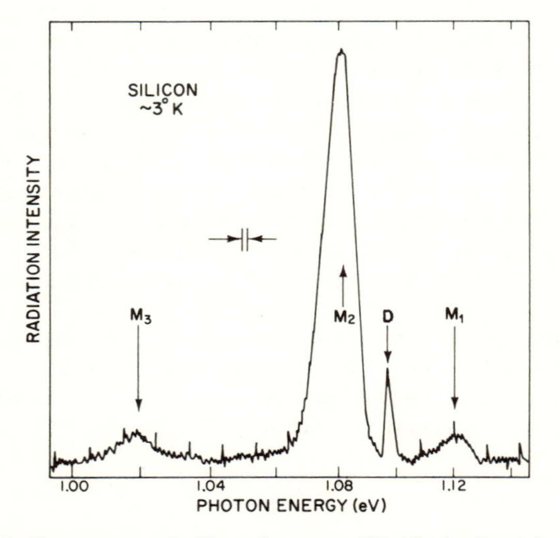

Fig. 6-12 Spectrogram of a Si specimen at ~3°K. The horizontal axis is the energy of the emitted photons in eV. The vertical response is nearly proportional to the number of photons per unit energy interval.[20]

6-D-1-d Excitonic Molecule

The excitonic molecule consisting of two electrons and two holes has been found by emission in silicon.[20] This is shown in Fig. 6-12. In this process, as an electron–hole pair recombines, a photon is emitted with an energy $h\nu = E_g - 2E_x - E_{x_2} - E_{p_m}$, where E_{x_2} is the binding energy of two excitons and E_{p_m} the energy of a phonon of mode m. The remaining electron and hole of the quartet are either ejected into the conduction and valence bands and form a new exciton or remain as an exciton. In Fig. 6-12, D is again an exciton recombination with emission of a TO-phonon. Because of the lower temperature, the processes ABE of Fig. 6-11 are not observed. The lines due to the excitonic molecule (M_1, M_2, and M_3), however, fall about 10 meV below the positions expected for the phonon replicas of the free exciton at 3°K—this is the sum of the binding energy E_{x_2} holding two excitons in a paired complex and the binding energy of one exciton (a relation that will become evident by referring back to Fig. 1-13). The most significant evidence

[20] J. R. Haynes, *Phys. Rev. Letters*, **17**, 860 (1966). The excitonic molecule has also been found recently in germanium: C. Benoit-à-la-Guillaume, F. Salvan, and M. Voos, *Bull. Am. Phys. Soc.* **14**, 867 (1969).

that this emission is correctly interpreted as due to an excitonic molecule is that its intensity increases as the square of the exciton line: this dependence shows that two excitons are needed to produce the molecular complex. Had the emission been due to the recombination of a complex consisting of two electrons bound to one hole or two holes bound to one electron, its intensity would grow as the $\frac{3}{2}$ power of the exciton line.

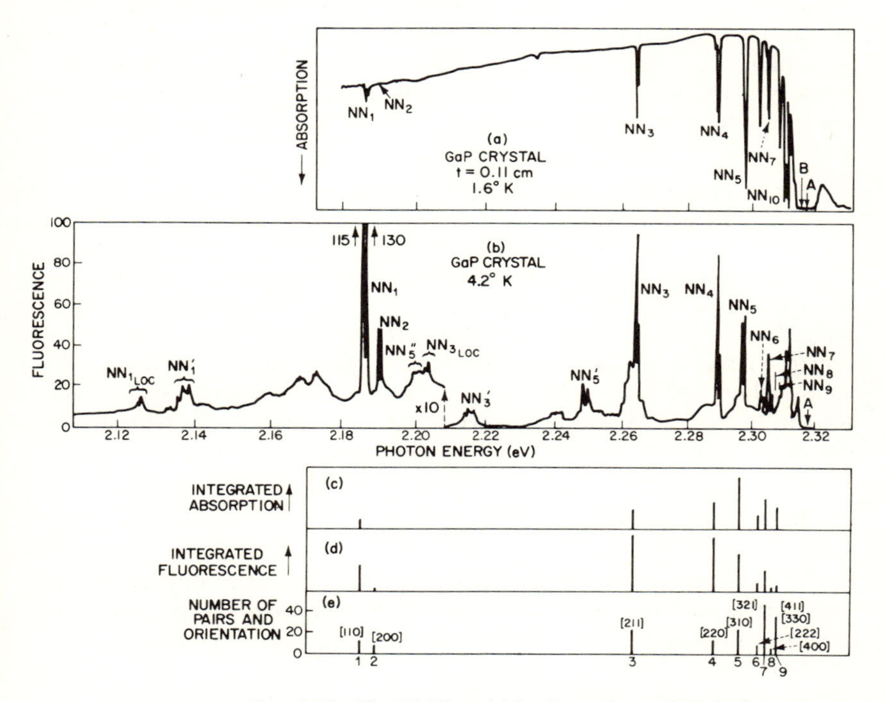

Fig. 6-13 The NN lines (a) in absorption and (b) in fluorescence in crystals containing about $10^{19} N$ atoms/cc. Lattice phonon replicas have been labeled NN_5', NN_5'', etc. Local modes are labeled $NN_{1\,loc}$, etc. In more lightly doped crystals there is a sharp A line in absorption and fluorescence; here the A line (and its higher energy phonon wings) dominates the absorption, but is very weak in emission. (c) and (d) show the approximate relative strengths of the lines in absorption and fluorescence. (e) shows the expected relative intensities of the lines if it is assumed that the nitrogen atoms are arranged randomly and that relative numbers alone control the transition intensities. The orientations of the pairs are also indicated.[21]

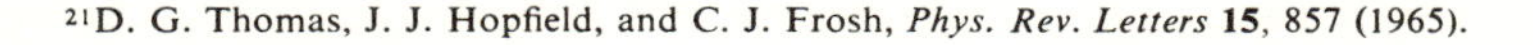
[21] D. G. Thomas, J. J. Hopfield, and C. J. Frosh, *Phys. Rev. Letters* **15**, 857 (1965).

6-D-1-e Excitons Bound to Isoelectronic Traps

It will be recalled from Sec. 3-D that isoelectronic traps (substitution of an atom with the same valence as the replaced host crystal atom) form bound excitons. The radiative recombination of these excitons produces a characteristic spectrum of narrow emission lines. The separation between these emission lines decreases as the photon energy approaches the gap energy of the semiconductor. Furthermore, the intensity of these lines, as well as their separation, is predicted from the statistics of the number of possible isoelectronic pairings along various crystallographic directions. Such a spectrum for nitrogen substituting on phosphorus sites in GaP is shown in Fig. 6-13.[21]

6-D-2 CONDUCTION-BAND–TO–VALENCE–BAND TRANSITIONS

Although exciton states represent lowest-energy states for electron–hole pairs, excitons are formed easily only in the purest materials and at low temperatures. But, in general, a fraction of the excited electron–hole pairs remain as free carriers occupying band states. This is certainly so at temperatures such that $kT > E_x$, and also in less pure or less perfect crystals, where local fields tend to break up the exciton into free carriers. The free carriers can then recombine radiatively in a band-to-band transition.

6-D-2-a Direct Transitions

In a direct-gap semiconductor, momentum-conserving transitions connect states having the same k-values (Fig. 6-14). Therefore, as in the corresponding absorption process of Sec. 3-A-1, the emission spectrum is given by

$$L(\nu) = B(h\nu - E_g)^{1/2} \tag{6-17}$$

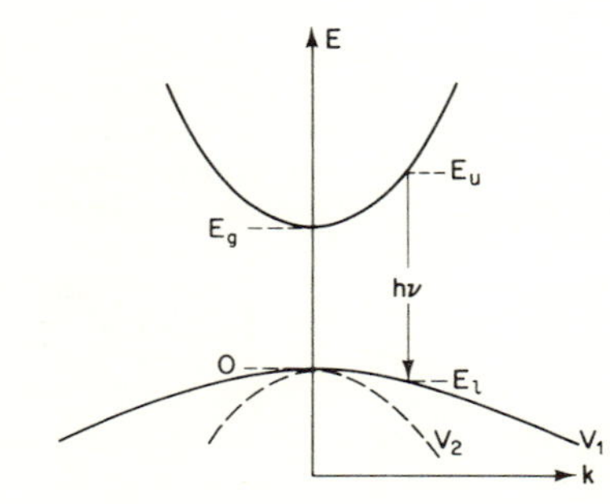

Fig. 6-14 Direct band-to-band radiative transition.

in which the coefficient B can be evaluated from[22]

$$B = \frac{2q^2(m_r^*)^{3/2}}{nch^2m_e^*} \tag{6-18}$$

where m_r^* is the reduced mass: $1/m_r^* = 1/m_e^* + 1/m_h^*$. Hence the emission should have a low-energy threshold at $h\nu = E_g$. As the excitation rate increases and also as the temperature increases, states deeper in the band become filled, permitting emission at higher photon energies. Therefore, free-carrier recombination is characterized by a temperature-dependent high-energy tail, while the low-energy edge is abruptly cut off at $h\nu = E_g$. At low excitation, the half-width of the emission peak is then approximately equal to $0.7kT$ (this is the behavior of the E_g peak of Fig. 6-9).

Note that transitions could terminate at the light-hole subband. The population of the light-hole subband equilibrates quickly with that of the heavy-hole subband, making the density of final states indifferent to the subband. But, because [as per Eq. (6-18)] the transition probability is proportional to $(m_r^*)^{3/2}$ for direct transitions, and since $m_{h_2}^* < m_{h_1}^*$, the transition probability to the light-hole subband is considerably lower than a transition to the heavy-hole states.

An example of free-electron–free-hole radiative recombination is shown in Fig. 6-15, which presents the photoluminescent spectrum of n-type InAs with various doping levels.[23] The shift of the emission peak and of the high-

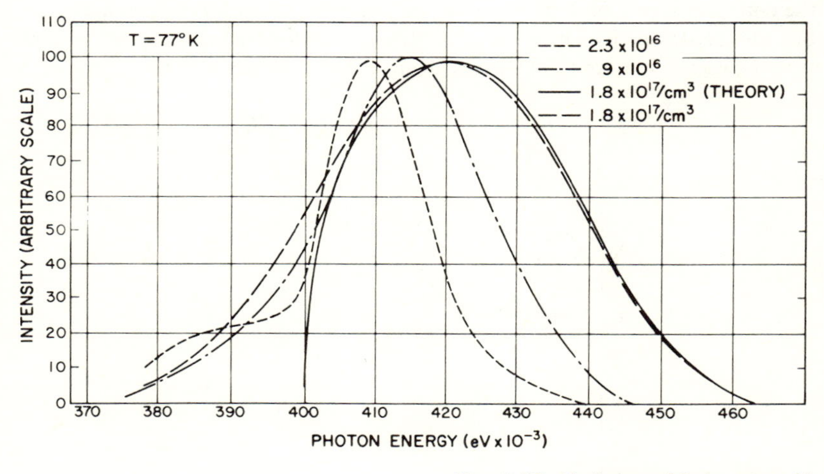

Fig. 6-15 Emission of InAs n-type.[23]

[22] J. Bardeen, F. J. Blatt, and L. H. Hall, *Proc. of Atlantic City Photoconductivity Conference*, 1954, J. Wiley and Chapman and Hall (1956), p. 146.

[23] A. Mooradian and H. Y. Fan, "Photoluminescence of Indium Arsenide," *Radiative Recombination in Semiconductors, 7th Int. Conf. Phys. of Semicond., Paris*, 1964, Dunod, (1965) p. 39.

energy edge to higher photon energies as the doping increases is due to the penetration of the Fermi level into the conduction band. The structure at lower photon energies is due to other transitions (involving tails of states and impurity bands), which we shall discuss later.

6-D-2-b Indirect Transitions

In an indirect-gap semiconductor all the occupied upper states connect to all the empty lower states. But the transition must be mediated by an intermediate process which conserves momentum (Fig. 6-16). Phonon emission is the most likely intermediate process. Another momentum-conserving process, phonon absorption, which forms a readily distinguishable structure in the absorption spectrum, becomes negligible during emission. As we have seen in Sec. 3-A-3, the number of phonons available for absorption is small and is rapidly decreasing at lower temperatures, whereas the emission of phonons by electrons which are already at a high-energy state is very probable. Furthermore, an optical transition assisted by phonon emission occurs at a lower photon energy than the gap energy, $h\nu_{\min} = E_g - E_p$; whereas phonon absorption results in a higher photon energy of at least $E_g + E_p$, which can be more readily reabsorbed by the semiconductor.

In an indirect-gap material, as a higher excitation rate increases the population of the bands, driving the quasi-Fermi levels deeper into the bands, the emission spectrum comprises all the possible transitions between any pair of states separated by a given $h\nu$, regardless of the difference in crystal momentum between initial and final states. Hence the spectrum can be described by

$$L(\nu) = \int_{E_g - E_p}^{\infty} B' n(E)\, p(E)\, dE$$

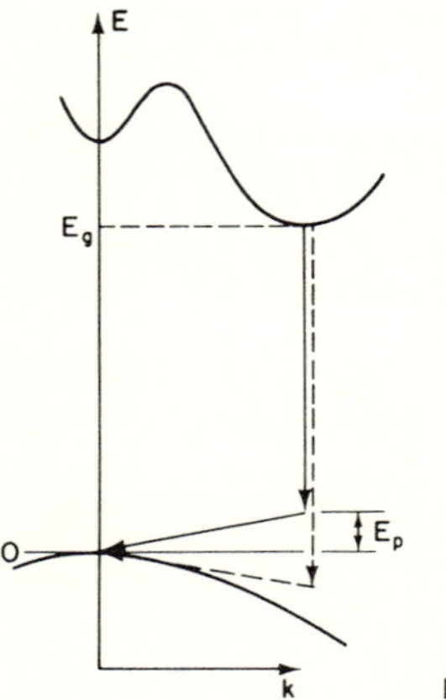

Fig. 6-16 Radiative indirect transition.

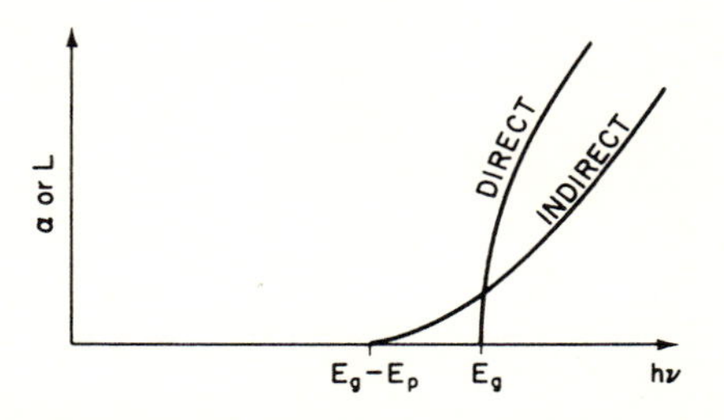

Fig. 6-17 Comparison of intensities of direct and indirect transitions.

which, by analogy with the derivation for the absorption coefficient [Eq. (3-11)] and assuming unit probability for phonon emission, can be written

$$L(\nu) = B'(h\nu - E_g + E_p)^2 \tag{6-19}$$

Comparing direct and indirect transitions, it appears that the indirect transition rises much faster (quadratically) with the excess energy above a threshold for emission compared to the square-root dependence [Eq. (6-17)] of the direct process (Fig. 6-17). However, the transition probability, the coefficient B', is much smaller for the indirect transitions.

6-D-2-c Self-absorption

Self-absorption affects the shape of the emission spectrum. If a spectrum $L_0(\nu)$ is emitted at some point at a distance d from the exit surface of reflectance R and the absorption coefficient is $\alpha(\nu)$, then the radiated spectrum is given by

$$L(\nu) = (1 - R)\, L_0(\nu)\, e^{-\alpha(\nu)d} \tag{6-20}$$

If, on the other hand, the radiative recombination occurs uniformly inside the specimen of thickness t, the spectrum radiated externally in one direction is

$$L(\nu) = (1 - R)\frac{L_0(\nu)}{t}\int_0^t e^{-\alpha x}\, dx$$

$$L(\nu) = (1 - R)\, L_0(\nu)\, \frac{1 - e^{-\alpha t}}{\alpha t} \tag{6-21}$$

Germanium is an indirect-gap semiconductor. However, the conduction band has a direct valley 0.15 eV above the bottom of the lowest valley. It is possible to excite electrons into both valleys, thus obtaining both direct and indirect radiative recombination (see the data in Fig. 6-18).[24] Although the electrons will rapidly relax to the lowest energy—i.e., to the indirect valley—and in spite of the high self-absorption for the higher photon energy of the

[24] J. R. Haynes, *Phys. Rev.* **98**, 1866 (1955).

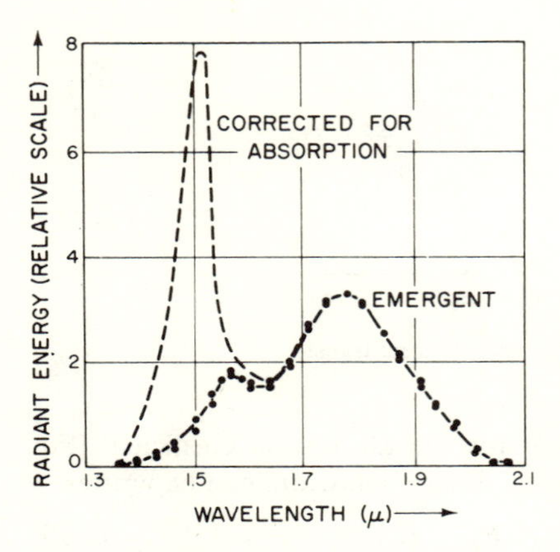

Fig. 6-18 Plot of experimental data obtained using a germanium sample 1.3×10^{-3} cm thick showing measured emergent radiation as a function of wavelength. The dotted curve was obtained from the solid line by correcting for absorption in the sample.[24]

direct transition, some of the higher-energy emission can be detected in very thin specimens. This observation of the direct transition under such adverse conditions is possible only because of the relatively high probability for the direct transition. Figure 6-18 shows also how the correction for self-absorption, using Eq. (6-21), reconstructs the expected emission peak for the direct transition at 1.5 μ.

6-D-2-d Case of a Heavily Doped Semiconductor

If the material is heavily doped, the Fermi level lies somewhere inside the band (conduction band for n-type material, valence band for p-type material). Let us neglect, for the time being, tails of states and impurity states and assume a perfect parabolic-band model, taking as an example a p-type semiconductor (Fig. 6-19).

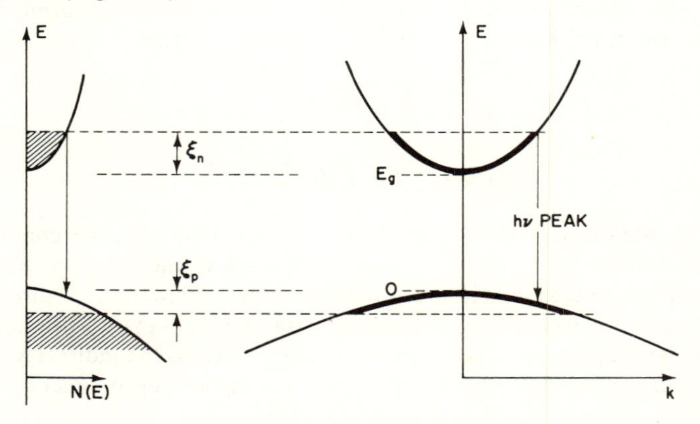

Fig. 6-19 Direct transition in the degenerate case.

Introducing a few electrons (n cm^{-3}) in the conduction band should give an emission spectrum with a lower-energy threshold at E_g. The spectrum should grow as $(h\nu - E_g)^{1/2}$. At 0°K, the peak position will occur at

$$h\nu_{\text{peak}} = E_g + \left(1 + \frac{m_e^*}{m_h^*}\right)\xi_n \tag{6-22}$$

where

$$\xi_n = (3\pi^2 n)^{2/3} \frac{\hbar^2}{2m_e^*} \tag{6-23}$$

The m_e^*/m_h^* factor of Eq. (6-22) expresses the restriction imposed by momentum conservation.

If the momentum-conservation restriction is relaxed by processes such as electron–electron scattering or electron–impurity scattering, all the occupied states of the conduction band can launch a radiative transition to any of the empty states of the valence band. Then the peak position at 0°K will occur at

$$h\nu_{\text{peak}} = E_g + \xi_n + \xi_p \tag{6-24}$$

where ξ_n is again as given by Eq. (6-23) and ξ_p is given by

$$\xi_p = [3\pi^2(p_0 + \Delta p)]^{2/3} \frac{\hbar^2}{2m_h^*} \tag{6-25}$$

Here, p_0 is the hole concentration due to doping (before excitation) and Δp is the increase in the hole concentration due to excitation. But now, because the emission spectrum becomes a convolution of all the upper states and all the lower states that can be linked by the same $h\nu$, the spectral distribution can be described by

$$L(\nu) = B(h\nu - E_g)^2 \tag{6-26}$$

Equations (6-24) and (6-26) are also valid in the case of an indirect-gap semiconductor when momentum conservation does not involve phonons.

6-D-2-e Emission With Carrier Interaction[25]

In Sec. 3-K we explored the possibility of a three-body interaction—an electron, a photon, and a "hot" electron—and we saw the possibility of completing the band-to-band transition with a photon of energy lower than the gap energy. Figure 6-20 illustrates the converse process: the first electron makes a transition to a virtual state a at a potential ΔE below the conduction band and excites an electron to a higher-energy state inside the band (this causes a change in momentum) and the first electron completes the transition from state a to the valence band by emitting a photon $h\nu$. Note that momentum-conservation rules make this process difficult to observe in pure direct-gap semiconductors, since an additional phonon may be needed (as shown in Fig. 6-21); then the transition becomes a three-step process having a very low

[25]S. M. Ryvkin, *Phys. Stat. Sol.* **11**, 285 (1965).

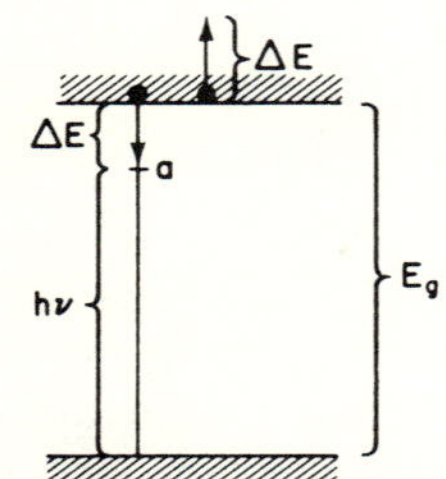

Fig. 6-20 Emission of a photon with an energy deficit $\Delta E = h\nu - E_g$ by heating an electron such that $kT_e = \Delta E$.[25]

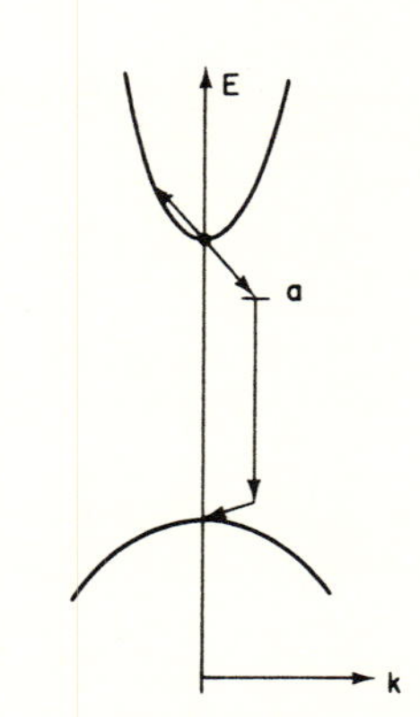

Fig. 6-21 Third-order recombination in a direct-gap semiconductor.

probability of occurrence. In impure materials other processes would mask such a transition.

In indirect-gap semiconductors, on the other hand, the usual radiative recombination is already a two-step process; therefore, the interaction with another carrier reduces the transition probability to a still-observable level. As illustrated in Fig. 6-22, the recombining electron "heats up" another electron in the conduction band and thus changes its energy by ΔE and its momentum by Δk. To complete the transition to the valence band, the recombining electron emits a photon $h\nu$ and a phonon E_p selected from the phonon-dispersion curves (the acoustic phonon is considered in this illustration). Hence the photon energy emitted is

$$h\nu = E_g - E_p - \Delta E$$

Experiments with germanium have shown a spectrum which seems to agree with this model. The data for the low photon-energy emission edge of germanium is shown in Fig. 6-23. Increasing the density of injected carriers causes a shift of this emission edge to lower energies—this is attributed to a shrinkage of the energy gap due to a coulomb interaction of the carriers. But what is significant for the present model of recombination with carrier heating is the presence of a low-energy tail which grows with the rate of excitation.

6-D-2-f Transitions Between Tails of States

In Sec. 1-C we discussed how doping and imperfections cause the density of states of the conduction and valence bands to tail off exponentially into the "forbidden" gap. Experimental evidence for these tails via absorption has

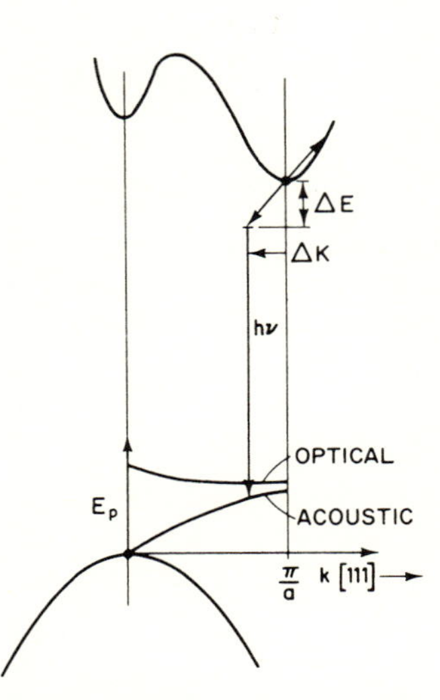

Fig. 6-22 Illustration of the indirect emission due to carriers and phonons in Ge. [$h\nu$ photon energy; ΔE energy delivered to an electron; E_p energy delivered to the lattice (phonon energy)].[25]

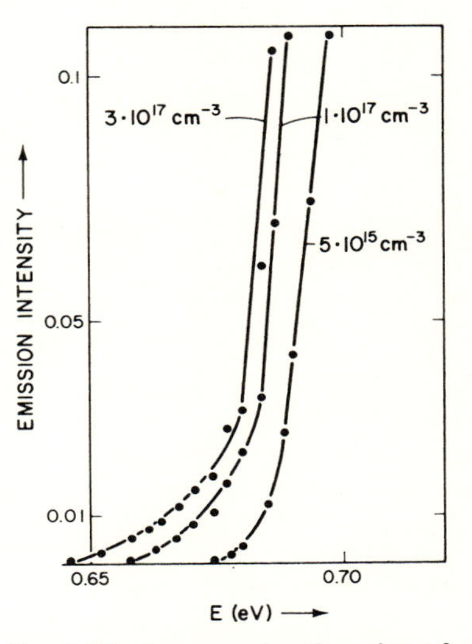

Fig. 6-23 Long-wavelength region of radiative recombination spectra of Ge at 80°K. (The figures on the curves indicate the injected carrier density.)[25]

been presented in Sec. 3-A-5. The radiative transition between the tails could, therefore, be observed. Since tails of state result from fluctuations of the band edges while the energy gap is usually constant, the lowest states of the conduction band are located at a different position than the highest states of the valence band. Therefore, a tail-to-tail transition requires that the electron tunnels from one tail to the other to change position; then it can radiate to change its energy. However, because of the complex nature of this "fundamental" process, it is best to defer its discussion as part of Sec. 6-F-2 after we have considered donor–acceptor transitions.

6-E Transition Between a Band and an Impurity Level

6-E-1 SHALLOW TRANSITIONS

The shallow transitions to neutralize ionized donors or acceptors are shown in Fig. 6-24. It is conceivable that these transitions could be radiative in the far infrared. Clever measurements[26-28] have been made to detect these

[26]S. H. Koenig, *Phys. Rev.* **110**, 988 (1958).
[27]G. Ascarelli and S. C. Brown, *Phys. Rev.* **120**, 1615 (1960).
[28]S. H. Koenig and R. D. Brown III, *Phys. Rev. Letters* **4**, 120 (1960).

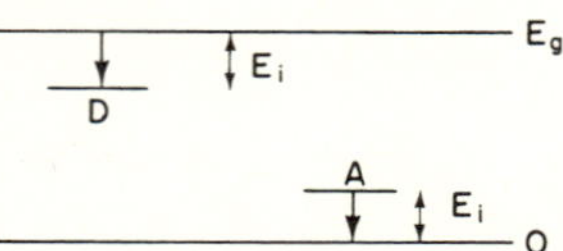

Fig. 6-24 Shallow transitions.

transitions, but the validity of the results has been always open to question.† The arguments have been settled by a calculation[29] of the probabilities for transition by photon emission and by phonon emission. The capture cross-section for photon emission is

$$\sigma_t = 1.71 \times 10^{-18}\epsilon^{1/2}\left(\frac{m}{m^*}\right)^2 E_i/T$$

where E_i is in electron volts and σ_t in cm^2. For n-type germanium at 4°K, $\sigma_t = 4 \times 10^{-19}$ cm^2. For phonon emission, the capture cross-section is

$$\sigma_n = 256\pi(E_1)^2\hbar^5(c_s)^3/\rho(a^*)^6(E_i)^5kT$$

where E_1 is the deformation potential, c_s the longitudinal sound velocity, ρ the density of the semiconductor, and a^* the effective radius of the first Bohr orbit. For n-type germanium at 4°K, $\sigma_n = 6 \times 10^{-15}$ cm^2. Therefore, the phonon-emission process is much more probable. Hence, taking a donor as an example, the recombination process consists of, first, trapping an electron in an excited state of the donor; then the electron cascades to lower-lying states, emitting a phonon at each step. When more detailed considerations are taken into account (such as the temperature dependence of electron distribution in the conduction bands and the probability for reionization by phonons), a capture cross-section of $\sigma_n \approx 10^{-12}$ cm^2 is obtained for n-type Ge at 4°K and the temperature dependence becomes approximately $T^{-2.5}$. This temperature dependence and $\sigma_n \approx 10^{-12}$ are in good agreement with the experimental data.

6-E-2 DEEP TRANSITIONS

By deep transition we shall mean either the transition of an electron from the conduction band to an acceptor state or a transition from a donor to the valence band (Fig. 6-25). Such transitions emit a photon $h\nu = E_g - E_i$ for direct transitions and $h\nu = E_g - E_i - E_p$ if the transition is indirect and involves a phonon of energy E_p.

A quantum-mechanical calculation[30] compared the transition probabilities of the deep transitions with those of the band-to-band transitions. A

† Quite recently, the radiative transition from the conduction band to the donor has been seen in GaAs by I. Melngailis, G. E. Stillman, J. O. Dimmock and C. M. Wolfe, reported in *Phys. Rev. Letters* **23**, 1111 (1969).

[29] G. Ascarelli and S. Rodriguez, *Phys. Rev.* **124**, 1321 (1961).

[30] W. P. Dumke, *Phys. Rev.* **132**, 1998 (1963).

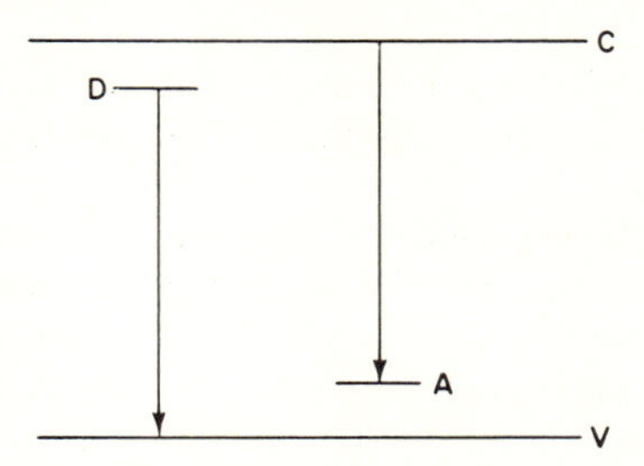

Fig. 6-25 Radiative transition between a band and an impurity state.

direct gap between parabolic bands was assumed and the impurities were taken to be shallow, discrete, and nonoverlapping. Under these assumptions, the results of the calculation of the maximum transition probabilities are as follows: donor–to–valence-band transition probability at $k = 0$:

$$\frac{1}{\tau_{p_0}} = 64\sqrt{2}\,\pi n \frac{q^2\hbar^2\omega\,|\overline{P_{vc}}|^2}{c^3m^2(m_cE_D)^{3/2}}n_D \tag{6-27}$$

And, for conduction-band–to–acceptor transition probability at $k = 0$;

$$\frac{1}{\tau_{n_0}} = 64\sqrt{2}\,\pi n \frac{q^2\hbar^2\omega\,|\overline{P_{vc}}|^2}{c^3m^2(m_AE_A)^{3/2}}p_A \tag{6-28}$$

where $|\overline{P_{vc}}|$ is the averaged interband-matrix element of the momentum operator; E_D and E_A are the impurity ionization energies; n_D and p_A are the concentrations of electrons in donors, and of holes in acceptors, respectively, per unit volume; m_c is the mass of the electron in the conduction band; and m_A, an equivalent effective mass for acceptors, is 5 to 10 times greater than m_c. Higher-energy carriers (at $k \neq 0$) have a lower transition probability.

For GaAs, Eqs. (6-27) and (6-28) become

$$\frac{1}{\tau_{p_0}} = 0.8 \times 10^{-8} n_D \text{ cm}^3/\text{sec}$$

$$\frac{1}{\tau_{n_0}} = 0.43 \times 10^{-9} p_A \text{ cm}^3/\text{sec}$$

Therefore for a doping on the order of 10^{18} cm^{-3}, the carrier lifetime would be of the order of several nanoseconds for this type of transition.

The band-to-band transition probability, on the other hand, is given by[30]

$$\frac{1}{\tau_n} = \frac{16nq^2\omega}{c^3m^2\hbar}|\overline{P_{vc}}|^2 \tag{6-29}$$

which gives, for GaAs, $\tau_n = 0.31$ ns. The minimum lifetime is roughly inversely proportional to the energy gap. Hence it is possible to find an approximate lifetime for direct band-to-band recombination in other semiconductors.

From these calculations one would conclude that, if there are electrons in the conduction band and in the donor states, and holes in the acceptors levels and in the valence band, the band-to-band transition would be about four times more probable than the transition between the impurity and the band.

The experimental observation of radiative transitions between impurity levels and the more distant band edge is limited to semiconductors with a relatively low concentration of impurities. When the impurity concentration is large enough to form an impurity band which merges with the nearest intrinsic band, the interpretation of the process becomes ambiguous: is the initial state a high-energy donor level or a low-energy conduction-band state? We shall see later that this dilemma can be resolved. For the purer materials, exciton recombination may be more probable than the deep transition. The exciton recombination can be distinguished by its very narrow emission spectrum.

In most direct-gap semiconductors the electron effective mass is considerably smaller than the hole effective mass. Therefore, the donor ionization energy E_D is lower than the acceptor ionization energy E_A. Hence the deep transitions can be distinguished as (1) a conduction-band–to–acceptor transition which produces an emission peak at $h\nu = E_g - E_A$, and (2) a donor–to–valence-band transition which produces an emission peak at the higher photon energy $h\nu = E_g - E_D$. In other semiconductors where $E_A \approx E_D$, the identification of the transition requires a knowledge of the conductivity type of the semiconductor, and the emission intensity must be correlated with the impurity concentration.

Figure 6-26 shows the energy of the emission peak (or peaks) in GaAs as a function of the impurity concentration.[31] Two sets of transitions are found at low impurity concentrations ($<10^{18}$ cm^{-3}). One set occurs at $h\nu = E_g - E_D$, where $E_D = 0.006$ eV agrees with estimates based on the hydrogenic model; this transition is the donor–to–valence-band recombination. The other set occurs at $h\nu = E_g - E_A$, where $E_A = 0.03$ eV; this is the conduction-band–to–acceptor transition. The deviations which occur at higher doping ($>8 \times 10^{17}$ cm^{-3}) are due to impurity banding and tailing of states, which we shall discuss later. The careful reader may have noticed that although the two most lightly doped samples are *n*-type, they exhibit a transition terminating on an acceptor level. This can be tentatively explained as due to the presence of uncontrolled residual acceptors in these materials (most acceptors: Zn,[32] Cd,[33] Be,[34] Si[35] in small concentrations are located about 30 meV above the valence band).

Similar data for emission peak energy vs. doping has also been obtained

[31] D. A. Cusano, *Solid State Communications* **2**, 353 (1964).
[32] M. I. Nathan and G. Burns, *Appl. Phys. Letters* **1**, 89 (1962).
[33] D. E. Hill, *Bull. Am. Phys. Soc.* **8**, 202 (1963).
[34] A. E. Yunovich et al., *Solid State Phys.* **6**, 1900 (1964).
[35] E. W. Williams and D. M. Blacknall, *Trans. AIME* **239**, 387 (1967).

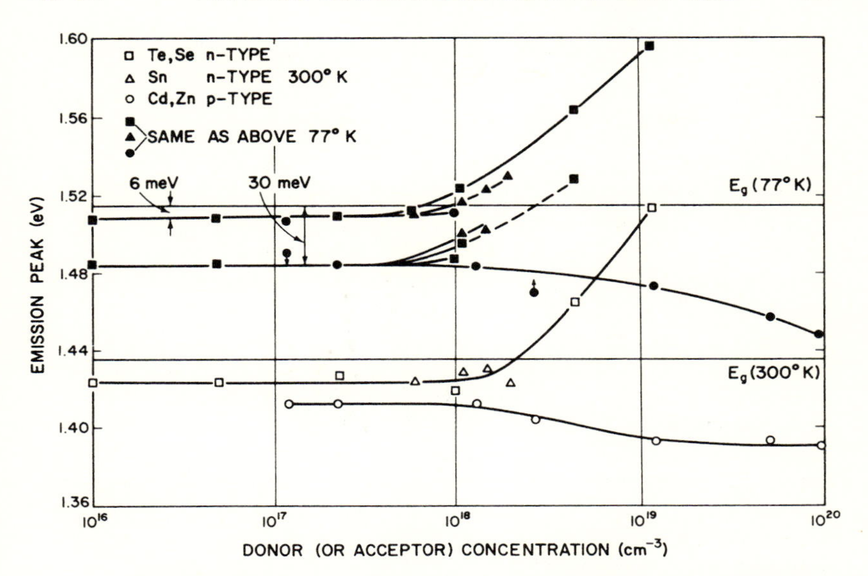

Fig. 6-26 Peak positions in cathodoluminescence at 77 and 300°K as a function of doping in n-type and in p-type GaAs crystals.[31]

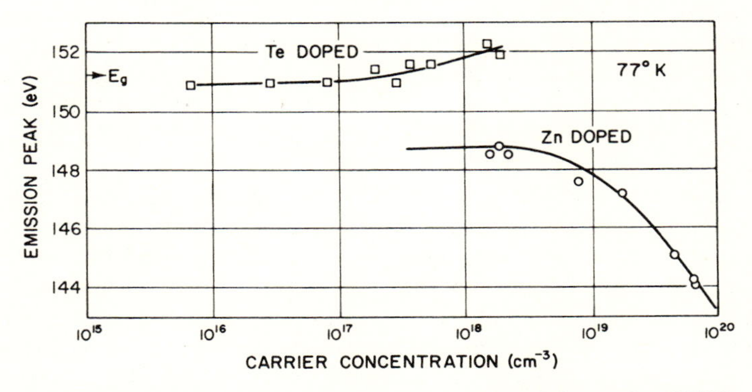

Fig. 6-27 Position of the photoluminescence edge emission peak at 77°K in n-type Te-doped and in p-type Zn-doped crystals as a function of doping.[32]

by photoluminescence (Fig. 6-27).[32] This data also shows that when the donor concentration is less than about 5×10^{17} cm^{-3}, the transition is from donor states to valence band; when the acceptor concentration is less than 2×10^{18}, the transition is from the conduction band to the acceptor states.

It is interesting to note that, as the impurity concentration increases, the evolution of the impurity levels into a band of states can be readily followed

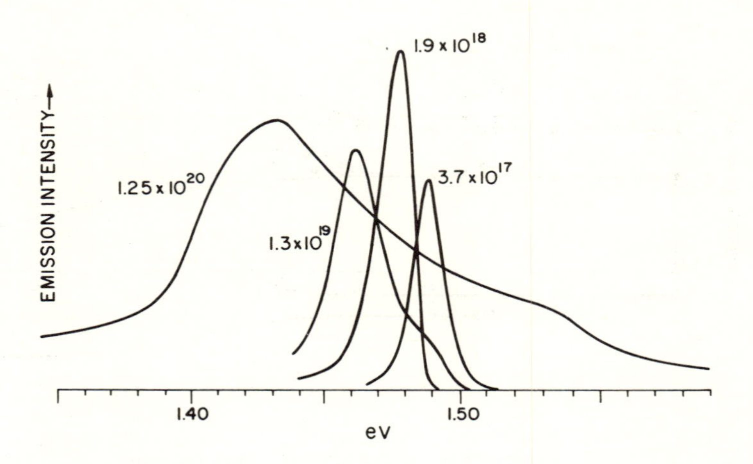

Fig. 6-28 Cathodoluminescence spectra of Zn-doped GaAs at 4.2°K.[36]

via the emission spectrum. Figure 6-28 shows the broadening of the emission spectrum in *p*-type GaAs as the acceptor concentration is increased. The shift of the spectral peak is due to the apparent shrinkage of the energy gap induced by the heavy doping. The broadening of the emission spectrum with increased doping is reproduced in Fig. 6-29 for *p*-type GaAs and in Fig. 6-30 for *n*-type GaAs.

In indirect-gap semiconductors the donor is associated with the lowest valley of the conduction band. Therefore, at low donor concentrations a phonon must assist the transition to the top of the valence band. A phonon is also needed for a transition from the indirect valley of the conduction band to an acceptor. Figure 6-31[38] shows the emission spectrum for impurity transitions in Si; the phonons used for conserving momentum are identified. For Si, the values of E_p are 0.016 eV and 0.055 eV for the TA- and TO-phonons, respectively. The impurity ionization energies are, in order of increasing values (in eV): $E_i = 0.046$(B); 0.069(Bi), 0.071(Ga), and 0.16(In). The longer vertical marker indicates $h\nu = E_g - E_i - E_p$ (TO); the shorter marker indicates $h\nu = E_g - E_i - E_p$ (TA). In the In-doped sample, and to a lesser degree in the Bi-doped sample, transitions with lower values of E_p and phononless transitions occur also, contributing emission at the higher photon energies.

[36]J. I. Pankove, *J. Phys. Soc.* (Japan) **21**, supplement, 298 (1966).

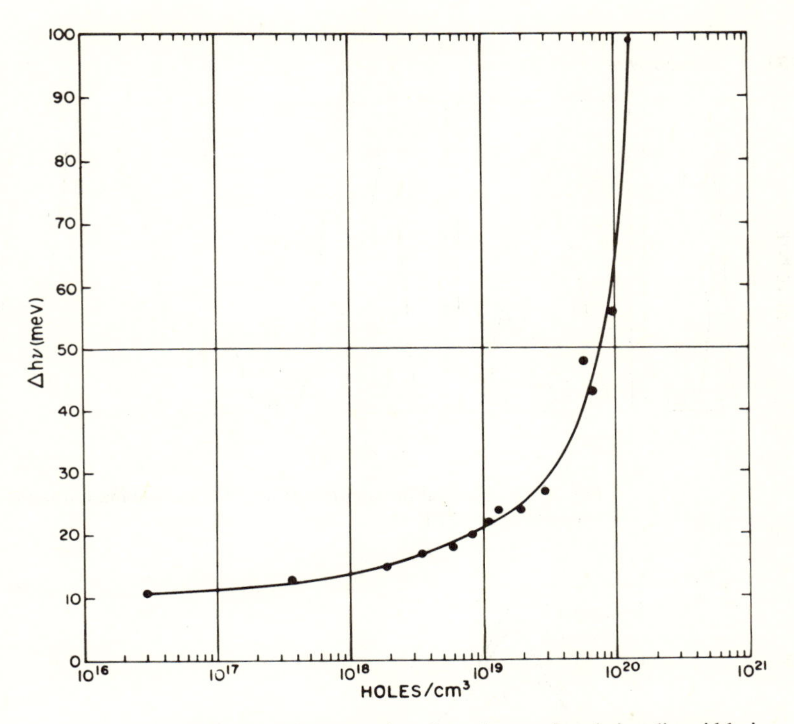

Fig. 6-29 Concentration dependence of emission linewidth in *p*-type GaAs at 4.2°K.[36]

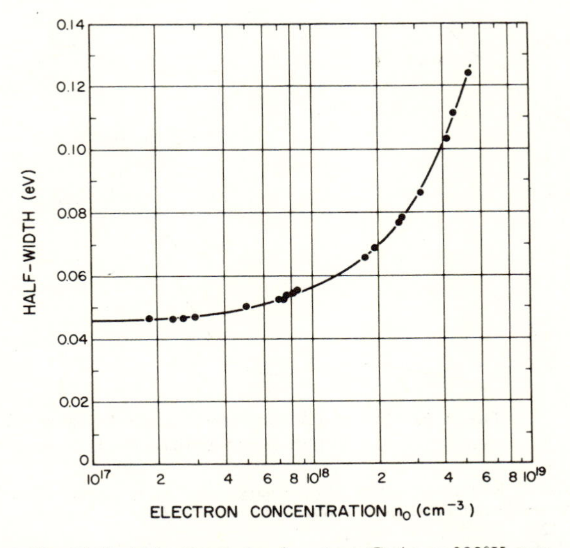

Fig. 6-30 Half-width of emission in *n*-type GaAs at 300°K as a function of the free-electron concentration.[37]

[37] H. C. Casey and R. H. Kaiser, *J. Electrochem Soc.* **114**, 149 (1967).

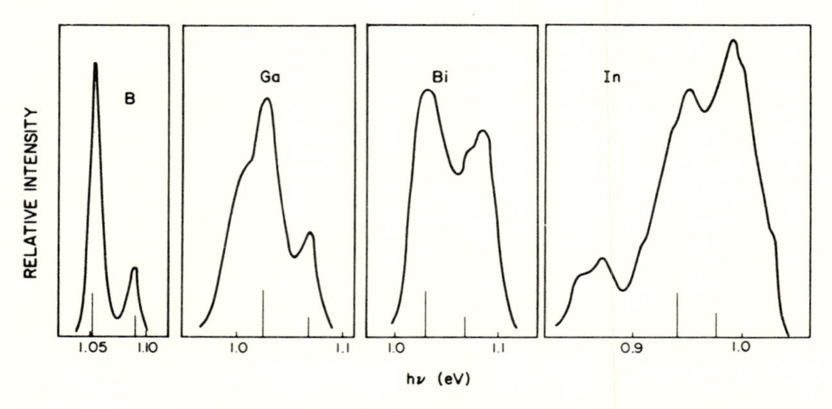

Fig. 6-31 Spectral distribution of impurity recombination radiation in Si at ~20°K.[38]

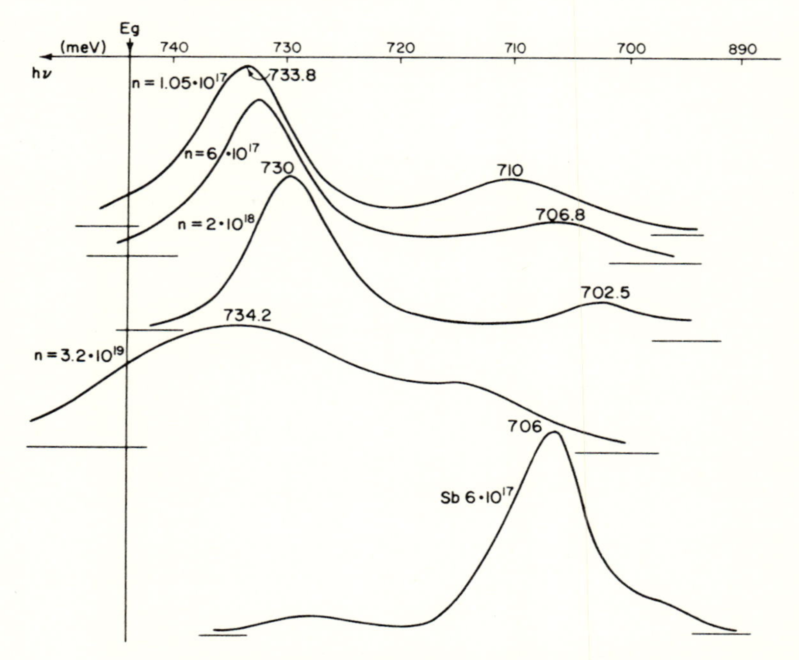

Fig. 6-32 Spectral distribution of donor-to-valence-band recombination in As- and Sb-doped germanium at 20°K.[39]

[38] Y. E. Pokrovsky, "Radiative Capture of the Charge Carriers by Impurity Atoms in Si and Ge," *Radiative Recombination in Semiconductors, 7th Int. Conf. Phys. of Semicond., Paris, 1964*, Dunod (1965), p. 129.

[39] C. Benoit-à-la-Guillame and J. Cernogora, "Recombinaison radiative dans le germanium fortement dopé et dans le germanium compensé," *Radiative Recombination in Semiconductors, 7th Int. Conf. Phys. of Semicond., Paris, 1964*, Dunod (1965), p. 121.

The phononless transition is especially evident in Ge moderately doped with As (Fig. 6-32).[39] The LA-phonon-assisted transition produces an emission peak below 0.71 eV. Note that in Sb-doped Ge the phononless transition is very weak, further illustrating that various donor atoms have different properties. As the impurity concentration increases, the emission spectrum shifts to lower energies, reflecting the effective gap shrinkage due to greater perturbation of the band edges. At very large doping, the impurity band broadens and merges with the tail of the conduction band; this results in a very broad emission spectrum.

6-E-3 TRANSITIONS TO DEEP LEVELS

Some impurities have large ionization energies; therefore, they form deep levels in the energy gap. Radiative transitions between these states and the band edge emit at $h\nu = E_g - E_i$. Thus transition elements form deep acceptors in GaAs: $E_i = 0.36$ eV (Fe); 0.35 eV (Ni); 0.345 (Co).[40] The corresponding emission spectra are shown in Fig. 6-33. Copper is known to produce

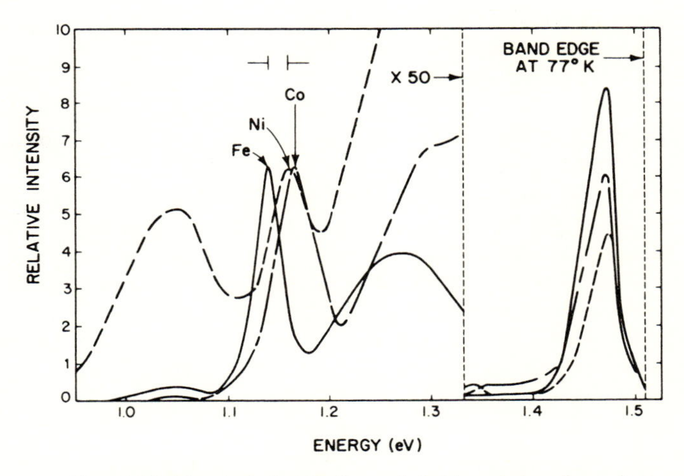

Fig. 6-33 Spectral distribution of light emission from iron-, cobalt-, and nickel-doped GaAs.[40]

deep acceptor levels at 0.18 and 0.41 eV (in addition to a shallow donor level at 0.07 eV).[41] An example of the emission spectrum attributed to copper in GaAs is reproduced in Fig. 6-34. Oxygen, which acts as a neutral impurity (neither donor nor acceptor) in GaAs is responsible for a 0.65-eV emission peak, the intensity of which increases with the concentration of oxygen in the

[40] H. Strack, *Trans. Metallurgical Soc. AIME* **239**, 381 (1967).
[41] T. N. Morgan, M. Pilkuhn, and H. Rupprecht, *Phys. Rev.* **138**, A1551 (1965).

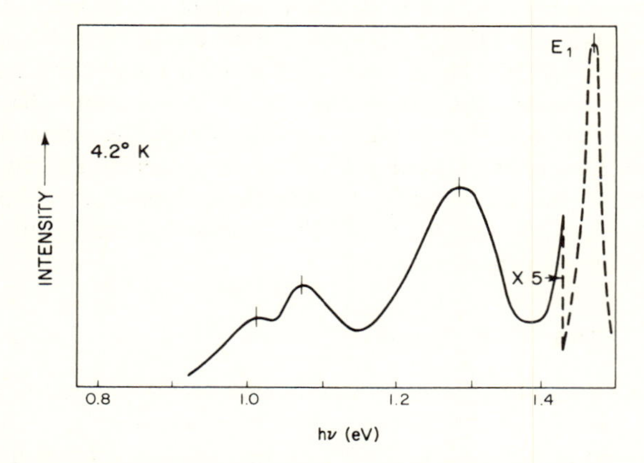

Fig. 6-34 Emission due to copper in GaAs (solid line); E_1 is due to "near gap" transition.[41]

crystal.[42] Although the nature of this transition is uncertain, its association with defects is likely. Emission due to defects has been identified. Thus a donor–Ga-vacancy complex in GaAs forms a deep level, or band of levels, which gives a broad emission at about 1.20 eV (Fig. 6-35).[43] When the material is grown in a Ga-rich ambient, minimizing the formation of Ga-vacancies, this emission is absent. It is also absent in *p*-type GaAs, where there is no donor to complex with.

A correlation has been found between the depth of the impurity level and the Franck–Condon shift.[44] The orbit of the electron or of the hole becomes more localized as the ionization energy of the impurity increases. In the configuration diagram, the increased localization of the charge associated with the center leads to a stronger interaction with neighboring ions. Hence the Franck–Condon shift should increase with ionization energy. Figure 6-36 demonstrates this correlation for five impurities in GaAs. In this figure, the Franck–Condon shift, plotted along the abscissa, is taken as the difference between the ionization energy determined optically and the thermal-activation energy from electrical-conductivity measurements; the corresponding activation energies labeled "optical" and "thermal," respectively, are plotted along the ordinate. The square-law correlation between the ionization or activation energy ΔE and the Franck–Condon shift d is remarkable, although no theoretical reason for this functional relation has been suggested.

From our discussion of the configuration diagram (Sec. 6-C), a larger Franck–Condon shift implies a stronger phonon interaction. Indeed, the

[42] W. J. Turner, G. D. Pettit, and N. G. Ainslie, *J. Appl. Phys.* **34**, 3274 (1963).
[43] E. W. Williams and D. M. Blacknall, *Trans. TMS-AIME* **239**, 387 (1967).
[44] E. W. Williams, *Brit. J. Appl. Phys.* **18**, 253 (1967).

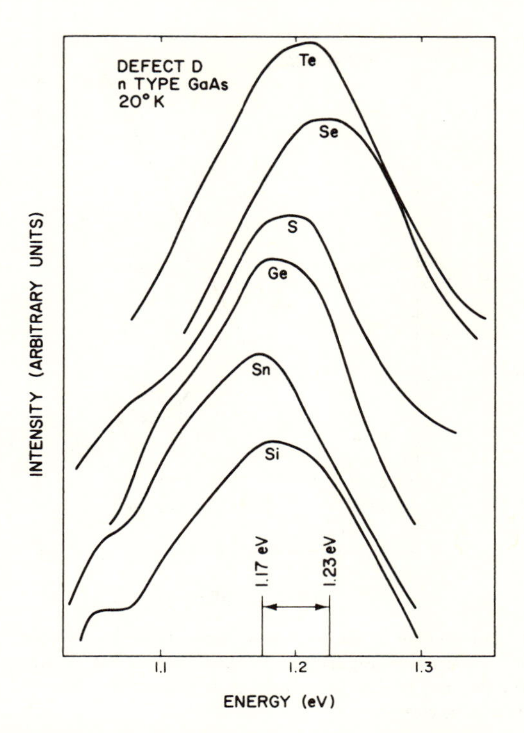

Fig. 6-35 The photoluminescence emission peaks seen in GaAs doped to concentration greater than 1×10^{18} cm^{-3} with six donor impurities.[43]

emission spectrum shows a series of equally spaced lower-energy peaks corresponding to recombination with phonon emission. As shown in Fig. 6-37, the energy spacing between peaks is equal to that of the LO-phonon, and the phonon peaks are stronger for impurities having larger activation energies.

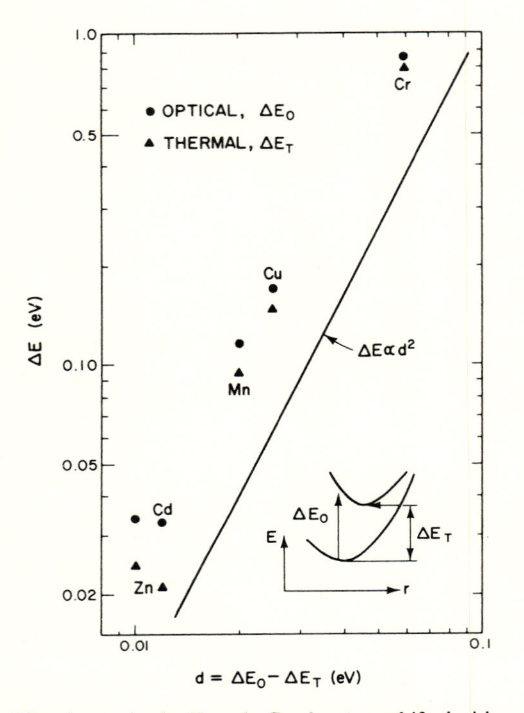

Fig. 6-36 The change in the Franck–Condon-type shift d with the activation energy of impurities in GaAs (ΔE_o is the optical and ΔE_T the thermal activation energy). A configuration diagram is shown in the inset.[44]

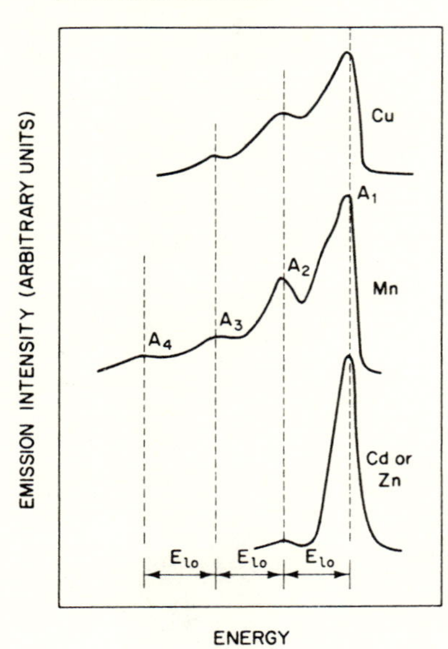

Fig. 6-37 Phonon emission observed in the photoluminescence spectra at 20°K of GaAs doped with cadmium or zinc, manganese and copper. The curves are displaced vertically for clarity and displaced horizontally to compare the longitudinal optical phonon energies E_{lo}.[44]

6-F Donor–Acceptor Transitions†

6-F-1 SPECTRAL STRUCTURE

Donor-to-acceptor transitions are illustrated in Fig. 6-38(a). We have seen in Sec. 1-E that when both donor and acceptor impurities are present in a semiconductor, coulomb interaction between donor and acceptor modifies the binding energies (compared to the isolated-impurity case) such that the energy separating the paired donor and acceptor states [Fig. 6-38(b)] is

$$h\nu = E_g - E_A - E_D + \frac{q^2}{\epsilon r} \tag{6-30}$$

For distant pairs the coulomb interaction term is very small and the lowest possible photon energy is obtained—except, of course, when phonon emission can occur.

For impurities separated by distances greater than the effective Bohr radius, the transition is assisted by a tunneling process. A transition between distant pairs is less probable than a transition between nearer pairs; therefore, the emission intensity should increase as the pair separation decreases. However, the number of possible pairings decreases as r decreases. Consequently, the emission intensity must go through a maximum as the separation r is varied. Since r varies in a discrete fashion, the emission spectrum should exhibit a fine structure. At large values of r ($r > 40$ Å) the emission lines overlap, forming a broad spectrum. The discrete line structure can be resolved only for pair separations in the range of 10 to 40 Å. Such spectra have been observed in GaP (Fig. 6-39).[45] The peak of the broad emission occurs at

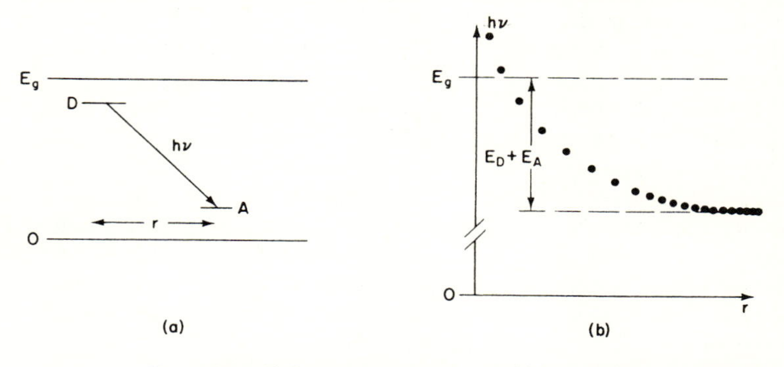

Fig. 6-38 (a) Donor-to-acceptor transition; (b) effect of coulomb interaction on emission energy (r is the donor–acceptor separation).

†Review paper: F. Williams, *Phys. Status Solidi* **25**, 493 (1968).
[45]D. G. Thomas, M. Gershenzon, and F. A. Trumbore, *Phys. Rev.* **133**, A269 (1964).

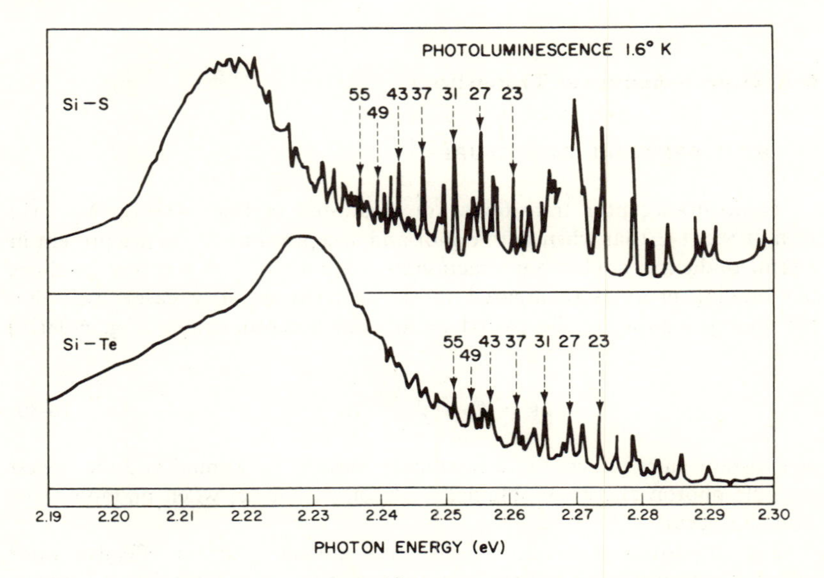

Fig. 6-39 Photoluminescence spectra of GaP crystals at 1.6°K, showing the isolated pair lines and the broad pair edge emission band corresponding to Si–S and Si–Te acceptor–donor combinations. Some of the shell numbers are given.[45]

donor–acceptor separations of about 50 Å, and the broad band cuts off at pair separations of about 200 Å. Actually, the low-energy edge of the emission spectrum may extend over a large range, and consist of a series of phonon-emission replicas of the main band. Figure 6-40 illustrates a partial overlap of up to seven LO-phonon-emission replicas of the main band.

Note that in Fig. 6-39 the spectral positions of the broad peak and of the line structure depend on the impurity. In both curves of Fig. 6-39 the acceptor is Si on a P-site and the donor is S for the upper curve and Te for the lower curve.

Let us recall that different pairings will occur depending on the sites occupied by the impurities. If the donor and the acceptor occupy similar substitutional sites—e.g., both on P-sites in GaP—a type-I pairing is said to occur. If, on the other hand, the donor and the acceptor occupy opposite sites (one on a Ga-site, the other on a P-site), a type-II pairing is obtained. The number of possible pairings N_r and the coulomb-interaction term can be calculated as a function of r for each of the two types (Fig. 6-41). Then the measured fine structure can be compared to the theoretical prediction, as shown in Fig. 6-42. Note that the fit between theory and experiment is excellent for the more distant pairs. The poorer fit for the nearer pairs is due to a neglect of higher-order interactions. The line separation due to coulomb interaction should be independent of the impurity for the same type of

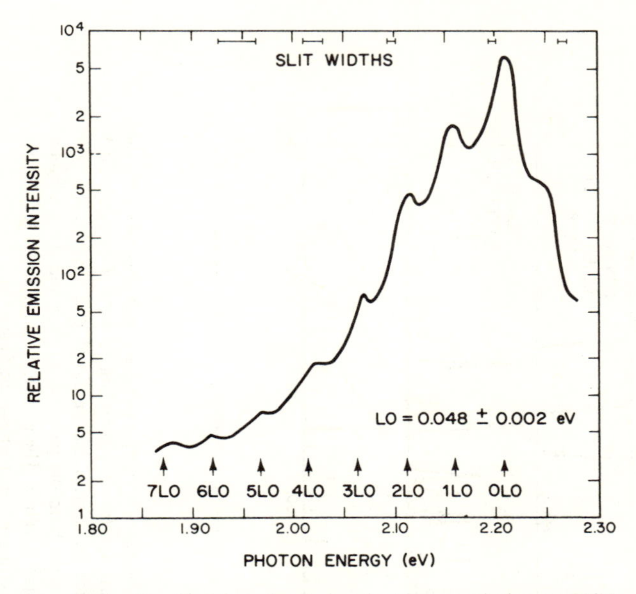

Fig. 6-40 The Si–S broad pair band at low resolution at 20°K, showing the simultaneous emission of multiple LO phonons. The sharp pair lines are unresolved, but lie in the high-energy shoulder above the 0 LO peak.[45]

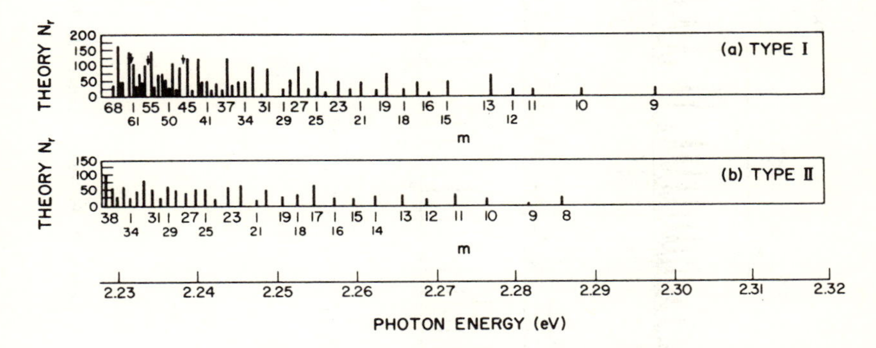

Fig. 6-41 The predicted distribution of pairs as a function of pair separation for donor and acceptor on the same lattice sites (type I) and on opposite lattice sites (type II). The shell numbers (m), measuring these separations, are given. The distribution is converted to one in energy by Eq. (6-30) and is related to the real photon energy of the pair lines by the arbitrary additive term $E_g - (E_D + E_A)$. This term is deduced by sliding the energy scale to fit the measured Si–S pair data (type I, upper spectrum) and the measured Zn–S pair data (type II, lower spectrum) and thereby identifying these lines and determining these parameters for the two sets of pairs.[45]

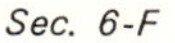

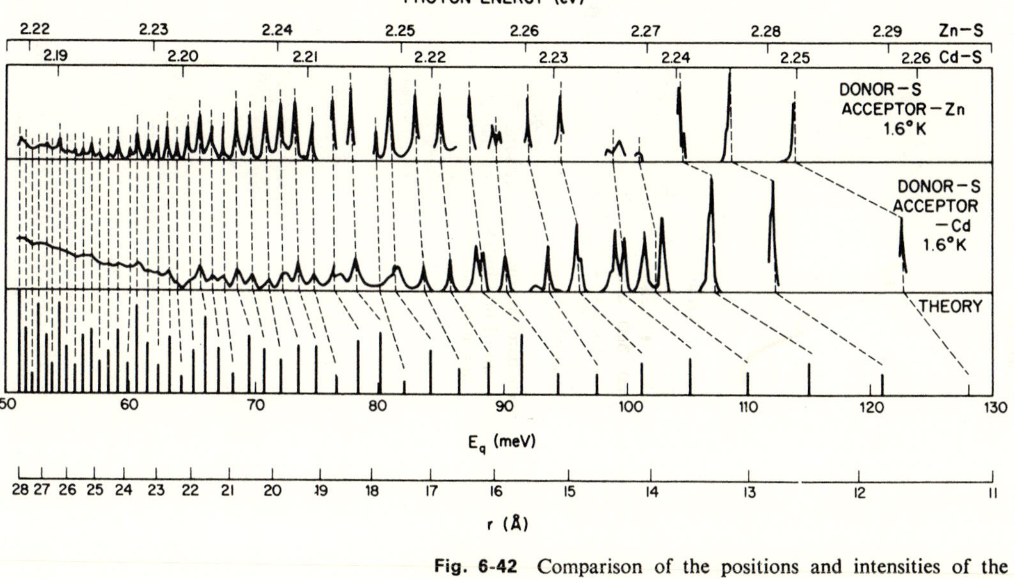

Fig. 6-42 Comparison of the positions and intensities of the sharp line pair spectra corresponding to both Zn–S and Cd–S acceptor–donor pairs with the predicted pair distribution. The lower scales show both the pair separation *r* and the coulombic energy E_q derived from *r*. The emission energy scales for the two measured spectra are shown above the figure.[46]

[46]M. Gershenzon, R. A. Logan, D. F. Nelson, and F. A. Trumbore, *Int. Conf. on Luminescence*, Budapest, 1966. Hungarian Academy of Sciences, ed., G. Szigetti (1968), p. 1737.

substitution. Only the position of the whole structure needs to be shifted to agree with the value $E_A + E_D$. Hence if either E_A or E_D is known for one impurity, the binding energy of the companion impurity can be deduced, and subsequently, by substituting other impurities, their binding energies can be determined in turn.

The following donor–acceptor pairs have been examined in GaP:[47]

Type I: Si–S, Si–Te, Si–Se—donor and acceptor on P-site (Si being the acceptor).

Type II: Zn–S, Zn–Te, Zn–Se, Cd–S, Cd–Te, Cd–Se—donor on P-site, acceptor on Ga-site.

Donor–acceptor-pair structures have also been seen in ZnO,[48] BP,[49] and CdS.[50]

In GaAs, Ge, Si, and other semiconductors where $E_A + E_D$ is small, only the distant pairs contribute to the emission spectrum. The coulomb-interaction term for near pairs drives the donor and acceptor levels out of the energy gap. A recombination between donor and acceptor states located inside the intrinsic bands would compete with band-to-band transitions; since the bands have a high density of states, the donor–acceptor transition would need to be extremely efficient to contribute a characteristic line emission.

6-F-2 TRANSITION PROBABILITY

A calculation has been made for the transition probability between donors and acceptors neglecting the coulomb interaction and the tunneling factor.[51] The donors and acceptors were assumed to form Gaussian bands of states with a characteristic distribution width E_0. The transition probability turns out to be proportional to $1/E_0$, which, for small E_0, can make the donor-to-acceptor coupling stronger than either the band-to-band or band-to-impurity transitions. In what follows we shall demonstrate by experimental evidence that donor-to-acceptor transitions are the dominant radiative-recombination mechanism.

The dominantly high probability for donor-to-acceptor transitions is apparent in cathodoluminescent studies of GaAs.[36,52] The electron-beam-excitation technique allows the generation of a high concentration of electron–hole pairs (about 1.5×10^{18} cm^{-3}). Let us consider the case of moderately

[47] M. Gershenzon, "Radiative Recombination in the III–V Compounds," *Semiconductors and Semimetals*, ed. R. K. Willardson and A. C. Beer, Academic Press (1966), p. 289.

[48] D. C. Reynolds, C. W. Litton, Y. S. Park, and T. C. Collins, *J. Phys. Soc. Japan* **21**, Supplement, 143 (1966).

[49] F. M. Ryan and R. C. Miller, *Phys. Rev.* **148**, 858 (1966).

[50] C. H. Henry, R. A. Faulkner, and K. Nassau, *Phys. Rev.* **183**, 798 (1969).

[51] J. Callaway, *J. Phys. Chem Solids* **24**, 1063 (1964).

[36] J. I. Pankove, *J. Phys. Soc.* (Japan) **21**, supplement, 298 (1966).

[52] J. I. Pankove, *J. Appl. Phys.* **39**, 5368 (1968).

doped p-type GaAs. If the radiative transitions occurred from the conduction band to the acceptor states, the emission peak would shift to higher energies as the excitation is increased, driving the quasi-Fermi level for electrons deeper into the conduction band (Fig. 6-43). As the quasi-Fermi level for holes penetrates deeper into the valence band, a shifting band-to-band emission peak should also be observed. A shift of more than 60 meV is expected from an excitation generating 1.5×10^{18} electrons/cm^3, assuming a parabolic conduction band with a distribution of states increasing with energy E above the conduction band edge at E_c as $(E - E_c)^{1/2}$. In more heavily doped GaAs the exponential tail of states would give an even larger shift of the quasi-Fermi level. However, no shift is observed! This behavior leads to the

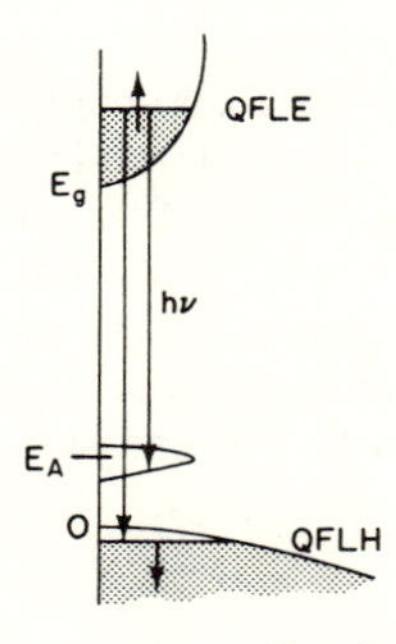

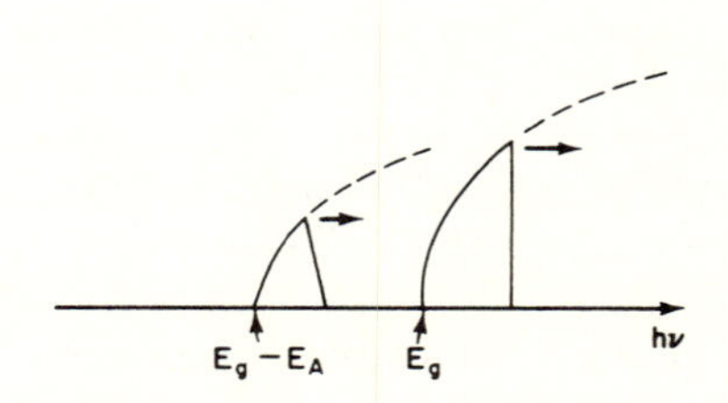

Fig. 6-43 Conduction-band–to–valence-band transition and conduction-band–to–acceptor transition with corresponding expected emission spectrum in p-type GaAs. Short arrows indicate motion of quasi-Fermi levels (QFL) and of emission peaks as excitation is increased.

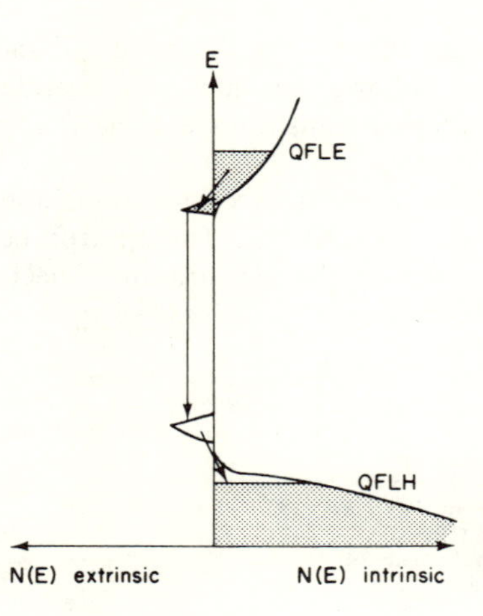

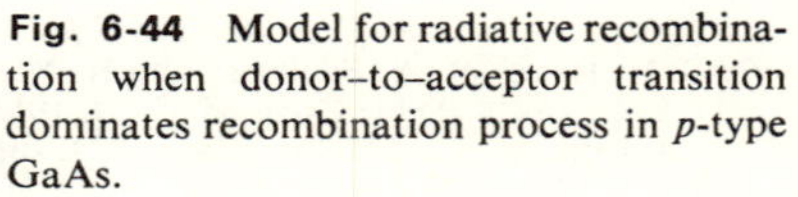
Fig. 6-44 Model for radiative recombination when donor–to–acceptor transition dominates recombination process in p-type GaAs.

model of Fig. 6-44, which includes donor states due to uncontrolled residual impurities, probably Si on Ga-sites. The donor states are close to the conduction-band edge, and in a density-of-states diagram overlap the tail of the conduction band. At excitation rates sufficient to observe an emission, the quasi-Fermi level for electrons is above the donor states. If the transition probability between donors and acceptors is greater than the conduction-band–to–acceptor transition probability, the dominant radiative recombination will be from donors to acceptors. The electrons will leak out of the conduction band through the donors and the quasi-Fermi level for electrons will not affect the emission spectrum.

A similar conclusion was reached for *n*-type GaAs,[52] where the dominant terminal state for radiative recombination is a residual acceptor, probably Si on As-sites. As shown in Fig. 6-45, in moderately doped *n*-type GaAs the donor band overlaps the tail of the conduction band. The acceptor levels are about 30 meV above the valence band. As the quasi-Fermi level for holes moves deeper into the valence band under increased excitation, a donor–to–valence-band transition would produce an emission spectrum with a threshold at about $E_g - E_D$ and the emission peak would shift to higher energies. Although this shift would be smaller than was expected in *p*-type material because of the higher effective mass of the valence band, a shift of about 7 meV was expected for a generation of 1.5×10^{18} holes/cm^3. But, here also, no shift was observed and the emission peak in moderately doped *n*-type GaAs occurred at about $E_g - 30$ meV $= E_g - E_A$. This behavior is again interpreted as a donor-to-acceptor radiative transition, which is independent of the position of the quasi-Fermi levels.

In contrast to the above described behavior of moderately and heavily doped samples, in closely compensated GaAs the emission spectrum shifts

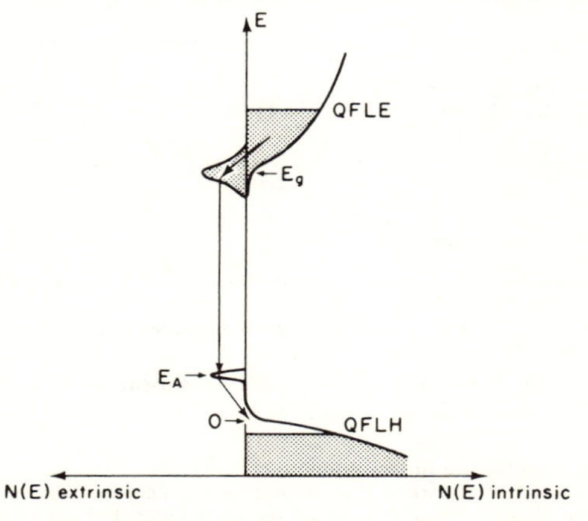

Fig. 6-45 Model for radiative recombination between donors and residual acceptors in *n*-type GaAs.

to higher energies as the excitation rate increases. Such shifts are observed under electron-beam excitation[53] as well as under photoluminescence.[54] As shown in Fig. 6-46, the energy of the emission peak of closely compensated n-type GaAs increases logarithmically with the emission intensity (and nearly logarithmically with the excitation rate). In the observed behavior of the shifting peak $L(\nu) = L_0 \exp (h\nu/E_0)$, the coefficient E_0 depends on the compensation and becomes very small in noncompensated samples, in which case there is no shift of the emission peak.

If the peak shift were due to a simple filling of band tails, one would expect the low-energy edge of the emission spectrum to represent a convolution of all the tail states linked by the same photon energy and, therefore, the low-energy edge to be described by an exponential—this is in accord with the experiment. However, the high-energy edge of the emission spectrum at low temperature should be an abrupt cutoff at a photon energy corresponding to the separation between the two quasi-Fermi levels.

Radiative transitions to an acceptor band would give the same behavior while the quasi-Fermi level for holes sweeps across the acceptor band. When the acceptor band is completely filled with holes, the emission spectrum would reproduce approximately the shape of the acceptor band. This shape would remain constant while the peak of the emission shifts to higher energies with increasing excitation, as the quasi-Fermi level for electrons moves deeper into the conduction band.

The experiment, however, shows that with increasing excitation the emission spectrum changes: although the low-energy edge maintains the same shape, the high-energy edge becomes more abrupt. This observation leads to the conclusion that the radiative transition is not a simple localized recombination at an acceptor center.

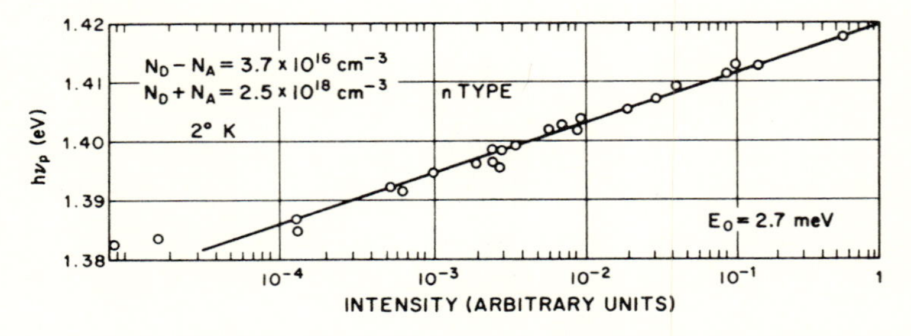

Fig. 6-46 Peak photon energy vs. peak light intensity for a compensated GaAs sample.[54]

[53] J. I. Pankove, unpublished observations.

[54] M. I. Nathan and T. N. Morgan, "Excitation Dependence of Photoluminescence in n- and p-type Compensated GaAs," *Physics of Quantum Electronics*, P. L. Kelley, B. Lax, and P. E. Tannenwald, McGraw-Hill (1966), p. 478.

An acceptable model is that the emission is due to tunnel-assisted transitions between spatially separated upper and lower states. Figure 6-47 shows the position-dependent energy-level diagram for a band structure perturbed by the heavy doping. As the quasi-Fermi levels move through the upper and lower states, first the most distant sets of states can participate in radiative recombination with tunneling over a large distance; then, at higher excitation rates, higher-energy photons can be emitted with a shorter tunneling range. Eventually, at still higher excitation, localized recombination (without tunneling) can occur. Since less tunneling is needed for transitions

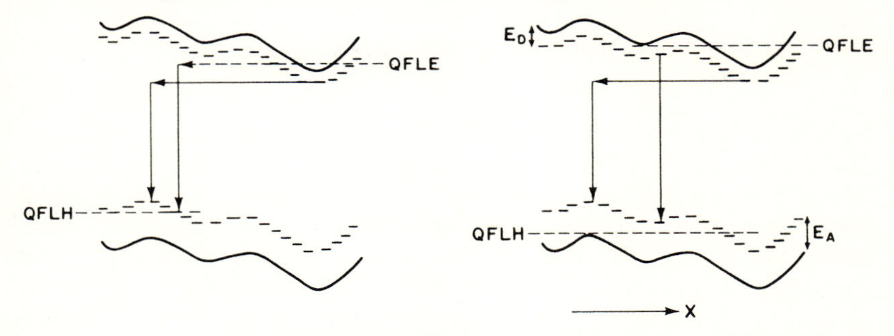

Fig. 6-47 Tunneling-assisted emission in compensated GaAs at low and high excitation rates.

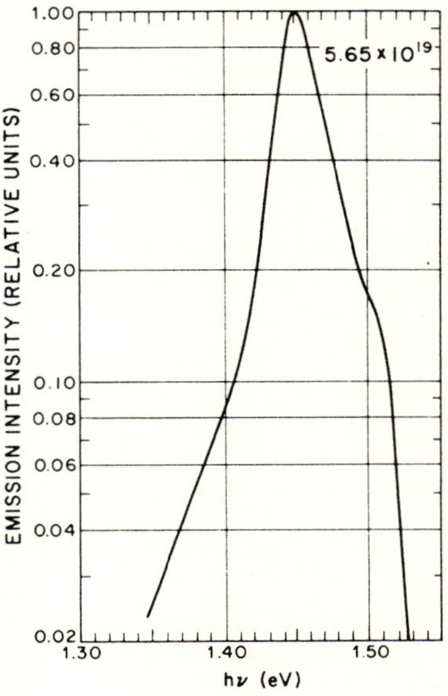

Fig. 6-48 Logarithmic recording of emission spectrum at 4.2°K in *p*-type GaAs.[36] Note the abrupt cutoff at the high energy edge, i.e., at the Fermi level where the acceptor band merges with the valence band.

generating the higher-energy photons, the high-energy edge of the emission spectrum sharpens considerably. Note that the transitions occur between donors and acceptors; hence the sharpest high-energy edge is defined by the convolution of those edges of the banded donors and acceptors which are away from midgap. It is only when the impurity band merges with the intrinsic band edge (for example the acceptor band merging with the valence-band tail) that the emission spectrum cuts off abruptly at the high-energy edge (Fig. 6-48). This abrupt cutoff disappears when the temperature is raised; then the high-energy edge tapers gradually.

6-F-3 TIME DEPENDENCE OF DONOR-TO-ACCEPTOR TRANSITIONS

A transition probability is expressed by the reciprocal of the pair lifetime. The transition probability decreases exponentially with increasing distance between donor and acceptor. Hence the recombination at low photon energies will be slower than the high-energy transition—the more distant pairs live longer. This time dependence of the recombination process is strikingly demonstrated in Fig. 6-49. This figure shows the emission spectrum at various times after the excitation of a GaP sample. The high-energy edge of the emission spectrum decays much faster than the low-energy edge. The rate of decay has been calculated on the basis of the tunneling model. The agreement between theory and experiment is as remarkable for the time dependence of the spectral decay as it was for the fine structure of the spectrum. Donor-to-acceptor tunneling "afterglow" has also been observed in ZnS,[56] InP,[57] and Si.[58]

In GaAs containing self-compensating Si, a time-dependent photoluminescence which behaves like the above tunneling afterglow has been observed;[59] however, there is no independent evidence that donor-to-acceptor photon-assisted tunneling occurs in this case. In Si-doped GaAs with a concentration of about 5×10^{18} Si/cm^3, both the rise time of photoluminescence and its decay time exhibit a monotonic increase as the photon energy decreases through the broad emission band which peaks at 1.37 eV. Thus at 1.45 eV the time constant is about 20 nanoseconds for rise and decay, while at 1.32 eV the rise time is 0.7 microseconds and the decay-time constant is about 2 μs. The time-resolved emission spectra show the corresponding peak shifts toward lower energies at longer times. Similar results are obtained in Si-doped samples which have a net excess of electrons or of holes. On the

[56] N. Riehl, G. Baur, L. Mader, and P. Thoma, "Afterglow in ZnS due to Tunneling," *II–VI Semiconducting Compounds*, ed. D. G. Thomas, Benjamin (1967), p. 724.

[57] U. Heim, *Solid State Communication*, **7**, 445 (1969).

[58] R. C. Enck and A. Honig, *Phys. Rev.* **177**, 1182 (1969).

[59] D. Redfield, J. I. Pankove, and J. P. Wittke, *Bull. Am. Phys. Soc.* **14**, 357 (1969).

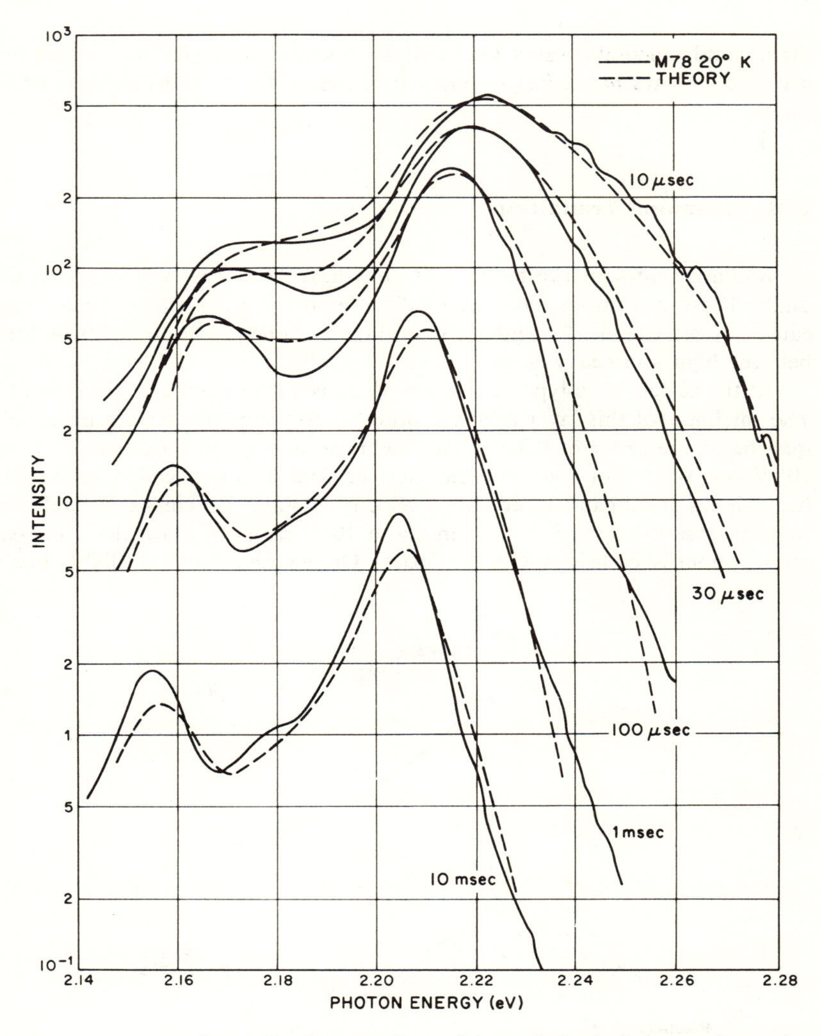

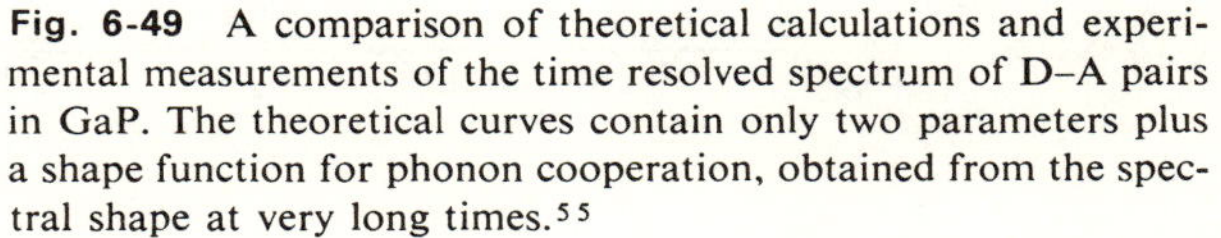
Fig. 6-49 A comparison of theoretical calculations and experimental measurements of the time resolved spectrum of D–A pairs in GaP. The theoretical curves contain only two parameters plus a shape function for phonon cooperation, obtained from the spectral shape at very long times.[55]

[55]J. J. Hopfield, "Radiative Recombination at Shallow Centers," *II–VI Semiconducting Compounds*, ed. D. G. Thomas, Benjamin (1967), p. 796.

other hand, uncompensated GaAs which contains only shallow donors or acceptors always has a fast response time (less than the 10 ns experimental limit).

6-G Intraband Transition

Radiative intravalence-band transitions have been observed in germanium.[60] These transitions have been called "emission by hot light holes" because an electric field is used to produce a nonequilibrium distribution between light and heavy holes.

At the surface of *n*-type germanium there is a inversion layer (Fig. 6-50). The thickness of this layer is of the order of 10^{-5} cm; the surface potential may be of the order of 0.3 eV. This yields an average electric field of 3×10^4 V/cm. When holes are injected into the field of the inversion layer, the heavy holes of subband V_1 acquire a velocity of about 10^6 cm/sec. Therefore, they traverse the inversion layer in about 10^{-11} sec (which, on the average, allows about 10 collisions with the lattice). On the other hand, the light holes,

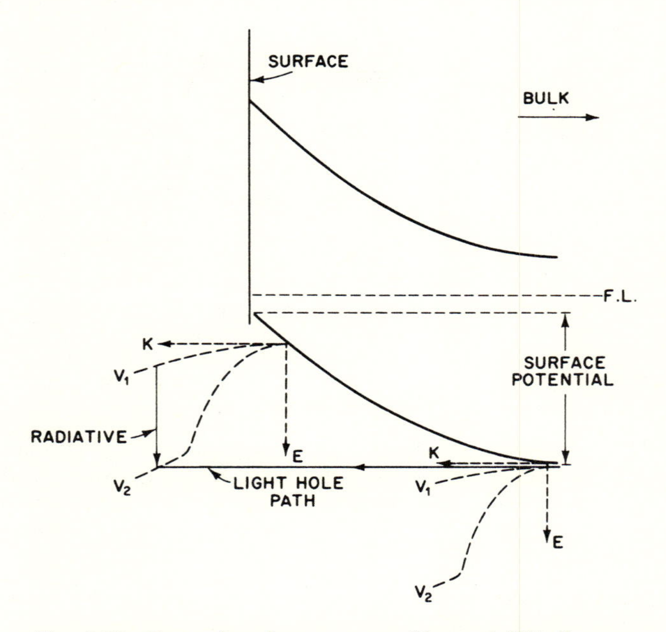

Fig. 6-50 Composite of energy vs. position and $E(k)$ diagrams showing the excitation of a light hole.

[60] J. I. Pankove, *J. Phys. Chem. Solids* **6**, 100 (1958) and *Annales de Phys.* **6**, 331 (1961).

being 8 times more mobile than the heavy holes,[61] suffer fewer collisions. Hence light holes are accelerated to greater kinetic energies than heavy holes. Figure 6-50 shows the case of a light hole which has not suffered a collision: the light hole maintains its total energy with respect to the Fermi level in the inversion layer. Because of the curvature of the bands (i.e., the internal field), the light hole effectively gains kinetic energy with respect to the top of the valence band. An electron from the heavy-hole subband can make a radiative direct transition into the hot light hole. The resulting emission spectrum (Fig. 6-51) peaks at about 6.6 μ (0.19 eV), corresponding to transitions in the region of $E(k)$-space where the two subbands V_1 and V_2 become parallel.

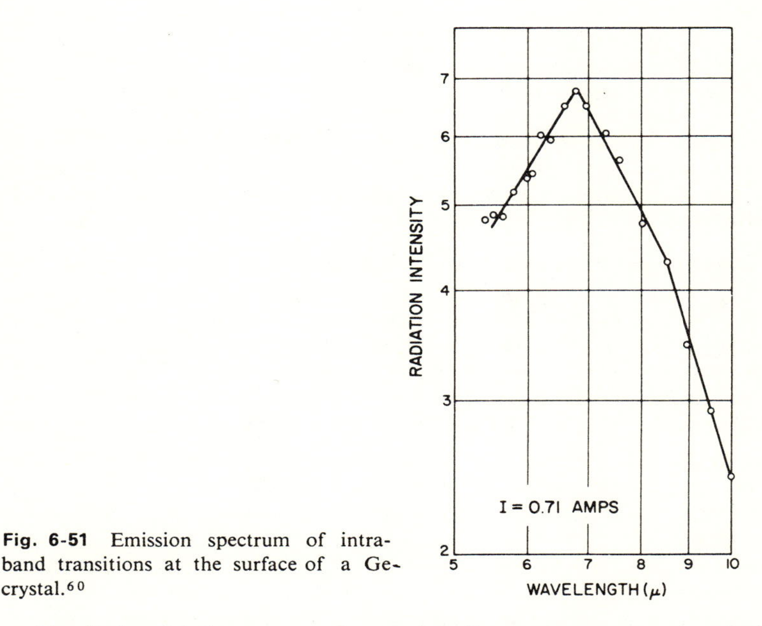

Fig. 6-51 Emission spectrum of intraband transitions at the surface of a Ge-crystal.[60]

To distinguish between surface and bulk processes, several techniques were utilized:

1. The Suhl effect was used: a magnetic field parallel to the surface and transverse to the current can deflect carriers toward the surface or away from it, i.e., toward the bulk, depending on the polarity of the magnetic field.
2. Since the present emission process requires a high local field to accelerate the light holes, the field-producing surface depletion layer can be replaced by the depletion layer of a *p–n* junction (see Sec. 8-A).

[61] R. K. Willardson, T. C. Harman, and A. C. Beer, *Phys. Rev.* **96**, 1512 (1954).

3. An optical-imaging technique allowed the possibility to distinguish between two regions: (a) the region where the holes were injected, producing strong fundamental transition but a weak hot-light-hole emission because of the low local field; (b) the adjacent surface region, or a nearby *p–n* junction, where hot-light-hole emission dominates.

Problem 1. A perfect Ge crystal 1 mm thick is illuminated at 0°K with 0.82-eV photons (where $\alpha = 10\ \text{cm}^{-1}$). The light flux is 10^{18} photons/cm²-sec. Assume that the transition probability is independent of energy, that the processes consist of indirect transitions with phonon participation ($E_{\text{phonon}} = 0.03$ eV), and that the radiative-recombination time is 0.1 sec. The energy gap is $E_g = 0.72$ eV; the effective masses are $m_e^* = 0.2m$ (4 valleys) and $m_h^* = 0.3m$; the index of refraction is $\boldsymbol{n} = 4$.
Find:

1. the initial and steady-state values of the transmitted flux;
2. the shape of the steady-state emission spectrum.

Problem 2. In view of the reciprocity between absorption and emission, explain why absorption due to transitions from acceptors to donors is not observed in heavily doped *n*-type or *p*-type semiconductors, although in emission the donor-to-acceptor transition is a dominant process.

Problem 3. The compounds GaAs and GaP form a continuous set of solid solutions, which have an energy gap varying from that of GaAs to that of GaP. Whereas GaAs is a direct-band-gap semiconductor having its lowest valley at $k = \langle 000 \rangle$, GaP is indirect, having its lowest valley at $k = \langle 100 \rangle$. Both have the valence-band maximum at $k = 000$. At 77°K, the direct valley of the $GaAs_{1-x}P_x$ system varies from 1.5 eV at $x = 0$ to 2.9 eV at $x = 1$. The $\langle 100 \rangle$ valleys, inside the Brillouin zone, vary from 1.9 to 2.2 eV in GaAs and GaP, respectively.

(a) Write a general expression for the concentration of electrons in an *n*-type alloy assuming (a) parabolic bands with density-of-state effective masses $m_{000}^* = 0.07$ and $m_{100}^* = 0.34$; (b) temperature-independent separation between the two sets of valleys.

(b) When holes are injected into *n*-type $GaAs_{1-x}P_x$, both radiative and nonradiative recombinations occur. If the radiative lifetime of the electron is 10^{-9} sec in the direct valley and 10^{-6} sec in the indirect valley, and the lifetime due to nonradiative recombination is 10^{-8} sec, describe the dependence of the emission efficiency on the alloy composition x.

Problem 4. Consider two discrete energy levels E_i and E_f such that

$$E_i - E_f = \hbar\omega$$

The probability per unit time that an electron in the state ψ_i will make a transition to the state ψ_f with the emission (or absorption $f \longrightarrow i$) of radiation of energy $\hbar\omega$ is given by

$$\mathscr{P} = \frac{2\pi}{\hbar} |\langle \psi_i | \frac{e}{mc} \bar{A} \cdot \bar{P} | \psi_f \rangle|^2$$

where $\bar{A}$ is the vector potential of the radiation and $\bar{P}$ is the momentum operator. (For this result consult any book on quantum mechanics.) Apply this to a semiconductor and show that the matrix element vanishes unles $\bar{k}_i = \bar{k}_f + \bar{q}$, where $\bar{q}$ is the wave vector of the radiation; $\bar{k}_i$ and $\bar{k}_f$ are the wave vectors associated with the initial and final Bloch states, respectively; and

$$\psi_i = U_{\bar{k}_i}(\bar{r})e^{i\bar{k}_i\cdot\bar{r}}$$
$$\psi_f = U_{\bar{k}_f}(\bar{r})e^{i\bar{k}_f\cdot\bar{r}}$$

Hints: 1. Note that $U_{\bar{k}}(\bar{r})$ is finite over a unit cell only and has the periodicity of the lattice:

$$U_{\bar{k}}(\bar{r} + \bar{R}_l) = U_{\bar{k}}(\bar{r})$$

where $\bar{R}_l$ is a lattice-translation vector.

2.
$$\sum_{\bar{R}_l} e^{i(\bar{k}-\bar{k}')\cdot\bar{R}} = 0$$

unless $\bar{k} = \bar{k}'$ (see, e.g., C. Kittel, *Introduction to Solid State Physics*, 3rd ed., John Wiley & Sons, Inc., New York, 1966, p. 50, in the limit of very large M). Note also that, in fact, the above sum is the discrete-sum representation of the delta function:

$$\int_{-\infty}^{+\infty} e^{ikx}\,dx = 2\pi\delta(k)$$

Problem 5. From the form of the transition rate given in Prob. 4, what can you deduce about the spin-selection rule in an optical transition allowed by first-order perturbation theory? Assume that $\psi = \phi\chi$, where ϕ is an orbital wave function and χ represents a pure spin state.

Problem 6. Taking into account Probs. 4 and 5, derive expressions for the optical density of states in the following cases:

(a) Band–band transitions in which the $\bar{k}$-selection rule applies. Neglect photon wave vector.
(b) Valence-band–donor transitions. Give answer in terms of $E = \hbar\omega$ and E_D, the donor ionization energy.
(c) Band–band transitions in which the $\bar{k}$-selection rule does not apply. An example of this is a strongly degenerate semiconductor. Can you give another example? The heavy doping grossly perturbs the states near the band edges and the result of Prob. 4 cannot be deduced.

In parts a and b, assume that the bands are parabolic. In each case explain clearly how you take into account momentum- and spin-selection rules, when applicable. Also assume that $\psi = \phi\chi$, as in Prob. 5. Omit occupation probabilities in parts a and b, but include them in part c.

Problem 7. Consider the van Roosbroeck–Shockley relation (Eq. 6-5). Given the absorption spectrum, discuss under what conditions you can use this relation to derive the emission spectrum. If applicable, what portion of the absorption spectrum should be used?

Problem 8. Consider the spontaneous emission in nondegenerate n-type GaAs when the emission is due to band–band transitions and when it is due to donor–valence-band transitions. Assume a parabolic density-of-state distribution.

Calculate the approximate energy of the emission peak in the above two cases when the valence band is also nondegenerately occupied. Assume $E_D \ll KT$, where E_D is the donor ionization energy measured from the conduction-band edge, and that the selection rule is $\bar{k}_u = \bar{k}_l$.

Problem 9. From the results of Prob. 8, discuss the temperature dependence of the peak energy and the line width of the emission spectrum in a nondegenerate n-type direct-gap semiconductor. Consider both band–band and donor–valence-band transitions.

Problem 10. Consider band–band transitions in a p-type material with the band diagram shown in the Fig. 6P-10. Draw and discuss the line shape at an excitation level such that both conduction and valence bands are degenerately occupied. Assume low temperature; F_u and F_l are the quasi-Fermi levels for electrons and holes, respectively.

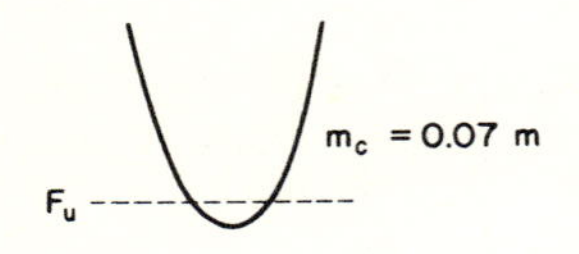

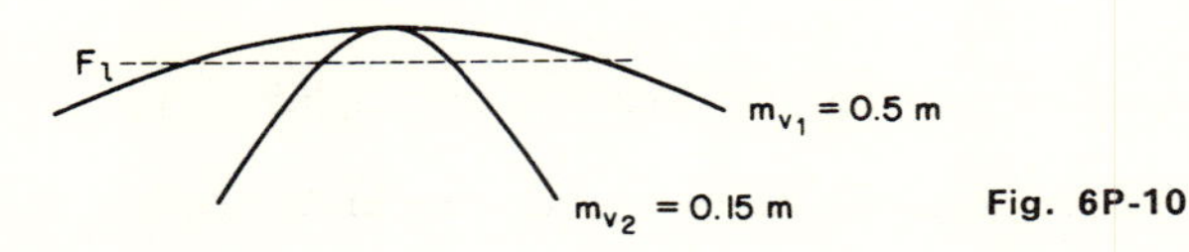

Fig. 6P-10

Problem 11. Hydrostatic pressure is applied to the material of Prob. 10 so that the valence-band degeneracy is split by Δ, as shown in the band diagram (Fig. 6P-11). Assuming the excitation level is the same as in Prob. 10, draw and discuss the shape of the emission spectrum. Compare the peak energies and line widths with the corresponding values found in the preceding problem. Do this for $S_{p_2} < \Delta$ and for $S_{p_2} > \Delta$, where S_{p_2} is the range of levels in the light-hole valence subband accessible to direct transitions from the conduction band. Comment on how the position of F_l affects the spectrum.

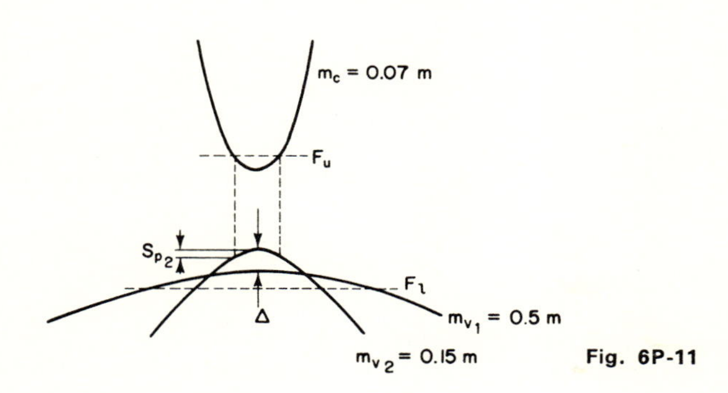

Fig. 6P-11

NONRADIATIVE RECOMBINATION

7

An electron–hole pair can recombine nonradiatively. In fact, in many semiconductors the nonradiative transition is the dominant process. Thus in pure germanium the radiative-transition probability calculated from the van Roosbroeck–Shockley relation (see Sec. 6-A) corresponds to a radiative lifetime of about one second, yet the measured minority-carrier lifetime is at most of the order of a millisecond and often less than a microsecond. Hence the nonradiative-recombination process in germanium is at least a thousand times more probable than the radiative transition.

Nonradiative recombination presents a conceptual difficulty. The statement "radiative recombination" is definitively explicit—a photon is emitted in the transition process—whereas "nonradiative recombination" is indefinite, vaguely indicating any process that does not emit a photon and leaving to one's ingenuity and imagination the construction of plausible models. The experimental study of nonradiative processes is, then, extremely difficult because the mechanism manifests itself only by the absence of the expected by-product. The only measurable parameters are emission efficiency, carrier lifetime, and kinetics of the recombination process in response to increased temperature or carrier concentration. Finally, the judgement is guided by circumstantial evidence rather than by definitive proof.

In this chapter we shall consider several recombination processes which do not result in external photon emission: Auger effect, surface recombination, and phonon emission. Although the Auger effect may sometimes appear to describe the process quite well, it is not a unique solution. Therefore, the field of nonradiative recombination is still open to further investigation.

7-A Auger Effect[1]

In the Auger effect, the energy released by a recombining electron is immediately absorbed by another electron which then dissipates this energy by emitting phonons. Thus this three-body collision, involving two electrons and a hole, results in no net photon emission. A large number of Auger processes can take place, depending on the nature of the possible transitions and on the carrier concentration. Some of these transitions are illustrated in Fig. 7-1. In Fig. 7-1(a) and (b), a band-to-band transition is considered; the second electron transforms the energy released in the recombination of the first electron into kinetic energy by being excited deep into the conduction band in n-type material, or from deep in the valence band in p-type material (in the latter case, one can consider the process as a collision between one electron and two holes). In Fig. 7-1(c)–(e) the transition occurs from a donor to the valence band; the recombination energy can be taken up by an electron in another donor or by a conduction-band electron, or, if many holes are present, by a second hole (process e). Auger effects associated with transitions from the conduction band to acceptors are illustrated in Fig. 7-1(f)–(h); and those associated with donor-to-acceptor transitions are shown in Fig. 7-1(i)–(l). Note that none of these processes replaces the electron making the first step of the collision. A process like that of Fig. 7-1(m) is not an Auger effect because the second electron could dissipate its energy radiatively. The process of Fig. 7-1(m) is called a "resonant absorption."

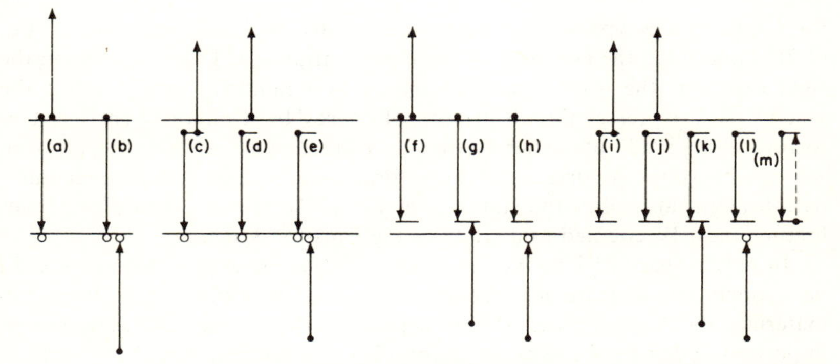

Fig. 7-1 Diagram of Auger processes (a) to (l) in a semiconductor, and resonant absorption (m). Processes (a), (c), (d), (f), (i), and (j) would be expected to occur in n-type semiconductors, while (b), (e), (g), (h), (k), and (l) would occur in p-type materials.

[1] R. Peierls, *Ann. Phys.* **13**, 905 (1932).

It is obvious that a process which depends on carrier–carrier interaction should become more intense as the carrier concentration increases. Therefore, as the temperature increases, the carrier concentration increasing proportionately to $(kT/E_g)^{3/2} \exp(-E_g/kT)$, the Auger effect should increase accordingly. However, both energy and momentum must be conserved. This leads to the theoretical prediction[2] that the electron–hole pair lifetime τ_A follows the proportion:

$$\tau_A \propto \left(\frac{E_g}{kT}\right)^{3/2} \exp\left(\frac{1+2M}{1+M}\frac{E_g}{kT}\right) \tag{7-1}$$

where M is the ratio of the electron and hole effective masses; $M = m_e/m_h$ if $m_e < m_h$ and τ_A is dominated by electron–electron collision, and $M = m_h/m_e$ if $m_h < m_e$ and τ_A is dominated by hole–hole collision.

The measured carrier lifetime in pure tellurium can be accounted for by a combination of radiative and Auger processes (Fig. 7-2).[3] The agreement is quite good at high temperatures; at low temperatures a third process which depends weakly on temperature seems to dominate.

Let us consider again that a process which depends on carrier–carrier interaction should increase as the carrier concentration increases. This dependence can be expressed in terms of the minority-carrier lifetime τ as

$$\left.\begin{aligned} \frac{1}{\tau} &= Anp + Bn^2 \\ \text{or} \qquad & \\ \frac{1}{\tau} &= Anp + Bp^2 \end{aligned}\right\} \tag{7-2}$$

for n-type and p-type semiconductors, respectively. In the relations of Eq. (7-2), n and p are the electron and hole concentrations. The first term on the right expresses the transmission of energy to a minority carrier, while the second term expresses the absorption of energy by a majority carrier. Unfortunately, a study of carrier lifetime as a function of carrier concentration will not prove the occurrence of Auger processes, because radiative recombination might also obey the statistics of Eq. (7-2). Hence a study of carrier lifetime must be coupled to a study of the emission kinetics.

In narrow-gap semiconductors, such as InSb, the Auger process should be strongly temperature-dependent, as expressed by Eq. (7-1). In large-gap materials, the Auger process should depend on the doping level and become important in degenerate semiconductors. The low-temperature photoluminescence efficiency of semiconductors generally decreases rapidly when the concentration of neutral donors or acceptors is increased above some thresh-

[2] A. R. Beattie and P. T. Landsberg, *Proc. Roy. Soc.* (London) **A249**, 16 (1958).

[3] J. S. Blakemore, *Proc. Intern. Conf. on Semiconductor Phys. Prague*, **1960**, Czechoslovak Academy of Sciences (1961), p. 981.

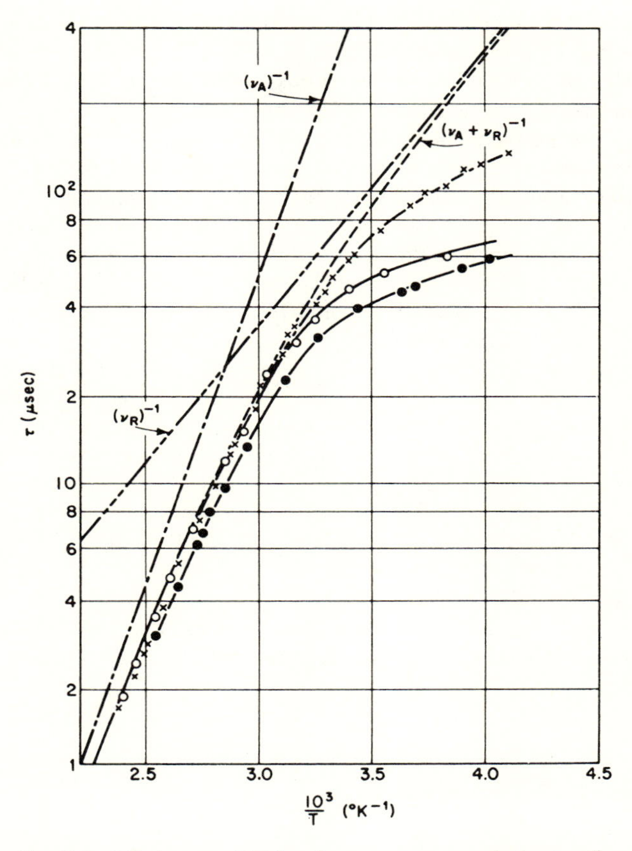

Fig. 7-2 Lifetime vs. $1/T$ for three very pure and structurally perfect samples, compared with the theoretical expectations of radiative and Auger recombinations.[3]

old concentration. Thus in GaP the threshold concentration N_T is about 10^{18} cm^{-3}. At low impurity concentrations and up to N_T the luminescence efficiency is constant, but at 10^{19} cm^{-3} the luminescence efficiency drops by a factor of 10^4 from its value at low concentrations.[4]

At concentrations beyond N_T, the electron wave functions of adjacent impurities overlap, thus delocalizing the electrons (or the holes) and increasing the chances for an Auger excitation of the delocalized carrier. The energy of the excited carrier would then be dissipated in a cascade emission of optical phonons to conserve energy and momentum.

[4] P. J. Dean, J. C. Tsang, and P. T. Landsberg, *Bull. APS* **13**, 404 (1968).

Note that although a critical concentration N_T may be predicted on theoretical grounds for the onset of the Auger effect, at impurity concentrations of the order of 10^{19} cm^{-3} other mechanisms can also set in, such as the formation of precipitates discussed in Sec. 7-C.

It must be pointed out that although the Auger effect accounts for the loss of photons expected to be generated in the dominant radiative recombination, the Auger effect can be responsible for other phenomena requiring energetic or "hot" carriers, since a hot carrier results from an Auger interaction. Thus a hot electron could be emitted from the semiconductor, or the hot electron could recombine radiatively to produce a more energetic photon. Similarly, a hot hole could overcome internal barriers in a semiconductor and produce a detectable event beyond that barrier, or it could participate in a second radiative recombination at a higher photon energy than in the first recombination. The latter effect has been observed in the electroluminescence of germanium, where in addition to the band-to-band radiative recombination, high-energy photon emission has been observed, extending to photon energies of twice the gap energy.[5] In this case, an across-the-gap transition (0.7 eV) excites a hole to the split-off valence subband, where it can recombine with a conduction-band electron, emitting up to about 1.3-eV photons. If this emission occurs near the surface, it is not strongly absorbed and, therefore, can be detected.

In the final analysis of "nonradiative" losses, one can be sure of the Auger effect only when one has definitive evidence of the resultant hot carrier.

7-B Surface Recombination

A surface is a strong perturbation of a lattice, creating many dangling bonds which can absorb impurities from the ambient. Hence a high concentration of deep and shallow levels can occur, and these may act as recombination centers. Although there is no definitive evidence for a uniform distribution of states, when a uniform distribution is assumed, the distribution of surface states is $N_s(E) = 4 \times 10^{14}$ cm^{-2} eV^{-1}, in good agreement with experimental estimates.[6]

Figure 7-3 shows a model of a continuous distribution of states at the surface of a semiconductor. It is evident that when electrons and holes are within a diffusion length of a surface, they will recombine, and that a transition through a continuum of states is readily nonradiative. This model can

[5] R. Conradt and W. Waidelich, *Phys. Rev. Letters* **20**, 8 (1968).

[6] A. Many, Y. Goldstein, and N. B. Grover, *Semiconductor Surfaces*, North-Holland Publishing Co. (1965), p. 434.

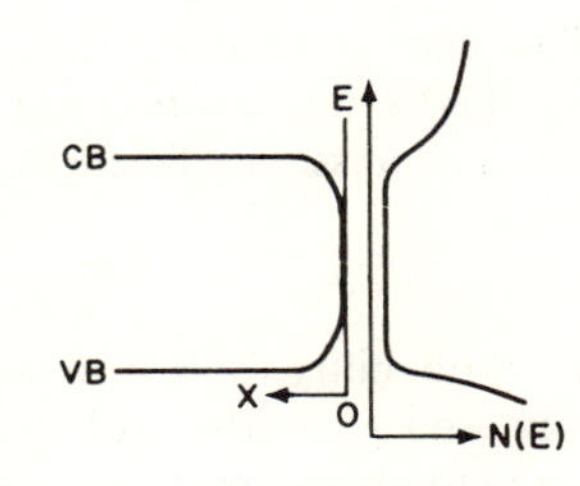

Fig. 7-3 Model of a continuous distribution of surface states.

be adapted to the concept of an internal surface, which we shall call a defect or an inclusion.

7-C Recombination Through Defects or Inclusions

Consider a localized defect which, like a microscopic internal surface or metallic inclusion, produces a continuum of states (Fig. 7-4). Electrons and holes which are within a diffusion length $\mathscr{L}$ from the edge of the defect, which may have an effective radius r, will be drawn to this trap, where they will recombine nonradiatively via the continuum of states. Such a model has no strong temperature dependence and may explain values of internal emission efficiencies lower than 1 which are observed at low temperatures in "pure" materials.

In n-type GaAs doped with Se or Te, precipitates of Ga_2Se_3 or Ga_2Te_3 have been detected when the donor concentration exceeds about $3 \times 10^{18}\,\text{cm}^{-3}$. The radiation efficiency drops by an order of magnitude when, at this donor concentration, the precipitates become detectable by transmission-electron microscopy.[7] A uniform distribution of small precipitates may affect the emission efficiency far more severely than a few large precipitates. If the same number of minority-phase particles aglomerate into a few large clusters, depleting the surrounding material of small precipitates, the efficiency around each cluster will be higher than in the regions where the precipitates are still

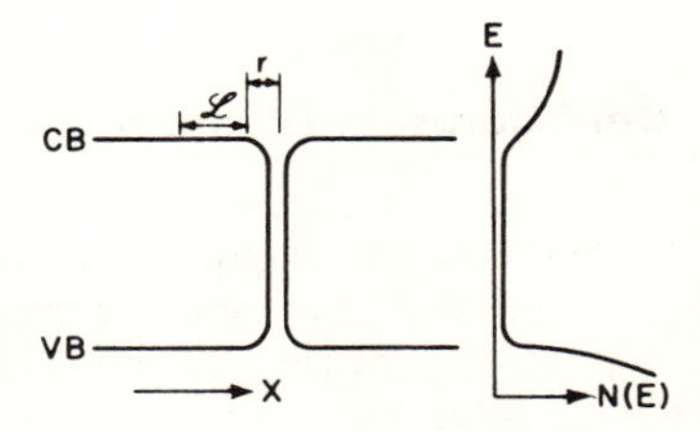

Fig. 7-4 Model of a defect or inclusion contributing a localized continuum of states.

[7] H. Kressel, F. Z. Hawrylo, M. S. Abrahams, and C. J. Buiocchi, *J. Appl. Phys.* **39**, 5139 (1968).

dispersed. This may account for the appearance of the cathodoluminescent pattern in Fig. 11-14, where a brighter region surrounds the darkest patches.

In general, the emission efficiency can be described by

$$\eta = \frac{P_r}{P_r + P_{nr}} \tag{7-3}$$

where P_r, the probability for a radiative transition, is assumed independent of temperature, and P_{nr} is the probability for a nonradiative transition. The term P_{nr} has the temperature dependence

$$P_{nr} = P_{nro}\exp(-E^*/kT) \tag{7-4}$$

where E^* is some activation energy to be described shortly, P_{nro} is a coefficient independent of temperature. Hence the temperature dependence of the radiative efficiency would be of the form

$$\eta = \frac{1}{1 + C\exp(-E^*/kT)} \tag{7-5}$$

where $C = P_{nro}/P_r$ is a constant.

A microscopic defect or inclusion could induce a deformation of the band structure as shown in Fig. 7-5. The deformations (a) and (b) could be due to local trapped charges and deformation (c) could be due to local strains. In either case, the deformation produces a barrier (of height E^*) around the recombination center. Only hot carriers with sufficient energy to overcome this barrier ($kT_e \geq E^*$) can recombine. Increasing the temperature will then increase the nonradiative transitions.

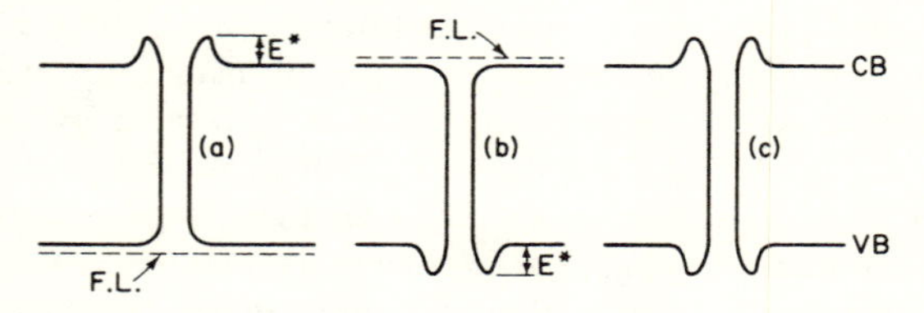

Fig. 7-5 Nonradiative recombination center surrounded by a barrier.

7-D Configuration Diagram

The configuration diagram provides a convenient, though hypothetical, model for describing nonradiative transitions. Figure 7-6 shows a configuration diagram with the electron in the excited states at A. At low temperature, the excited system has a configurational position r_A. The electron can make a radiative transition $A \longrightarrow A'$ and then the system will relax to the ground-state equilibrium at B. At high temperatures, atomic vibrations can move the

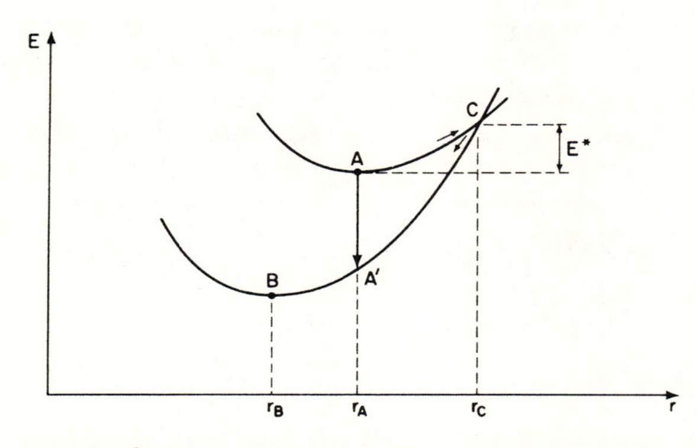

Fig. 7-6 Configuration diagram showing a radiative transition A ⟶ A′ and a nonradiative transition via C.

atom from a configurational position r_A to position r_C. At C, the electron can make a nonradiative transition to the ground state. When the system relaxes to position r_B, the electron is at the lowest-energy state without having emitted a photon. Of course, as the atom relaxes from position r_C to position r_B, many phonons are emitted. If the energy separation between A and C is E^*, the nonradiative-transition probability will have the temperature dependence of Eq. (7-4). As an example, transitions involving either defects or a copper centers in germanium have a thermal-activation energy $E^* = 0.14$ eV.[8]

In an efficient photoluminescent material, absorption must raise the electron to an energy in the excited state lower than that of point C, otherwise the "hot" carrier may cross over to the ground state via a vibrational level passing through C.

Another possible mechanism for making the luminescence efficiency temperature-dependent is discussed in Sec. 17-D as "luminescence quenching." Since this is not a model for nonradiative recombination, we shall not dwell on it in this chapter.

7-E Multiple-phonon Emission

It is conceivable for a nonradiative transition to occur by emitting a cascade of phonons. Since the energy of a phonon is much smaller than the expected loss of energy during recombination, the present process requires that a very large number of phonons be emitted with a high probability. Since in the present case we are not assuming a continuum of states bridging

[8]C. Benoit-à-la-Guillaume, *Annales de Physique*, *Ser. 13* **4**, 1187 (1959), also *J*, *Phys. Chem. Solids* **8**, 150 (1959).

the gap, the multiphonon process must be a high-order transition which would have very low probability. We are touching on this subject, however, because this concept occurs occasionally in the literature.

References are sometimes made to observations of multiphonon processes in germanium[9] (Fig. 7-7) or in silicon.[10] However, these are not recombina-

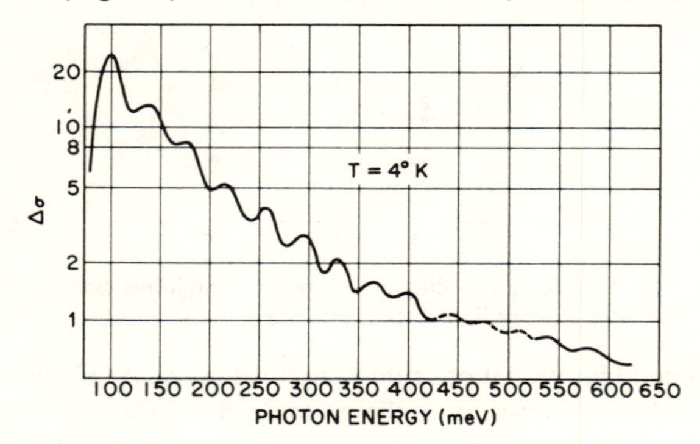

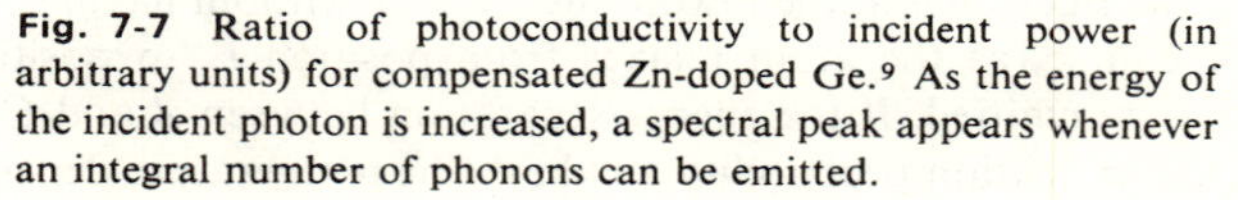

Fig. 7-7 Ratio of photoconductivity to incident power (in arbitrary units) for compensated Zn-doped Ge.[9] As the energy of the incident photon is increased, a spectral peak appears whenever an integral number of phonons can be emitted.

tion processes but rather dissipation of excess energy when an electron is excited from deep in the valence band to an acceptor level. In this case optical phonons are emitted by transition along the continuum of states formed by the valence band.

Problem 1. The following photoluminescence data was obtained by irradiating GaN with a constant flux of photons having a greater-than-gap energy.

Temperature (°K)	*Light Intensity* (*Arbitrary Units*)	*Temperature* (°K)	*Light Intensity* (*Arbitrary Units*)
20	160	95	50
40	155	105	37
50	150	125	25
55	140	165	15
70	110	220	10
80	80	295	8

Find the activation energy for the non-radiative recombination process.

[9]C. Benoit-à-la-Guillaume and J. Cernogora, *J. Phys. Chem. Solids* **24**, 383 (1963).
[10]I. V. Kryukova, *Soviet Phys. Solid State* **7**, 2060 (1966).

Problem 2. This problem deals with the identification of Auger electrons. Suppose that in n-type GaAs hot electrons having a kinetic energy $E_k > 0.5$ eV can be collected and suppose that their energy in excess of 0.5 eV can be measured. Let the crystal be excited by 2.0-eV photons. Assume that the probability of finding a hot carrier at an energy E is $1 - \exp[(E - E_{max})/E_o]$ where E_{max} is the maximum energy to which an electron can be excited and E_o is the energy lost per collision (typically 0.05 eV). The energy gap of GaAs is 1.4 eV at 300°K.

Find the approximate distribution of hot electrons whose kinetic energy is greater than 0.5 eV and identify the three regions of the distribution curve. How might one distinguish Auger electrons from other hot electrons?

PROCESSES IN *p-n* JUNCTIONS

8

The *p–n* junction is the vital component of most semiconductor devices. In this chapter we shall develop the concept of the *p–n* junction and characterize some of its parameters. We shall see that this singularity in the semiconductor crystal is endowed with important properties, such as a potential barrier to electrical transport of majority carriers through the junction and a strong local field which assists the migration of minority carriers across the junction into a region where they become majority carriers. We shall also see that the junction is a region of intense optical activity in many forms.

8-A Nature of the *p–n* Junction

When *n*-type and *p*-type regions of a semiconductor are joined to form a continuous crystal, the carriers redistribute themselves in such a way as to equalize the Fermi level throughout the semiconductor (Fig. 8-1). In the neighborhood of this juncture electrons from donors transfer to nearby acceptors, and a dipole layer is formed consisting of positive empty (ionized) donors on the *n*-type side and negative occupied (ionized) acceptors on the *p*-type side. The dipole generates an electric field which would drive a conduction-band electron to the *n*-type side and a valence-band hole to the *p*-type side. The junction proper is defined by that location where the Fermi level is in the middle of the gap. The dipole layer extends on either side of the junction, and the total extent of this "depletion layer" is called the junction thickness.

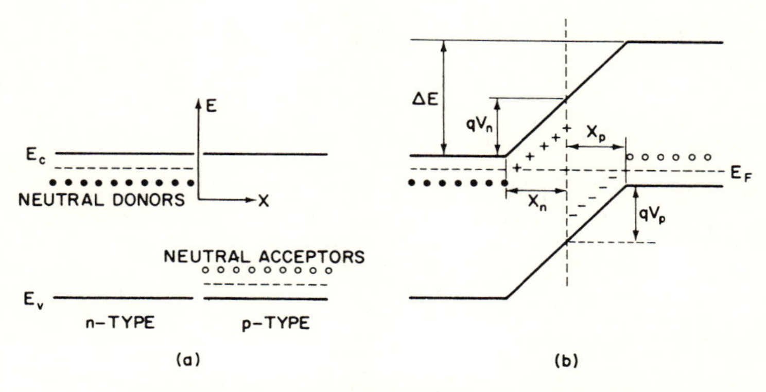

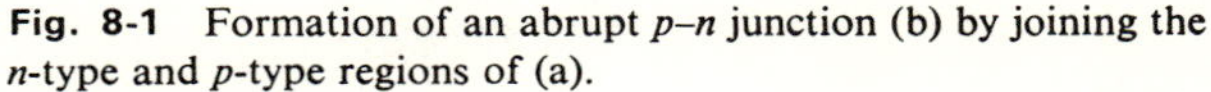

Fig. 8-1 Formation of an abrupt *p–n* junction (b) by joining the *n*-type and *p*-type regions of (a).

8-A-1 THE DEPLETION LAYER

The junction thickness can be related to the impurity concentration as follows.

First, we equate the charges on either side of the junction resulting from the donor-to-acceptor transfer while forming the junction:

$$N_d X_n = N_a X_p \tag{8-1}$$

where X_n and X_p are the lengths of the depletion regions into the *n*-side and *p*-side, respectively. The net charge across the whole junction is then zero. Next, we evaluate the resulting change in potential by solving Poisson's equation in one dimension:

$$\left.\begin{aligned} \frac{d^2V_n}{dX^2} &= \frac{qN_d}{\epsilon} \\ \frac{d^2V_p}{dX^2} &= \frac{qN_a}{\epsilon} \end{aligned}\right\} \tag{8-2}$$

The solution of Eq. (8-2) is

$$\left.\begin{aligned} V_n &= \frac{qN_d}{2\epsilon}(X_n)^2 \\ V_p &= \frac{qN_a}{2\epsilon}(X_p)^2 \end{aligned}\right\} \tag{8-3}$$

Hence the conduction- and valence-band edges change their potential energy by the total amount

$$\Delta E = q(V_n + V_p) \tag{8-4}$$

$$= \frac{q^2}{2\epsilon}[N_d(X_n)^2 + N_a(X_p)^2] \tag{8-5}$$

From Eq. (8-3), the junction thickness is

$$X_n + X_p = \left(\frac{2\epsilon}{q}\right)^{1/2}\left[\left(\frac{V_n}{N_d}\right)^{1/2} + \left(\frac{V_p}{N_a}\right)^{1/2}\right] \tag{8-6}$$

Let us insert $V_n + V_p$ from Eq. (8-4):

$$\begin{aligned} X_n + X_p &= \left(\frac{2\epsilon}{q}\right)^{1/2}\left(\frac{\Delta E}{q}\right)^{1/2}\left\{\left[\frac{V_n}{(V_n + V_p)N_d}\right]^{1/2} + \left[\frac{V_p}{(V_n + V_p)N_a}\right]^{1/2}\right\} \\ &= \frac{(2\epsilon\,\Delta E)^{1/2}}{q}\left\{\left[\frac{1}{N_d\left(1 + \frac{V_p}{V_n}\right)}\right]^{1/2} + \left[\frac{1}{N_a\left(1 + \frac{V_n}{V_p}\right)}\right]^{1/2}\right\} \end{aligned}$$

Note that Eq. (8-3) gives

$$\frac{V_p}{V_n} = \frac{(X_p)^2}{(X_n)^2}\frac{N_a}{N_d}$$

Substituting X_p/X_n from Eq. (8-1),

$$\frac{V_p}{V_n} = \frac{N_d}{N_a} \tag{8-7}$$

Therefore,

$$\left.\begin{aligned} X_n + X_p &= \frac{(2\epsilon\,\Delta E)^{1/2}}{q}\left\{\left[\frac{1}{N_d\left(1 + \frac{N_d}{N_a}\right)}\right]^{1/2} + \left[\frac{1}{N_a\left(1 + \frac{N_a}{N_d}\right)}\right]^{1/2}\right\} \\ &= \frac{(2\epsilon\,\Delta E)^{1/2}}{q(N_a + N_d)^{1/2}}\left[\left(\frac{N_a}{N_d}\right)^{1/2} + \left(\frac{N_d}{N_a}\right)^{1/2}\right] \end{aligned}\right\} \tag{8-8}$$

If the p-type region is much more heavily doped than the n-type region, $N_a \gg N_d$; then because of Eq. (8-1), $X_n \gg X_p$, i.e., most of the depletion extends into the n-type side of the junction. Since

$$X_n \approx \frac{(2\epsilon\,\Delta E)^{1/2}}{q(N_d)^{1/2}} \tag{8-9}$$

the thickness of the depletion region decreases with increasing doping. The above is correct only for abrupt junctions between uniformly doped n-type and p-type regions. If the impurity concentration varies gradually across the

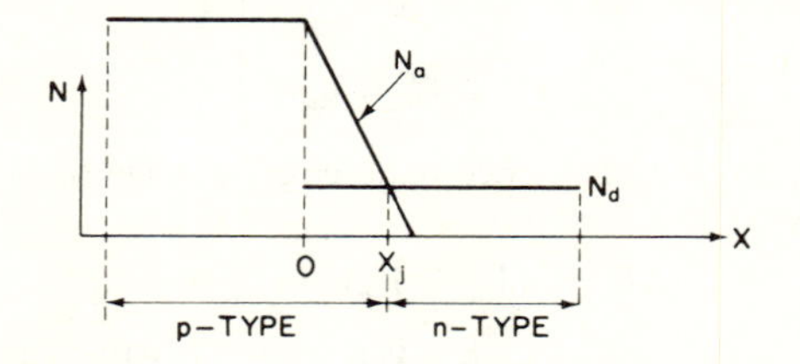

Fig. 8-2 Distribution of impurity concentration in linearly graded p–n junction.

p–n junction, a slightly different functional dependence between doping, depletion, and change in potential is obtained. The exact dependence is found by replacing the constants N_d or N_a in Eq. (8-2) by $N(X)$.

Thus for a linearly graded junction with N_d constant and $N_a(X) > N_d$ at $X = 0$ (Fig. 8-2), the charge in the depletion region varies proportionally with X. The solution of Poisson's equation makes V_n proportional to X^3.

8-A-2 JUNCTION CAPACITANCE

A p–n junction forms a parallel-plate capacitor consisting of two conducting regions separated by the space-charge layer. In the space-charge layer there are no mobile carriers; therefore, the depletion region is insulating.

The capacitance per unit area is given simply by

$$C = \frac{\epsilon}{X_n + X_p} \tag{8-10}$$

For an abrupt junction between uniformly doped regions, if $X_n \gg X_p$, substituting Eq. (8-9) gives

$$C = q\left(\frac{\epsilon N_d}{2\,\Delta E}\right)^{1/2} \tag{8-11}$$

In the presence of a bias V, Eq. (8-11) becomes

$$C = \frac{(q\epsilon N_d)^{1/2}}{\sqrt{2}\left(\dfrac{\Delta E}{q} - V\right)^{1/2}} \tag{8-12}$$

Therefore, when V is positive (forward bias), the capacitance increases; and when V is negative (reverse bias), the capacitance decreases.

As shown in Fig. 8-3, it is customary to plot the capacitance vs. voltage dependence of an abrupt p–n junction in the form

$$\frac{1}{C^2} = \frac{2}{q\epsilon N_d}\left(\frac{\Delta E}{q} - V\right) \tag{8-13}$$

The slope of the straight line allows an evaluation of N_d, while the extrapolated intercept with the abscissa gives $\Delta E/q$. The latter quantity is usually

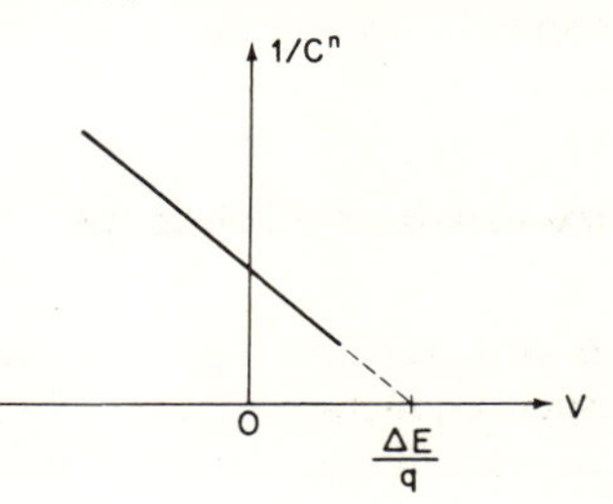

Fig. 8-3 Dependence of junction capacitance on voltage across the junction. The exponent n is 2 for an abrupt p–n junction and 3 for a linearly graded transition.

the forward voltage across the junction needed to flatten the band edges. However, in junctions between degenerately doped regions, $\Delta E/q$ is the forward voltage needed to inject an abundance of minority carriers—i.e., when $V = E_g + \xi$ (where ξ is the smaller of the two values ξ_n or ξ_p).

In the case of the linearly graded impurity distribution of Fig. 8-2, it can be shown that the capacitance has the following bias dependence:

$$\frac{1}{C^3} = \frac{12}{\epsilon^2 qa}\left(\frac{\Delta E}{q} - V\right) \tag{8-14}$$

where

$$a = \left.\frac{d(N_a - N_d)}{dX}\right|_{X=X_j} \tag{8-15}$$

is the impurity gradient at the junction proper.

8-A-3 ELECTRIC FIELD IN THE *p-n* JUNCTION

The average electric field $\bar{\mathscr{E}}$ in the *p–n* junction can be derived from the electrostatic potential V_n, assuming again an abrupt *p–n* junction with $N_a \gg N_d$:

$$\bar{\mathscr{E}} = \frac{V_n}{X_n}$$

Using Eq. (8-3), and then (8-9):

$$\bar{\mathscr{E}} = \left(\frac{N_d \Delta E}{2\epsilon}\right)^{1/2} \tag{8-16}$$

A typical value for V_n is 1 volt, and X_n ranges from 10^{-4} to 10^{-6} cm; hence local fields of 10^4 to 10^6 V/cm are possible in *p–n* junctions.

The solution of Poisson's equation (8-2) in the presence of an applied reverse bias V shows that X_n varies proportionately to $\sqrt{V}$ in abrupt junctions. Hence the application of a reverse bias will increase the electric field in the junction proportionately to the square root of the applied bias. This field will increase until breakdown sets in.

Note that with a linearly graded impurity distribution, the electric field grows as the $\frac{1}{3}$ power of the applied voltage.

8-B Forward-bias Processes

Let us consider first a *p–n* junction between two heavily doped regions, because such a junction exhibits the richest set of phenomena. Less heavily doped regions will make some of these processes too weak to be observed.

8-B-1 BAND-TO-BAND TUNNELING[1]

Figure 8-4 shows a *p–n* junction between two degenerately doped regions, with a small forward bias V_1 applied across the junction. If the junction is very narrow, the wave functions of conduction-band electrons overlap the wave functions of holes in the valence band. Then the electrons can tunnel at constant energy from the conduction band of the *n*-type region to the valence band of the *p*-type region. Obviously, only electrons in the energy range E_{F_n} to $E_{F_n} - qV_1$ can make transitions to the corresponding empty states between E_{F_p} and $E_{F_p} + qV_1$. As the overlap between the filled states in the conduction band and the empty states in the valence band increases, the tunneling current increases. This process manifests itself in portion 1 of the *I–V* characteristic of Fig. 8-5. The current reaches a maximum when the bias allows a maximum overlap of filled conduction-band states and empty valence-band states. The maximum occurs at $qV_1 = \xi$, where ξ is the smaller of the two values ξ_n and ξ_p. At larger biases the overlap decreases and, therefore, the tunneling transport decreases. This negative-resistance region includes tunneling between tails of state[2] and, therefore, can extend over a fair range of forward bias.

When a reverse bias is applied, filled valence-band states of the *p*-type region overlap empty states in the conduction band of the *n*-type region.

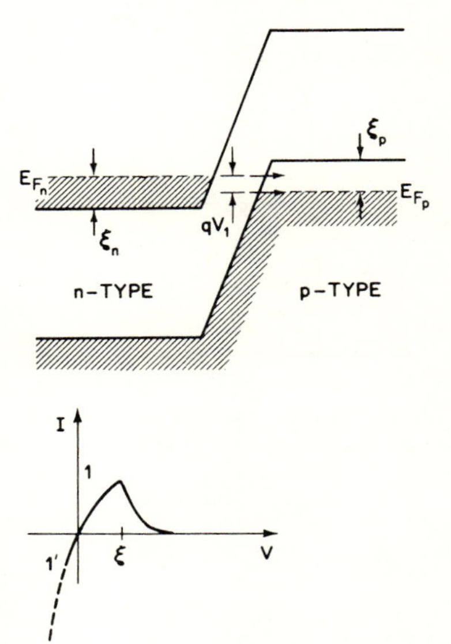

Fig. 8-4 Tunneling current between heavily doped regions with a forward bias V_1 applied across the junction.

Fig. 8-5 *I–V* characteristic of tunnel diode in the tunneling mode of transport.

[1] L. Esaki, *Phys. Rev.* **109**, 603 (1958).
[2] E. O. Kane, *Phys. Rev.* **131**, 79 (1963).

Hence in narrow junctions, electrons tunnel in a direction opposite to forward-bias tunneling.[3] This process contributes portion 1′ of the I–V characteristic in Fig. 8-5.

Note that in considering tunneling transitions at constant energy we have overlooked momentum conservation. In general, scattering processes would conserve momentum; but to be more rigorous, let us illustrate the momentum-conservation problem by considering the tunneling between two direct valleys, as in Fig. 8-6. The three-dimensional E–k diagram superposed on the E–X diagram shows that the momentum of an electron A can be decomposed into a longitudinal component and a transverse component. As the electron enters the junction, the longitudinal component decreases and then vanishes at the "turning point" X_2 under the restraining action of the local field at the junction. But since there is no force acting in a transverse direction, the transverse-momentum component remains unchanged. When the electron has tunneled into the valence band, its transverse momentum is conserved so that when it appears at X_3 its location in the three-dimensional E–k diagram is at a point B. Obviously, effective-mass considerations impose further constraints as to the range of initial and final states that can be coupled by this process.

When the conduction-band valley is not at the same crystal momentum as the valence band, and when scattering processes are not available to conserve momentum, no transport can occur until the bias V_1 allows a suitable phonon to be emitted.[4,5] The phonon then permits momentum conservation, but then the tunneling does not occur at a constant energy.

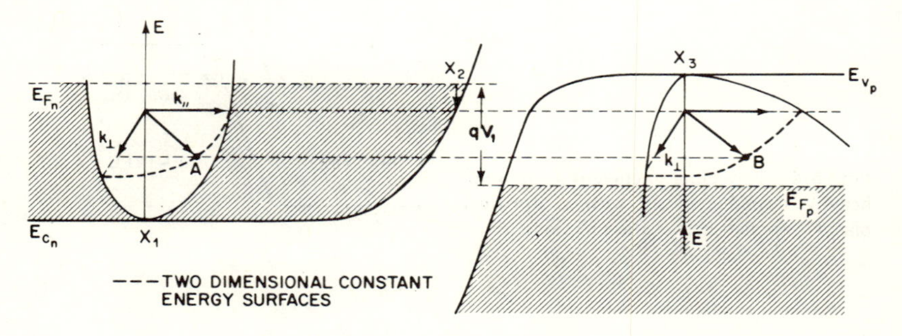

Fig. 8-6 Detail of tunneling process $A \longrightarrow B$ showing how conservation of energy and of transverse momentum $k_\perp$ determine the final state B at X_3.

[3]A. G. Chynoweth, W. L. Feldmann, C. A. Lee, R. A. Logan, G. L. Pearson, and P. Aigrain, *Phys. Rev.* **118**, 425 (1960).

[4]N. Holonyak, I. A. Lesk, R. N. Hall, J. J. Tiemann, and H. Ehrenreich, *Phys. Rev. Letters* **3**, 167 (1959).

[5]E. O. Kane, *J. Appl. Phys.* **32**, 83 (1961)

It is appropriate here to mention that a varying reverse bias allows scanning the distribution of the density of states in the conduction band. When more valleys can receive electrons from the valence band of the *p*-type region, the current increases, and the conductance dI/dV exhibits a step. These steps have been used to find the positions of the valleys as the impurity concentration increases.[6,7] Tunneling spectroscopy under forward bias has also been used to find the distribution of states inside the energy gap;[8] but great care is needed to interpret these results, since with fewer states the tunneling is weak and can be masked by other effects.

8-B-2 PHOTON-ASSISTED TUNNELING

When the forward bias is large enough to uncross the bands (Fig. 8-7), an electron can make a transition across the junction to the valence band by a two-step process: tunneling and photon emission[9]—hence the name "photon-assisted tunneling" from a transport point of view, or "tunneling-assisted photon emission" from a luminescence point of view. This transport process contributes the so-called "excess current,"[10] which is portion 2 of the I–V characteristic in Fig. 8-8. According to a simple model with direct parabolic bands, for a given bias V_2 the minimum photon energy to be emitted is

$$h\nu_{\min} = qV_2 - \xi_n - \xi_p \tag{8-17}$$

corresponding to a transition between the band edges. However, if the band edges are extended by an exponential tail, the lower energy of the emission

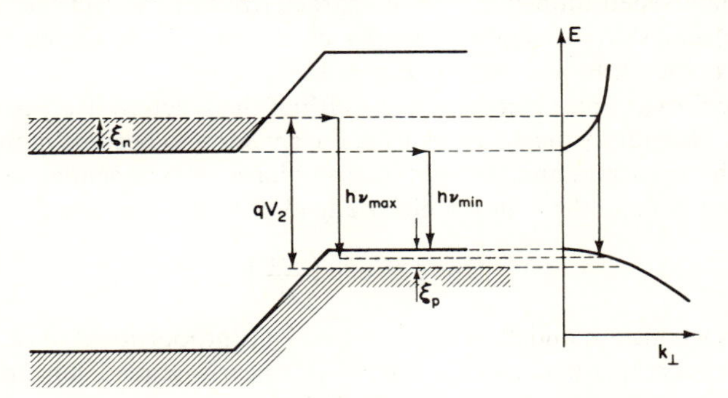

Fig. 8-7 Photon-assisted tunneling across a *p-n* junction with forward bias V_2.

[6] J. V. Morgan and E. O. Kane, *Phys. Rev. Letters* **3**, 466 (1959).
[7] W. N. Carr, *J. Appl. Phys.* **34**, 2467 (1963).
[8] P. V. Gray, *Phys. Rev. Letters* **9**, 303 (1962).
[9] J. I. Pankove, *Phys. Rev. Letters* **9**, 283 (1962); also *Annales de Physique* **6**, 363 (1961); also *J. Electrochemical Soc.* **108**, 998 (1961).
[10] A. G. Chynoweth, W. L. Feldmann, and R. A. Logan, *Phys. Rev.* **121**, 684 (1961).

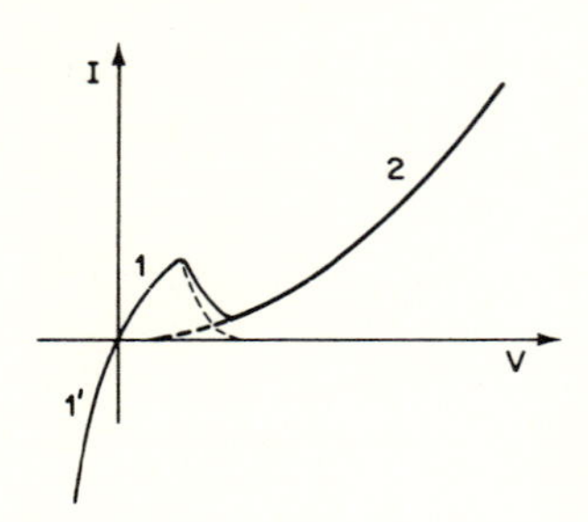

Fig. 8-8 I–V characteristic of a tunnel diode showing tunneling current 1, 1′ and the photon-assisted tunneling mode, 2.

spectrum does not have an abrupt cutoff. The maximum photon energy emitted at 0°K is

$$h\nu_{\max} = qV_2 - \xi_p + \frac{m_c}{m_v}\xi_n \tag{8-18}$$

where the $[-\xi_p + (m_c/m_v)\xi_n]$ term is imposed by the conservation of transverse momentum, as is evident from inspection of Fig. 8-7. Of course, if a scattering means is available to change momentum at constant energy, the maximum emitted energy is

$$h\nu_{\max} = qV_2 \tag{8-19}$$

On the other hand, in an indirect-gap semiconductor, where a phonon E_p is emitted to conserve momentum,

$$h\nu_{\max} = qV_2 - E_p \tag{8-20}$$

Photon-assisted tunneling is then characterized by the fact that the emission spectrum shifts to higher energies as the bias V_2 is increased.[11] This behavior is shown by the data of Fig. 8-9.

As the forward bias increases, the width of the junction decreases and the tunneling probability increases, causing the emission to become more intense. Thus both the current and the light intensity have an exponential dependence on the bias voltage through the electric field $\mathscr{E}$ at the junction:[12]

$$L \propto I \propto \exp\left(\frac{-a}{\mathscr{E}}\right) \tag{8-21}$$

where a is a constant and $\mathscr{E} = (E_g - V)/X(V)$ is proportional to $(E_g - V)^{1/2}$ in abrupt junctions and to $(E_g - V)^{2/3}$ in linearly graded junctions. With increasing bias, then, the intensity of all the transitions increases, including those between the tails of states beyond the band edges. However, if the band edges form a sharp cutoff in the distribution of states, the emission spectrum should exhibit, in addition to a shifting peak, a shifting low-energy edge (it can be noticed in the lower curves of Fig. 8-9). The shifting low-energy edge

[11] H. C. Casey and D. J. Silversmith, *J. Appl. Phys.* **40**, 241 (1969).
[12] T. N. Morgan, *Phys. Rev.* **148**, 890 (1966).

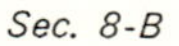

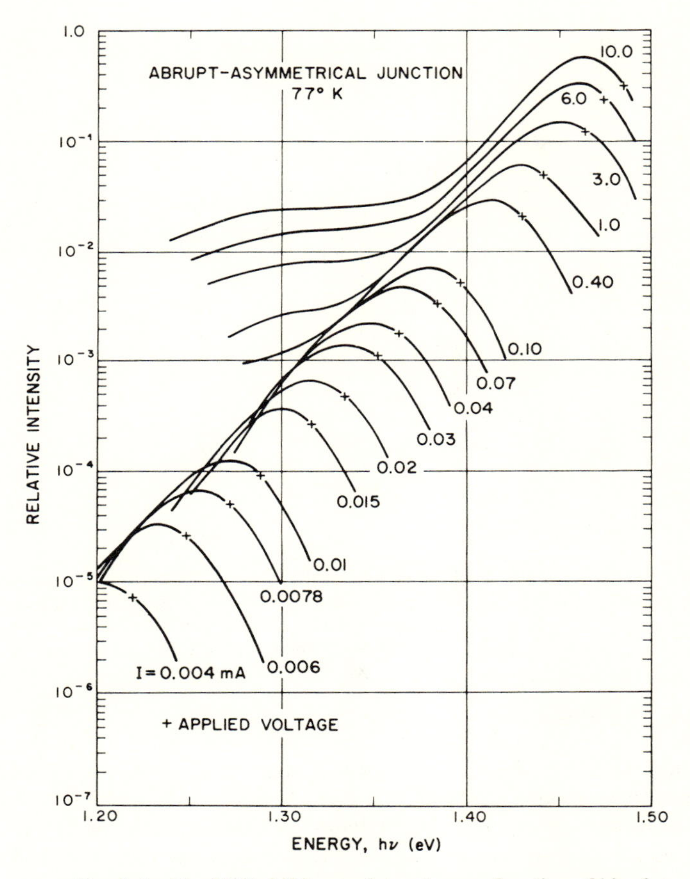

Fig. 8-9 The 77°K shifting-peak spectra as a function of bias for an abrupt-asymmetrical junction with $n = 1.74 \times 10^{18}$ electrons/cm^3.[11]

is seldom observed, because the tailing of states is usually pronounced; however, it has been reported,[9] and it is clearly shown in Fig. 8-10(a).[13] In this case, as the bias increases the emission intensity at some low photon energy first increases, goes through a maximum, and then decreases as the energy separating the band edges exceeds the observed photon energy [Fig. 8-10(b)].

For a detailed theoretical treatment of photon-assisted tunneling we recommend references 11 and 12.

[13]A. A. Rogachev and S. M. Ryvkin, *Fizika Tverdovo Tela* **6**, 3188 (1964).

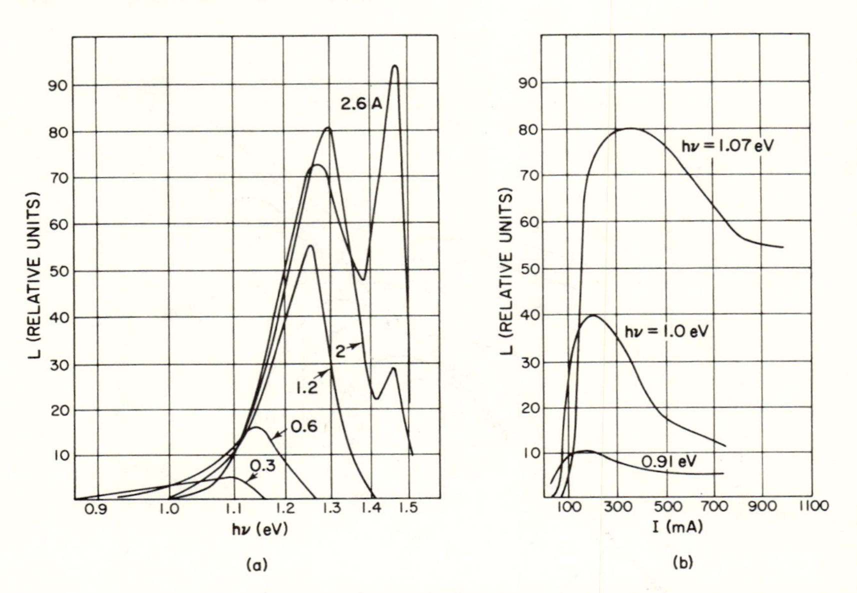

Fig. 8-10 (a) Emission spectrum for the photon-assisted tunneling process in a GaAs diode at 77°K and at several currents from 0.3 to 2.6 A. (b) Bias dependence of emission intensity at three wavelengths in the low-energy edge of the emission spectrum.[13]

8-B-3 INJECTION

When the forward bias is sufficiently large to allow the propagation of electrons throughout the conduction band beyond the junction [Fig. 8-11(a)] (or of holes throughout the valence band), the current assumes the injection mode, which appears as portion 3 of the I–V characteristic of Fig. 8-11(b). In the injection mode, direct band-to-band recombination becomes possible in direct-gap semiconductors (in indirect-gap materials, phonon-assisted recombination can occur). The injection current and the corresponding emission increase rapidly with the bias voltage V_3 according to the "diode equation":

$$I = I_0\left(\exp\frac{qV_3}{kT} - 1\right) \tag{8-22}$$

where I_0 is a constant having the dimensions of current. Implicit in I_0 is a term

$$\exp\left(\frac{-E_g - \xi}{kT}\right)$$

where ξ is the smaller of ξ_n or ξ_p; $E_g + \xi_n$ is the barrier which holes must

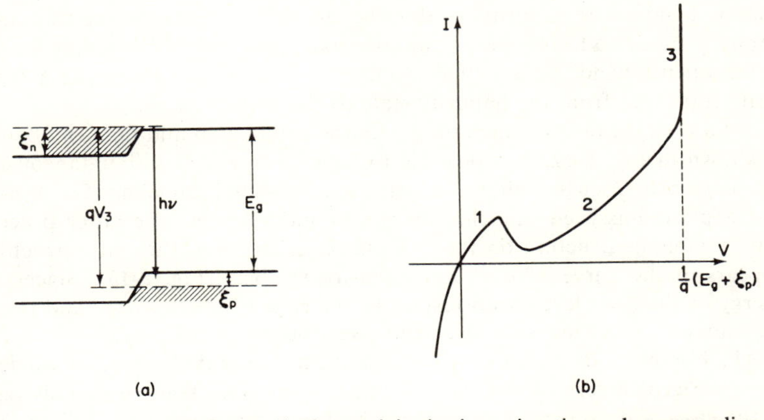

Fig. 8-11 (a) Electron injection in p–n junction and corresponding I–V characteristic "3" in (b).

overcome to be injected into the n-type region. Similarly, $E_g + \xi_p$ is the barrier that electrons must overcome to enter the p-type region by injection. Hence for the case illustrated in Fig. 8-11, the injection current rises rapidly with voltage when $qV_3 > E_g + \xi_p$. Later, we shall see that injection luminescence can result in lasing.

It is important to realize that in the injection mode, in addition to the band-to-band recombination, it is possible to get recombination, radiative and nonradiative, via a variety of levels: conduction band to acceptors, donors to valence band, donor to acceptor, free and bound excitons—in brief, all the processes covered in Chapter 6.

8-B-4 TUNNELING TO DEEP LEVELS

Let us assume now that a deep impurity band is present in the energy gap (Fig. 8-12). With a small forward bias—which, however, is greater than that at which band-to-band tunneling occurs—electrons can tunnel from the con-

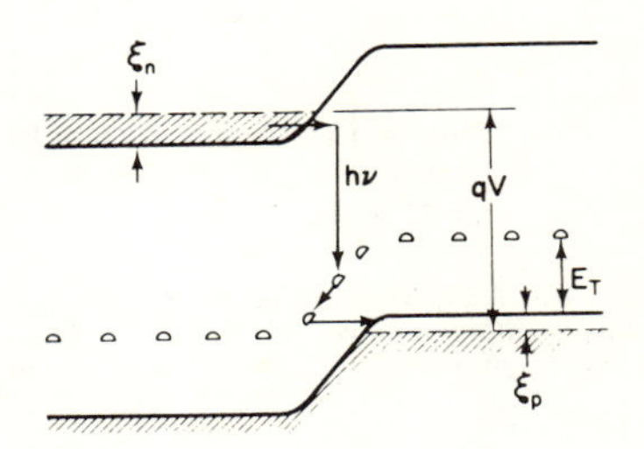

Fig. 8-12 Photon-assisted tunneling to a deep impurity band.

duction band to the impurity band in the junction. After cascading from one impurity state to a lower-energy impurity state, the electron can finally tunnel to the valence band of the p-type region. The electron can also make a "vertical" transition from the impurity state to the valence band.

At a forward bias V which uncrosses the conduction and impurity bands as shown in Fig. 8-12, it is possible to have two processes simultaneously: the previously discussed photon-assisted band-to-band tunneling (Sec. 8-B-2) and a photon-assisted tunneling to the impurity levels. The latter process causes a bump structure on the I–V characteristic, and the same structure appears in the curve of emission intensity vs. bias (Fig. 8-13). Since the energy of the deep levels with respect to the edge of the valence band is E_T, the emission spectrum should extend over the range $(qV - E_T - \xi_n - \xi_p)$ to qV. However, it is also possible to have a substantial traffic of carriers through deep centers via nonradiative transitions. This seems especially true in the case of deep centers introduced by fast-particle irradiation.[15] In this case the "excess current" increases with the irradiation dosage (Fig. 8-14), but no emission is observed. Deep centers introduced by radiation damage are usually sites of efficient nonradiative recombination.

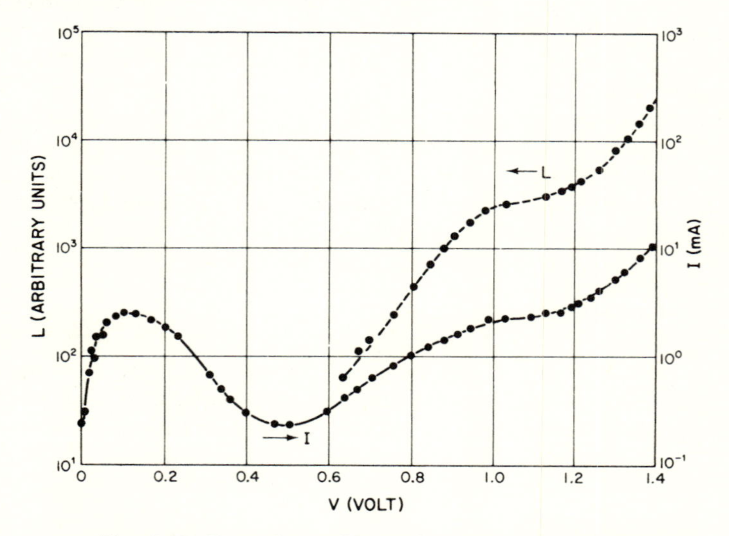

Fig. 8-13 Dependence of forward current I, and integrated light intensity L, on the voltage V, applied across a GaAs tunnel diode at 77°K.[14]

[14] A. I. Imenkov, M. M. Kozlov, S. S. Meskin, D. N. Nasledov, V. N. Ravitch, and B. V. Tsarenkov, *Fizika Tverdovo Tela* **7**, 634 (1965).

[15] A. G. Chynoweth, W. L. Feldmann, and R. A. Logan, *Phys. Rev.* **121**, 684 (1961).

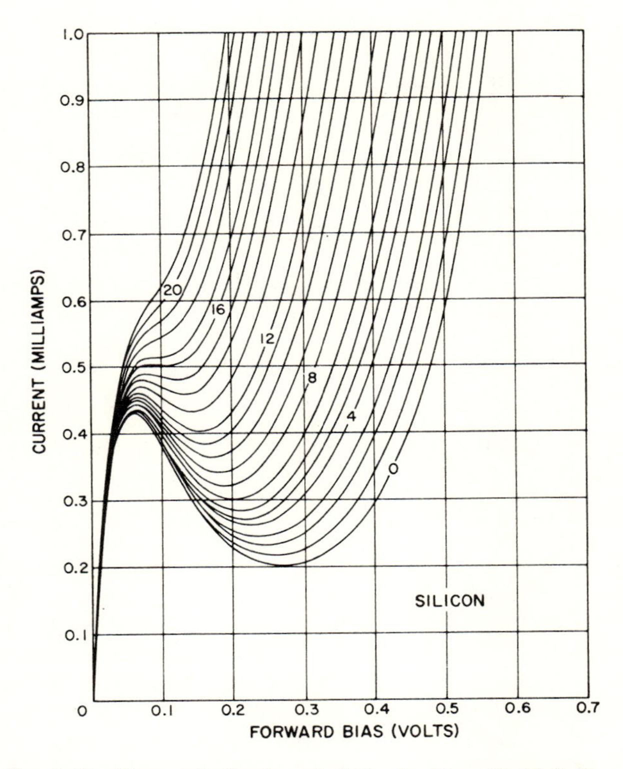

Fig. 8-14 Changes in the characteristic of a silicon Esaki diode, brought about by bombardment with 800-keV electrons. Curve 0 refers to the junction before bombardment and the series of curves up to curve 21 represent the junction characteristics at various stages of the bombardment.[15]

8-B-5 DONOR-TO-ACCEPTOR PHOTON-ASSISTED TUNNELING

We saw in Chapter 6 that in a number of semiconductors (e.g., GaAs, GaP) the dominant radiative transition at low temperatures is from donor to acceptor. In fact, controlled experiments with GaAs *p–n* junctions demonstrate that radiative recombination is more efficient in those junctions where there is a deliberate overlap of donors and acceptors.[16] In GaAs the net impurity binding energy is $E_D + E_A \approx 30$ to 50 meV depending on the doping

[16]N. N. Winogradoff and H. K. Kessler, *Solid State Communications* **2**, 119 (1964).

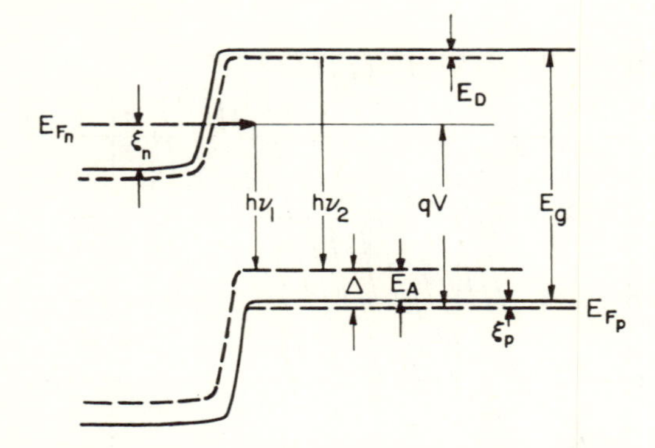

Fig. 8-15 Energy diagram for a *p–n* junction illustrating two modes of radiative transport.

level. Hence, as shown in Fig. 8-15 with a voltage V applied across the *p–n* junction, the emission peak corresponding to donor-to-acceptor transitions should occur at 0°K (and at other "low" temperatures) at a photon energy

$$h\nu_1 = qV - \Delta \tag{8-23}$$

where $\Delta \geq E_A$. Here we neglect two factors: (1) the coulomb-interaction term which allows emission of higher-energy photons for near pairs; and (2) transitions inside the depletion layer of the *p–n* junction, which allow the emission of photon energies as high as qV. The radiative recombination occurs mostly on the *p*-type side of the *p–n* junction; the highest probability for electrons tunneling out of the donors occurs at the quasi-Fermi level E_{F_n} (where the field is highest and the barrier is lowest); the highest density of acceptor final states occurs at E_A above the valence-band edge. Hence the emission peak should occur at

$$h\nu_1 = qV - E_A - \xi_p = qV - \Delta \tag{8-24}$$

Experimental observation at low temperatures usually confirms this relation. Note that one usually measures the voltage applied across a diode, V_a, and not the voltage V across the junction. The difference between V_a and V is the internal Ir-drop:

$$V = V_a - Ir \tag{8-25}$$

Substituting Eq. (8-25) into Eq. (8-23) gives

$$qV_a - h\nu_1 - \Delta = qIr \tag{8-26}$$

It is found that Δ is equal to 30 ± 5 meV for acceptor concentrations in the ranges 2.5×10^{18} to 3×10^{19} cm^{-3}.[17] Hence a plot of $qV_a - h\nu_1 - \Delta$ vs. the current through the diode should show a linear dependence, the slope of which is the internal resistance.

[17]J. I. Pankove, *J. Appl. Phys.* **35**, 1890 (1964).

Figure 8-16 demonstrates such a dependence. The rather high internal resistance ($r = 1$ ohm) exhibited by Fig. 8-16 agrees with a direct measurement of the diode's internal resistance.

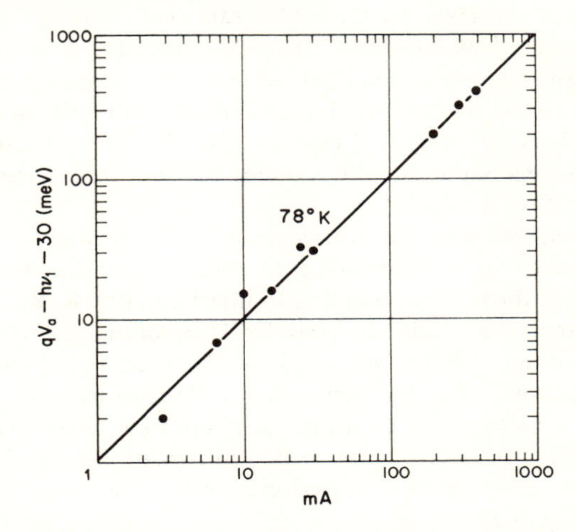

Fig. 8-16 Dependence of ($qV_a - h\nu_1$ − 30 meV) on current through the junction for the photon-assisted tunneling mode of transport.[17]

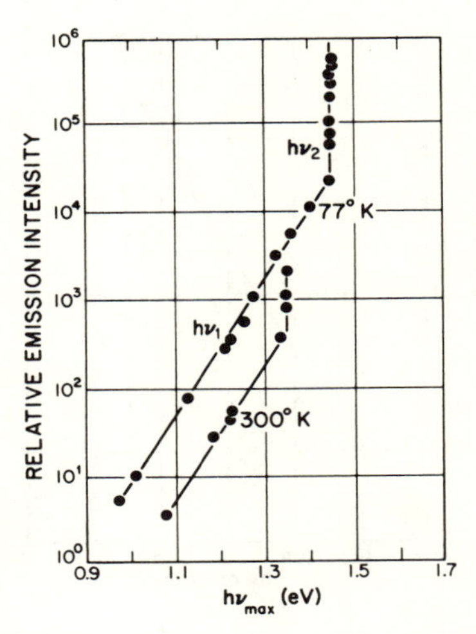

Fig. 8-17 Emission intensity vs. energy of the emission peak in GaAs at 77° and 300°K.[14] $h\nu_1$ is the shifting emission peak; $h\nu_2$ is the "stationary peak."

[14]A. I. Imenkov, M. M. Kozlov, S. S. Meskin, D. N. Nasledov, V. N. Ravitch, and B. V. Tsarenkov, *Fizika Tverdovo Tela* 7, 634 (1965).

As the temperature is increased, the electron distribution populates levels above the quasi-Fermi level E_{F_n}. The tunneling probability for electrons in these higher-energy states is higher than for those at E_{F_n} because the barrier height is lower at this level. Hence for the same bias, raising the temperature shifts the emission peak to higher energies, decreasing Δ in Eq. (8-23) and eventually causing the emission peak to occur at energies greater than qV. Of course, the above dependence [Eq. (8-23)] can hold only until the onset of injection at a bias $qV = E_g + \xi$ (see Sec. 8-B-3). Then "localized" donor–acceptor transitions can occur with emission of the corresponding nonshifting peak $h\nu_2$ (Figs. 8-15 and 8-17).

Here we shall digress with a discussion of the emission peak $h\nu_2$, which is observed under injection luminescence and, in some diodes, also at biases much lower than those corresponding to injection (Fig. 8-18). The peak $h\nu_2$, often referred to as the "stationary peak,"[18,19] has been attributed to an Auger effect.[19] Accordingly, energy $h\nu_1 < qV$ is absorbed by an electron in the conduction band near the junction; the hot electron migrates to the p-type region, where it recombines, emitting the photon $h\nu_2 > qV$. The hypothesis of simple injection of electrons far in the Boltzmann tail can be discounted, because the $h\nu_2$ emission is observed even when $(h\nu_2 - qV) > 1000kT$, as evident from Fig. 8-18.

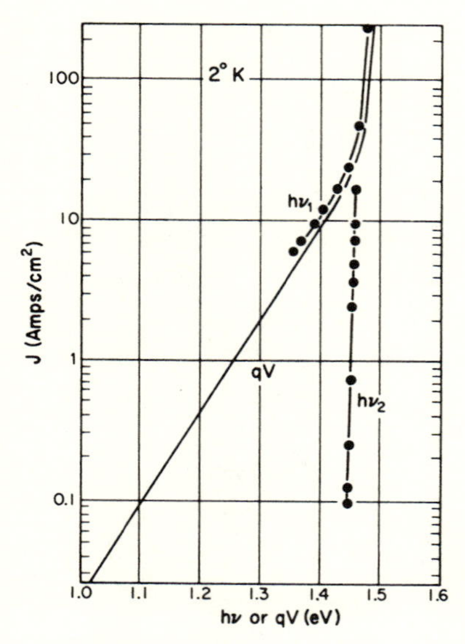

Fig. 8-18 Current density vs. voltage and photon energies $h\nu_1$ or $h\nu_2$ for a GaAs diode at 2°K.[19] See Fig. 8-15 for diagram of processes $h\nu_1$ and $h\nu_2$.

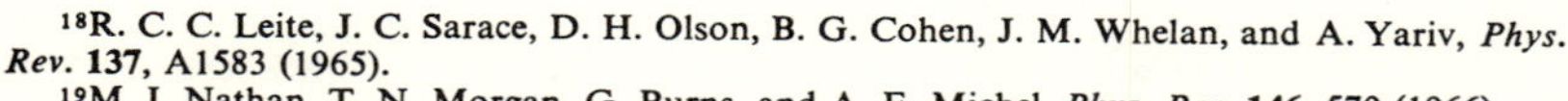

[18] R. C. C. Leite, J. C. Sarace, D. H. Olson, B. G. Cohen, J. M. Whelan, and A. Yariv, *Phys. Rev.* **137**, A1583 (1965).

[19] M. I. Nathan, T. N. Morgan, G. Burns, and A. E. Michel, *Phys. Rev.* **146**, 570 (1966).

It is interesting to note that the transition $h\nu_2$ terminates also at acceptors, giving the relation $h\nu_2 = E_g - \Delta$. In this case it is found that $\Delta = 34 \pm 5$ meV,[17] which agrees with the sum of $E_A = 30$ meV and $E_D = 5$ meV in GaAs.

Finally we must dwell on the dependence of the photon-assisted tunneling process on the impurity gradient. The shifting emission peak has the following exponential relationship to the current or to the emission intensity:

$$L \propto I \propto \exp \frac{h\nu_1}{E_0} \tag{8-27}$$

where E_0 is the slope of the semilogarithmic plot of $h\nu_1$ vs. I or L (Fig. 8-19); E_0 is related to the tunneling probability, which increases with the field at the junction. The field at the junction increases with the impurity gradient, as we saw in Sec. 8-A-3. The impurity gradient can be determined from capacitance vs. bias measurements. Figure 8-20 shows a correlation between E_0 and the impurity gradient.

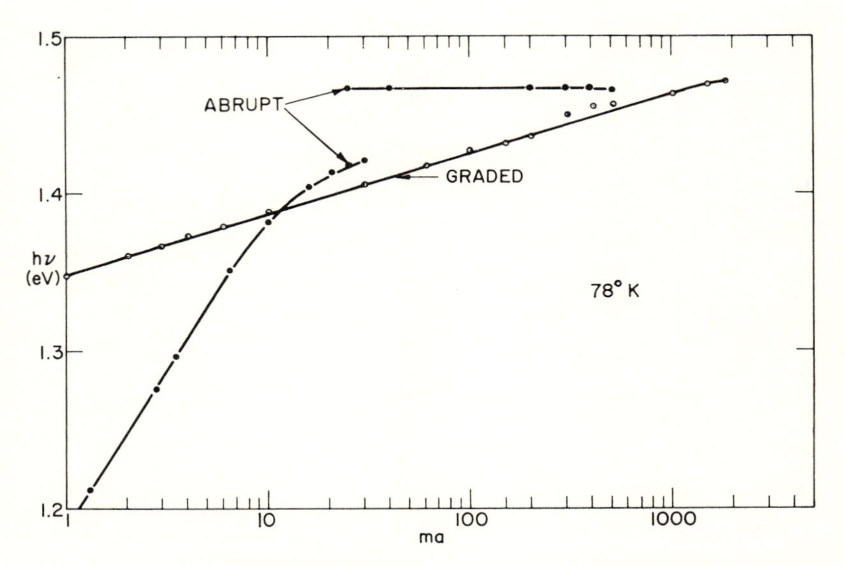

Fig. 8-19 Variation of emission peak with current through the junction for two diodes made of the same material. The abrupt junction was obtained by solution growth (alloying); the graded junction was obtained by a subsequent diffusion treatment of a similar solution-grown p–n junction.[17]

[17] J. I. Pankove, *J. Appl Phys.* **35**, 1890 (1964).

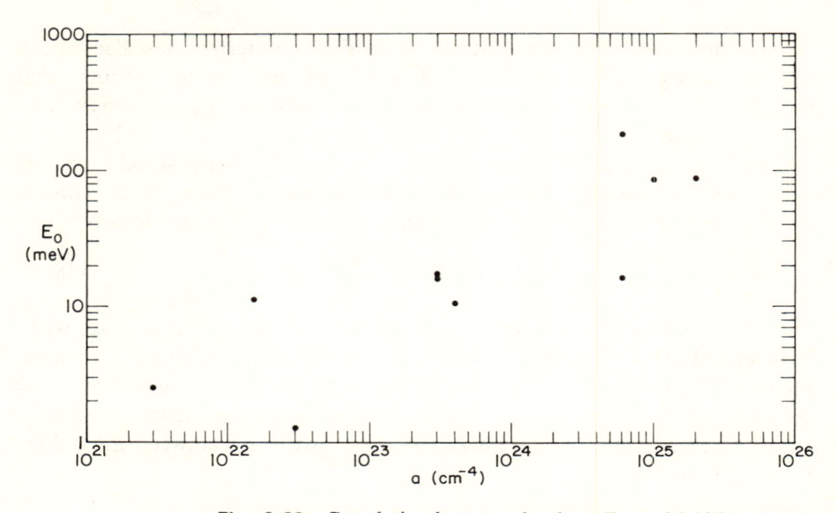

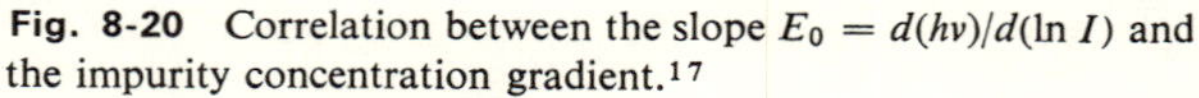
Fig. 8-20 Correlation between the slope $E_0 = d(h\nu)/d(\ln I)$ and the impurity concentration gradient.[17]

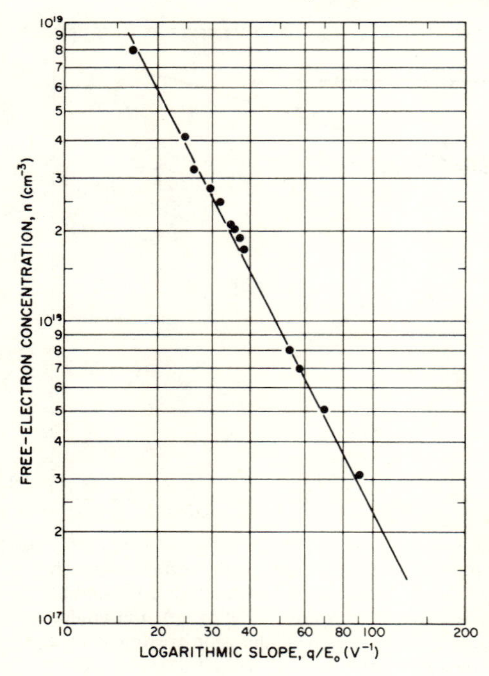

Fig. 8-21 Variation of the logarithmic slope constant as a function of the free-electron concentration. The experimental values are given by the small solid circles. The values of q/E_0 calculated from Morgan's theory[20] are represented by the solid line.[22]

[17] J. I. Pankove, *J. Appl. Phys.* **35**, 1890 (1964).

Although theoretical treatments of photon-assisted tunneling have been moderately successful by assuming a uniform field in the junction,[20,21] excellent agreement between theory and experiment is obtained when a parabolic depletion region is assumed.[22] For an abrupt junction, the parabolic depletion region, which is given by Eq. (8-9), depends only on the impurity concentration in the less heavily doped region. The coefficient E_0 can be calculated from first principles. Comparison with experimental data (Fig. 8-21) shows excellent agreement.

8-B-6 BAND FILLING

In this section we shall develop the concept of band filling[23] and then demonstrate that it is only a special case of photon-assisted tunneling.

In some *p–n* junctions the emission peak shifts at a slower rate when the transport process approaches the injection mode. There is still an exponential dependence between the current through the junction (or the emission intensity) and the energy of the emission peak:

$$I = I_0 \exp \frac{h\nu_2}{E_0} \tag{8-28}$$

Whereas in the photon-assisted tunneling mode the value of E_0 ranges typically between 25 and 200 meV (fast-shifting peak), in the present case there is a break in the dependence (8-28) as E_0 assumes a new value, typically in the range from 2 to 15 meV (slow-shifting peak). When injection sets in, $h\nu_2$ stops shifting. Sometimes there is a range of bias over which both fast and slow peaks appear simultaneously. As the bias is increased, the fast peak disappears into the low-energy edge of the slow peak.

In GaAs diodes this slow peak has been attributed to a band-filling mechanism, illustrated in Fig. 8-22.[18,24] According to the band-filling hypothesis, above some critical bias, electrons would either diffuse or tunnel from the *n*-type region to the tail of states of the conduction band in the *p*-type region. The subsequent recombination would be the result of "direct" tail-to-tail or tail-to-valence-band (or tail-to-acceptor) transitions. The filling of the tails of states, and not the tunneling probability, would then determine the shape of the emission spectrum. Thus with a distribution of tail states proportional to $\exp(E_{F_n}/E_0)$ (E_0 being a constant characterizing the exponential distribution of states and E_{F_n} being the energy up to which the tail is filled), the emission spectrum would have a low-energy edge the shape of which is proportional to $\exp(h\nu/E_0)$ (assuming a narrow set of terminal states—narrow

[20] T. N. Morgan, *Phys. Rev.* **148**, 890 (1966).
[21] A. E. Yunovich and A. B. Ormont, *JETP* (USSR) **51**, 1292 (1966).
[22] H. C. Casey and D. J. Silversmith, *J. Apply. Phys.* **40**, 241 (1969).
[23] J. I. Pankove, *Phys. Rev. Letters* **4**, 20 (1960).
[24] D. F. Nelson, M. Gershenzon, A. Ashkin, L. A. D'Asaro, and J. C. Sarace, *Appl. Phys. Letters* **2**, 182 (1962).

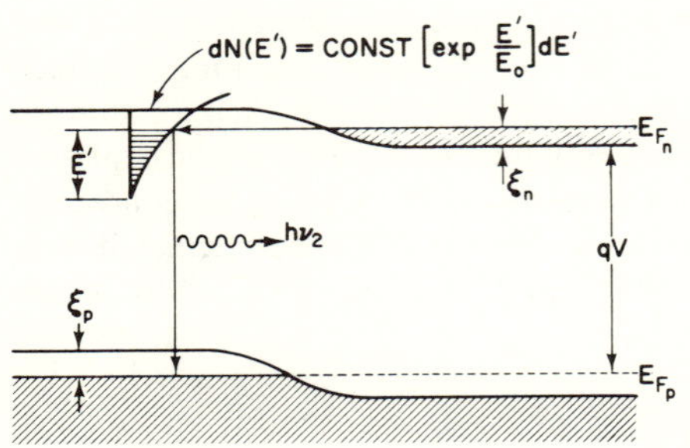

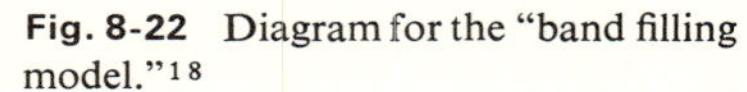
Fig. 8-22 Diagram for the "band filling model."[18]

compared to E_0). The position of the slowly shifting peak $h\nu_2$ would be determined by the position of the quasi-Fermi level for electrons:

$$h\nu_2 = E_{F_n} - E_f$$

where E_f is the energy of the densest set of final states, E_{F_p} (or E_A, as in the previous section).

The slope of $h\nu$ vs. log of current (or of light intensity) is equal to E_0. For several years this method of finding E_0 has been widely accepted as a means of determining the distribution of states in the tail.[18,24-28] However, we saw in Sec. 1-F that the origin of the tails can be ascribed to a perturbation of the band edges such that the energy gap remains locally constant. Then the bottom of the conduction band has a different energy at different positions. Therefore, tunneling from the Fermi level in the n-type region to various states in the tail of the conduction band in the p-type region involves different transition probabilities, depending on the distance of the state from the junction. Once all these states are filled, the radiative-recombination process is not a localized direct transition but rather a tunneling-assisted photon emission. Hence this is a complex process where localized states are filled by tunneling[29] and then emptied by photon-assisted tunneling. The distinction between this mode of radiative transport and the mode responsible for the fast-shifting peak is that the slow peak is due to photon-assisted tunneling in the bulk near the junction, whereas the fast peak is due to photon-assisted tunneling within the junction.

Each of these two processes should become observable at a different bias threshold, and each process should exhibit a different rate of shift with current (or emission intensity) and also a different emission spectrum.

[18] R. C. C. Leite, J. C. Sarace, D. H. Olson, B. G. Cohen, J. M. Whelan, and A. Yariv, *Phys. Rev.* **137**, A1583 (1965).

[25] G. C. Dousmanis, C. W. Mueller, and H. Nelson, *Appl. Phys. Letters* **3**, 133 (1963).

[26] R. J. Archer, R. C. C. Leite, A. Yariv, S. P. S. Porto, and J. M. Whelan, *Phys. Rev. Letters* **10**, 483 (1963).

[27] G. Lucovsky, *Proc. Conf. Phys. of Quantum Electronics*, McGraw-Hill (1966), p. 467.

[28] L. W. Aukerman and M. Millea, *J. Appl. Phys.* **36**, 2585 (1965).

[29] V. S. Bagaiev, Y. N. Berozashvili, L. V. Keldysh, A. P. Shotov, B. M. Vul, and E. I. Zavaritskaya, *Radiative Recombination in Semiconductors, Paris* (1964), Dunod (1965), p. 149.

Photon-assisted tunneling in the junction can occur as soon as the bias is applied, although at low voltage it is masked by direct tunneling and the emission occurs in the far infrared. Photon-assisted tunneling in the bulk is noticeable only when the number of bulk tail and donor states accessible to the quasi-Fermi level E_{F_n} exceeds the number of donors in the junction at E_{F_n}.

In both cases the emission spectrum should shift with the applied voltage because the emission peak is derived from E_{F_n} as the initial level for the radiative transition. However, the rates of shift with current or radiation intensity differ for the two processes. Tunneling in the junction couples fewer states, since the active volume is determined by the thickness of the depletion layers. With increasing bias, although the active volume decreases because the thickness decreases, the tunneling probability increases so that the radiative transport increases exponentially with voltage. On the other hand, tunneling in the bulk extends over a wider spatial range and, therefore, can couple a greater number of initial and final states. Hence the corresponding radiation intensity increases much faster with voltage (and with $h\nu_2$).

The emission spectra of the two processes may also be different, especially in the low-energy edge. We saw that with increasing voltage photon-assisted tunneling in the junction can shift the cutoff of the low-energy edge to higher energies so that a family of emission spectra with increasing bias would give a set of intersecting low-energy edges [Figs. 8-9 and 8-10(a)]. However, when band tailing is extensive, the low-energy cutoff is less noticeable. Photon-assisted tunneling in the bulk, on the other hand, for a given photon energy, always couples the same set of states. Therefore, in the low-energy edge, the emission intensity at a given photon energy does not change with bias. Hence a family of spectra with increasing bias exhibits a saturated low-energy edge (Fig. 8-23).

Note that in a very gradual junction, such as is obtained by a diffusion treatment, the active region can be very wide and, therefore, exhibit a behavior very much like bulk photon-assisted tunneling.

Experiments with abrupt and with graded junctions made of the same material show a fast-shifting peak in the abrupt junction, and first a fast peak and then a slow peak in the graded junction.[30] Theoretical calculations[30,31] have shown that photon-assisted tunneling accounts for both types of shifting peaks and for the corresponding emission spectra.

There is another important argument against true band filling: since emission occurs in the *p*-type region, the thickness of the active region should extend on the order of an electron diffusion length, i.e., about one micron. An active region of one micron can, in fact, be obtained, but only in the injection mode, when the emission peak stops shifting. However, while the emission peak shifts, the width of the active region is less than 600 Å.[30]

[30] H. C. Casey and D. J. Silversmith, *J. Appl. Phys.* **40**, 241 (1969).
[31] T. N. Morgan, *Phys. Rev.* **148**, 890 (1966).

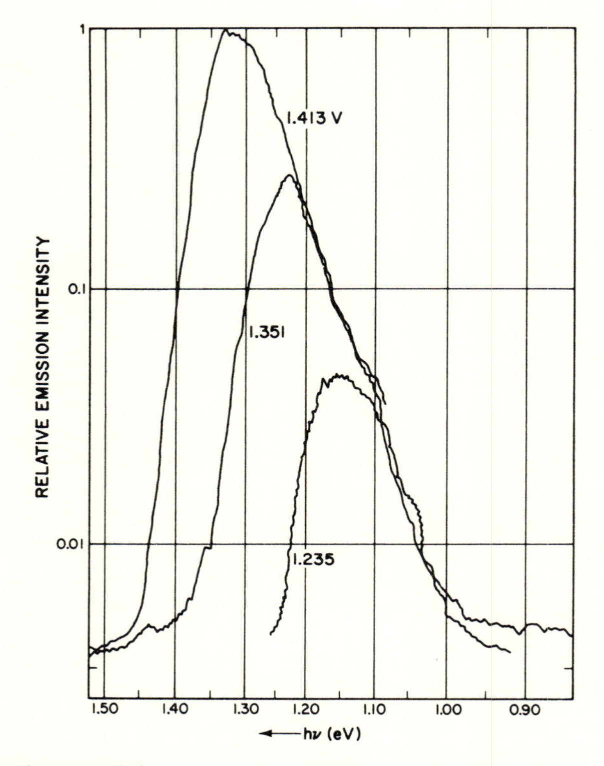

Fig. 8-23 Emission spectra of InP diode at 77°K for three different biases.[21]

To summarize this section, the "band-filling" process consists of photon-assisted tunneling not in the junction proper, but rather in the bulk within about 1000 Å from the junction.

8-B-7 INJECTION LUMINESCENCE IN LIGHTLY DOPED JUNCTIONS

In all the cases we have treated above, the junction was formed between heavily doped *p*-type and *n*-type regions of a semiconductor. Lightly doped materials are seldom used for making light-emitting diodes because the high resistivity of the material leads to large I^2r-losses and generates heating problems. It is possible, however, to make thin layers of pure epitaxial GaAs on

[21]A. E. Yunovich and A. B. Ormont, *JETP* (USSR) **51**, 1292 (1966).

a heavily doped substrate which provides a sturdy low resistance support.

Radiative recombination has been studied under forward bias in thin regions where the donor concentration is $N_d \leq 5 \times 10^{16}$ cm^{-3}.[32] These are abrupt junctions which emit three peaks at, respectively, 1.35, 1.48, and 1.505 eV at 78°K. The 1.35-eV peak is attributed to recombination at a deep level; the 1.48-eV peak corresponds to the donor-to-acceptor transition; the 1.505-eV peak is attributed to the radiative decay of bound excitons. It is observed only at large forward bias where the junction field is considerably reduced, allowing the formation of bound excitons. These excitons have a short radiative lifetime.

8-B-8 OPTICAL REFRIGERATION[33,34]

In Sec. 8-B-6 we saw that it is possible to obtain radiation with a photon distribution that peaks at an energy substantially larger than the energy qV corresponding to the voltage applied across the junction. In a number of diodes, only one emission peak has been reported, this peak occurring at $h\nu > qV$ over a large range of bias, as shown in Fig. 8-24.

Although the efficiency of this emission process is not well documented, it is interesting to speculate on the possibility that the energy difference

$$\Delta = h\nu - qV \tag{8-29}$$

comes from the lattice heat rather than from an Auger effect.

Let Q be the heat removed from the lattice by a recombining electron–hole pair which emits a photon of energy $h\nu$ while the battery supplies an energy qV. Therefore,

$$h\nu \leq qV + Q \tag{8-30}$$

The equality sign applies when the heat removal is a 100% efficient process; then, $\Delta = Q$.

The quantity Q can be estimated from the change in entropy. When the carriers absorb a heat Q from the lattice at a temperature T, their entropy changes by $-Q/T$. The photon-emission process, on the other hand, increases the entropy by $h\nu/T^*$, where T^* is the effective temperature of the radiation field. Then,

$$Q = T\frac{h\nu}{T^*} \tag{8-31}$$

In other words, Q transfers energy from a reservoir at a temperature T to another reservoir at a temperature T^*, from the lattice to the radiation field.

[32] D. K. Wilson, *Appl. Phys. Letters* **3**, 127 (1963).
[33] R. T. Keyes and T. M. Quist, *Proc. IRE* **50**, 1822 (1962).
[34] G. C. Dousmanis, C. W. Mueller, H. Nelson, and K. G. Petzinger, *Phys. Rev.* **133**, A316, (1964). In this section we shall follow Dousmanis's derivation.

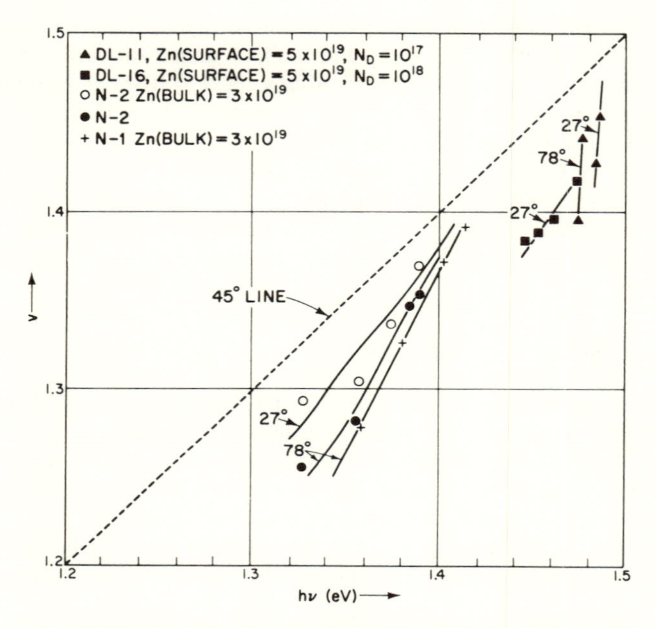

Fig. 8-24 Applied voltage vs. $h\nu$ in different GaAs diodes at 27° and 78°K. The DL diodes were fabricated by diffusion, the N diodes were grown epitaxially on a p-type substrate.[34]

The temperature T^* is related to the number of photons $\rho(h\nu)$ in an electromagnetic mode by

$$\rho(h\nu) = \frac{1}{\exp\left(\dfrac{h\nu}{kT^*}\right) - 1}$$

Hence

$$\frac{h\nu}{kT^*} = \ln\left(1 + \frac{1}{\rho(h\nu)}\right) \approx \ln\frac{1}{\rho(h\nu)} \tag{8-32}$$

since $\rho(h\nu) \ll 1$ for spontaneous emission, as we shall see in the next chapter. Inserting Eq. (8-32) into (8-31) gives

$$Q = kT \ln\frac{1}{\rho(h\nu)} \tag{8-33}$$

At low currents, $\rho(h\nu)$ is small and Q is largest; as the current increases, $\rho(h\nu)$ grows and Q tends to zero. According to Eq. (8-31), as Q tends to zero, T decreases while T^* increases in response to this heat-pumping process.

Now, to evaluate the efficiency of the optical-refrigeration process, we must relate Q to the electrical-power input. For efficient diodes the emission intensity is proportional to the current I:

$$\int_0^\infty \rho(h\nu)\, d\nu = \eta I \tag{8-34}$$

where η is the quantum efficiency in photons per electron.

Since the shape of the emission spectrum does not change over a large range of currents, Eq. (8-34) can be restated as

$$\rho(h\nu) = a\eta I \tag{8-35}$$

where a is a constant. Then Eq. (8-33) can be rewritten as

$$\frac{dQ}{dI} = -\frac{kT}{I} - \frac{kT}{\eta}\frac{d\eta}{dI} \tag{8-36}$$

In the diodes of Fig. 8-25, the emission intensity was found to be proportional to the current and the efficiency appeared high, a circumstance under which the efficiency is usually independent of the current. Hence the last term in Eq. (8-36) can be neglected.

Recall from Eqs. (8-27) and (8-28) that there is an exponential relationship between the emission peak and the current:

$$I = I_0 \exp \frac{h\nu}{E_0} \tag{8-37}$$

Substituting Eq. (8-37) into (8-35) and then the resulting expression for $\rho(h\nu)$ into Eq. (8-33) gives

$$Q = kT\left[\ln\left(\frac{1}{a\eta I_0}\right) - \frac{h\nu}{E_0}\right] \tag{8-38}$$

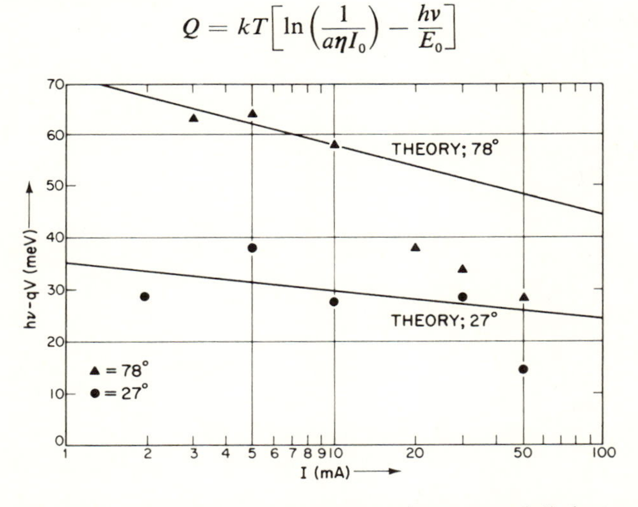

Fig. 8-25 $(h\nu - qV)$ vs. current averaged over several diodes at 27° and 78°K. The two straight lines are theoretical curves. The value of $a\eta$ of (8.35) has been adjusted for best fit to the data.[34]

Hence substituting Eq. (8-38) into (8-30) yields, for the equality limit,

$$V = \frac{kT}{q}\left[h\nu\left(\frac{1}{kT} + \frac{1}{E_0}\right) + \ln\,(a\eta I_0)\right] \tag{8-39}$$

which predicts a linear dependence between V and $h\nu$; and since

$$\frac{dV}{d(h\nu)} = \frac{1}{q}\left(1 + \frac{kT}{E_0}\right) \tag{8-40}$$

a greater slope of $V(h\nu)$ is expected at higher T and at lower E_0 (the latter case corresponds to graded junctions). The data of Fig. 8-24 are in accord with these conclusions.

The efficiency $\ni$ of the refrigerator is the ratio of heat extracted by all the electrons to the electrical energy supplied to the diode:

$$\ni = \frac{(|Q|\,I)/q}{IV} = \frac{|Q|}{qV} \tag{8-41}$$

Substituting Eqs. (8-38) and (8-39) into Eqs. (8-41) gives

$$\ni = \frac{\ln\,(a\eta I_0) + \dfrac{h\nu}{E_0}}{\ln\,(a\eta I_0) + \dfrac{h\nu}{E_0} + \dfrac{h\nu}{kT}} \tag{8-42}$$

Equation (8-42) expresses the fact that the refrigeration efficiency decreases as the photon energy increases and as the temperature decreases.

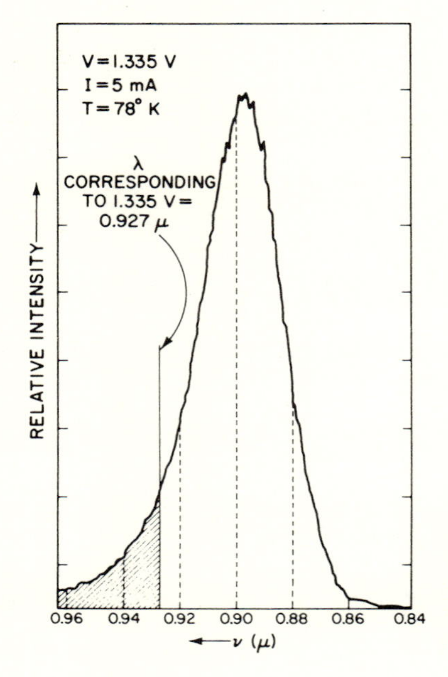

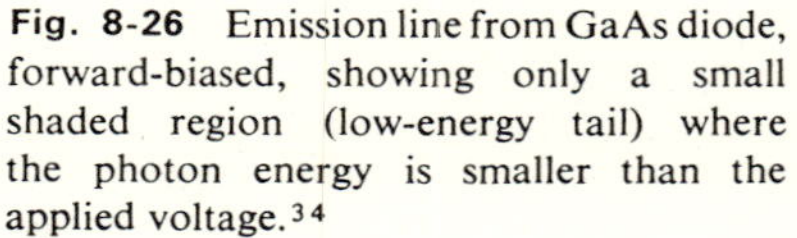

Fig. 8-26 Emission line from GaAs diode, forward-biased, showing only a small shaded region (low-energy tail) where the photon energy is smaller than the applied voltage.[34]

By way of a quantitative example, consider the emission spectrum of Fig. 8-26. There, $h\nu - qV = 60$ meV. Hence $Q/h\nu$ is about 0.03. In other words, 3% of the radiated energy is derived from lattice heat. If the emission efficiency η were low, a fraction $1 - \eta$ of the electrical energy would be generating heat and no net cooling might result. However, if all the electrical energy supplied by the battery is radiated and forms 97% of the emitted energy, then refrigeration is obtained. On the other hand, if the quantum efficiency is $\eta = 0.99$, 1% of the electrical energy from the battery generates heat; hence only 2% of the emitted energy effectively extracts heat from the lattice. In the diode of Fig. 8-26, the I^2r-loss is equal to 2% of the emitted energy at a current of 30 mA. Therefore, effective refrigeration occurs only at currents lower than 20 mA. At 10 mA, the cooling rate is 3×10^{-4} W, while the I^2r-loss is 5×10^{-5} W.

8-C Heterojunctions

When the n-type and p-type regions on either side of the junction consist of different semiconductors, the transition is called a heterojunction. A fine distinction can be made when the two semiconductors are miscible and the transition is gradual. The latter transition is sometimes called "quasi-homojunction.[35] Thus a gradual transition between n-type ZnSe and p-type ZnTe, or between p-type GaAs and n-type GaP is a quasi-homojunction. In this section we shall use the more general term heterojunction to include both the abrupt and the graded transitions.

One purpose for resorting to heterojunctions is to obtain a high injection efficiency of the minority carrier into the lower-gap semiconductor.[36] This property of heterojunctions can be readily understood from a consideration of Fig. 8-27: when the forward bias flattens the valence band edge, holes are injected into the n-type region. However, the barrier $\Delta E = E_{g_1} - E_{g_2}$ blocks the injection of electrons from the n-type to the p-type material.

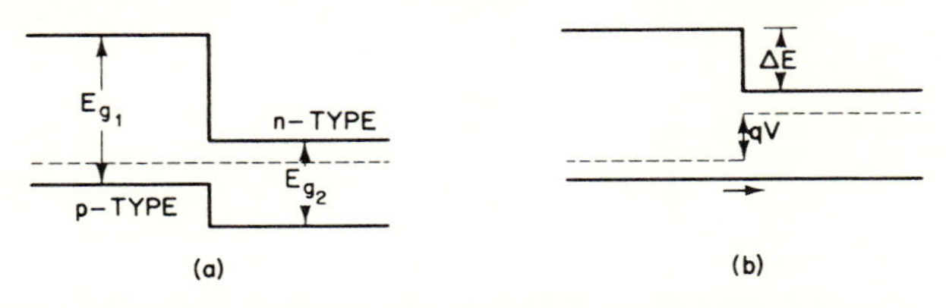

Fig. 8-27 Idealized band structure of heterojunction (a) at equilibrium, (b) with an applied forward bias V.

[35] A. G. Fisher, "Electroluminescence in II–VI Compounds," *Luminescence of Inorganic Solids*, ed. P. Goldberg, Academic Press (1966), p. 541.

[36] H. Kroemer, *Proc. IRE* **45**, 1535 (1957).

Obviously, radiative recombination will then occur in the lower-gap region. Thus in a GaAs–GaSb heterojunction, injection luminescence occurs at about 0.7 eV,[37] near the gap energy of GaSb. The larger-gap material, in general, is transparent to the radiation generated in the lower-gap material and, therefore, is sometimes used as a window for transmitting the radiation.

However, in practice, heterojunctions suffer from an interfacial difficulty: the Fermi level is pinned at the interface by surface states.[38] Hence instead of flattening one band edge, as in Fig. 8-27, there is usually a Schottky barrier, as in Fig. 8-28.

Because the Schottky barrier can rectify, its presence is evident in the

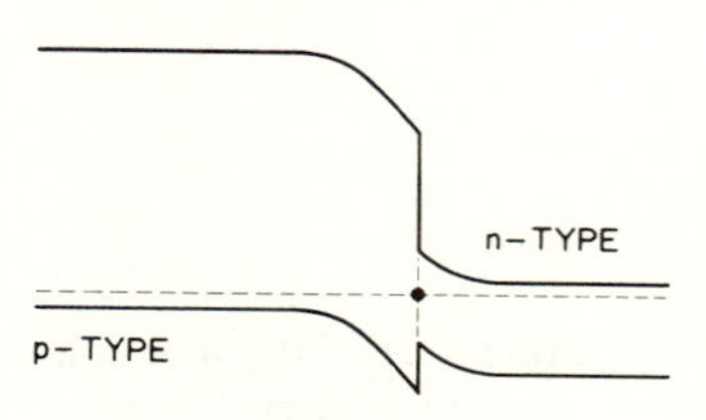

Fig. 8-28 The change in electric field (slope of the band edges) at the interface arises from interfacial charges on interface states produced by the lattice mismatch. In this case, a Schottky barrier for holes is obtained.

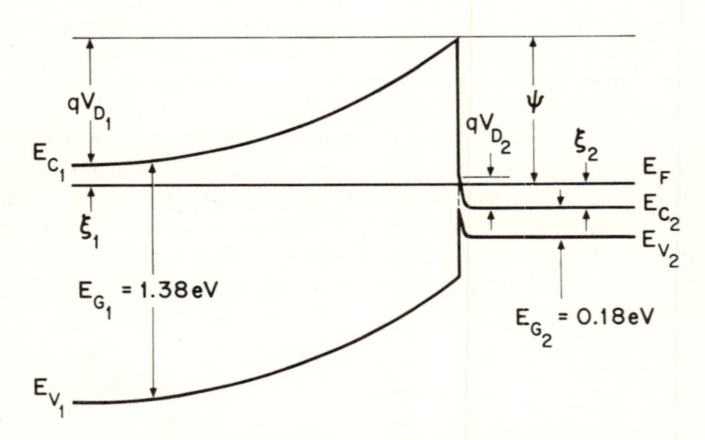

Fig. 8-29 Energy band profile of a GaAs–InSb *n–n* heterotransition. For different purity materials the ξ values will change as well as the depletion layer widths.[39]

[37] R. H. Rediker, S. Stopek, and J. H. R. Ward, *Solid State Electronics* **7**, 621 (1964).
[38] J. Bardeen, *Phys. Rev.* **71**, 717 (1947).
[39] E. D. Hinkley, R. H. Rediker, and D. K. Jadus, *Appl. Phys. Letters* **6**, 144 (1965).

n–n heterostructure (a transition between two different semiconductors which are both *n*-type, as in Fig. 8-29). Thus a GaAs–InSb *n–n* heterostructure can give a rectification ratio of 10^8 at 0.5 V.[39]

Diode theory predicts[39] a forward-current density j:

$$j = A^* T^2 e^{-\psi/nkT}(e^{qV/nkT} - 1) \tag{8-43}$$

where $A^* = A(m^*/m) = 120 \times 0.072\ \mathrm{A \cdot cm^{-2} \cdot {}^\circ K^{-2}}$ is the effective Richardson constant for GaAs; ψ is the metal-to-semiconductor barrier, which is approximately equal to 0.8 eV; and $n \approx 1$.

The reverse current in a Schottky barrier does not saturate as it would in a lightly or moderately doped *p–n* junction. The image force lowers the Schottky barrier and causes the reverse-bias current density j_r to rise as[39]

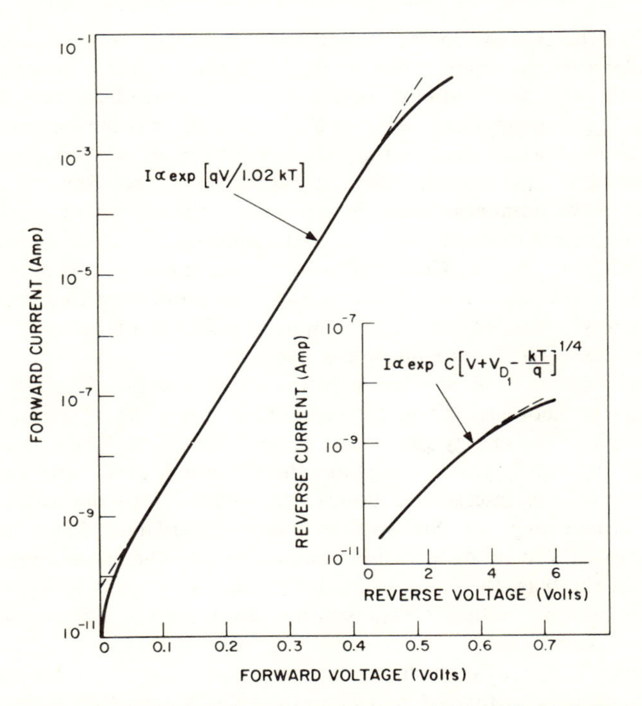

Fig. 8-30 Room-temperature *I–V* characteristics of GaAs–InSb *n–n* heterotransition. Net donor densities are $3.9 \times 10^{15}\ \mathrm{cm^{-3}}$ and $1.0 \times 10^{18}\ \mathrm{cm^{-3}}$ for the GaAs and InSb, respectively. The solid lines represent *x–y* recorder data. In the forward characteristic the dashed lines are extrapolations of $I = I_0 \exp(qV/1.02kT)$. In the reverse characteristic (insert) the dashed line is an extrapolation of $I = I_0 \exp[C(V + V_{D_1} - kT/q)^{1/4}]$. Junction area is $5.1 \times 10^{-3}\ \mathrm{cm^{-2}}$.[39]

$$j_r = B \exp\left\{C\left[V + V_{D_1} - \frac{kT}{q}\right]^{1/4}\right\} \qquad (8\text{-}44)$$

where V_{D_1} is the GaAs diffusion potential. Both the forward- and the reverse-bias experimental data seem to agree with theory, as shown by Fig. 8-30.

The above two equations apply to heterotransitions between two *n*-type or two *p*-type regions. In *p–n* heterojunctions, on the other hand, the temperature independence of the *I–V* characteristic indicates that carrier transport occurs by tunneling through the Schottky barrier. The tunneling probability for electrons from a wide-gap *n*-type material to a narrow-gap *p*-type material is given by[40]

$$\mathscr{T} \approx \exp\left\{-\frac{2}{\hbar}\left(\frac{\epsilon m^*}{qN_d}\right)^{1/2} E_{b(\max)}\right\} \exp\left\{\frac{2}{\hbar}\left(\frac{\epsilon m^*}{qN_d}\right)^{1/2} \alpha V\right\} \qquad (8\text{-}45)$$

where m^* is the effective mass; N_d is the net impurity concentration in the wide-bandgap *n*-type semiconductor; $E_{b(\max)}$ is the maximum height of the barrier with respect to the edge of the conduction band for zero applied voltage [$E_{b(\max)}$ corresponds to qV_{D_1} of Fig. 8-29]; and α is the fraction of the applied voltage V which changes the potential of the wide-gap semiconductor. The tunneling current is the product of the tunneling probability $\mathscr{T}$ of Eq. (8-45) and of the number of electrons arriving at the barrier per second.

Heterojunctions made between CdS and a wider-gap *p*-type semiconductor are of particular interest; CdS is always *n*-type, and *p–n* junctions in this material have not been made to date in spite of more than two decades of effort by many research teams. The compound CdS has a large direct-energy gap ($E_g \approx 2.5$ eV) and can emit blue–green light.

Heterojunctions have been made between CdS and SiC. The latter is an indirect-gap semiconductor which can be made *n*-type or *p*-type at will by suitable doping. The energy gap of SiC varies from 2.7 to 3.3 eV, depending on the polytype; polytypism designates the rotational periodicity by which successive layers of interatomic bonds are ordered.[41] In one case, *n*-type CdS was grown on *p*-type SiC so that under forward bias, holes could be injected into CdS to produce visible luminescence.[42] The emission spectrum shifted with the bias, the color of the light changing gradually from red to green. This behavior indicates a transition to deep levels or a photon-assisted tunneling process.

The compound Cu_2S which is *p*-type has also a larger energy gap than CdS. Heterojunctions formed by evaporating Cu_2S onto CdS have yielded red injection luminescence the intensity of which varied linearly with the current.[43] This process was presumably a recombination at deep centers.

[40] R. H. Rediker, S. Stopek, and E. D. Hinkley, *Trans. Metal. Soc., AIME* **233**, 463 (1965).

[41] H. Jagodzinski and H. Arnold, "Anomalous SiC Structures," *Silicon Carbide*, ed. J. R. O'Connor and J. Smiltens, Pergamon (1960), p. 136.

[42] E. A. Salkov, *Fizika Tverdovo Tela* **7**, 289 (1965).

[43] P. N. Keating, *J. Phys. Chem. Solids* **24**, 1101 (1963).

8-D Reverse-bias Processes

8-D-1 SATURATION CURRENT AND PHOTOCONDUCTIVITY

We saw in Sec. 8-B-1 that a reverse bias applied to a junction between heavily doped n- and p-type regions permits the tunneling of electrons from the p-type side to the empty conduction-band states in the n-type region. If, on the other hand, one or both regions on either side of the junction are not heavily doped, the depletion region is too wide to allow tunneling when a small reverse-bias voltage is applied. Then, the only contribution to the current across the junction is the migration of minority carriers toward the junction. Holes are thermally generated in the n-type region at a rate p_n/τ_h per unit volume, where p_n is the hole concentration in the n-type region and τ_h is the hole lifetime. Similarly, the rate of thermal generation of electrons in the p-type region is n_p/τ_e. The holes will diffuse toward the junction from a distance of about L_h, the hole diffusion length, so that the hole-current density is $qL_h p_n/\tau_h$ (see Fig. 8-31). Similarly, the electron-current density contributed by the p-type region is given by $qL_n n_p/\tau_e$. We can replace L/τ by D/L where D_h is the diffusion constant for holes, which is related to L_h and τ_h by $L_h = \sqrt{D_h \tau_h}$ (and similarly, for electrons, $L_e = \sqrt{D_e \tau_e}$). Then the total current density flowing through the p–n junction is

$$J_s = q\left[\frac{p_n D_h}{L_h} + \frac{n_p D_e}{L_e}\right] \tag{8-46}$$

Ideally, over some range of applied voltage the reverse bias affects only the extent of the depletion region. In many semiconductors the depletion region is much smaller than either diffusion length. Since only minority carriers within a diffusion length from the edge of the depletion region contribute appreciably to the current, if the concentration of minority carriers is uni-

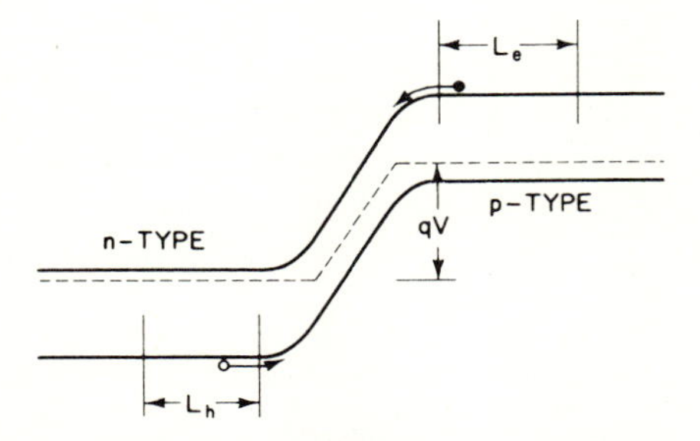

Fig. 8-31 Carrier flow in a reverse-biased p–n junction.

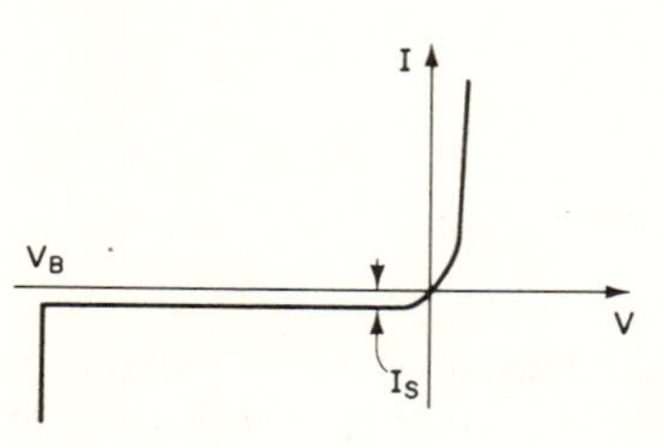

Fig. 8-32 I–V characteristic of an ideal p–n junction.

form in each region, changing the bias does not change the current. Therefore, the reverse current is saturated, as shown in Fig. 8-32, and remains constant until breakdown at V_B.

In practice, the current is not quite constant, but increases slightly with bias. The slight increase in reverse current—the "leakage" current—is due mainly to two contributions: (1) the surface generation of minority carriers, which occurs where the junction intersects the surface of the semiconductor; and (2) the generation of minority carriers inside the depletion layer.[44]

It is possible to increase the concentration of minority carriers by optical generation of electron–hole pairs. If the optical excitation is produced within a diffusion length of the junction, the saturation current is increased. This mechanism is called "photoconductivity." Photoconductivity is the change in resistance due to optical generation of extra carriers in a semiconductor or in an insulator. This effect is not unique to p–n junctions; however, it becomes measurable only in the presence of a driving electric field. In a bulk photoconductor, a material which contains no junction, the transport properties are strongly governed by carrier-trapping phenomena; whereas in a p–n junction the internal field is usually sufficient to empty the traps. Traps will be discussed in Chapter 17.

We shall not dwell on photoconductivity, since several excellent books have already been written on this subject.[45] Photoconductivity is a combination of optical excitation and transport phenomena. The optically generated carriers are moved by an applied electric field. Carrier generation has already been discussed under "absorption" in Chapter 3. The transport of the carriers is complicated by a variety of scattering processes and trapping effects. However, since light can also excite carriers out of traps, the simultaneous excitation of a semiconductor with photons of different energies can result in a variety of effects: instabilities, oscillations, photoresistance, delayed conductance, etc.

[44]C. T. Sah, R. N. Noyce, and W. Shockley, *Proc. IRE* **43**, 1228 (1957).

[45]A. Rose, *Concepts in Photoconductivity and Allied Problems*, Wiley (1963); S. M. Ryvkin, *Photoelectric Effects in Semiconductors*, Consultants Bureau (1964); R. H. Bube, *Photoconductivity of Solids*, Wiley (1960).

8-D-2 ZENER BREAKDOWN

In a strong electric field, electrons can tunnel through the energy gap.[46] The onset of this tunnel current is called "Zener breakdown." We have seen that in a heavily doped semiconductor a strong electric field exists in the *p–n* junction, and that a small forward or reverse bias applied across the junction to cause a deviation from equilibrium permits electrons to tunnel across the junction. This transport across the junction between heavily doped regions of a semiconductor is, therefore, Zener tunneling. For a theoretical treatment of this process, the reader is referred to references 47 and 48.

In junctions between more lightly doped regions of a semiconductor, the equilibrium electric field inside the junction is too low to allow Zener tunneling. However, as the reverse bias V is increased, the thickness of the depletion layer increases as $V^{1/2}$ or $V^{1/3}$ for abrupt and graded junctions, respectively. Therefore, the corresponding field inside the junction increases as $V^{1/2}$ or $V^{2/3}$. Eventually, the internal field is large enough (10^7 V/cm)[49] to permit Zener tunneling. Then, the current increases suddenly and grows very abruptly with increased bias. This is shown at V_B in Fig. 8-32. In abrupt junctions between heavily doped and moderately doped regions, the Zener breakdown is usually observed below about 5 volts.

Zener breakdown is so reliable a process, V_B is so stable, and the change in impedance is so large, that this phenomenon is commonly used in "Zener diodes" to establish a voltage reference which is stable with changing currents. Zener diodes are also placed across delicate circuits to prevent damage due to high excursions of voltage, such as lightning surges. It must be pointed out that many commercial so-called "Zener diodes" operate on the avalanche-breakdown mode, to be discussed next.

8-D-3 AVALANCHE BREAKDOWN

Although Zener breakdown would be the ultimate cause for the sudden increase in reverse current as the reverse bias is increased, another mechanism, the avalanche breadkown,[50] usually occurs first in the more lightly doped diodes.

8-D-3-a The Mechanism of Avalanche Breakdown

When a minority carrier—e.g., an electron—which contributes to the saturation current is accelerated by the electric field in the junction to a kinetic

[46] C. Zener, *Proc. Roy. Soc.* **A145**, 523 (1934).
[47] L. V. Keldysh, *Sov. Phys. JETP* **6**, 763 (1958).
[48] E. O. Kane, *J. Phys. Chem. Solids* **12**, 181 (1959).
[49] A. G. Chynoweth, W. L. Feldmann, C. A. Lee, R. A. Logan, G. L. Pearson, and P. Aigrain, *Phys. Rev.* **118**, 425 (1960).
[50] K. G. McKay, *Phys. Rev.* **94**, 877 (1954).

energy equal to or greater than $\frac{3}{2}E_g$, it can transfer some of its energy to a valence-band electron and ionize it into the conduction band. In this event, an electron–hole pair is generated while the primary electron thermalizes toward the bottom of the conduction band. Now, there are two electrons and one hole which can be heated by the field in the junction. When these in turn acquire a kinetic energy of about $\frac{3}{2}E_g$, they can each create further pairs, and the process repeats itself many times in an avalanching fashion. Several stages of this breakdown process are shown in Fig. 8-33.

The avalanche breakdown is often localized at a few spots called "microplasmas," which can be visible because of the radiative recombination which occurs whenever electron–hole pairs are generated. Each microplasma operates intermittently, carrying current in the form of pulses.[50] There is much evidence that microplasmas form at sites of dislocations, precipitates, or inhomogeneities within the depletion layer.[51] By way of example, a precipitate of SiO_2 in Silicon forms a region of lower dielectric constant; the presence of such a precipitate in the space-charge region causes a local enhancement of the electric field around the precipitate.[52]

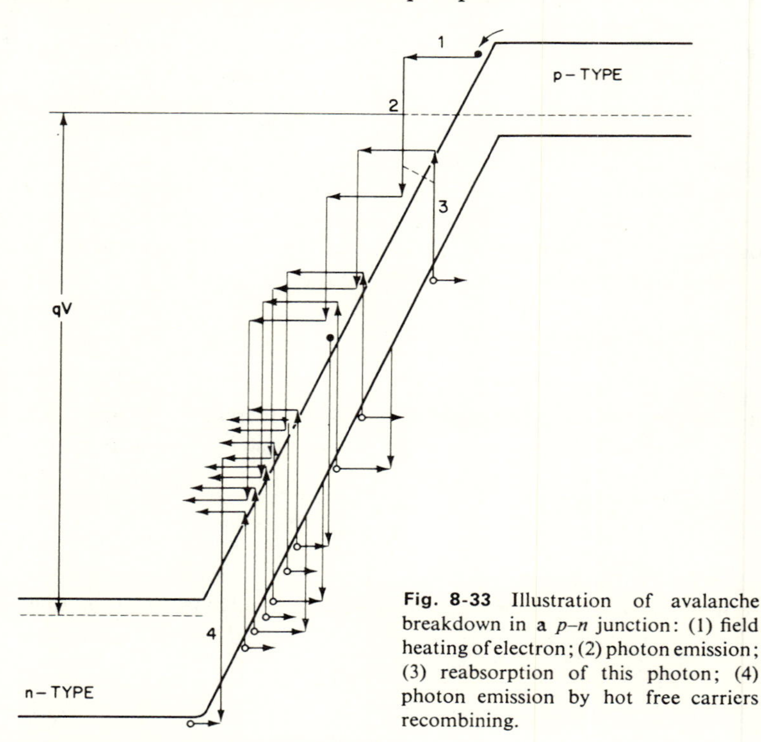

Fig. 8-33 Illustration of avalanche breakdown in a *p–n* junction: (1) field heating of electron; (2) photon emission; (3) reabsorption of this photon; (4) photon emission by hot free carriers recombining.

[51] H. Kressel, *RCA Review* **28**, 175 (1967) (a survey).
[52] W. Shockley, *Solid State Electronics* **2**, 35 (1961).

Avalanche breakdown occurs in random bursts, and the summation of these bursts results in an intense broad-spectrum noise.[53]

The avalanche breakdown is characterized by an I–V characteristic in which the current is proportional to some power of the voltage, this power ranging between 3 and 6. The I–V characteristic curve exhibits a more gradual break than in the case of the Zener breakdown. The reverse current I_R is given by the empirical relation

$$I_R = MI_s$$

where I_s is the saturation current and M is the multiplication factor:[54]

$$M = \frac{1}{1 - (V_R/V_B)^n} \tag{8-47}$$

where V_R is the reverse-bias voltage, V_B is the breakdown voltage at the junction (when $M \to \infty$), and n ranges from 3 to 6. The dependence of V_B on the carrier concentration in the most lightly doped region of the junction is shown in Fig. 8-34. Its dependence on the impurity gradient in the junction is shown in Fig. 8-35.

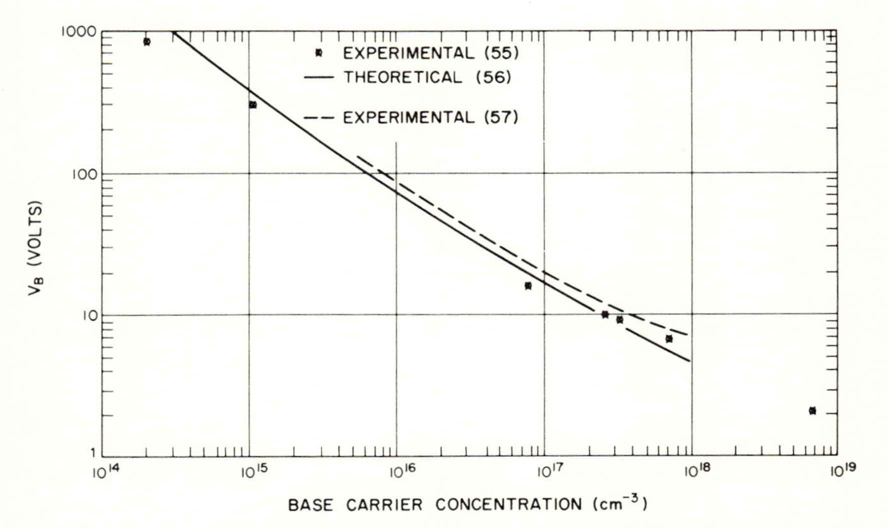

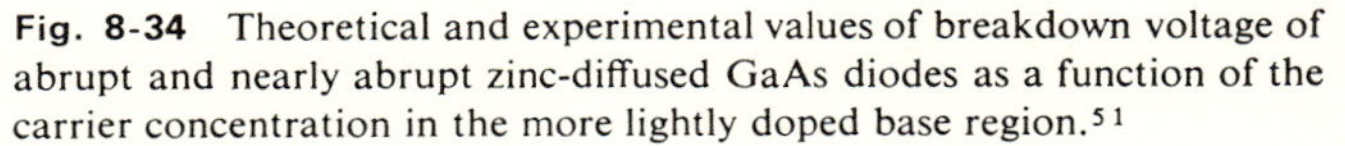

Fig. 8-34 Theoretical and experimental values of breakdown voltage of abrupt and nearly abrupt zinc-diffused GaAs diodes as a function of the carrier concentration in the more lightly doped base region.[51]

[53]K. S. Champlin, *J. Appl. Phys.* **30**, 1039 (1959); R. E. Burgess, *Can. J. Phys.* **37**, 730 (1959); J. L. Moll, A. H. Uhlir, and B. Senitzky, *Proc. IRE* **46**, 306 (1958).

[54]S. L. Miller, *Phys. Rev.* **105**, 1246 (1957).

[55]M. Weinstein and A. J. Mlavsky, *Appl. Phys. Letters* **2**, 97 (1963).

[56]S. M. Sze and G. Gibbons, *Appl. Phys. Letters* **8**, 111 (1966).

[57]H. Kressel, A. Blicher, and L. H. Gibbons, Jr., *Proc. IRE* **50**, 2493 (1962).

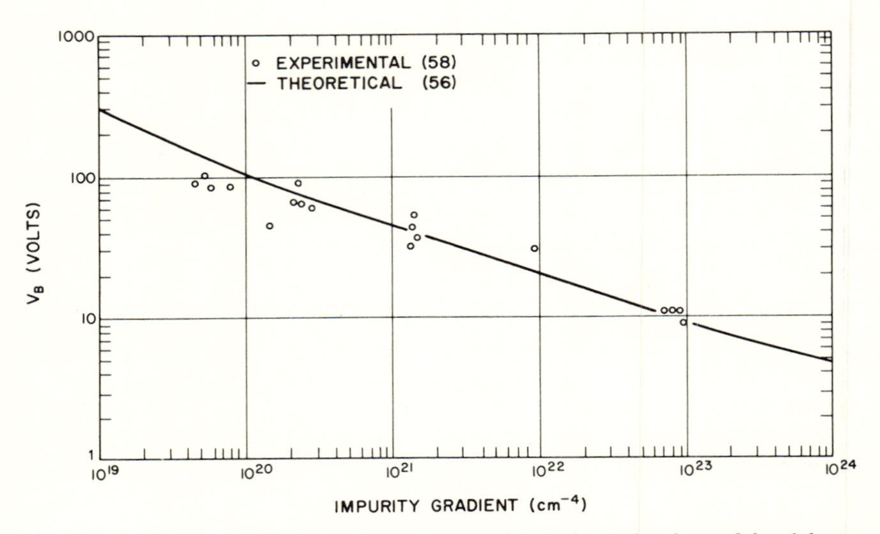

Fig. 8-35 Theoretical and experimental values of breakdown voltage of graded GaAs diodes as a function of impurity gradient.[51]

8-D-3-b The Solid-state Photomultiplier (or Multiplying Photodetector)

The saturation current I_s can be increased by illumination. Therefore, at V_R close to V_B, where the multiplication factor is large, the photocurrent is amplified. This principle has been used to make a fast and sensitive light detector.[59]

For maximum sensitivity the multiplication factor must be made as large as possible. Hence the junction must be prepared free of such imperfections as dislocations and precipitates. Furthermore, the edge effects must be avoided: where the junction emerges to the surface, the leakage current is likely to be large. The edge effects are minimized by using a guard-ring structure, as shown in Fig. 8-36. The active region (p^+n) is surrounded by a p–n junction in which the p-type region is less heavily doped, so that the electric field in the peripheral region is lower than in the active area. Hence the avalanche breakdown will occur in the central region rather than at the periphery. The bias dependence of the sensitivity of the device of Fig. 8-36 is shown in Fig. 8-37. Multiplication factors exceeding 300 have been obtained.

[51]H. Kressel, *RCA Review* **28**, 175 (1967).
[58]H. Kressel and A. Blicher, *J. Appl. Phys.* **34**, 2495 (1963).
[59]K. M. Johnson, *IEEE Trans. Electr. Devices* **ED12**, 55 (1965).

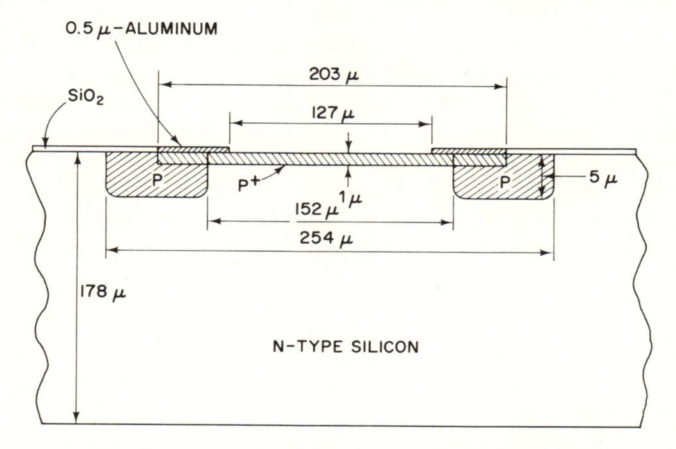

Fig. 8-36 Cross-section of silicon guard-ring photo-diode.[51]

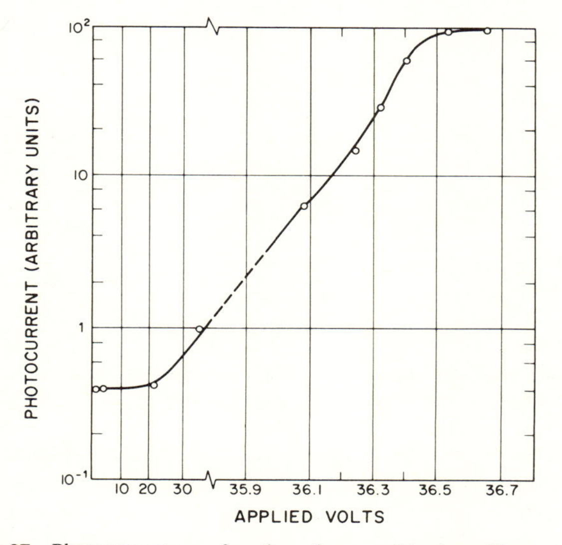

Fig. 8-37 Photocurrent as a function of reverse bias in a silicon guard-ring diode.[57]

8-D-3-c Luminescence During Avalanche Breakdown

Since electron–hole pairs are produced during avalanche breakdown, some radiative recombination occurs. Both the electrons and the holes can be heated by the electric field. The radiative transition between hot carriers emits

[57]H, Kressel, A. Blicher, and L. H. Gibbons, Jr., *Proc. IRE* **50**, 2493 (1962).

photons larger than the energy gap. Hence the luminescence during avalanche breakdown is characterized by a broad emission spectrum. This spectrum extends to $h\nu \approx 3E_g$, which represents the energy separating the hottest electron from the hottest hole (recall that the energy for impact ionization by a hot carrier is about $\frac{3}{2}E_g$). An example of the high-energy edges of avalanche-breakdown luminescence is shown in Fig. 8-38. The low-energy edge of the emission spectrum, on the other hand, extends to energies lower than the gap energy, due to the tunneling-assisted photon emission.

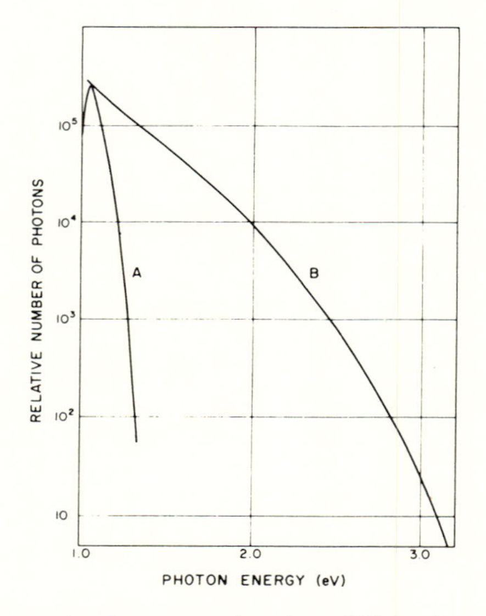

Fig. 8-38 Emission spectra from a *p–n* junction of Si in the forward direction A and the reverse direction B.[60]

An accurate description of the avalanche-breakdown luminescence spectrum must take into account the details of collisional losses by the hot carriers. As shown in Fig. 8-39, the shape of the high-energy edge of the emission spectrum fits the following description:[61]

$$U(\nu) = A\left(1 - \operatorname{erf}\frac{h\nu}{C}\right) \tag{8-48}$$

where A is a constant and

$$C = \left(\frac{3\pi}{8}\right)^{1/2} \frac{\mu \mathscr{E} k T_e}{v_s}$$

[60]A. G. Chynoweth and K. G. McKay, *Phys. Rev.* **102**, 369 (1956).
[61]J. Shewchun and L. Y. Wei, *Solid State Electronics* **8**, 485 (1965).

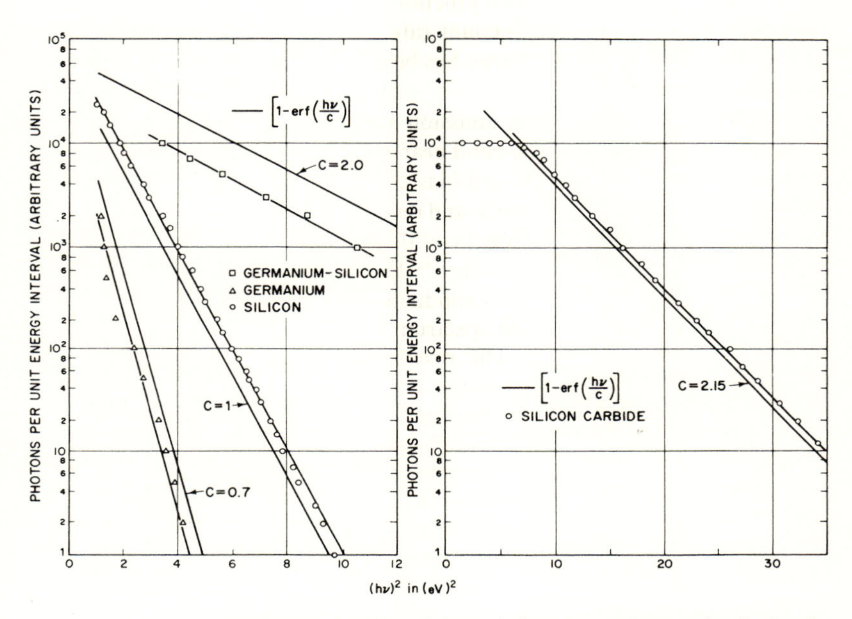

Fig. 8-39 Replot of the emission spectra from avalanche breakdown in silicon,[60] germanium,[62] germanium–silicon heterojunctions,[63] and silicon carbide[64] against $(h\nu)^2$. The equation $1 - \text{erf}\,(h\nu/C)$ is also shown for various values of the parameter C.[61]

where μ is the ohmic mobility, $\mathscr{E}$ is the electric field, kT_e is the energy of the hot carriers, and v_s is the sound velocity. Equation (8-48) was derived from first principles assuming that the hot carriers lose their energy by interacting with the coulomb field of charged centers, the energy being radiated and the momentum being transferred to the ion.[65]

The emission, especially the higher-energy edge, can be readily reabsorbed. Therefore, it is seen only in shallow junctions specially prepared to minimize self-absorption. The breakdown luminescence is usually spotty, occurring at microplasmas which develop at imperfections in the junction.[66] There is evidence that the light emitted at one microplasma can trigger avalanching at other sites where new microplasmas develop.[67] For example,

[62]A. G. Chynoweth and H. K. Gummel, *J. Phys. Chem. Solids* **16**, 191 (1960).
[63]J. Shewchun and L. Y. Wei, *J. Electrochem. Soc.* **111**, 1145 (1964).
[64]G. F. Kholuyanov, *Soviet Phys. Solid State* **3**, 2405 (1962).
[65]T. Figielski and A. Torun, "On the Origin of Light Emitted from Reversed-Biased p-n Junctions," *Int. Conf. Phys. Semiconductors, Exeter*, (*1962*), Institute of Physics and The Physical Society, London (1962), p. 863.
[66]A. G. Chynoweth and G. L. Pearson, *J. Appl. Phys.* **29**, 1103 (1958); M. Kikuchi, *J, Phys. Soc.* (Japan) **15**, 1822 (1960).
[67]I. Ruge and G. Keil, *J. Appl. Phys..* **34**, 3306 (1963).

increasing the separation of two junctions operating in the avalanche-breakdown mode reduces the pulse-count rate with the square of the distance.[68] Uniform breakdown luminescence has been observed only in dislocation-free *p–n* junctions.[69]

Although broad-spectrum emission can be observed in GaAs during avalanche breakdown, narrow-band near-gap emission has also been reported and found comparable to forward-bias injection luminescence.[70] Even in the alloy $GaAs_{1-x}P_x$, similar forward- and reverse-bias emission spectra are found over that range of composition ($x < 0.45$) where the direct valley of the conduction band is the lowest valley.[71] However, at compositions $x > 0.45$, where the $\langle 100 \rangle$ valley of the conduction band is lower than the direct valley, the breakdown-emission spectrum is very different from the forward-bias luminescence spectrum. The reverse-bias emission spectrum contains the fundamental near-edge emission corresponding to indirect transitions (also observed under forward bias) and a higher-energy emission which is not seen under forward bias. The high-energy edge of the emission spectrum cuts off abruptly at the energy corresponding to direct transitions (Fig. 8-40). The abrupt cutoff is attributed to strong self-absorption (the absorption coefficient is very high for direct transitions). The cutoff, then, varies with the composition x. A plot of the cutoff energy for the reverse-bias emission as a

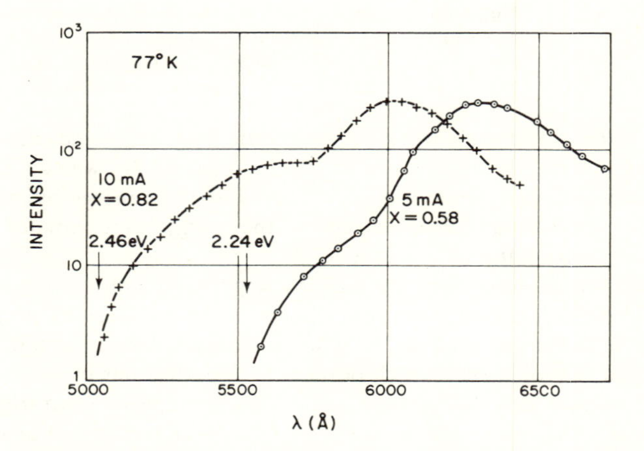

Fig. 8-40 Reverse-bias emission spectra of $GaAs_{1-x}P_x$ near and above bandgap for two compositions at which the minimum of the conduction band is an indirect valley.[71]

[68] R. H. Haitz, *Solid State Electronics*. **8**, 417 (1965).

[69] A. Goetzberger, B. McDonald, R. H. Haitz, and R. H. Scarlett, *J. Appl. Phys*. **34**, 1591 (1963).

[70] A. E. Michel and M. I. Nathan, *Bull. Am. Phys. Soc*. **9**, 269 (1964).

[71] M. Pilkuhn and H. Rupprecht, *J. Appl. Phys*. **36**, 684 (1965).

function of alloy composition in Fig. 8-41 shows that the cutoff energy follows the expected variation of the direct gap with x (Fig. 8-42). Figure 8-41 also shows (as crosses) the forward-bias emission peak which follows the variation of the lowest-energy peak. This data supports independent deter-

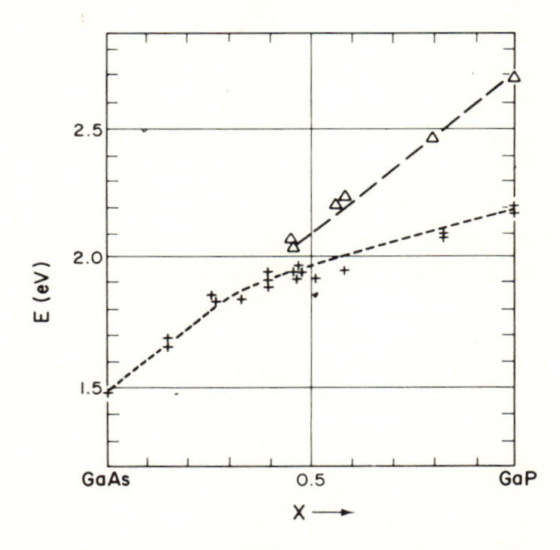

Fig. 8-41 For the system $GaAs_{1-x}P_x$, plot of the cutoff energy of the reverse-bias emission above bandgap (△) and of the peak of the near-edge emission in forward bias (+) as a function of composition x at 77°K.[71]

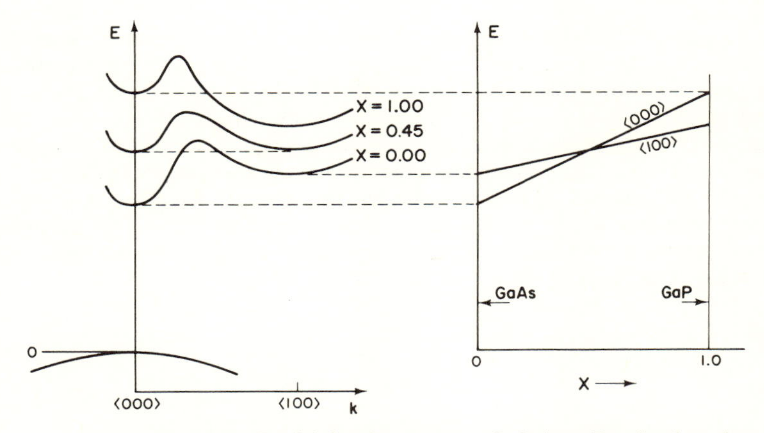

Fig. 8-42 Partial band structure of $GaAs_{1-x}P_x$ showing the variation of the direct and indirect valleys of the conduction band as a function of alloy composition x.

minations of the cross-over composition at which the $\langle 100 \rangle$ and $\langle 000 \rangle$ valleys lie at the same potential. Note, however, that the emission data must be interpreted in the context of donor-to-acceptor transitions. At the $\langle 000 \rangle$ direct valley ($x < 0.45$) the donor ionization energy is $E_D \approx 5$ meV; but for the indirect $\langle 100 \rangle$ valleys ($x > 0.45$) the donor ionization energy is $E_D \approx 100$ meV. Hence, the energy of the emission peak shifts to lower values by about 0.1 eV near and above the composition for valley crossover. This displacement of the emission peak in turn shifts the break of the emission curve to a lower value of x (by about 0.1) than the composition at which the valley crossover actually occurs.

Although the avalanche-breakdown luminescence is not an efficient process (its quantum efficiency is about 10^{-4} in GaAs),[70] it can be useful as a small light source with a fast response time and a broad emission spectrum.

Problem 1. In considering emission or absorption problems, it is usually assumed that $|\bar{q}| \ll |\bar{k}|$, and the selection rule simply becomes $\bar{k}_i = \bar{k}_f$ (see Prob. 4 in Chapter 6). Examine this assumption in the case of GaAs and discuss the conditions under which it is a satisfactory assumption. Note that the minimum of the conduction band and the maximum of the valence band occur at $\bar{k} = 0$ (where, therefore, $|\bar{k}| < |\bar{q}|$). Calculate the total number of states which have $|\bar{k}| \leq |\bar{q}|$. How does this number compare with the number of electrons injected into the p-side of a forward-biased GaAs p–n junction evaluated at the junction edge as stated below? Perform this calculation for room temperature (take $kT = 25$ meV) and assume, for convenience, that the Fermi level coincides with the band edge on each side of the junction in thermal equilibrium. Take the applied voltage about $8KT$ less than the bandgap $E_g = 1.4$ eV. The emission peak wavelength is at about $\lambda = 0.9\ \mu$m; $N_c = 5 \times 10^{17}$ cm^{-3}.

Problem 2. In some moderately doped GaAs electroluminescent diodes it is observed that, in the "injection mode," light intensity has the voltage dependence $\exp(qV/kT)$ and is proportional to the diffusion current. In addition, it has been established that most of the light originates in the p-region. One can conclude, therefore, that the electron-diffusion current in the p-region gives rise to the observed radiation. With this information in mind, discuss the question of whether one can optimize the external efficiency by changing the length of the p-region. If so, calculate the optimum length. Treat the problem as one-dimensional. The light intensity is measured from the p-side face of the device. Also assume that the applied potential qV is several kT less than the bandgap.

[70] A. E. Michel and M. I. Nathan, *Bull. Am. Phys. Soc.* **9**, 269 (1964).

STIMULATED EMISSION

9

In this chapter we shall consider the relationship between spontaneous and stimulated emission and develop the criteria for lasing in semiconductors.

9-A Relationship Between Spontaneous and Stimulated Emission[1]

A photon of energy $h\nu$ traversing a semiconductor can stimulate a transition between two levels E_1 and E_2 whose energy difference is $E_2 - E_1 = h\nu_{12}$ (Fig. 9-1). The transition can proceed either way, $1 \rightarrow 2$ or $2 \rightarrow 1$, provided there are electrons in the initial state and holes in the final state. The $(1 \rightarrow 2)$-transition is the absorption or pair-generation process; the $(2 \rightarrow 1)$-transition is the recombination process. If the recombination occurs in response to a passing photon, it is a "stimulated" recombination; if it occurs without apparent provocation, it is a "spontaneous" recombination.

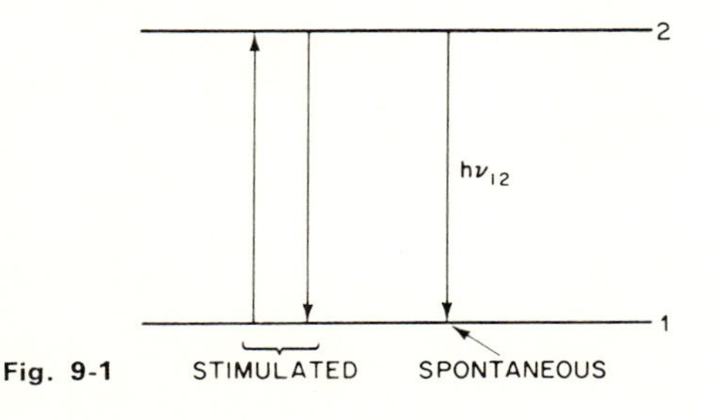

Fig. 9-1

[1] A. Einstein, *Phys. Zeit.* **18**, 121 (1917).

The density of photons in a system at a temperature T is given by Planck's distribution:

$$\rho(h\nu) = \frac{8\pi n^3 \nu^2}{c^3} \frac{1}{\exp\left(\frac{h\nu}{kT}\right) - 1} \tag{9-1}$$

The quantity $\rho(h\nu_{12})$ is, then, the density of photons in the mode $h\nu_{12}$. The pair-generation rate is

$$G_{12} = A_{12} N_1 f_1 N_2 (1 - f_2) \rho(h\nu_{12}) \tag{9-2}$$

where $A_{12}\rho(h\nu_{12})$ is a probability factor for stimulated transitions from 1 to 2; N_1 and N_2 are the densities of states at levels 1 and 2, respectively; and f is the electron-distribution function:

$$f(E) = \frac{1}{1 + \exp\left(\frac{E - E_F}{kT}\right)} \tag{9-3}$$

Hence $N_1 f_1$ is the density of electrons in the lowest state and $N_2(1 - f_2)$ is the density of holes in the upper state.

The recombination rate is

$$R_{21} = A_{21} N_2 f_2 N_1 (1 - f_1) \rho(h\nu_{12}) + B_{21} N_2 f_2 N_1 (1 - f_1) \tag{9-4}$$

The first term on the right side of Eq. (9-4) is the stimulated recombination, which depends on the photon field present; the second term is the spontaneous recombination, which does not depend on the photons already present in the system—B_{21} is the probability coefficient for the spontaneous recombination.

At thermal equilibrium, the pair-generation rate is equal to the recombination rate; therefore, equating Eq. (9-2) and (9-4),

$$A_{12}\rho(h\nu_{12}) f_1 (1 - f_2) = [A_{21}\rho(h\nu_{12}) + B_{21}][f_2(1 - f_1)] \tag{9-5}$$

Usually, $A_{12} = A_{21}$, and it can be shown that

$$\frac{f_1(1 - f_2)}{f_2(1 - f_1)} = \exp\left(\frac{E_2 - E_1}{kT}\right) \tag{9-6}$$

Therefore,

$$A_{21}\rho(h\nu_{12})\left[\exp\left(\frac{E_2 - E_1}{kT}\right) - 1\right] = B_{21} \tag{9-7}$$

and substituting Eq. (9-1),

$$\frac{B_{21}}{A_{21}} = \left[\exp\left(\frac{E_2 - E_1}{kT}\right) - 1\right] \frac{8\pi n^3 (\nu_{12})^2}{c^3} \frac{1}{\exp\left(\frac{h\nu_{12}}{kT}\right) - 1} \tag{9-8}$$

Since we chose $E_2 - E_1 = h\nu_{12}$, the ratio of the coefficients for stimulated and spontaneous emission is

$$\frac{B}{A} = \frac{8\pi n^3 \nu^2}{c^3} \tag{9-9}$$

which depends only on the frequency of the photon involved and on the index of refraction of the semiconductor.

Equation (9-9) shows the ratio of the coefficients for spontaneous and stimulated recombinations; but the probabilities per electron–hole pair for the two recombinations are in the ratio

$$\frac{P_{sp}}{P_{st}} = \frac{B}{A\rho(h\nu_{12})}$$

or, substituting Eqs. (9-1) and (9-9),

$$\frac{P_{sp}}{P_{st}} = \exp\left(\frac{h\nu_{12}}{kT}\right) - 1 \tag{9-10}$$

A stimulated recombination generates a photon which has the same frequency, direction of propagation, and phase as the stimulating photon. A spontaneous recombination generates photons propagating in random directions and random phases, although their frequency is also ν_{12} (for our two-state system). One can consider the spontaneous recombination as being triggered by "zero-point" photons, a quantum-mechanical concept which corresponds to the minimum energy of the system (at absolute zero)—their density is $8\pi n^3\nu^2/c^3$. This is an acceptable point of view, since a stimulating photon is not affected by the stimulation or triggering process. The stimulated recombination, on the other hand, is triggered by thermal photons, those determined by the temperature of the material and given by Planck's distribution Eq. (9-1). At room temperature, the density of thermal photons in the infrared and in the visible is many orders of magnitude lower than the density of zero-point modes. Hence at equilibrium the recombination is mostly spontaneous.

When the system is not at equilibrium, the density of electrons in the upper state and of holes in the lower state having been increased, the recombination rate increases, contributing additional photons to the field of thermal photons already present. Hence the rate of stimulation grows with the photon density. However, the generation rate (absorption) increases also, thus reducing the photon density. We shall now consider under what conditions the rate of stimulated recombination can exceed the rate of absorption. This will lead to a self-sharpening of the emission spectrum and eventually to lasing.

9-B Criteria For Lasing in a Semiconductor

An obvious approach to making the recombination rate greater than the absorption rate is possible at absolute zero. Let us fill the upper states of a direct-gap semiconductor to a quasi-Fermi level ξ_n and introduce holes to a

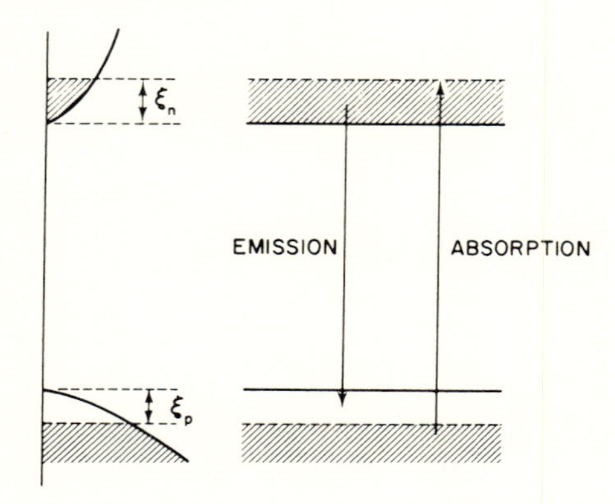

Fig. 9-2 Degenerate band occupation prevents band-to-band absorption at the emission frequency.

quasi-Fermi level ξ_p in the lower states (Fig. 9-2). Under these conditions, a population inversion is achieved; all the states involved in the emission process are not available for absorption.

A quantum-mechanical treatment[2] of this problem leads to the conclusion that, for the simple system considered thus far, the stimulated recombination rate will exceed the absorption rate if $\xi_n + \xi_p$ is greater than about $2kT$. In an indirect-gap semiconductor, if the radiative recombination occurs with phonon emission at low temperature, and competing nonradiative recombinations do not occur (an unrealistic ideal condition), the absorption process would be negligible because phonon absorption would be required to complete the upward transition.[3,4] This is illustrated in Fig. 9-3.

As the recombination rate increases, the density of photons builds up. These stimulate further recombinations. Therefore, starting with a spontaneous-emission spectrum at 0°K and an inverted population $np > (n_i)^2$,

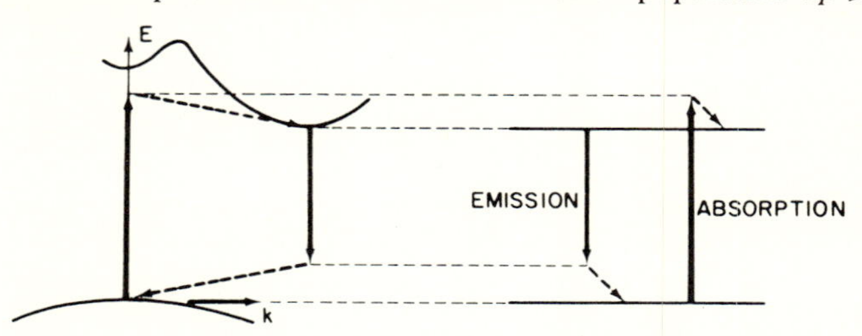

Fig. 9-3 In an indirect gap semiconductor at low temperature, the energy of the emitted photon is too low for its reabsorption.

[2]M. G. A. Bernard and G. Duraffourg, *Physica Status Solidi* **1**, 699 (1961).

[3]P. Aigrain, *Abstracts of Int. Congress on the Physics of Solid State, Brussels, June, 1958*, Société Belge de Physique, p. 1 (1960).

[4]N. G. Basov, O. N. Krohkin, and Yu. M. Popov, *Sov. Phys. Uspekhi* **3**, 702 (1961).

the initial field of photons will stimulate further recombinations. These, like the spontaneous emission, propagate in all directions with random phases. The photons at the peak of the distribution are more numerous and, therefore, stimulate more transitions at that frequency than those in the wings of the spectrum. Hence the emission spectrum sharpens up considerably. This growth of the emission peak is called the "gain." If the intensity of the emission peak grows superlinearly with the excitation, the process is "superradiant." Superradiance results in spectral narrowing, but the radiation remains incoherent. Superradiance amplifies the intensity of photons of all phases.

In order to use the recombination process in a laser, two conditions must be met: (1) that the gain at least be equal to the losses, and (2) that the radiation be coherent.

Let us consider coherence first. Coherence can be obtained by placing the source of radiation in a cavity which will favor the growth of one frequency and one phase. This selective amplification is the result of positive feedback for those electromagnetic waves which form a standing pattern in the cavity. Now the gain must be made equal to the losses.

Consider the cavity of Fig. 9-4. It comprises two partly reflecting surfaces, of respective reflectances R_1 and R_2, which are plane, parallel, and spaced by a distance l. Consider now a point in the center of this cavity emitting an intensity L_0 toward 1. A fraction R_1 of the radiation is reflected at 1 toward 2, and there, another fraction R_2 is reflected toward the original source. The system is characterized by a gain per unit length g and a loss per unit length α. Hence the light intensity at the center of the cavity after this $2l$-travel is given by[5]

$$L = L_0 R_1 R_2 \exp(2gl - 2\alpha l) \tag{9-11}$$

In a laser the gain is at least equal to the losses. In Eq. (9-11), in addition to the distributed loss α there is also an effective end loss (characterized by $R_1 R_2$) which is due to the escape of photons out of the cavity.

When the gain is equal to the sum of all the losses, the intensity of the

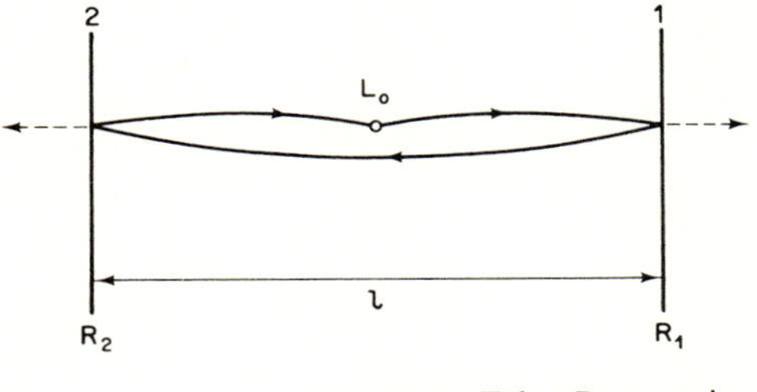

Fig. 9-4 Fabry-Perot cavity.

[5]G. Lasher, *IBM J. Res. Devel.* 7, 58 (1963).

radiation at the center of the cavity remains unchanged after the $2l$-path. Hence $L = L_0$ and Eq. (9-11) becomes

$$1 = R_1 R_2 \exp(2gl - 2\alpha l)$$

or, taking the natural logarithm of the above,

$$gl - \alpha l - \frac{1}{2} \ln \frac{1}{R_1 R_2} = 0 \tag{9-12}$$

The first term is the gain, the second term represents the distributed losses, and the last term represents the end losses. The gain is given by the following expression[5]:

$$g = \frac{c^2 j\eta}{8\pi q n^2 \nu^2 \, \Delta\nu \, d} \tag{9-13}$$

where j is the excitation rate (current density in the case of a p–n junction); η is the radiative-recombination efficiency (not all electron–hole pairs recombine radiatively); n is the index of refraction; ν is the frequency of the photons emitted in a spontaneous spectral half-width $\Delta\nu$; and d is the thickness of the active region in a direction transverse to the cavity length l; ν is determined by the choice of the material ($h\nu \approx E_g$). Hence the smaller-gap semiconductor should have the higher gain, everything else being constant. The half-width $\Delta\nu$ of the spontaneous emission depends on the doping. Thus in pure materials, exciton recombination should give the narrowest spectrum and, therefore, the highest gain. Pure materials, however, present a drawback for injection lasers in their inherently high internal resistance. The thickness of the active region is usually determined by the carrier diffusion length. In a forward-biased p–n junction, the injected minority carriers penetrate a distance which on the average equals a diffusion length $L = \sqrt{D\tau}$, where D is the diffusion coefficient and τ the carrier lifetime. In optical excitation, the thickness of the active region depends on the mean free path $1/\alpha$ of the exciting radiation; but when $1/\alpha$ is smaller than the diffusion length, the thickness of the active region is still determined by the diffusion length. Similarly, in electron-beam excitation the thickness of the active region is either the penetration depth of the incident electrons or the diffusion length, whichever is greater.

Let us rewrite Eq. (9-13) as

$$g = \beta j$$

and calculate the threshold current density for lasing, when the gain becomes equal to the losses. Then, Eq. (9-12) becomes

$$\beta j_{\text{Th}} = \alpha + \frac{1}{2l} \ln \frac{1}{R_1 R_2}$$

hence

$$j_{\text{Th}} = \frac{\alpha}{\beta} + \frac{1}{2\beta l} \ln \frac{1}{R_1 R_2} \tag{9-14}$$

If $R_1 = R_2 = R$, Eq. (9-14) becomes

$$j_{\text{Th}} = \frac{\alpha}{\beta} + \frac{1}{\beta l} \ln \frac{1}{R} \tag{9-15}$$

A plot of j_{Th} vs. $1/l$, as in Fig. 9-5, results in a straight line which extrapolates to an intercept with the j_{Th} axis at α/β. Its slope is $(1/\beta) \ln (1/R)$. The intercept and the slope allow a determination of α and β. In practice, the loss coefficient α ranges between 10 and 100 cm^{-1} (lowest values are most desirable), and the gain factor β ranges from 10^{-2} to 10^{-4} cm/A (the highest value being preferred).

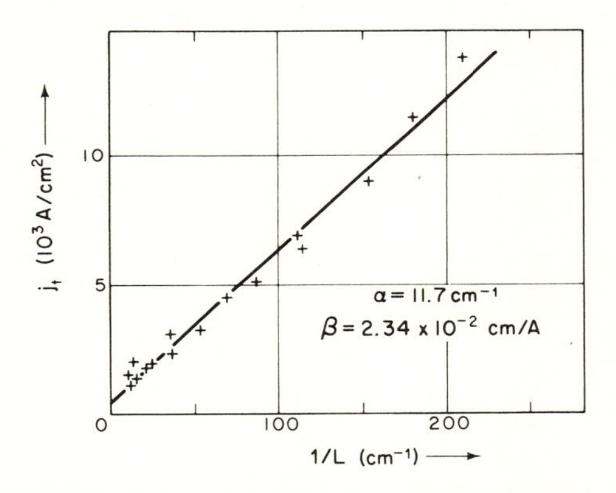

Fig. 9-5 Threshold current density as a function of reciprocal laser length at 77°K. Zn-diffused GaAs Fabry-Perot lasers ($N_d = 2 \times 10^{18}$ cm^{-3}).[6]

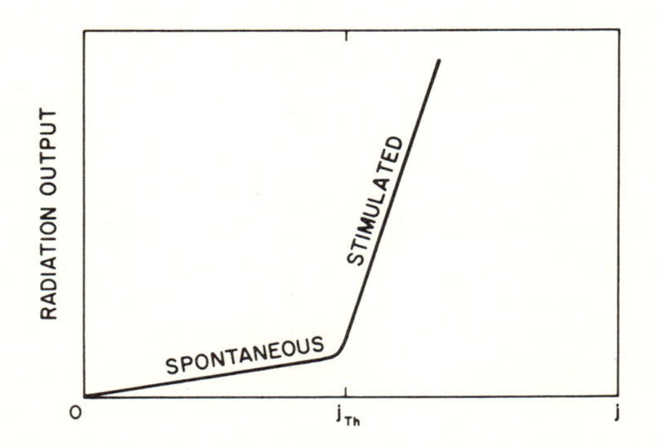

Fig. 9-6 Radiant output from a semiconductor laser vs. excitation rate.

[6]M. Pilkuhn and H. Rupprecht, *Trans. AIME* **230**, 282 (1964).

We shall see later that, although Fig. 9-5 suggests that lasers should be made as long as practical, other considerations, such as power dissipation, impose an optimum length.

The emission from a laser depends on the excitation, as shown in Fig. 9-6. Below threshold, the emission is mostly spontaneous. Above threshold, the stimulated emission dominates and, therefore, the intensity of the output rises sharply. Furthermore, the radiation comes out as a directional beam because the cavity favors the stimulation of certain modes. The final criterion for a laser is that the emission be coherent. A coherent radiation consists of identically phased photons. The degree of coherence is determined by the quality of the cavity. The properties of spontaneous, superradiant, and lasing radiation are summarized and compared in Table 9-1. The directionality of the stimulated radiation depends only on the geometry of the active region: the output will be most intense in the direction having the largest net gain. In a rectangular cavity with zero reflectance at the edges, the favored direction would be a diagonal; with reflecting ends, a longitudinal propagation is favored. Below the threshold for lasing, the stimulated radiation usually is not directional. Above threshold, the only distinction between superradiance and lasing is the coherence.

Table 9-1

Comparison of spontaneous, superradiant, and lasing radiations

Radiation	*Intensity*	*Spectral linewidth*	*Directional output*	*Coherence*
spontaneous	weak	broad	no	no
stimulated superradiant	intense	narrow	not necessarily	no
stimulated lasing	intense	very narrow	yes	yes

Note from Figs. 9-5 and 9-6 that for a finite maximum excitation, a short laser, having high threshold, will emit a smaller proportion of stimulated emission than a long laser.

To lower the lasing threshold, besides increasing the length of the laser, one can try to reduce the losses α, increase the gain factor β, or increase the reflectance R. One technique for simultaneously increasing R and L is to make a totally internally reflecting cavity (for example, a cube)—a seemingly trivial solution because no radiation comes out. However, in practice, imperfections in the structure allow some radiation to emerge and to demonstrate that lasing at lower threshold has been obtained.

Problem 1. Show that for laser emission involving band-to-band transitions in a laser diode, the potential difference applied across the junction must be larger than the bandgap (consider a direct-gap material).

Problem 2. Is the restriction of Prob. 1 essential for laser emission in an indirect-gap material? Discuss the reason why an indirect-gap material is unsuitable for laser applications.

SEMICONDUCTOR LASERS

10

10-A Cavity and Modes

The laser is usually shaped in the form of a Fabry–Perot cavity, a rectangular parallelepiped with one pair of opposite facets that are made perfectly plane and parallel. Radiation propagating perpendicularly to this pair of plane facets forms standing waves in the cavity. Standing waves occur whenever the cavity contains an integral number of half-wavelengths. For a cavity length l, this condition is expressed by

$$\mathbf{m}\frac{\lambda}{2n} = l \tag{10-1}$$

when **m** is an integer and λ/n is the wavelength of the radiation in the semiconductor. The spacing between modes is given by[1]

$$\Delta\lambda = \frac{\lambda^2}{2l\left(n - \lambda\dfrac{dn}{d\lambda}\right)} \tag{10-2}$$

Figure 10-1 shows how the longitudinal modes of a cavity modulate the emission spectrum.

In addition to longitudinal modes, it is possible to propagate modes having a transverse component if the other facets of the parallelepiped are also perfectly plane and parallel. Figure 10-2 illustrates several such possibilities. Note that because of refraction at the facet (Snell's law), the spatial distribution of radiation intensity outside the laser (the "far-field pattern") shows a larger angular deviation from the longitudinal axis than that of the internal propagation. The far-field pattern—reproduced in Fig. 10-3(a) [and in more detail in Fig. 10-3(b)]—has been correlated to the complex modes of Fig. 10-2.

[1]M. I. Nathan, A. B. Fowler, and G. Burns, *Phys. Rev. Letters* **11**, 152 (1963).

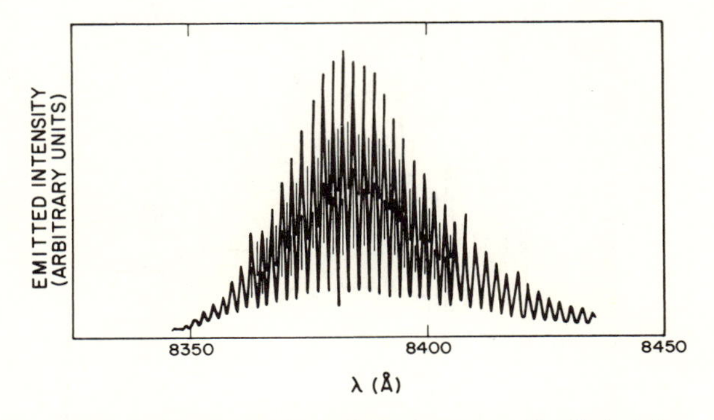

Fig. 10-1 Oscillations in the emission spectrum of a GaAs injection laser below lasing threshold at 2°K.[1]

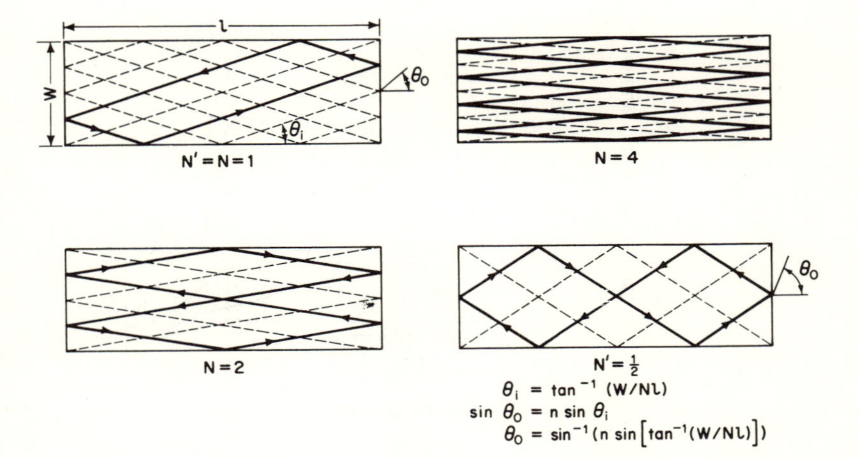

Fig. 10-2 Schematic representation of some simple geometric standing modes with easy closure.[2]

[2] R. A. Laff, W. P. Dumke, F. H. Dill, Jr., and G. Burns, *IBM J. Res. and Devel.* **7**, 63 (1963).

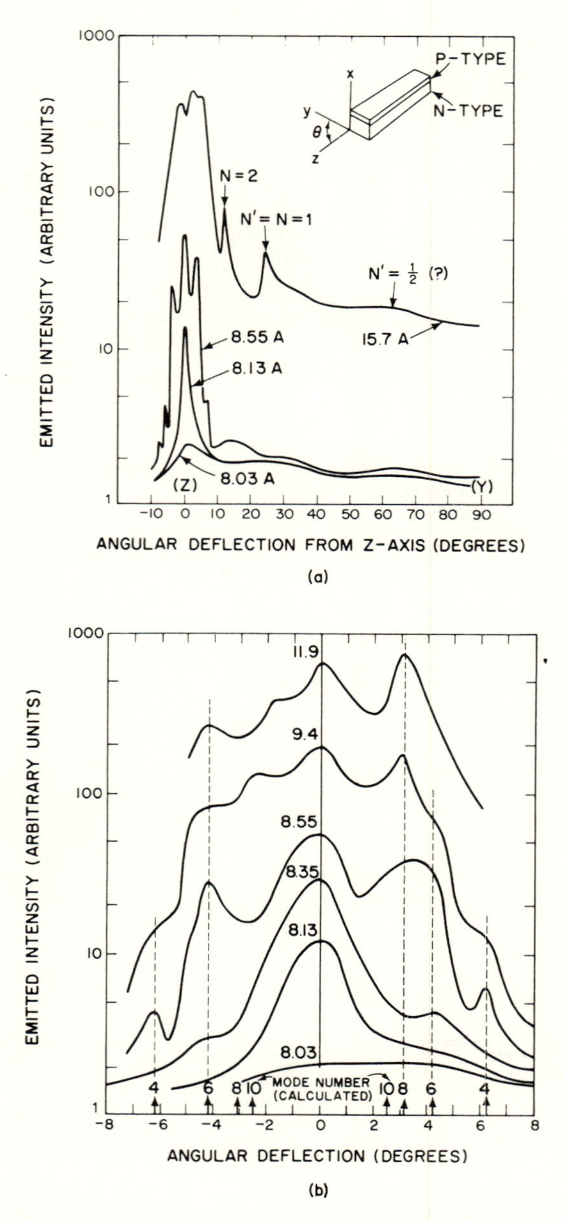

Fig. 10-3 (a) Angular dependence of emitter intensity in the junction plane. (b) Detailed structure of emitted intensity near length axis. The numbers on the curves are the excitation rate (in amps).[2]

10-B Waveguiding Properties of the Active Region

The active region, where stimulated radiative recombination occurs, has a higher index of refraction than the adjacent regions (Fig. 10-4). In an injection laser, the active region is a thin layer on the *p*-type side of the *p–n* junction. In an optically pumped laser or in an electron-beam-excited laser, the active region is bound by air or vacuum on one side ($n_1 = 1$) and by the passive part of the semiconductor on the other side.

By virtue of the Kramers–Kronig relation (Sec.4-C), it can be shown that the index of refraction is larger in the active region than in the adjacent layers. Let us consider the case of a *p–n*-junction laser, for example. Figure 10-5 shows that the absorption edge of the *n*-type region is the most abrupt one, and that it occurs at the highest energies. Consequently, the index of refraction of the *n*-type region begins increasing at the absorption edge. In the *p*-type region, the absorption edge starts at a lower energy and grows more slowly with photon energy. The index of refraction for the *p*-type region thus starts increasing at somewhat lower energies than the index of refraction of the *n*-type region. In the compensated, active region, on the other hand, the absorption edge starts at even lower energies and, therefore, the index of refraction grows at correspondingly lower energies. At the lasing photon

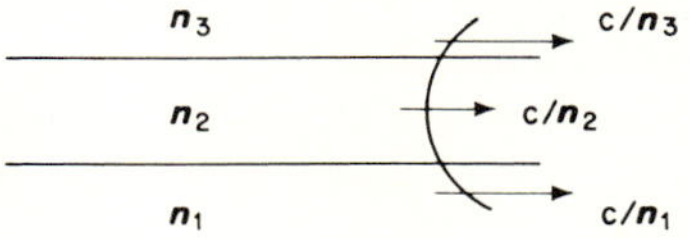

Fig. 10-4 Model of semiconductor laser having an active region with a refraction index n_2 such that $n_2 > n_1$ and $n_2 > n_3$.

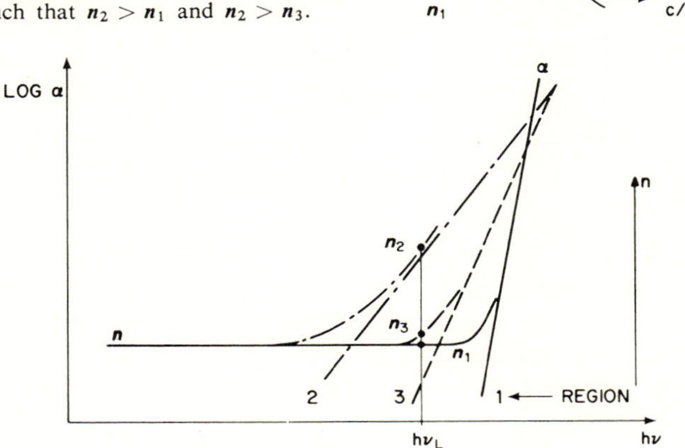

Fig. 10-5 Application of the Kramers–Kronig relation to the determination of the index of refraction in the three regions of Fig. 10-4. $h\nu_L$ is the lasing photon energy.

energy $h\nu_L$, the three regions have different indices of refraction $n_2 > n_1$ and n_3.

The propagation velocity in the three regions is then different; hence $c/n_2 < c/n_1$ or c/n_3. In other words, a wavefront of constant phase (Fig. 10-4) travels more slowly in the center of the active region than in the fringing regions. Therefore the wavefront must be concave in the direction of propagation. This shaping of the wavefront is the focusing or waveguiding property of the three-layer structure. The evidence for this property has been demonstrated in the experiment of Fig. 10-6 in which GaP was used.[3] Although GaP is not a lasing material, it is transparent to visible radiation, which makes the light-guiding property easily observable. Light incident at an angle with respect to the plane of a *p–n* junction in GaP breaks up into three components, as shown in Fig. 10-6: the radiation is partly reflected by the top of the crystal, the rest of the beam is refracted by the bulk of the crystal, and a portion of the refracted beam is guided along the *p–n* junction. The high efficiency of the semiconductor laser is in great part due to the containment of the radiation inside the active region.

Note that the index of refraction is increased in the active region as a result of carrier injection. The resulting radiation confinement is evident in the case of homogeneous semiconductors pumped either optically or with an electron beam where there is only one obvious discontinuity, that of the air interface. A mechanism for increasing the index of refraction by carrier injection may be the occurrence of negative loss which makes a positive contribution to the Kramers–Kronig integral (4-24).

A closer confinement of the coherent radiation in the active region of a GaAs laser is obtained by making the p^+ region of a wider-gap material: $Ga_{1-x}Al_xAs$.[4] This structure insures a large dielectric discontinuity between the p^+ region and a thin *p*-type GaAs layer. The resulting greater confinement

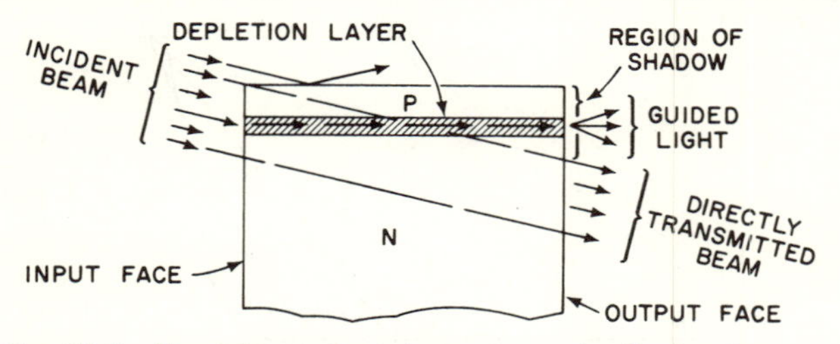

Fig. 10-6 Geometry used to observe internal reflection from a *p–n* junction (optical waveguide action). Refraction at the surface is omitted.[3]

[3] A. Ashkin and M. Gershenzon, *J. Appl. Phys.* **34**, 2116 (1963).

[4] H. Kressel and H. Nelson, *RCA Review* **30**, 106 (1969); H. Kressel, H. Nelson and F. Z. Hawrylo, *J. Appl. Phys.* **41**, 2019 (1970); M. B. Panish, I. Hayashi, S. Sumski, *IEEE J. Quantum Electronics*, **5**, 210 (1969); I. Hayashi, M. B. Panish, and P. Foy, *IEEE J. Quantum Electronics*, **5**, 211 (1969).

of radiation reduces an important absorption-loss mechanism by preventing the radiation from fringing into the p^+ region as it does in the all-GaAs laser. The reduction of absorption loss lowers substantially the lasing threshold. Furthermore, the barrier at the heterotransition blocks the injection of electrons into the wider gap p^+ region, thus confining the carriers to a narrow active region and thereby increasing its gain. Improved performance has also been obtained by making both the n^+ and the p^+ regions of the wider-gap material $Ga_{1-x}Al_xAs$ on either side of a thin GaAs active layer.[5]

10-C Far-field Pattern

Since the active region is thin (thickness d), the radiation is subjected to single-slit diffraction. Hence the radiation emerges from the thin cavity as a beam which spreads in a plane transverse to the active layer within an angle θ at half-maximum such that

$$\theta = \frac{\lambda}{d} \tag{10-3}$$

In GaAs at 78°K, $\lambda = 8400$ Å and $d \approx 1.5\ \mu$m; hence $\theta = 0.56$ radians = 32 degrees. In the plane of the active layer, the width of the beam at half-maximum should be $\phi = \lambda/w$, where w is the width of the active region. Since w is typically about 100 μm, the expected ϕ should be about 0.5 degrees—i.e., the radiated beam should consist of a narrow fan. However, the beam is usually considerably thicker, about 1 degree, and is acompanied by many lobes (Fig. 10-7). The thickness of the beam correlates with the observation that only part of the active region lases. The surface of the laser looks beady (Fig. 10-8); the size and number of beads increase with the excitation. From the width of the beam one can conclude that the size of the lasing region is about 10 μm wide. The lobes of the far-field pattern are, then, due to the interference pattern of two or more beady sources. The bright spot of the laser is the termination of a lasing filament, and one can observe the corresponding bright spot on the opposite facet of the Fabry–Perot cavity. The filamentary nature of the laser has been attributed to material inhomogeneity, and in some cases has been well correlated with the presence at edge dislocations of precipitates which themselves form filaments.[8] Sometimes the far-field pattern contains an interference structure within each fan-shaped lobe.[9] This interference structure is due to the radiation diffracted upon reflection at the back

[5]Zh. I. Alferov, *Bull. Am. Phys. Soc.* **14**, 875 (1969).
[8]M. S. Abrahams and J. I. Pankove, *J. Appl. Phys.* **37**, 2596 (1966).
[9]C. Deutsch, *Physics Letters* **24A**, 467 (1967).

facet and transmitted through the n-type region. The radiation emerging from the n-type region subsequently interferes with the radiation emerging from the junction.

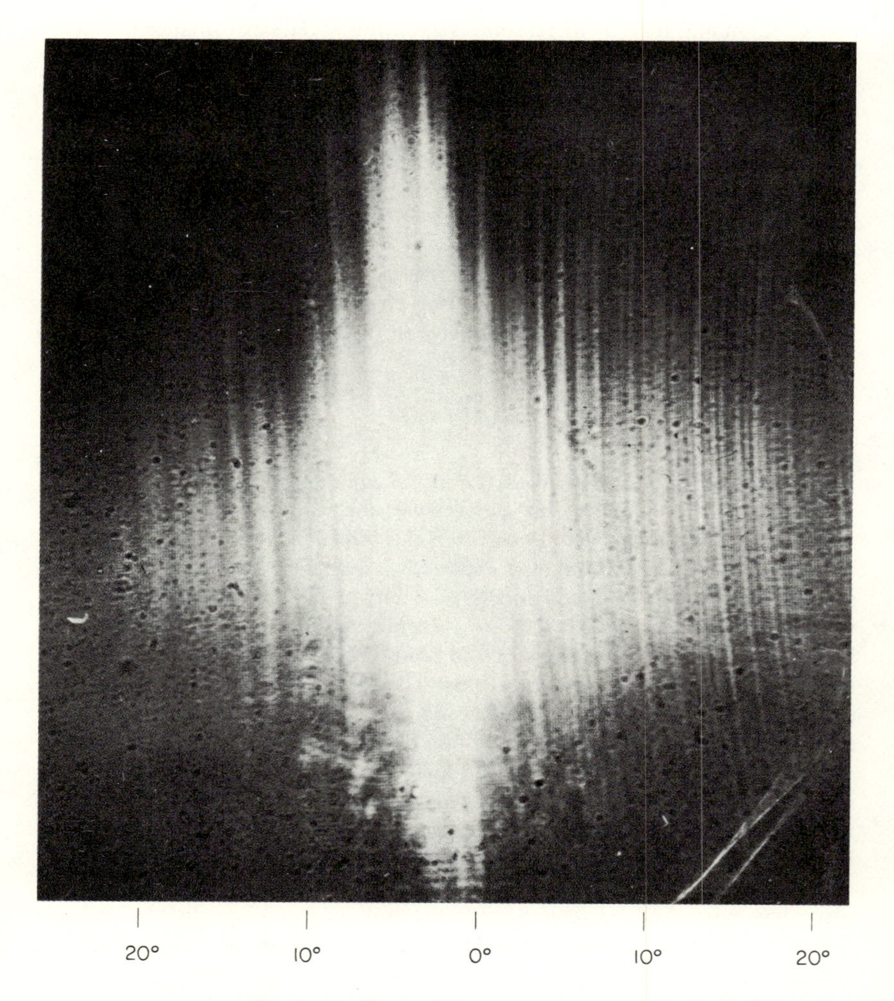

Fig. 10-7 The radiation pattern of a GaAs injection laser as detected by an infrared sensitive photographic plate with no optics between the diode and plate.[6]

[6]G. E. Fenner and J. D. Kingsley, *J. Appl. Phys.* **34**, 3204 (1963).

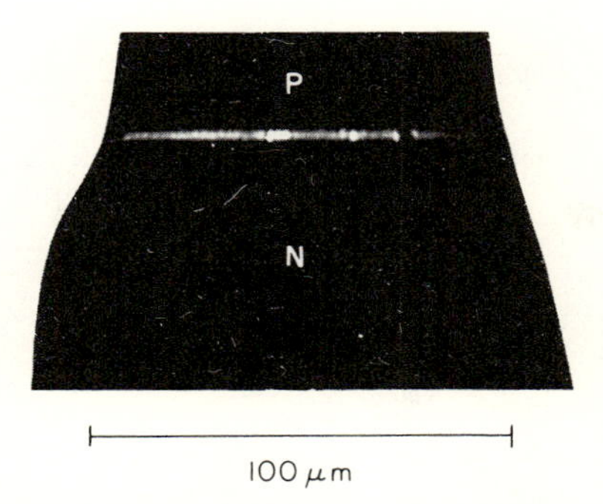

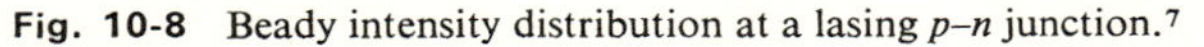

Fig. 10-8 Beady intensity distribution at a lasing *p–n* junction.[7]

10-D Temperature Dependence

10-D-1 EFFECT OF THE CAVITY

The radiative transition responsible for lasing is a donor-to-acceptor transition in heavily doped GaAs lasers. In lightly doped GaAs, in CdS, and in several other semiconductor lasers, the lasing transition is due to an exciton recombination. Hence in most cases the spontaneous-emission peak has the same temperature dependence as the energy gap (Fig. 10-9). The lasing peak, however, must be a mode allowed by the cavity. The cavity modes depend on temperature through the length of the cavity (via the thermal-expansion coefficient, which is very small) and through the temperature dependence of the index of refraction. It is, therefore, possible to predict the temperature dependence of the emission spectrum. As the temperature increases, the emitted peak jumps from one mode to the next one at a lower photon energy. At some temperatures two modes are allowed, and the emission is shared by both; as the temperature varies, one mode becomes more intense than the other.

10-D-2 TEMPERATURE DEPENDENCE OF LOSSES, OF EFFICIENCY, AND OF THRESHOLD CURRENT DENSITY

We saw in Eq. (9-14) that the threshold current density is proportional to the bulk losses α. A large portion of bulk losses is due to free-carrier absorption. Since the threshold current density increases with temperature, one can readily see that the increased carrier density will cause increased losses.

[7]Courtesy H. S. Sommers, Jr.

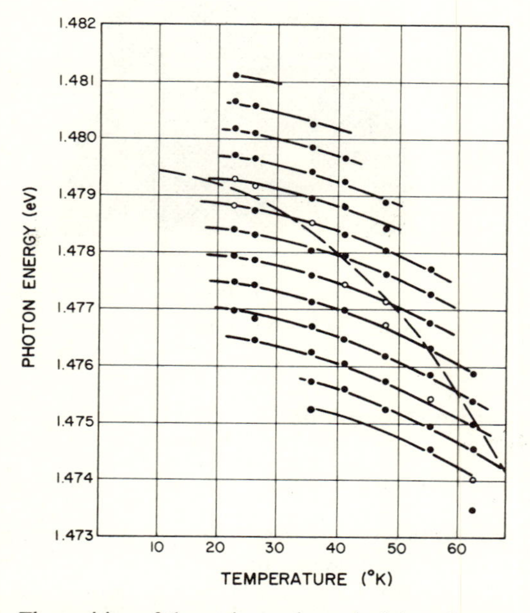

Fig. 10-9 The position of the cavity modes and of the coherent output. The open circles are the coherent lines and the dashed curve is the band gap of pure GaAs less 41.6 mV, i.e., the spontaneous emission peak: $E_g - (E_D + E_A)$.[10]

On the other hand, the gain factor shows, through Eq. (9-13), that the threshold current density must vary with the spontaneous line width $\Delta\nu$, which increases with temperature. The threshold current density j_{Th} must increase with the index of refraction, which increases slightly with temperature. Also, j_{Th} must increase with the thickness of the active region, which increases slightly with temperature; and j_{Th} must increase with decreasing internal efficiency, which may drop by a factor of 2 between 77°K and 300°K. Although a theoretical calculation of the temperature dependence of the threshold current gives results in agreement with experiment,[11] the formulation does not express in an explicit way the empirical dependence[12] observed in most semiconductor lasers (Fig. 10-10):

$$J_{\mathrm{Th}} = J_0 \exp \frac{T}{\theta} \tag{10-4}$$

where J_0 is the threshold current density at $T = 0$ and θ is a coefficient which

[10] W. E. Engeler and M. Garfinkel, *J. Appl. Phys.* **34**, 2746 (1963).
[11] F. Stern, *Phys. Rev.* **148**, 186 (1966).
[12] J. I. Pankove, *IEEE J. Quantum Electronics* **4**, 119 (1968).

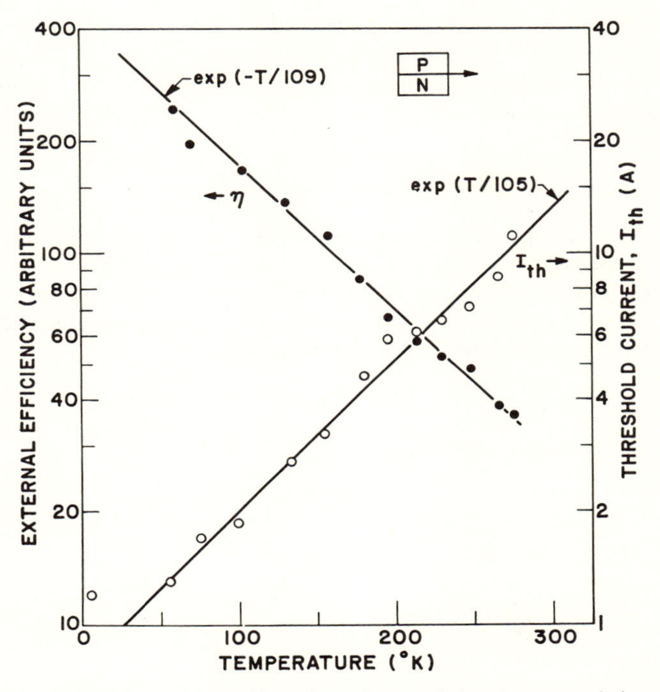

Fig. 10-10 Temperature dependence of incoherent emission efficiency and of threshold current for a GaAs diode.[12]

ranges from 50°K to 110°K; the value of θ depends on the doping, the larger values being obtained in the more heavily doped materials.

The temperature dependence of Eq. (10-4) is rather unique behavior. Most physical processes vary with the T in the denominator of an exponent, or else they vary as some power of T.

It is interesting to note that the external efficiency for incoherent emission (below threshold) also has an exponential dependence (Fig. 10-10):

$$\eta_{\text{ext}} = \eta_0 \exp\left(-\frac{T}{\theta}\right) \tag{10-5}$$

Usually, though not always, the coefficients θ of Eqs. (10-4) and (10-5) are nearly equal—this may be only a coincidence. The temperature dependence of the subthreshold external efficiency can be readily explained in terms of self-absorption.

The absorption edge of GaAs has an exponential dependence $\alpha = \alpha_0$

$\exp(h\nu/E_0)$, as shown in Fig. 10-11, and shifts to lower energies as the temperature increases. This shift is more rapid than the shift of the energy gap. In fact, the differential shift $[E_g(T) - h\nu(T, \alpha)]$ at constant α seems to vary linearly with temperature. Figure 10-11 shows that as a consequence of this temperature dependence of the absorption edge relative to the slower dependence of the emission peak, the self-absorption increases from α_0 to a new value:

$$\alpha = \alpha_0 \exp\frac{aT}{E_0} \tag{10-6}$$

When the light propagates along the active layer in a direction x, the facets being located at $x = 0$ and $x = l$, the light generated per unit length is $I(dx/l)\eta_i$, where I is the pair-excitation rate (the current) and η_i is the internal radiative-recombination efficiency. Self-absorption α and reflection R reduce the light generated per unit length to a value

$$(1 - R)I\frac{dx}{l}\eta_i e^{-\alpha(l-x)} \quad \text{beyond } l.$$

Hence the external efficiency is

$$\eta_{\text{ext}} = \frac{(1 - R)\eta_i}{l}\int_0^l e^{-\alpha(l-x)}\,dx$$

or

$$\eta_{\text{ext}} = \frac{(1 - R)\eta_i}{\alpha l}[1 - e^{-\alpha l}]$$

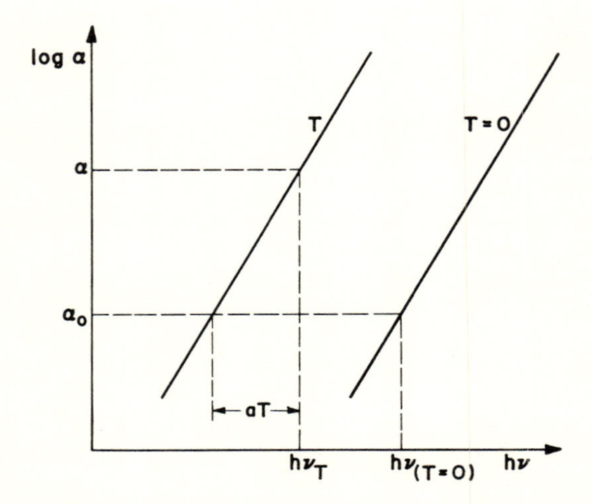

Fig. 10-11 Relationship between emission energy ($h\nu_0$ and $h\nu_T$) and the absorption edge at two different temperatures.

Since, in practice, αl is on the order of 2 or 3, we can neglect the exponential and omit the bracketed factor. In the above expression, the temperature dependence of η_{ext} would result from the temperature dependence of the absorption coefficient in the denominator. Hence, substituting Eq. (10-6) in the last expression,

$$\eta_{\text{ext}} \approx \frac{1-R}{\alpha_0 l}\eta_i \exp\left(-\frac{aT}{E_0}\right) \tag{10-7}$$

and we recognize a/E_0 as the coefficient $1/\theta$ of Eq. (10-5). We must stress again that the complementarity of Eqs. (10-4) and (10-5) may be purely coincidental, rather than phenomenologically correlated, since in the threshold-relation absorption is due largely to free carriers, whereas it is the absorption edge which affects the transmission of incoherent radiation. However, the radiation which fringes into the p^+ region, where it is strongly absorbed, represents a loss which, if dominant, could account for the observed temperature dependence.

10-D-3 POWER DISSIPATION

A large portion of the power supplied to the laser is dissipated internally as heat. For an injection laser, the power dissipated internally consists of: (1) joule losses I^2r; (2) that fraction of the power supplied to the p–n junction which is not converted into radiation; and (3) that fraction of the radiation which does not leave the laser. These last two components are accounted for by the external efficiency η_{ext}. Hence the power dissipated is

$$P_d = (1 - \eta_{\text{ext}})IV_j + I^2r$$

where η_{ext} is called the external quantum efficiency and represents the number of photons emitted per injected electron; V_j is the voltage across the junction. However, it is customary to quote the differential quantum efficiency which is the efficiency of emitting coherent radiation. The differential external quantum efficiency $\eta_{\text{ext D}}$ is the slope of a plot of the number of coherent photons emitted externally per injected electron for currents exceeding the threshold value. Typical values for $\eta_{\text{ext D}}$ are 0.5 at low temperature and 0.3 at room temperature, although values as high as 0.8 and 0.5 have been obtained at the two respective temperatures. The threshold current I_{Th} is typically 1 A at 77°K and 20 A at 300°K; and the internal resistance r is typically 0.05 ohm. The typical performance of a GaAs injection laser at an excitation rate twice as large as threshold is tabulated in Table 10-1.

The power output is given by

$$P_{\text{out}} = \eta_{\text{ext}}(I - I_{\text{Th}})V_j \tag{10-8}$$

The power dissipated is $P_d = P_{\text{in}} - P_{\text{out}}$.

As we shall see later, the power efficiency of electron-beam-excited lasers

Table 10-1

Typical performance of GaAs injection laser at $I = 2I_{Th}$

T (°K)	I_{Th} (A)	IV_j (W)	I^2r (W)	P_{in} (W)	P_{out} (W)	P_d (W)	η_P %	η_{ext} %	$\eta_{ext\,D}$ %
77	1	3	0.2	3.2	0.75	2.5	23	25	50
300	20	56	80	136	8.4	127.6	6.2	15	30

P_{in} input power $\quad$ η_P Power efficiency
P_{out} output power $\quad$ η_{ext} external quantum efficiency
P_d dissipated power $\quad$ $\eta_{ext\,D}$ Differential external quantum efficiency

can be much lower than in injection lasers because pair formation by electron impact requires about three times as much energy as the pair will be able to emit radiatively.

10-E Optimum Design for Injection Laser

The design of a laser may be governed by the desired operating wavelength which dictates the choice of a semiconductor having the required energy gap. Injection lasers have been made over the range of about 0.63 to 30 μ. Only direct-gap semiconductors have emitted coherent radiation; GaAs, the first semiconductor made to lase, has been the most extensively studied material.

Optimization means making compromises in the performance of the device. High power output requires a high dissipation rate, which to avoid excessive heating can be obtained only under pulsed conditions and at a low-duty cycle. The ultimate power output is determined by the threshold for degradation, which is estimated at about 10^7 W/cm^2.[13] At such a high power density, damage seems to occurs by piezoelectric stresses induced in the crystal. This catastrophic damage causes pitting of the Fabry–Perot facets.[14] However, in practice, the maximum reliable operating level is limited by a gradual degradation, which is not yet well understood. The maximum safe operating level is the lower of either 300 W/cm of active region about 1.5 μ thick, or 10^5 A/cm^2.[15] Lengthening the laser reduces its quantum efficiency.[16] Increasing the reflectance R, of the Fabry–Perot facets also reduces the exter-

[13] N. G. Basov, A. Z. Grasiuk, V. F. Efimkov, I. G. Zubarev, V. A. Katulin, and Ju. M. Popov, *J. Phys. Soc.* (Japan) **21**, Supplement, 277 (1966).
[14] H. Kressel and H. Mierop, *J. Appl. Phys.* **38**, 5419 (1967).
[15] H. Kressel and N. E. Byer, *Proc. IEEE* **57**, 25 (1969).
[16] G. Cheroff, F. Stern, and S. Triebwasser, *Appl. Phys. Letters* **2**, 173 (1963).

nal efficiency. Making the laser wide increases the power output, but when the width is greater than the length, the gain in a direction transverse to the longitudinal mode becomes larger than the longitudinal gain and the laser becomes superradiant in the transverse direction at the expense of the longitudinal mode.

The external quantum efficiency for coherent emission, i.e., the differential quantum efficiency above the excitation threshold for lasing, is given by[17]

$$\eta_{\text{ext D}} = \frac{\eta_i}{1 + \dfrac{\alpha l}{\ln(1/R)}} \tag{10-9}$$

Substituting Eqs. (10-9) and (9-15) into Eq. (10-8) gives an expression for the stimulated power per unit width of laser. This is best described by the plot of Fig. 10-12, which shows that lowering the reflectance of Fabry–Perot facets may optimize the performance of the laser.

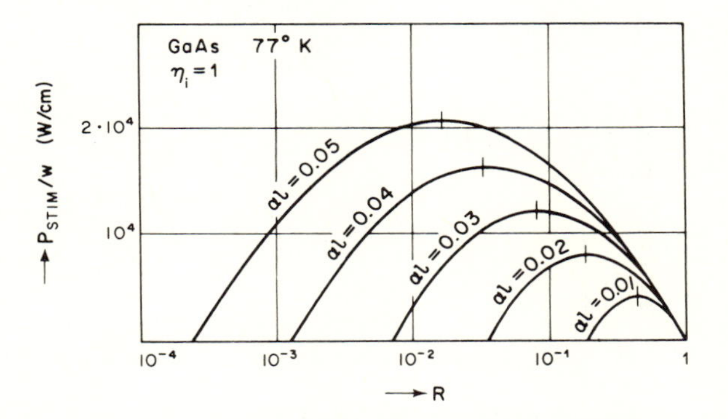

Fig. 10-12 Dependence of the stimulated light power per unit width on the reflection coefficient R for different laser lengths (pulsed operation). The following experimental data referring to GaAs lasers at 77°K were used: $\alpha = 15\ \text{cm}^{-1}$; $\beta = 2.5 \times 10^{-2}$ cm/A; V = 1.5 volts.[18]

Another performance characteristic of a semiconductor laser is its ability to operate continuously (cw). The limitation here is due to the temperature rise of the active region during operation. The temperature rise can be determined from the internal dissipation P_d and the thermal resistance of the device. Typically, the thermal resistance Ω is about 30°/watt and can be monitored through the shift of the emission spectrum with temperature. Hence the temperature rise of the laser is

$$\Delta T = \Omega[IV_j(1 - \eta_{\text{ext}}) + I^2 r] \tag{10-10}$$

[17] J. R. Biard, W. N. Carr, and B. S. Reed, *Trans. AIME* **230**, 286 (1964).
[18] M. H. Pilkuhn and G. T. Guettler, *IEEE J. Quantum Electronics* **4**, 132 (1968).

while the threshold current is given by

$$I_{Th} = I_0 \exp \frac{T + \Delta T}{\theta} \tag{10-11}$$

A solution of the set of simultaneous equations, Eqs. (10-10) and (10-11), gives the range of current and temperature where cw operation is possible.[19]

Still another design factor is the doping of the semiconductor. Thus if one is aiming for maximum external quantum efficiency, the donor and acceptor concentrations must be controlled. While the acceptor concentration is not critical below 2×10^{19} cm^{-3} (Fig. 10-13), the donor concentration

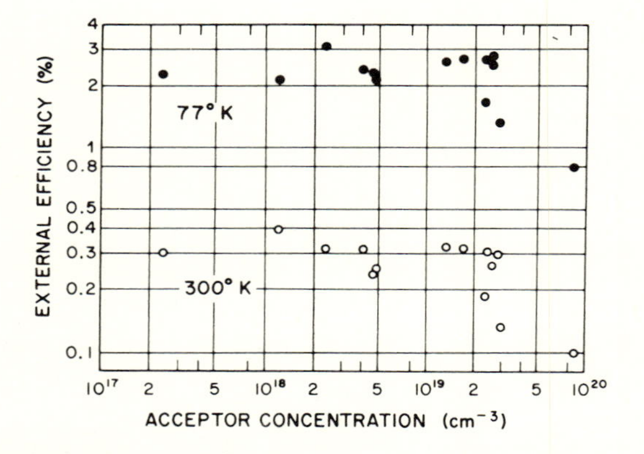

Fig. 10-13 External quantum efficiency for incoherent GaAs diodes at 77° and 300°K as a function of the acceptor concentration in the uppermost *p*-type epitaxial layer. Donor concentration is 5×10^{17} cm^{-3} in the lower epitaxial layer.[20]

seems to require an optimal value of 1 to 2×10^{18} cm^{-3} (Fig. 10-14).[20] The decrease in external quantum efficiency at donor concentrations greater than 2×10^{18} cm^{-3} is believed due to the formation of selenium or tellurium precipitates.[21]

Doping affects also the spontaneous linewidth and the energy of the emission peak—as shown in Fig. 10-15—and influences the light-guiding property of the active region.

[19]To solve this problem, an approximation, $I = I_0[1 + (\Delta T/\theta)]^3$, has been used instead of Eq. (10-11) by C. H. Gooch, *IEEE J. Quantum Electronics* **4**, 140 (1968).

[20]C. J. Nuese, J. J. Tietjen, J. J. Gannon, and H. F. Gossenberger, *Trans. MS AIME* **242**, 400 (1968).

[21]J. Vieland and I. Kudman, *J. Phys. Chem. Solids*. **24**, 437 (1963).

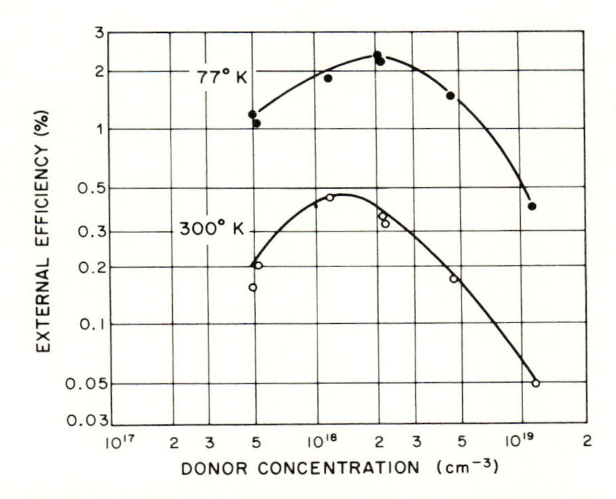

Fig. 10-14 External quantum efficiency for incoherent GaAs diodes at 77° and 300°K as a function of the donor concentration in the uppermost n-type epitaxial layer. Acceptor concentration is 1×10^{19} cm^{-3} in the lower epitaxial layer.[20]

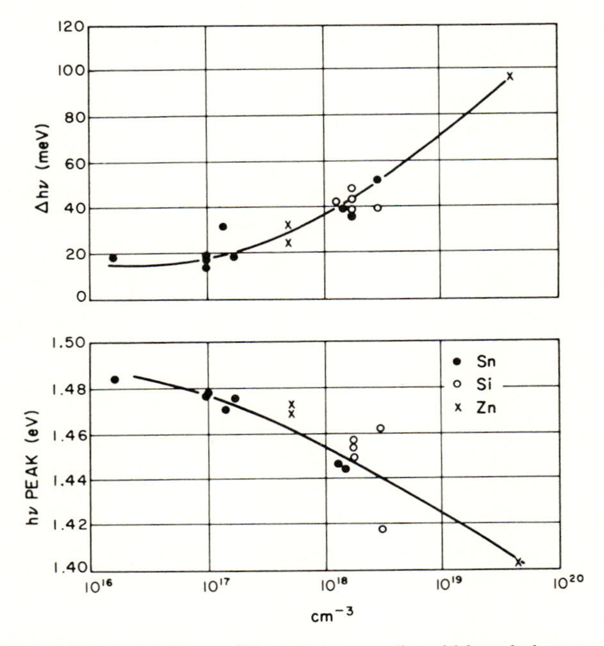

Fig. 10-15 Dependence of the spontaneous linewidth and photon energy as a function of both n- and p-type doping.[22]

[22]R. Braunstein, J. I. Pankove, and H. Nelson, *Appl. Phys. Letters* **3**, 31 (1963).

10-F Influence of a Magnetic Field

A magnetic field induces a Landau splitting of the energy levels. In the case of InSb, where the radiative recombination seems to be a band-to-band transition, the Landau splitting condenses band states into a narrower dis-

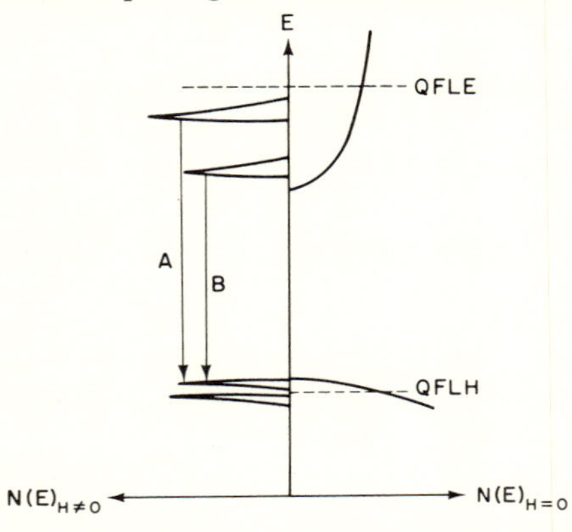

Fig. 10-16 Effect of a strong magnetic field on the distribution of states in InSb (Landau splitting). A and B are two sets of radiative transitions.

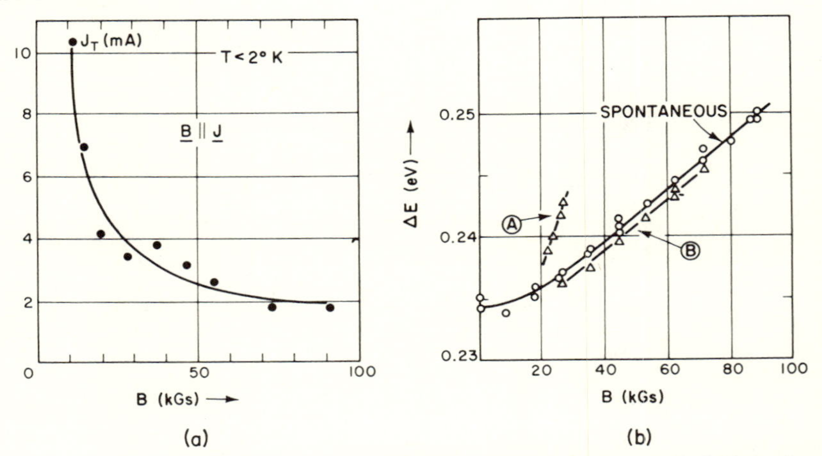

Fig. 10-17 (a) The effect of a longitudinal magnetic field on the laser threshold of an InSb diode. (b) The variation of peak emission energy with magnetic field for both coherent and incoherent emission data. A and B are the two coherent peaks.[23]

[23]R. J. Phelan, A. R. Calawa, R. H. Rediker, R. J. Keyes, and B. Lax, *Appl. Phys. Letters* **3**, 143 (1963).

tribution of states (Fig. 10-16). Consequently, the spontaneous line width is reduced and the threshold excitation rate can be considerably decreased [Fig. 10-17(a)]. Since the magnetic field moves the quantized states deeper into the bands, the emission peak (or peaks) shift to higher energies as the magnetic field is increased [Fig. 10-17(b)].

In InAs, however, the effect of the magnetic field is appreciable only when the field is transverse to the current. The threshold current is lowered by the transverse magnetic field as shown in Fig. 10-18. This behavior is explained in terms of a reduction of the thickness d of the active region [see Eq. (9-13)]; d is reduced because the cyclotron motion restrains the diffusive tendency of the injected minority carriers and holds them close to the p–n junction.

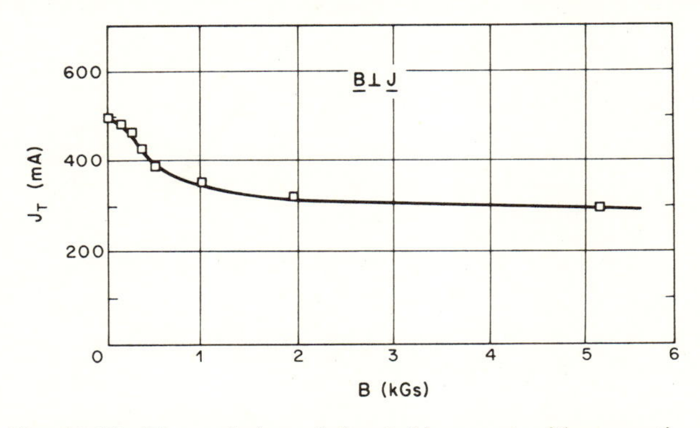

Fig. 10-18 The variation of threshold current with magnetic field for an InAs laser at 2.0°K.[24]

10-G Pressure Effects

Hydrostatic pressure causes a change in the energy gap of semiconductors and a corresponding change in the emission spectrum. In GaAs, the energy gap increases with pressure at a rate of 1.1×10^{-5} eV/bar, while the cavity length changes at the rate $(dl/dP)/l = -0.44 \times 10^{-6}$ bar^{-1}, and the index of refraction varies as $[d(\ln \boldsymbol{n})/dP)]_{77°K} = 2.4 \times 10^{-6}$ bar^{-1}.[25]

Within these restrictions, the laser emission consists of several modes which grow and decay as the pressure dependence of the emission spectrum intersects the pressure-dependent allowed modes.

In the lead salts, however, hydrostatic pressure causes a narrowing of the

[24] F. L. Galeener, I. Melngailis, G. B. Wright, and R. H. Rediker, *J. Appl. Phys.* **36**, 1574 (1965).

[25] J. Feinleib, S. Groves, W. Paul, and R. Zallen, *Phys. Rev.* **131**, 2070 (1963); also G. E. Fenner, *J. Appl. Phys.* **34**, 2955 (1963).

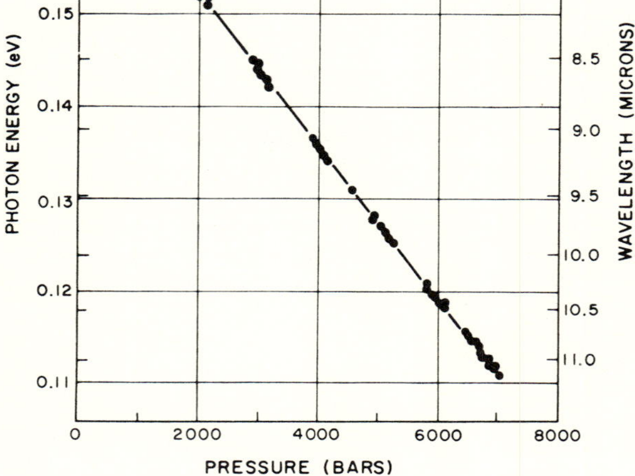
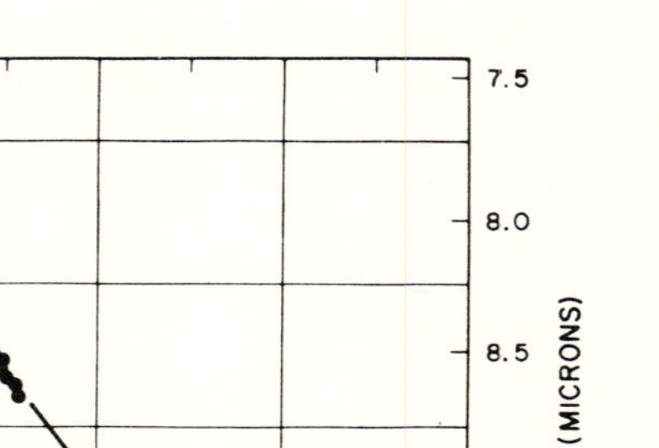

Fig. 10-19 Variation with pressure of the dominant emission modes from a PbSe diode laser at 77°K.[26]

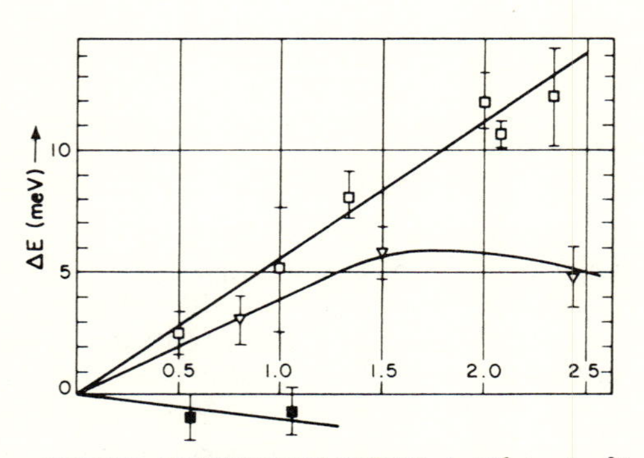

Fig. 10-20 Shift of the spontaneous emission peak with uniaxial stress for three different diodes.[28]

[26] J. M. Besson, J. F. Butler, A. R. Calawa, W. Paul, and R. H. Rediker, *Appl. Phys. Letters* **7**, 206 (1965).

[28] D. Meyerhofer and R. Braunstein, *Appl. Phys. Letters.* **3**, 171 (1963).

energy gap. Measurements on PbSe injection lasers show that the dominant emission mode shifts to lower energy with pressure at a rate of -8.5×10^{-6} eV/bar (Fig. 10-19).[26]

Uniaxial pressure, which is easier to apply than hydrostatic pressure, can cause a drop in the threshold current density,[27] but the shift of the emission spectrum may increase, decrease, or go through a maximum (Fig. 10-20).[28]

Problem 1. Demonstrate that, in a Fabry–Perot cavity of length L, the mode spacing is given by

$$\Delta\lambda = \frac{\lambda^2}{2L\left\{n(E) + E\dfrac{d[n(E)]}{dE}\right\}}$$

Problem 2. In studying the temperature dependence of the threshold current density J_{Th}, it has usually been assumed that α in Eq. (9-15) represents mainly free-carrier losses in the active region. Is this assumption sufficiently accurate? Draw a diagram of a laser diode showing the radiation-field distribution in the active and passive regions and discuss your answer in terms of the fringing field extending in the passive regions. Make use of the concept expressed by Eq. (6-21).

[27]G. M. Ryan and R. C. Miller, *Appl. Phys. Letters* 3, 162 (1963).

EXCITATION OF LUMINESCENCE AND LASING IN SEMICONDUCTORS

11

To excite luminescence in semiconductors, a high concentration of electron–hole pairs must be generated. This can be done electrically, by optical excitation, or with an electron beam.

11-A Electroluminescence

Electrical excitation is the most direct method of excitation, since electroluminescence converts electrical energy directly into radiation. Furthermore, this form of excitation lends itself readily to modulation.

11-A-1 FORWARD-BIASED p-n JUNCTION

A forward-biased *p–n* junction is capable of producing a very high density of electron–hole pairs very near the *p–n* junction. The current increases exponentially with the voltage across the *p–n* junction. If the internal resistance of the diode is negligible, most of the applied voltage appears across the junction. In a semiconductor laser, both the *p*-type and the *n*-type regions are heavily doped, thus minimizing the internal resistance.

In GaAs injection lasers, a high efficiency for injecting electrons into the *p*-type region is usually obtained. This is due to two factors: (1) different penetrations of the Fermi levels into the two bands, and (2) differential effective-gap shrinkages. Both these factors result in a barrier to holes, as we shall see presently.

Figure 11-1 shows an idealized *p–n* junction biased in the forward direction by a voltage $V_j = (E_g + \xi_p)/q$ just sufficient to start injecting electrons into the p-type region. For the same doping level, the Fermi level penetrates deeper into the conduction band of the n-type region (low effective mass) than it does into the valence band of the p-type region (large effective mass). Then $\xi_n > \xi_p$. At very low temperatures, with a bias V_j such that $qV_j = E_g + \xi_p$, electrons begin to flow into the p-type region, while the holes are blocked by a barrier equal to $\xi_n - \xi_p$.

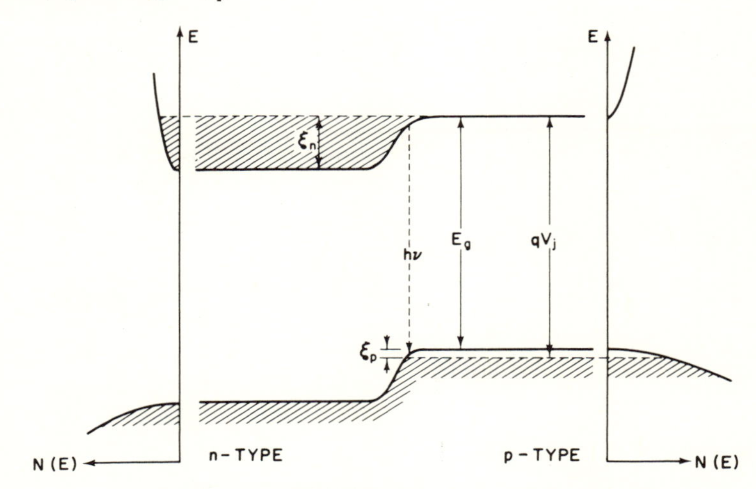

Fig. 11-1 Injection luminescence at a p-n junction with a V_j, forward bias.

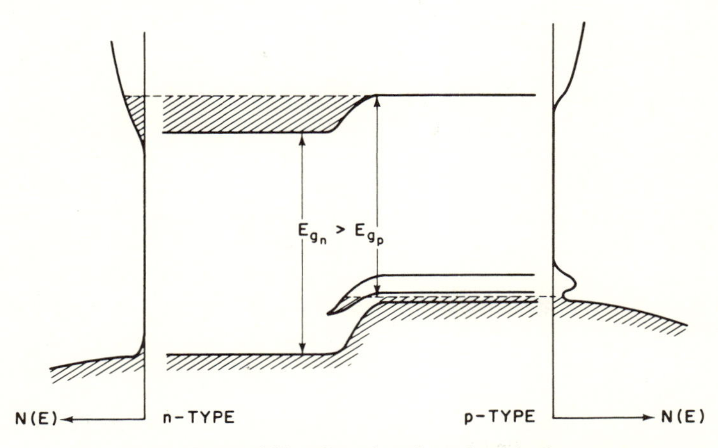

Fig. 11-2 Forward biased p–n junction where doping causes greater effective gap shrinkage of p-type region than of n-type region.

We saw in Sec. 1-C that heavy doping in semiconductors forms tails of states which effectively reduce the energy gap. The doping used in GaAs lasers, being one order of magnitude higher in the *p*-type than in the *n*-type region, causes a slightly greater effective shrinkage of the energy gap in the *p*-type side than in the *n*-type side. In other words, such a *p–n* junction is a quasi-hetero-junction, since the energy gap of the *n*-type region is effectively larger than that of the *p*-type region. Furthermore, in the *p*-type region the acceptors form a deep band, which causes the quasi-Fermi level for holes to be closer to the valence-band edge than if only valence-band states were available to accommodate the high concentration of holes. This is shown in Fig. 11-2.

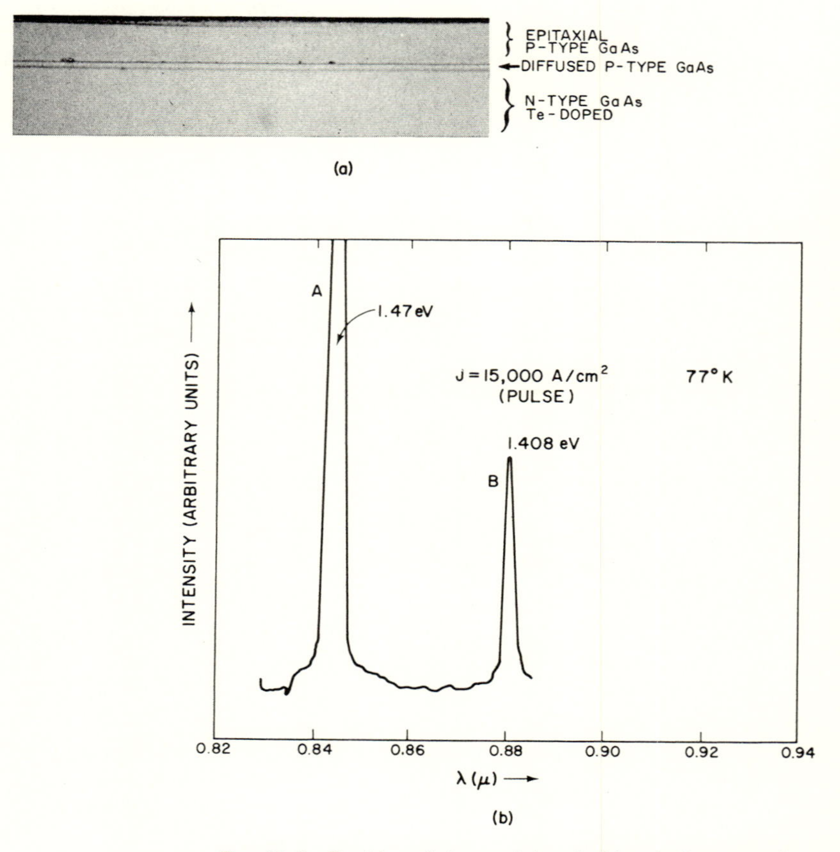

Fig. 11-3 Double emission peak laser[1]: (a) etched cross section; (b) emission spectrum.

[1]H. Nelson and G. C. Dousmanis, *Appl. Phys. Lett.* **4**, 192 (1964); also H. Kressel and F. Z. Hawrylo, *J. Appl. Phys.* **39**, 205 (1968).

For all the above reasons, luminescence and lasing usually occur only in the *p*-type side of the *p–n* junction. In some laser diodes [Fig. 11-3(a)] the *p*-type side consists of two regions: a very heavily doped one (p^+) and a less heavily doped region (p).[1] When the p^+p transition is close to the *p–n* junction radiative recombination and lasing can occur in the *p*-type region and, at higher excitation, in the p^+ region as well. The two regions are characterized by different transitions which emit photons of very different energies (1.47 eV and 1.408 eV, respectively). These two lasing emissions can occur simultaneously [Fig. 11-3(b)]. Of interest is the observation[2] of the simultaneous emission of the normal lasing peak and of a possibly coherent peak at a greater-than-gap photon energy in a variety of GaAs injection lasers at high current density (Fig. 11-4). The emission $h\nu > E_g$ is not yet understood.

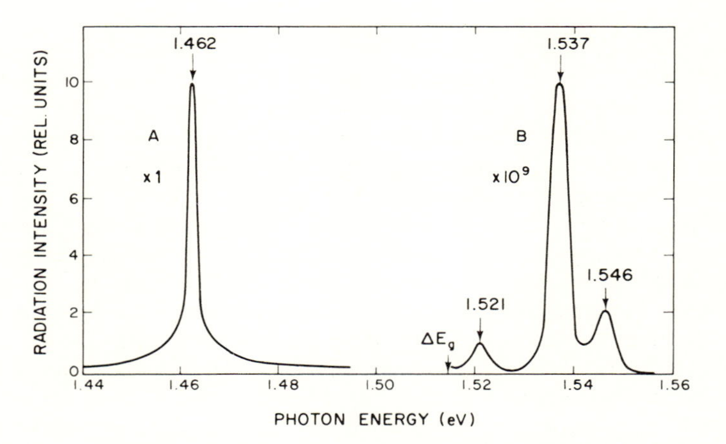

Fig. 11-4 Emission spectrum from a GaAs injection laser at 77°K at about 22,000 A/cm² (5 × threshold). Curve A is the envelope of the normal coherent radiation modes; B is the short-wavelength band.[2]

11-A-2 FORWARD-BIASED SURFACE BARRIER

The surface states of a semiconductor often induce an inversion layer as illustrated in Fig. 11-5. The inversion layer can be a source of minority carriers for the bulk when the surface is biased in such a way as to flatten the bands. The inversion layer then acts as a quasi-*p–n* junction.

A related technique has been used to generate luminescence in *n*-type and

[2] B. I. Gladkii, D. N. Nasledov, and B. V. Tsarenkov, *Sov. Phys.—Semiconductors*. **2**, 530 (1968).

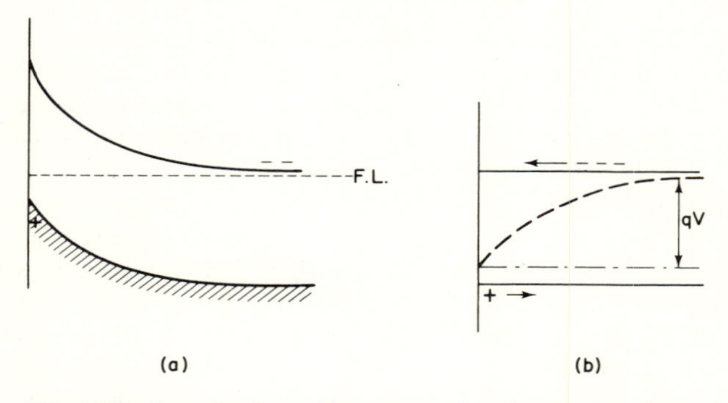

Fig. 11-5 Inversion layer in n-type semiconductor: (a) without applied bias; (b) with forward bias V.

in p-type GaAs[3] and also in GaP.[4] With the surface biased strongly enough with that polarity which depletes majority carriers at the surface, either avalanche breakdown or tunneling can be induced. Then the minority-carrier density is allowed to build up at the surface of the semiconductor. When the polarity of the applied voltage is changed into a forward bias, the minority carriers can recombine radiatively.

If the surface is driven by a sinusoidal voltage of frequency f, analysis[4] shows that the maximum density of minority carriers is obtained when $ft = \frac{1}{4}$. Beyond this quarter-period, the majority carriers flow back into the depletion layer, reducing the electric field and stopping the breakdown. Beyond $ft = \frac{1}{2}$, recombination occurs. Since the power radiated is proportional to the frequency, it is advantageous to use semiconductors in which the carriers have a short radiative lifetime τ_r. Another consideration in the design of such a device is that the majority carriers must be driven away from the surface in a time τ_d shorter than one-quarter of a period. Therefore, $1/4f$ must be greater than the larger of τ_d or τ_r. The input power is dissipated primarily in the insulator and in the semiconductor during breakdown. The input power increases also with frequency.

Experimental data with GaP[4] show that the power input increases as the cube of the applied voltage, whereas the radiated power is proportional to the voltage. Therefore, the power efficiency of such an electroluminescent system goes through a maximum which occurs at about double the breakdown voltage.

Note that luminescence occurs also during the avalanche breakdown, producing a broad emission spectrum which is visible to the eye. However, most of the radiation power is emitted during the "forward" bias.

[3] C. N. Berglund, *Appl. Phys. Letters*. **9**, 441 (1966).
[4] F. E. Harper and W. J. Bertram, *IEEE Trans. on Electron Devices*. **16**, 641 (1969).

11-A-3 TUNNELING THROUGH AN INSULATING LAYER[5]

A thin oxide layer separating a metallic electrode from the semiconductor can be used for tunneling electrons to or from the semiconductor (the latter direction is hole injection). Figure 11-6 shows such an "MIS" (metal–insulator–semiconductor) structure. With the metal made sufficiently positive, electrons can tunnel from the valence band of the semiconductor, leaving behind holes and allowing the radiative recombination of conduction-band electrons. This arrangement is especially useful in CdS, in which, in spite of a decade of efforts, *p–n* junctions have not been made. Such a structure has given very bright luminescence in CdS[6] and could possibly lead to a CdS electroluminescent laser. However, the MIS structure is an inefficient hole injector, because electrons from the conduction band of the semiconductor can also tunnel into the metal.

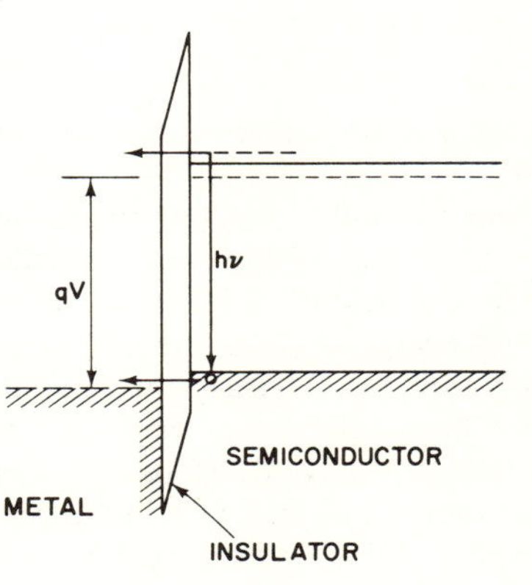

Fig. 11-6 MIS structure with bias V to extract electrons from (or inject holes into) *n*-type semiconductors.

11-A-4 BULK EXCITATION BY IMPACT IONIZATION

A high electric field ($\mathscr{E} > 2 \times 10^5$ V/cm) can cause across-the-gap ionization of electron–hole pairs. The carriers thus generated are accelerated by the electric field and interact with valence electrons via a collision process which causes an avalanche of electron–hole-pair generation.[7] Some of the pairs recombine radiatively, but because of their high kinetic energy, the emission spectrum is quite broad. However, after the electric field is turned off,

[5]For a variety of tunneling schemes, see the review paper by A. G. Fisher "Electroluminescence in II–VI Compounds," *Luminescence of Inorganic Solids*, ed. P. Goldberg, Academic Press (1966), p. 572.

[6]J. H. Yee and G. A. Condas, *Solid State Elec.* **11**, 419 (1968).

[7]K. G. McKay, *Phys. Rev.* **94**, 877 (1954). For a review of impact ionization in *p–n* junctions, see H. Kressel, *RCA Review* **28**, 175 (1967).

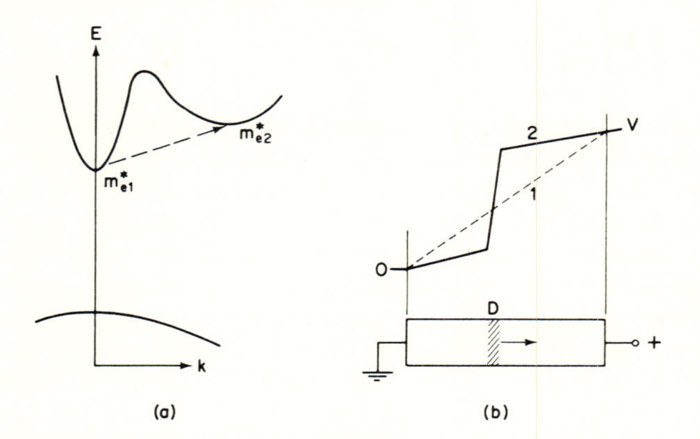

Fig. 11-7 Bulk excitation by Gunn domain: (a) transfer of hot electron to indirect low mobility valley; (b) potential distribution across specimen: 1—before and 2—after formation of traveling domain D.

the carriers thermalize to the band edges and recombine in a narrower spectrum.

One method of creating impact ionization in GaAs is to generate Gunn domains in lightly doped n-type material.[8,9] The electrons are accelerated by the electric field $\mathscr{E} > 2.2 \times 10^3$ V/cm. When the hot electrons have sufficient kinetic energy to populate the indirect valley where their effective mass is m^*_{e2} [See Fig. 11-7(a)], their mobility suddenly decreases to $\mu_2 = q\tau/m^*_{e2}$ and a higher resistivity region or "domain" results [Fig. 11-7(b)]. Most of the applied voltage develops across the high-resistivity domain. The domain being thin, the electric field in this region exceeds the impact-ionization value ($\mathscr{E} \approx 2 \times 10^5$ V/cm) and generates many electron–hole pairs. Their radiative recombination forms a light source which propagates with the Gunn domain.[8] When the field is turned off (or in the wake of the domain), the hot carriers thermalize to the edges of the bands and make a direct radiative transition. When a Fabry–Perot cavity is provided, coherent radiation is obtained.[10] Although this is a bulk process, the active regions are still small, as evidenced by the size of the conical lasing beam which emerges within a 7° angle from a normal to the surface.

Lasing after avalanche breakdown in GaAs has been obtained in a p^+pp^+ structure.[11] Here, the avalanche breakdown occurs in the high-resistivity p-type region. The resulting electrons are injected into the positively biased p^+ region, where they recombine radiating coherently.

[8] J. S. Heeks, *IEEE Trans. Electr. Devices* **13**, 68 (1966).
[9] P. D. Southgate, *J. Appl. Phys.* **38**, 6589 (1967).
[10] P. D. Southgate, *IEEE J. of Quantum Elec.* **4**, 179 (1968).
[11] K. Weiser and J. F. Wood, *Appl. Phys. Letters* **7**, 225 (1965).

11-B Optical Excitation

In optical excitation a photon is absorbed by the semiconductor, creating an electron–hole pair which then recombines, emitting another photon. This technique has the advantage that it can be used to excite materials in which contact or junction technology is not adequately developed, or in high-resistivity materials where electroluminescence would be inefficient or impractical. Optical excitation permits flexibility in the configuration of the excited region and in the choice of the location on the crystal. In some cases it is possible to select a source of radiation emitting excitation photons having an energy close to the gap energy ($h\nu_i \geq E_g$) and, therefore, close to the en-

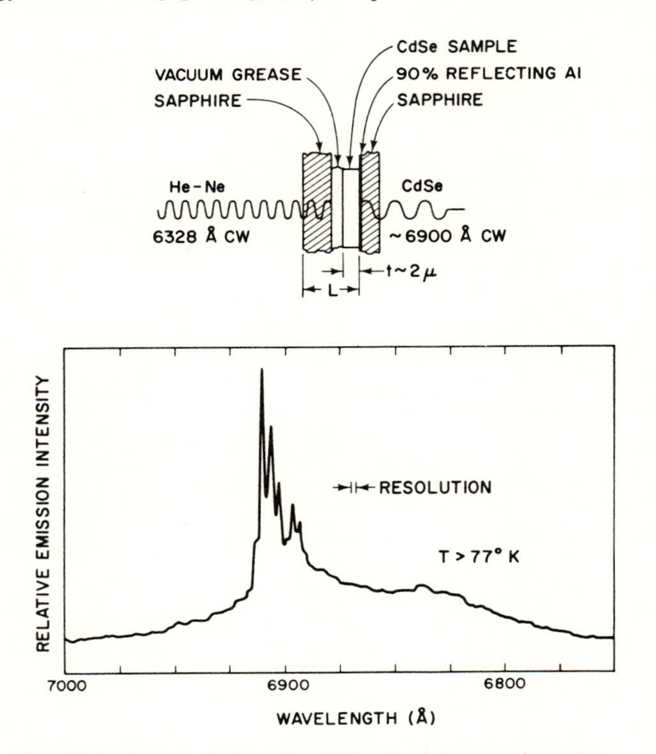

Fig. 11-8 Laser emission of a CdSe platelet pumped continuously well above threshold. Laser emission and mode structure occur on the low energy side of the spontaneous emission spectrum.[12]

[12]M. R. Johnson, N. Holonyak, Jr., M. D. Sirkis, and E. D. Boose, *Appl. Phys. Letters* **10**, 281 (1967).

ergy of the recombination radiation. Hence if the quantum efficiency for the absorption and emission processes is nearly unity, a high power efficiency obtains.

Thus a He–Ne laser has been used to pump a CdSe laser emitting cw visible light at 6900 Å (Fig. 11-8).[12]

As shown in Fig. 11-9, GaAs has been used in conjunction with a cylindrical lens to pump InSb[13] and InAs[14] lasers. In a simpler arrangement (Fig. 11-10), the material to be pumped is fastened with a thin layer of vacuum grease to one facet of the injection laser.[15] This provides optimum optical coupling. Since a peak power density of over 10^6 W/cm^2 can be obtained with a GaAs laser, it is a convenient source of optical-pump power for semiconductors of lower gap. For higher-gap semiconductors, one can use $GaAs_{1-x}P_x$ alloy lasers as optical pumps. Their emission can be extended to about 6400 Å, where peak powers of 10^6 W/cm^2 are still obtainable.[16] With the $GaAs_{1-x}P_x$ laser as a pump, other alloys of $GaAs_{1-x}P_x$ can be excited[15] provided the phosphide concentration x of the pump is higher than that of the target. Alloys of $CdSe_{1-x}S_x$ have also been made to lase by this technique.[15] Using an injection laser to drive another semiconductor laser offers the convenience of measuring the kinetics of the recombination process,

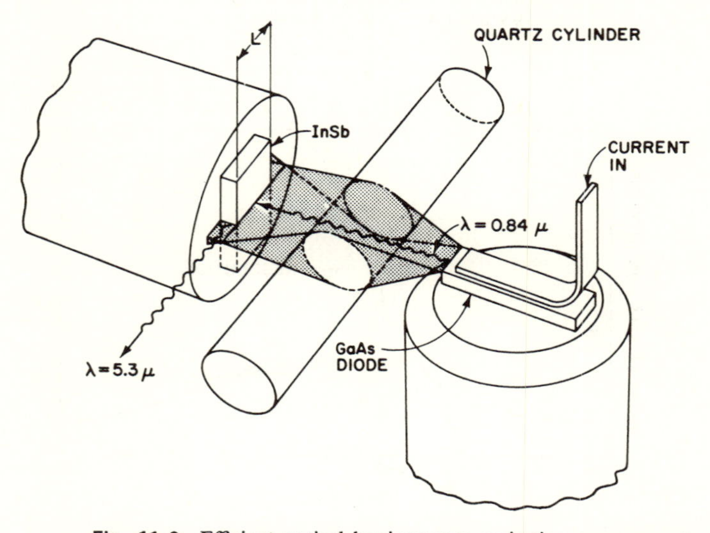

Fig. 11-9 Efficient optical luminescence excitation arrangement using a GaAs injection laser as source.[13]

[13] R. J. Phelan and R. H. Rediker, *Appl. Phys. Letters* **6**, 70 (1965).
[14] I. Melngailis, *IEEE J. of Quantum Elec.* **1**, 104 (1965).
[15] M. R. Johnson and N. Holonyak, Jr., *J. Appl. Phys.* **39**, 3977 (1968).
[16] J. I. Pankove and I. J. Hegyi, *Proc. IEEE* **56**, 324 (1968).

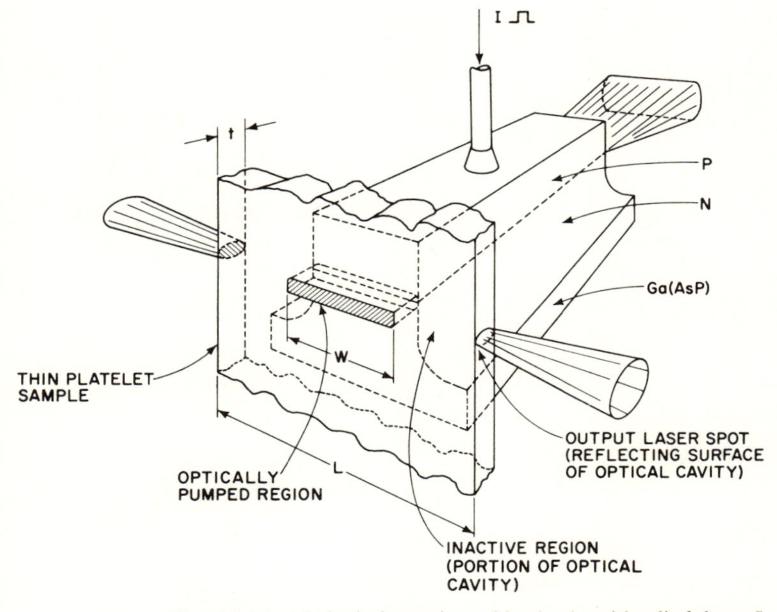

Fig. 11-10 Method of pumping a thin platelet with a diode laser. L may be greater or less than W, and the sample may be rotated to any angle relative to the plane of the diode junction. The platelet is heat sunk directly to the face of the diode, which in turn is heat sunk directly to the liquid nitrogen cold finger.[15]

because, with the possibility of a high pulse-repetition rate, one can use pulse-sampling techniques.

Extremely intense optical pumping can be obtained with Q-switched ruby lasers, intense enough to damage the semiconductor.[17] In a clever experiment (Fig. 11-11), the 30-MW output of a ruby laser is passed through liquid nitrogen, where vibrational excitation (Raman scattering) adds sidebands to the incident radiation. The first lower-frequency sideband (the first Stokes-shifted radiation) occurs at a wavelength of 8281 Å. This corresponds to a photon energy of 1.49 eV, which is weakly absorbed by GaAs with an energy gap of 1.51 eV at 77°K. The excitation is quite intense, because 8% of the ruby-laser output is converted into the 1.49-eV radiation. The GaAs sample lases with an output of 200 kW with a quantum efficiency of 47%. The GaAs laser is excited to a penetration depth of 0.5 mm; hence the cross-section of the active region is much larger than in an injection laser and the output beam is correspondingly narrower (divergence = 3°).

[17]N. G. Basov, A. Z. Grasiuk, V. F. Efimkov, I. G. Zubarev, V. A. Katulin, and Ju'. M. Popov, *J. Phys. Soc.* (Japan) **21**, supplement, 277 (1966).

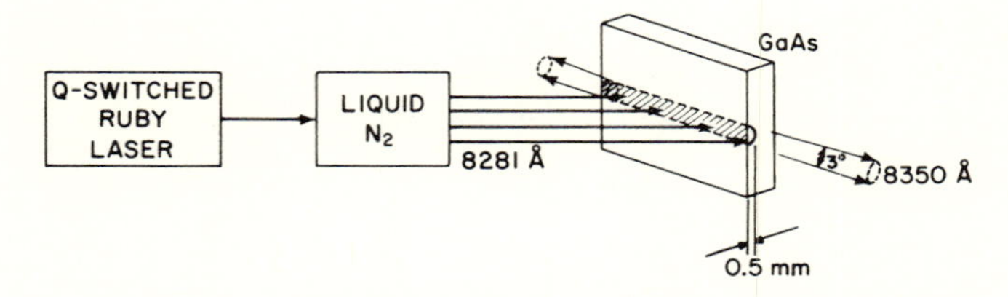

Fig. 11-11 Excitation of GaAs with powerful optical excitation derived from Raman scattering of ruby light.[17]

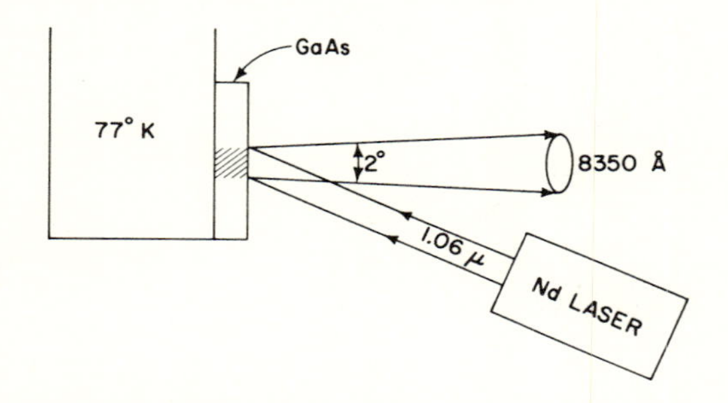

Fig. 11-12 Deep excitation of a GaAs laser by two-photon absorption.[17]

It is also possible to excite a semiconductor with radiation of much lower energy than gap energy, provided the incident energy is very intense. Nonlinear optical properties of the semiconductor allow two coherent photons to effectively add as a combined photon of doubled energy. The photon-doubling process has been observed with the 1.06-μ radiation from a Nd laser and has been used to pump GaAs.[17] The 1.06-μ radiation is very penetrating in GaAs, and by the doubling process can excite electron–hole pairs throughout the thickness of a thin slab mounted on a cold finger (Fig. 11-12). With a 6-MW/cm^2 incident 1.06-μ radiation, 60 kW/cm^2 of coherent 8350-Å radiation can be obtained from the GaAs. Because of the geometry used, the laser has a large radiation cross-section and the output emerges in a narrow beam (divergence = 2°).

11-C Electron-beam Excitation

Like optical excitation, electron bombardment permits the study of semiconductors for which electrode and junction technology is not sufficiently advanced to allow electroluminescence studies. Furthermore, the exciting electrons have far more energy than photons and, therefore, are suitable for the excitation of large-gap semiconductors. The most energetic electrons that can be used without inducing radiation damage have a kinetic energy of about 200 keV. However, in spite of radiation-induced damage, cathodoluminescence can be obtained with still higher excitations. The beam can be easily moved, focused, modulated, or pulsed. One of the drawbacks of electron-beam excitation is that the electron-density distribution across the beam is Gaussian. An aperture can partly solve this problem by transmitting only electrons near the peak of the distribution. The other drawback consists of the large losses, which we shall discuss soon.

Perhaps the most sophisticated equipment for electron-beam excitation of semiconductors is the scanning electron microscope.[18] This machine, schematically illustrated in Fig. 11-13, probes the semiconductor with a beam as small as 200 Å in diameter. As the beam scans a raster or a line across the specimen or a portion thereof, its effect can be monitored on a synchronously scanned kinescope. The display on the kinescope is a map of the semiconductor. Several types of mapping can be displayed:

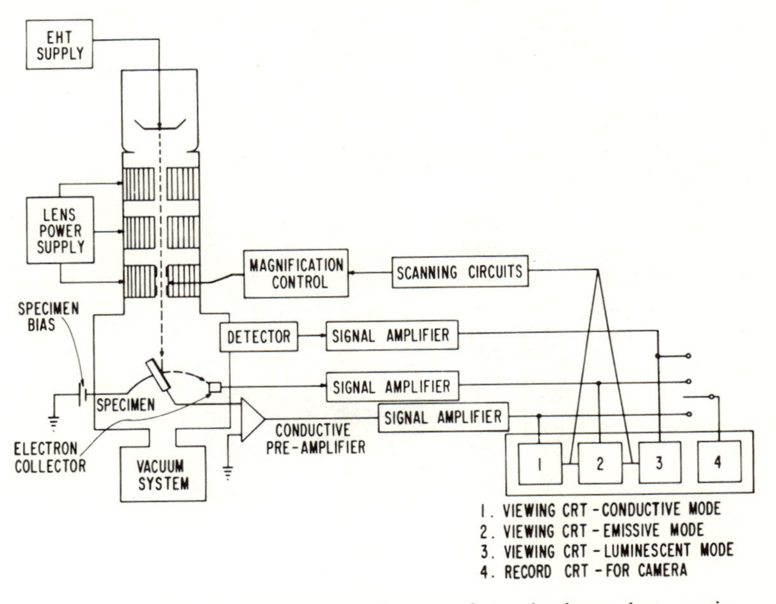

Fig. 11-13 Schematic diagram of scanning beam electron microscope. A bank of monitors allows a mapping of cathodoluminescence, electron emission and bombardment induced conductivity. The specimen can be translated and rotated.[19]

[18]C. W. Oatley, W. C. Nixon, and R. F. W. Pease, "Scanning Electron Microscopy," *Advan. in Electronics and Electron Phys.* **21**, 181 (1965), ed. L. Marton, Academic Press.
[19]Illustration courtesy of M. Coutts.

1. a cathodoluminescence display derived from a light detector;
2. a morphological display derived from scattered electrons and secondary emission measured by an electron collector;
3. a map of bombardment-induced conductivity by measuring the change in current through a biased specimen; and
4. with a machine equipped with an X-ray analyzer (this is called a "microprobe"), the chemical composition of the bombarded region can be determined via the characteristic X-ray emission spectrum.[20]

An example of cathodoluminescence mapping is shown in Fig. 11-14, which compares the luminescent pattern with etch-pit topography, taken later

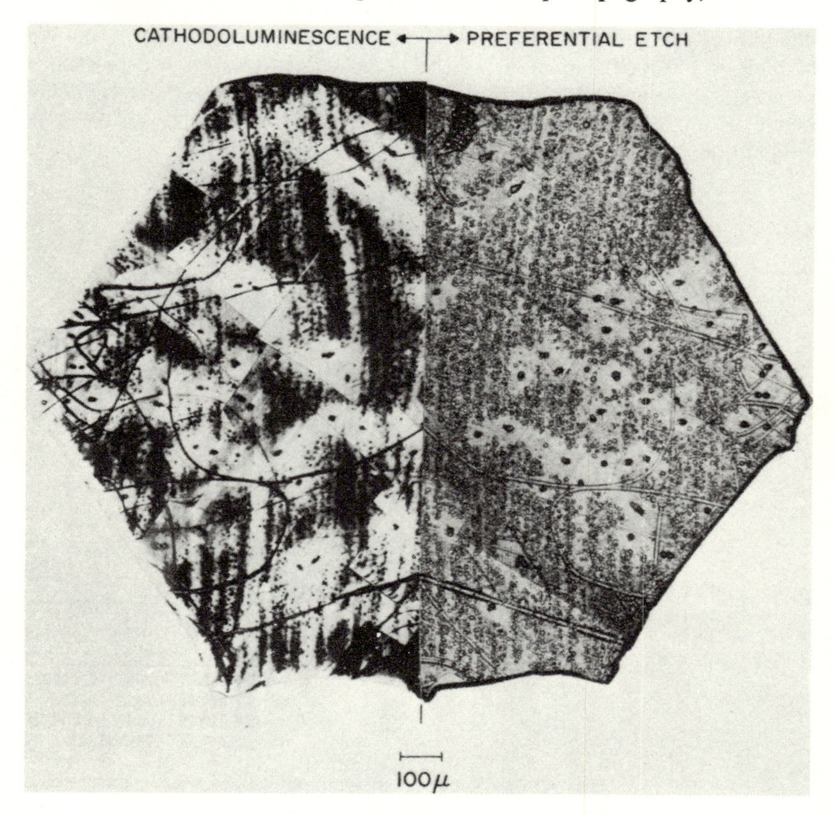

Fig. 11-14 Comparison of the cathodoluminescent and preferential etch features in GaAs. The free electron concentration is 2.2×10^{18} cm^{-3}.[21]

[20]D. B. Wittry, *J. Appl. Phys.* **29**, 1543 (1958).
[21]H. C. Casey, *J. Electrochem. Soc.* **114**, 153 (1967).

on the same specimen (note the matching scratch lines and the correspondence of dark bands with regions of high etch-pit density).

When the incident electron penetrates the semiconductor, it ionizes a copious supply of electron–hole pairs. In this process the incident (primary) electron loses kinetic energy and eventually comes to a stop. This "stopping power," "range," or "penetration depth" is of great interest to us. It can be defined as that distance in the semiconductor where the intensity of the primary electron beam reduces to zero. Experiments with a variety of materials indicate that the penetration depth d varies with the energy $\boldsymbol{E}_p$ of the primary electron as[22]

$$d \approx a\boldsymbol{E}_p^{3/2} \tag{11-1}$$

where a is a constant.

The ionization forms a spherical cloud of electron–hole pairs; the effective radius of this cloud is the "range of excitation" (Fig. 11-15). Some of the primary and secondary electrons are scattered back toward the gun, contributing a loss of excitation power; the rest excite electron–hole pairs.

The energy E_o needed to create an electron–hole pair is best determined by a semiempirical relation:[23]

$$E_o \doteq \tfrac{14}{5} E_g + E' \tag{11-2}$$

where E' represents some integral number of optical phonons lost in the process ($0.5 < E' < 1.0$ eV). The first term of E_o in Eq. (11-2) consists of the bandgap E_g and a "residual" kinetic-energy component equal to $\frac{9}{5}E_g$. Figure 11-16 shows the excellent fit of the semiempirical relation of Eq. (11-2) to available data.

With a knowledge of the ionization energy and of the distribution of

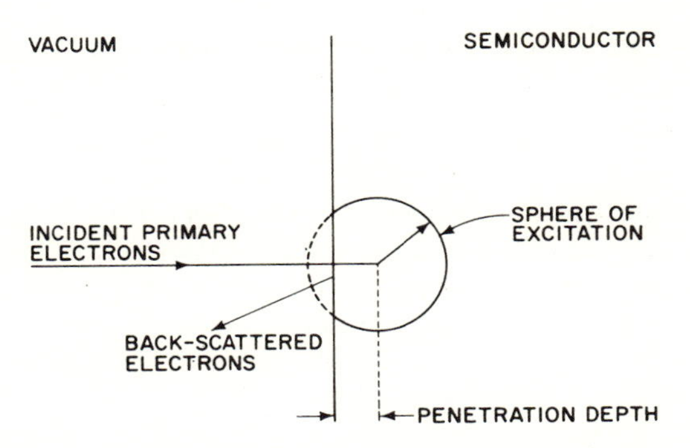

Fig. 11-15 Electron beam excitation of semiconductor.

[22]A. F. Makhov, *Sov. Phys.—Solid State* **2**, 1934 (1960). For various definitions of range, see V. E. Cosslett and R. N. Thomas, *British J. Appl. Phys.* **15**, 1283 (1964).
[23]C. A. Klein, *J. Appl. Phys.* **39**, 2029 (1968).

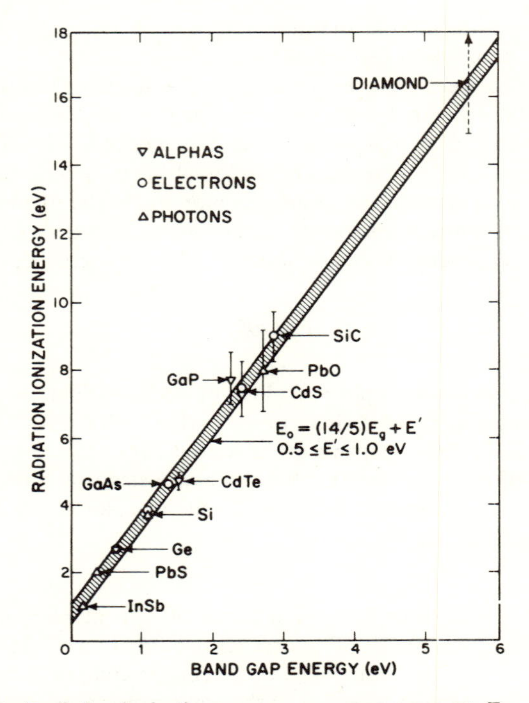

Fig. 11-16 Radiation-ionization energy, or average amount E_o of incident radiation energy (γ rays, fast electrons, or α particles) consumed per generated electron–hole pair, as a function of the bandgap energy E_g.[23]

energy loss, one can estimate the pair-excitation rate. Thus the pair density N_p is given approximately by

$$N_p = \frac{J}{1.6 \times 10^{-19}} \frac{E_p}{E_o} \frac{\tau}{d} \tag{11-3}$$

where J is the current density of the electron beam [$J/(1.6 \times 10^{-19})$ is the number of electrons/cm²/sec incident on the semiconductor]; E_p/E_o is the number of pairs created per incident electron; τ is the pair lifetime; and d is the effective-penetration depth of the primary electron.

Since the radiative recombination of the electron–hole pair emits a photon of energy $h\nu \approx E_g$, the maximum power efficiency to be expected from cathodoluminescence is

$$\eta = \frac{E_g}{2.8E_g + E'} \tag{11-4}$$

where η is of the order of 30% for GaAs. The energy which is not radiated is converted into heat. With the several hundreds of watts of coherent emission obtained in GaAs[24] and in CdS,[25] the burnout of the sample is avoided by using very short pulses.

Problem 1. Find under which conditions the kinetic energy of a carrier should be $\frac{3}{2}E_g$ in order to create an electron–hole pair by impact ionization across the energy gap E_g of a semiconductor.

[24]O. V. Bogdankevich, N. A. Borisov, I. V. Kryukova, and B. M. Lavrushin, *Sov. Phys. Semicond.* **2**, 845 (1969).
[25]C. E. Hurwitz, *Appl. Phys. Letters* **9**, 420 (1966).

PROCESSES INVOLVING COHERENT RADIATION

12

Here we shall consider a variety of phenomena which occur in response to excitation by coherent radiation. Some of these effects are due to a change in the recombination process, a change more readily achievable with laser light, because of its high intensity, than with incoherent radiation. Other effects are strongly dependent on the precise phase relationship between two oscillations. One of these oscillations is the incident coherent radiation; the other oscillation may be another electromagnetic wave or a coherent lattice vibration. We shall see how radiation can be amplified or attenuated and how its frequency can be shifted.

12-A Photon–Photon Interactions in Semiconductors

Two very different types of interactions between electromagnetic waves will be described: first, how a laser beam modifies the emission characteristics of an injection laser; second, how coherent photons can interact through the nonlinear optical properties of a semiconductor to generate optical harmonics and produce beat frequencies (heterodyning).

Our first category, which deals with the interaction between lasers, is of considerable practical interest. Great expectations have been placed on the use of semiconductor lasers in computers. The swift propagation and high resolving power of light promises fast processing of closely packed data. Semiconductor lasers, by virtue of their rapid response and small size, seem particularly suitable for computer applications. Of special interest is the ability of semiconductor lasers to perform logic operations, in which all the processing signals are in the form of optical energy, the electrical input being only a source of power rather than of information.

Laser digital devices have been developed in which amplification, saturable absorption, and quenching of lasing are the basic operant processes. Thus optical switches, gates, multivibrators, adders, half-adders, inverters, etc. have already been demonstrated.[1]

12-A-1 QUENCHING OF A LASER BY ANOTHER LASER[2,3]

Consider the structure of Fig. 12-1, where the output of laser 1 can traverse laser 2 in the plane of its junction, but transversely to the propagation direction of the lasing modes in laser 2. The population inversion in laser 2 is uniformly distributed, but the coherent radiation propagates in the direction for which the stimulation is greatest as provided by the feedback from the Fabry–Perot facets. When the coherent output of laser 1 traverses laser 2, it stimulates the latter to emit, in the same direction as the incident beam (which is transverse to laser 2's normal lasing mode.) When the intensity of the radiation from laser 1 is more intense than that of the normal mode, the inverted population in laser 2 is stimulated to emit in the transverse direction more strongly than in the normal longitudinal direction. Hence less inverted population is available for the normal lasing mode and laser 2 stops lasing. The power robbed from the lasing process is added to the quenching radiation, which then emerges from laser 2 amplified. The amplifying properties of this structure will be discussed in the next section.

To improve the ease in quenching the laser, the coupling between the two diodes must be optimized. This is achieved by using a common *p–n* junction for the two lasers as illustrated in Fig. 12-2.[4]

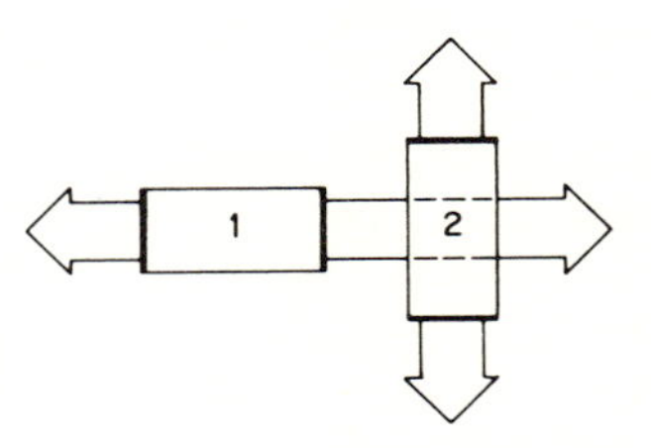

Fig. 12-1 Quenching of injection laser 2 by injection laser 1.

[1] N. G. Basov, V. V. Nikitin, and A. S. Semenov, *Sov. Phys.—Uspekhi*, **12**, 219 (1969); N. G. Basov, V. N. Morozov, V. V. Nikitin, and V. D. Samoilov, *Radio Engineering and Electronic Physics* **9**, 1409 (1969); V. V. Nikitin and V. D. Samoilov, *Sov. Phys.—Semiconductors* **2**, 1012 (1969); A. Kawaji, H. Yonezu, and T. Nemoto, *Proc. IEEE* **55**, 1766 (1967); W. F. Kosonocky, *Optical and Electro-Optical Information Processing*, MIT Press (1965), p. 269.

[2] A. B. Fowler, *Appl. Phys. Letters* **3**, 1 (1963).

[3] G. J. Lasher and A. B. Fowler, *IBM J. Res. & Dev.* **8**, 471 (1964).

[4] C. E. Kelly, *IEEE Trans. on Electron Devices* **ED-12**, 1 (1965).

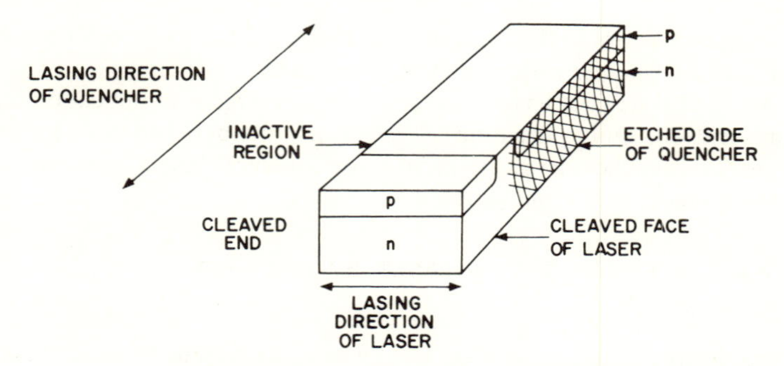

Fig. 12-2 Schematic diagram of laser-quencher pair.[4]

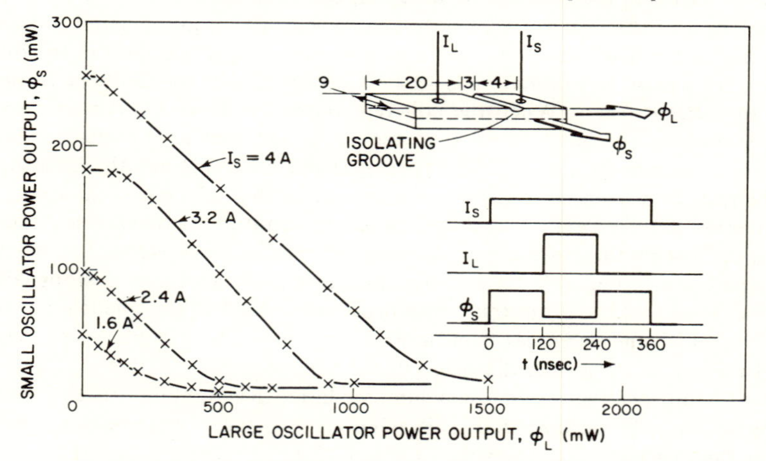

Fig. 12-3 Quenching curves, or transfer characteristics, for the structure shown in the inset. The numbers in the inset are dimensions in mils.[5]

In a laser–quencher pair, the output of the laser decreases linearly with increasing input. The corresponding tansfer characteristic curve is shown in Fig. 12-3. A laser–quencher pair can be designed to operate as an "inverter," a bistable device in computer technology.[6]

In another type of bistable operation, the junction either lases or does not during a constant electrical input.[7] This bistable operation is obtained in a dual-diode structure, where one diode pumps the other diode optically. Both

[5] W. F. Kosonocky and R. H. Cornely, *IEEE, 1968 Wescon Technical Papers*, Part 1, 16/4 (1968).

[6] W. F. Kosonocky, R. H. Cornely, and F. J. Marlowe, *Digest International IEEE Solid State Circuits Conference* (Philadelphia), 1965, p. 48.

[7] M. I. Nathan, J. C. Marinace, R. F. Rutz, A. E. Michel, and G. J. Lasher, *J. Appl. Phys.* **36**, 473 (1965).

diodes being biased at different current densities, the section which is least pumped electrically acts as a saturable absorber: its absorption decreases as the intensity of the radiation incident from the other section increases.[8] (Saturable absorption has been demonstrated in bulk materials: at a flux density of the order of 10^5 W/cm^2 at 1.47 eV and 77°K, the transmission of a slab of Mn-doped GaAs was increased by a factor of 14.)[9] In the dual-diode structure, for the same injection current the device has two stable states. In one state it emits coherent light, in the other it emits spontaneous radiation. The switching is obtained by a short trigger pulse which saturates the absorption. Once the unit has started lasing, the optical flux in the lasing direction is sufficient to keep the absorber saturated. Switching can also be obtained by optical pumping from an external light source.

12-A-2 AMPLIFICATION

An injection laser can be prevented from lasing by applying antireflection coatings on the Fabry–Perot facets [making $R = 0$ in Eq. (9-15)]. Lasing can be prevented also if one end facet is tilted by more than 3° so as not to retrap the reflected radiation into the active region. Without feedback to provide phase coherence, the emission of the forward-biased diode is at best superradiant. If externally generated coherent radiation at a wavelength close to that emitted spontaneously by this diode is introduced into the *p–n* junction, the coherent radiation becomes amplified as it stimulates radiative recombination

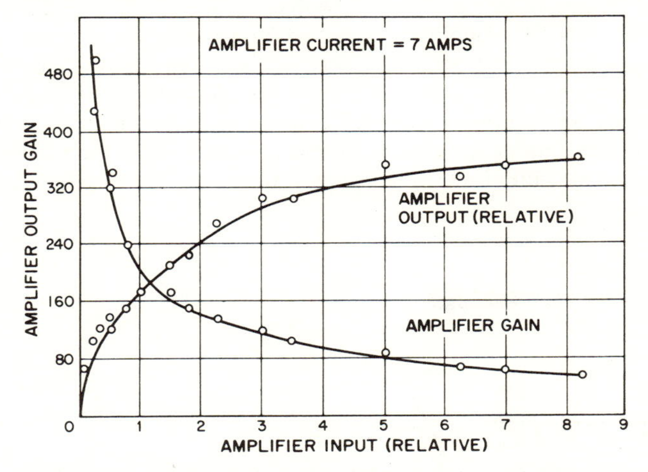

Fig. 12-4 Small-signal gain vs. input intensity and output vs. input for a GaAs diode.[10]

[8] G. J. Lasher, *Solid State Elec.* **7**, 707 (1964).
[9] A. E. Michel and M. I. Nathan, *Appl. Phys. Letters* **6**, 101 (1965).

in the junction. For very small inputs of coherent radiation, gains of more than a thousand have been obtained.[10] However, as the intensity of the input signal is increased, the gain tends to saturate (Fig. 12-4). At complete saturation, all the radiative recombination occurs in the amplified mode. Hence the level of the power output at saturation is comparable to that obtained if the amplifier were operating as a laser. Therefore, the level at which the output saturates increases with the current density through the junction of the amplifier.

Note that, just as in the laser–quencher, the power output at the amplified wavelength increases at the expense of other possible modes within the broad spontaneous spectrum of the amplifier. The reason for this rearrangement of the mode distribution is that the stimulation by the coherent input makes the pair recombination much more probable via the stimulated mode than via the multiplicity of possible spontaneous recombinations. This mode preference is equivalent to syphoning off the inverted population from other possible modes into the amplified one. When the input signal occurs at a wavelength different from that of the spontaneous-emission peak from the amplifier diode at a steady state, the intensity of the spontaneous peak, P_{sp}, decreases with increasing input power P_{in} according to[11]

$$P_{sp}(P_{in}) = P_{sp}(0) \exp\frac{-P_{in}}{P^*}$$

where P^* is a coefficient which ranges typically from 0.04 to 0.08 watts.

In practical application the main problem is the efficient coupling of the laser oscillator to the amplifier. A successful arrangement consists in using the same *p–n* junction for both functions, but isolating the oscillator from the amplifier by a narrow gap, as in Fig. 12-5. The resulting improved coupling between laser and amplifier imposes a mode structure on the superradiant emission of the amplifier. Hence with zero coherent input, the power output is a superradiant output which increases with the current through the ampli-

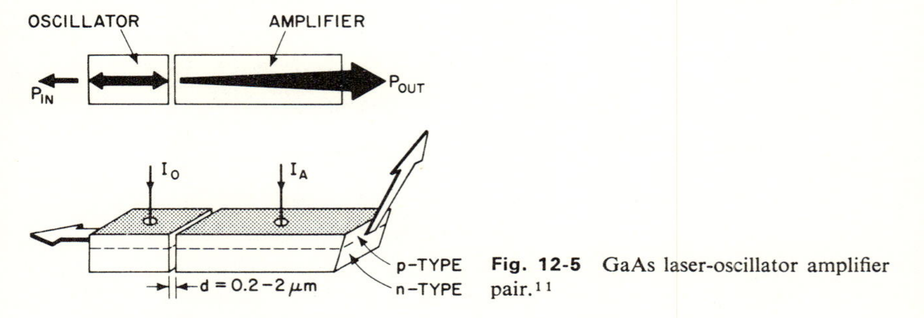

Fig. 12-5 GaAs laser-oscillator amplifier pair.[11]

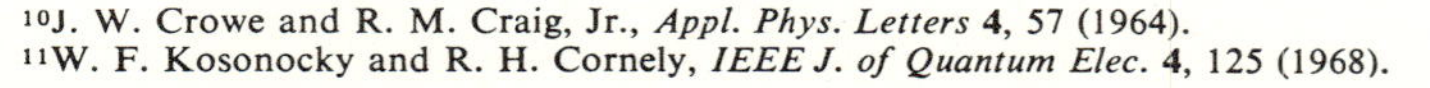

[10] J. W. Crowe and R. M. Craig, Jr., *Appl. Phys. Letters* **4**, 57 (1964).
[11] W. F. Kosonocky and R. H. Cornely, *IEEE J. of Quantum Elec.* **4**, 125 (1968).

fier and is represented in Fig. 12-6 by the intercept of the data with the P_{out} axis. Hence to find the gain of the amplifier, one must subtract from the output the superradiance obtained in the absence of any input and also the power transmitted passively by the amplifier when it is not energized (dashed data curve $I_A = 0$). The result of these subtractions is shown as the family of "calculated power curves" in Fig. 12-6. In practice, a gain of 150 or less is obtained at an input of 100 μW per micron of junction width, and the gain saturates at an input signal of about 400 μW per micron of junction.[11]

One may wonder about the wisdom of replacing a simple-diode laser by an oscillator–amplifier pair. Recall that the injection laser operated at high power density emits many modes simultaneously. The main advantage of the coupled pair is that it can achieve at a high power density the single-mode operation which the injection laser can produce only at low excitation.

The spontaneous output of the amplifier is "noise." A low-noise amplifier can be made by devising a structure which at steady state has practically no output in the desired direction. As shown in Fig. 12-7, the sides of the amplifier are made into Fabry–Perot facets, so that at the operating bias the amplifier lases in a direction perpendicular to the direction of desired gain. Compared to the amplifier structure of Fig. 12-5, the present arrangement has

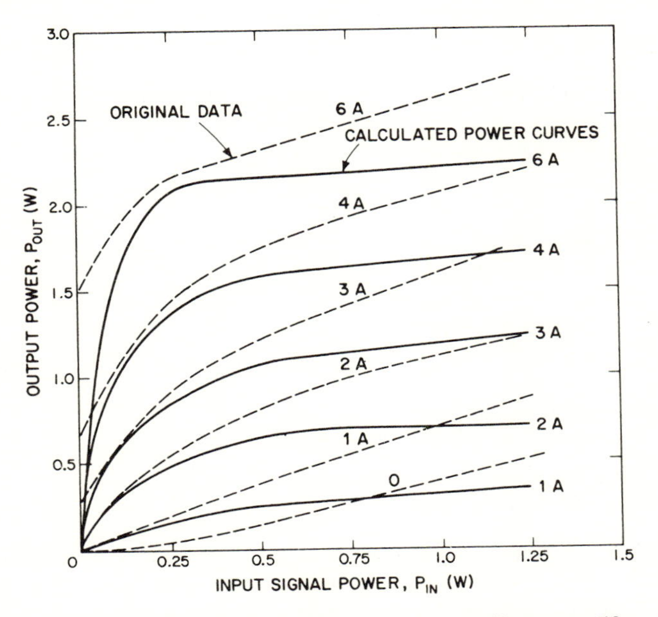

Fig. 12-6 Transfer characteristics of laser-oscillator amplifier pair of Fig. 12-5. The parameter labeling each curve is the current I_A through the amplifier.[11]

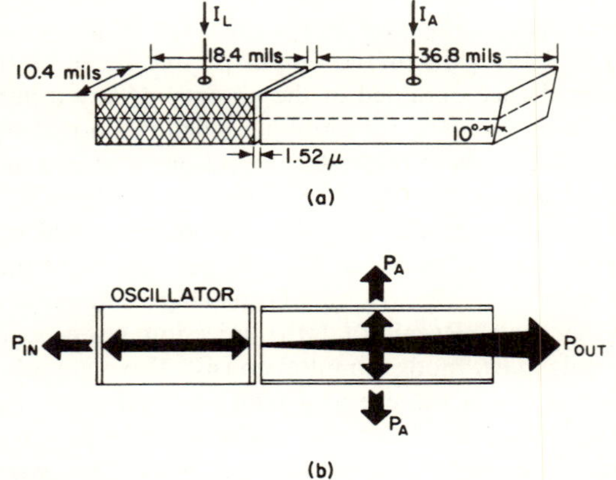

Fig. 12-7 Low-noise oscillator-amplifier pair. The double lines in diagram (b) represent Fabry–Perot facets.[11]

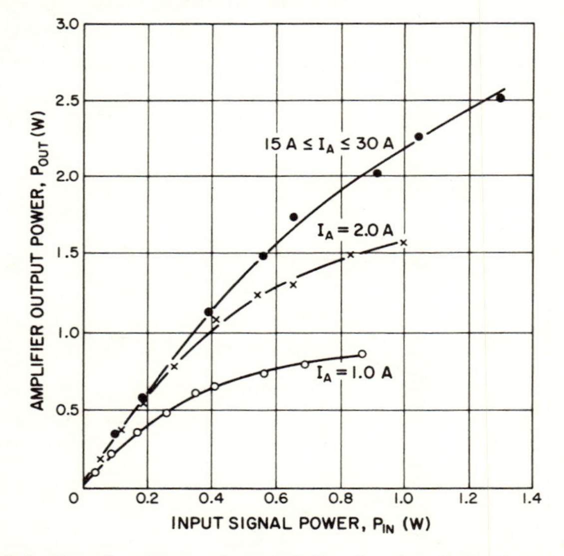

Fig. 12-8 Transfer power characteristics for the GaAs laser amplifier shown in Fig. 12-7.[11]

no steady-state output in the gain direction, since most of the recombination emits in the transverse lasing mode. When a coherent input is introduced from the oscillator section, the device operates as a laser–quencher: the coherent input is amplified at the expense of the transverse modes. The penalty for obtaining amplification at low noise is that a higher level of input is necessary than for the arrangement of Fig. 12-5 before coherent output is obtained. Figure 12-8 shows the performance of the amplifier of Fig. 12-7. A comparison with the data of Fig. 12-6 shows that there is no saturation over a large range of input power.

12-A-3 HARMONIC GENERATION

A material is said to be optically either linear or nonlinear depending on whether or not its polarization is proportional to the driving electric field. At high electric fields, the electrical susceptibility of a material and, therefore, its index of refraction are no longer constant. Thus when an intense sinusoidal electromagnetic wave from a laser interacts with a semiconductor, the wave shape becomes distorted as the electric vector goes through its maximum value. A distorted wave contains harmonics of the fundamental frequency.

To get a better feel for this process, recall that the electric field polarizes the semiconductor by aligning the electron–nucleus dipole at each atomic site. Evidently, as the electric field approaches the value of the local field binding the electron to the atom (about 10^7 V/cm), the polarization begins to saturate; eventually, the material breaks down into a plasma. Note that if the material is already partly polarized—as are II–VI compounds, and to a lesser extent III–V semiconductors—the polarization will saturate at a lower field when the electric vector is lined up with the natural polarization than when the electric vector is lined up in the opposite direction. Hence an intense coherent radiation will saturate its polarization every half-cycle. Such a rectification effect generates a second harmonic. In nonpolar semiconductors the saturation of the polarization is symmetric, flattening the sinusoidal wave at either polarity and thus generating third harmonics. It is conceivable that in slightly polar materials, both second and third harmonics could be generated.

By irradiating a GaAs surface with a laser beam the photons of which have a higher-than-gap energy, second harmonics have been produced, thus demonstrating that GaAs is a nonlinear optical material.[12] Optical nonlinearity has been found in several other semiconductors as well, and the dispersion of the nonlinear susceptibility could be measured at the doubled frequency over an appreciable range of photon energy.[13]

Second harmonics have been produced in GaAs[14,15] and InP[16] injection lasers by the intense electric field of the coherent radiation generated within the laser diodes. In view of the high absorbance of the semiconductor at

[12] J. Ducuing and N. Bloembergen *Phys. Rev. Letters* **10**, 474 (1963).
[13] R. K. Chang, J. Ducuing, and N. Bloembergen, *Phys. Rev. Letters* **15**, 415 (1965).
[14] J. A. Armstrong, M. I. Nathan, and A. W. Smith, *Appl. Phys. Letters* **3**, 68 (1963).
[15] L. D. Malmstrom, J. J. Schlickman, and R. H. Kingston, *J. Appl. Phys.* **35**, 248 (1964).
[16] A. W. Smith, M. I. Nathan, J. A. Armstrong, A. E. Michel, and K. Weiser, *J. Appl. Phys.* **35**, 733 (1964).

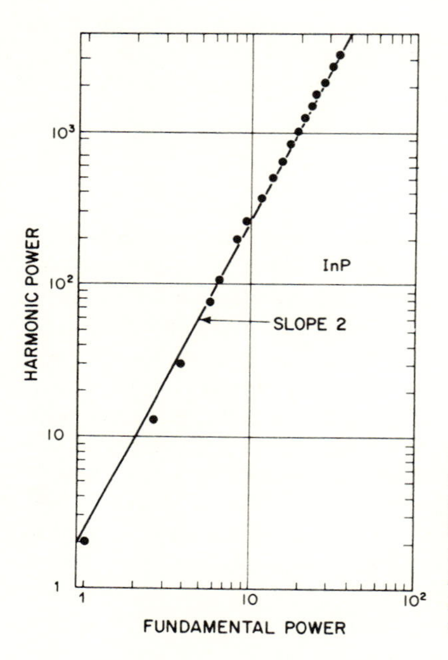

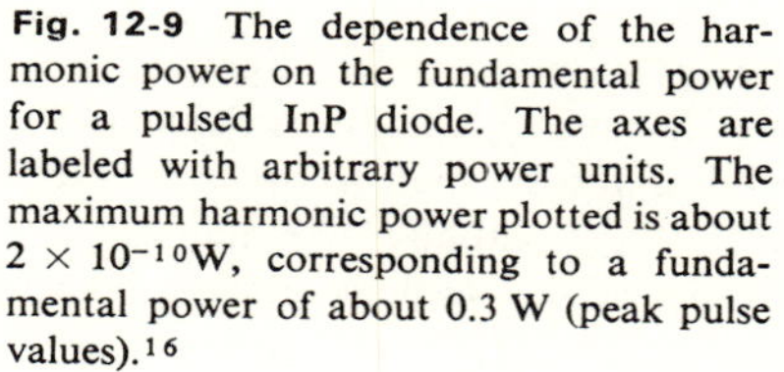
Fig. 12-9 The dependence of the harmonic power on the fundamental power for a pulsed InP diode. The axes are labeled with arbitrary power units. The maximum harmonic power plotted is about 2×10^{-10}W, corresponding to a fundamental power of about 0.3 W (peak pulse values).[16]

photon energies which have approximately twice the energy of the gap, the observed second harmonic must originate in a layer about 100 Å thick at the Fabry–Perot edge of the *p–n* junction. As shown in Fig. 12-9, the power in the harmonic varies as the square of the power in the fundamental—the harmonic may be construed as the product of two fundamental waves.

In nonlinear materials the electric fields of two coherent photons interact through the nonlinear dielectric tensor, which adds up vectorially the displacement vectors. Hence the interaction of the two sinusoidal fields at different frequencies results in the generation of sum and difference frequencies. This process is called "frequency mixing" or "heterodyning."

When the high-power cw emission from a GaAs laser is examined with a high-resolution spectrometer, it is found that the fundamental spectrum contains many modes [Fig. 12-10(a)]. The harmonic generation then clearly appears to be the result of heterodyning: the second-harmonic spectrum comprises the sum frequencies of all the fundamental modes taken in pairs, including a doubling of each fundamental line. Thus in Fig. 12-10(b), which shows the second-harmonic spectrum, each odd-numbered line has twice the frequency of the correspondingly lettered fundamental mode of Fig. 12-10(a) and also contains the sums of pairs of lines which are symmetric with respect to this fundamental. For example, line 9 comprixes H^2, GI, FJ, and EK. The even-numbered lines, however, are made up of an intermediate combina-

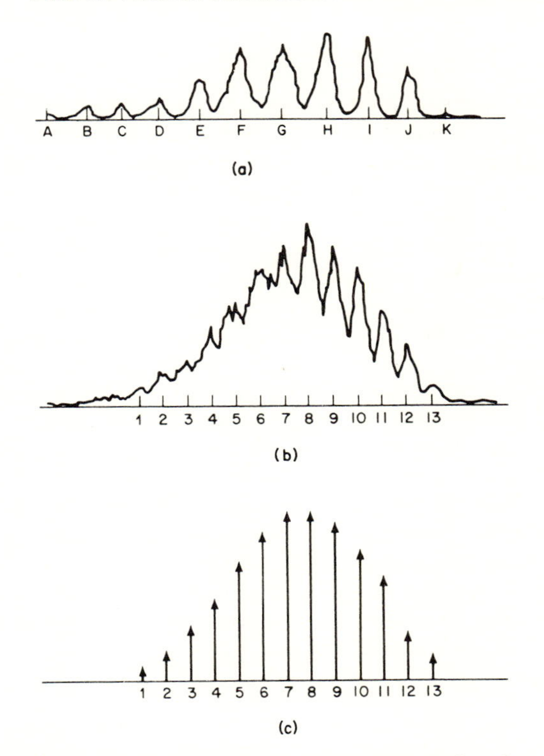

Fig. 12-10 Typical spectrometer traces of fundamental and harmonic spectra[17]: (a) fundamental; (b) observed harmonic; (c) calculated harmonic.

tion of pairs of fundamental lines. Thus line number 10 is made up of the combinations *HI*, *GJ*, and *FK*.

In an injection laser, the beat between adjacent modes generates difference frequencies which, for GaAs, are on the order of 10^{11} Hz. When more than two modes are present, these modes are not identically spaced, because the index of refraction varies with the photon energy $h\nu$; modes of higher frequencies are more closely spaced. Then the beat frequency between two successive pairs of fundamental modes results in the generation of two different frequencies in the 100-GHz region. These two difference frequencies can interact in turn to produce a second-order beat in the low-GHz range. This second order heterodyne interaction has been observed as a GHz modulation of the output of GaAs injection lasers.[18,19]

[17]M. Garfinkel and W. E. Engeler, *Appl. Phys. Letters* **3**, 178 (1963).
[18]L. A. D'Asaro, J. M. Cherlow, and T. L. Paoli, *IEEE J. Quantum Elec.* **4**, 164 (1968).
[19]T. L. Paoli and J. E. Ripper, *Phys. Rev. Letters* **22**, 1085 (1969).

12-A-4 TWO-PHOTON ABSORPTION

Two phase-coherent photons can cooperate in exciting an electron to twice the energy of a single photon. Thus a coherent radiation with $h\nu < E_g$, to which the semiconductor should be transparent, can be absorbed in the cooperative phenomenon. Therefore, electron–hole pairs can be generated by two-photon absorption. Pair generation can then be detected as a photocurrent or as a photoluminescence emitting photons of greater energy than that of the exciting photons.

The semiconductors GaAs, CdSe, and CdTe have been excited with 1.06-μ coherent radiation ($h\nu = 1.17$ eV $< E_g$) from a Q-switched Nd laser to emit near-gap-energy photons;[20] CdS has been optically pumped with the 6943-Å output of a ruby laser[20-22] as well as with the output of an injection laser: 6440 Å with $GaAs_{0.6}P_{0.4}$ and 8400 Å with GaAs.[23] In all of these experiments, the CdS emitted at about 4870 Å, and even coherent emission could be obtained when the excitation from the ruby laser exceeded 60 MW/cm²;[22] ZnS has also been made to lase with two-photon excitation by a ruby laser.[24] The experimental arrangement and results obtained with a GaAs injection laser are shown in Fig. 12-11.

The probability for a cooperative two-photon process to occur should increase quadratically with the photon intensity. In some specimens, the one-photon ($h\nu_1 > E_g$) photoluminescent output L_o has a power-law dependence on the excitation rate L_i:

$$L_o(L_{i_1}) = A(L_{i_1})^n \quad (\text{at } \lambda_1 = 3660 \text{ Å})$$

where n is a constant which is different for the various emission lines of CdS and A is a proportionality constant. A two-photon excitation ($h\nu_2 < E_g$) for the same set of emission lines results in a squared photoluminescent dependence on the excitation rate L_{i_2}; therefore,[21]

$$L_o(L_{i_2}) = B(L_{i_2})^{2n} \quad (\text{at } \lambda_2 = 6943 \text{ Å})$$

where B is another proportionality constant.

Thus far we have considered the two-photon absorption as a $2h\nu$ excitation of an electron. In this case the excitation might be viewed as a transition via a virtual state at an energy $h\nu$ above the initial state. Now let us consider another cooperative two-photon process resulting from a Franz–Keldysh effect.

[20] N. G. Basov, A. Z. Grasiuk, V. F. Efimkov, I. G. Zubarev, V. A. Katulin, and J. M. Popov, *J. Phys. Soc.* (Japan) **21**, Supplement, 277 (1966).

[21] R. Braunstein and N. Ockman, *Phys. Rev.* **134**, A499 (1964).

[22] N. G. Basov, A. Z. Grasiuk, I. G. Zubarev, and V. A. Katulin, *Sov. Phys.—Solid State* **7**, 2932 (1966).

[23] J. I. Pankove and A. H. Firester, unpublished results.

[24] S. Wang and C. C. Chang, *Appl. Phys. Letters* **12**, 193 (1968).

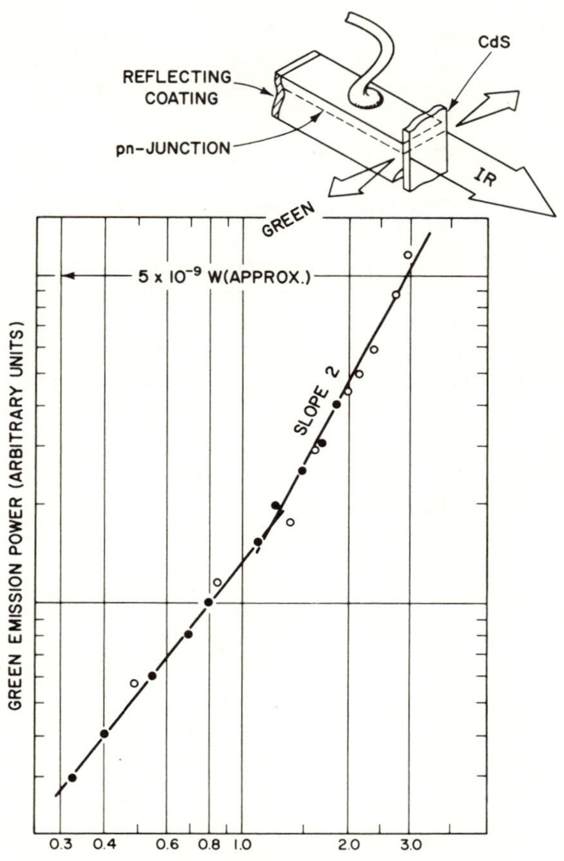

Fig. 12-11 Dependence of green emission intensity on intensity of two-photon excitation by the arrangement shown in inset. The dots show the spectrally integrated output in the green arrow direction; the circles were obtained with an IR cutoff filter to verify that the low-level signal was not due to scattered exciting radiation (the data has been adjusted for filter attenuation).[23]

In Sec. 3-A-6 we saw that a strong electric field shifts the absorption edge to lower energies. Accordingly, the intense electric field of a coherent electromagnetic wave can modulate the absorption edge at the optical frequency.[25] Then if two coherent radiations at different frequencies are incident on a semiconductor, the absorption of one frequency component is

[25]R. Braunstein, *Phys. Rev.* **125**, 475 (1962).

modulated by the oscillating electric field of the other frequency component. Of course, this modulation process requires that one of the two frequencies falls within the absorption edge of the semiconductor.

12-A-5 FREQUENCY MIXING

Note that in the heterodyning process, difference frequencies as well as sum frequencies are produced. Note also that the interaction process can be more complex than two-photon absorption with the consequent increase in carrier concentration. Thus not only photoluminescence can be obtained, but also phenomena pertinent to the presence of free carriers: increased conductivity, field dependence of conductivity, and free-carrier absorption.

Difference frequencies were obtained between the pure stable 10.6-μ

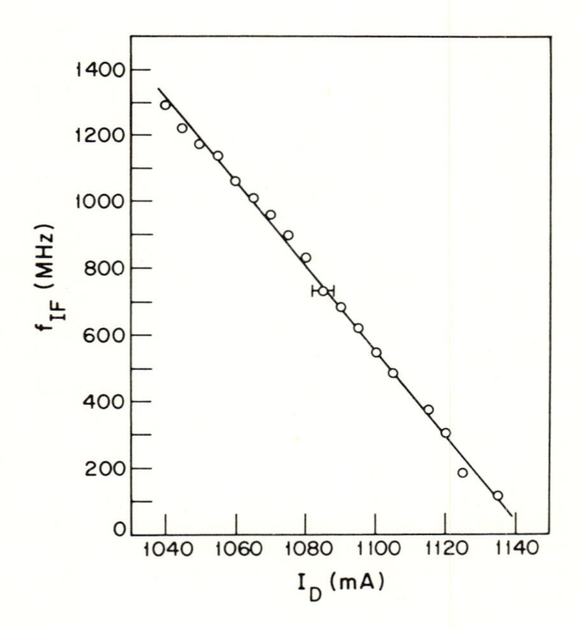

Fig. 12-12 Beat frequency as measured on a spectrum analyzer for various values of diode current. Slope of 12.8 MHz/mA corresponds to that of diode laser mode closest to the P-20 transition of the CO_2 laser. Heterodyning with the P-18 transition of the CO_2 laser results in beat frequencies tunable over the 300 to 3500 MHz range.[26]

[26]E. D. Hinkley, T. C. Harman, and C. Freed, *Appl. Phys. Letters* **13**, 49 (1968).

output of a CO_2 laser and the tunable output of a $Pb_{0.88}Sn_{0.12}Te$ injection laser by heterodyning these two outputs in a Ge detector.[26] The output of the semiconductor laser could be thermally tuned in the vicinity of 10.6 μ by adjusting the current through the diode. Since the two lasers operate at optical frequencies similar to within the fifth or sixth decimal place, their mixing produces a signal tunable over the 50-to-3500-MHz range (Fig. 12-12). Microwaves have also been produced in GaAs by beating two adjacent CO_2 vibrational lines.[27]

Third-order combinations are also possible. Thus the $\lambda_1 = 10.6$-μ and the $\lambda_2 = 9.2$-μ lines of a CO_2 laser have been mixed in InAs and in GaAs to produce the following third-order results:[28]

$$\begin{aligned} 11.8\ \mu &\quad \text{due to} \quad 2\nu_1 - \nu_2 = \nu_3 \\ 8.7\ \mu &\quad \text{due to} \quad 2\nu_2 - \nu_1 = \nu_3 \\ 3.53\ \mu &\quad \text{due to} \quad 3\nu_1 = \nu_3 \end{aligned}$$

When the same two CO_2 lines are mixed in Si or Ge, which have refractive indices different from those of InAs and GaAs, the third-order process yields[29]

$$\begin{aligned} 12.4\ \mu &\quad \text{due to} \quad 2\nu_1 - \nu_2 = \nu_3 \\ 8.2\ \mu &\quad \text{due to} \quad 2\nu_2 - \nu_1 = \nu_3 \end{aligned}$$

12-B Photon–Phonon Interactions in Semiconductors

A photon of energy $h\nu$ can interact with a set of oscillators which resonate at a lower frequency ν_o to produce beat frequencies. In semiconductors there are always two sets of oscillators with which photons can interact. These are the optical and the acoustical modes of lattice vibration. The interaction with optical phonons is called Raman scattering; the interaction with acoustic phonons results in Brillouin scattering.

In the scattering process [Fig. 12-13(a)], the incident photon gives part of its energy $h\nu_i$ to the lattice in the form of a phonon of energy $h\nu_o$ and emerges with a lower energy $h\nu_s$:

$$h\nu_s = h\nu_i - h\nu_o \tag{12-1}$$

This down-converted frequency shift is the Stokes-shifted scattering. Another way of viewing this interaction is that the incident photon excites a phonon, while the semiconductor re-emits what remains of the exciting photon. In general, this re-emission is isotropic. Hence a convenient method of obser-

[27]T. Y. Chang, N. Van Tran, and C. K. N. Patel, *Appl. Phys. Letters* **13**, 357 (1968).
[28]C. K. N. Patel, R. E. Slusher, and P. A. Fleury, *Phys. Rev. Letters* **17**, 1011 (1966).
[29]J. J. Wynne and G. D. Boyd, *Appl. Phys. Letters* **12**, 191 (1968).

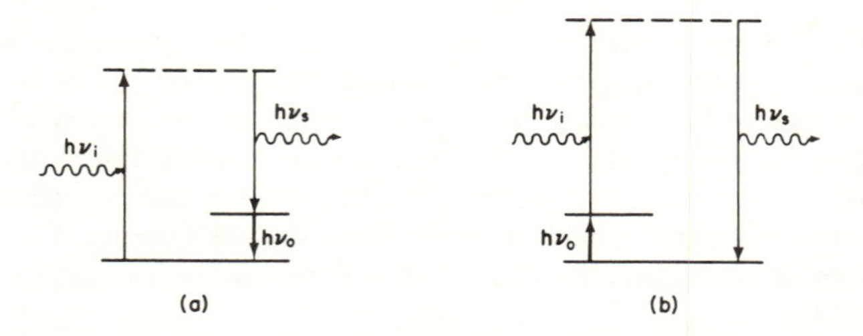

Fig. 12-13 Conservation of energy in photon–phonon scattering with (a) emission, and (b) absorption of phonon $h\nu_o$.

vation is to seek the scattered radiation at some angle to the exciting beam. However, as we shall see later, the directional dependences of Raman scattering and of Brillouin scattering have significant differences.

The exciting photon can raise an electron to a virtual level from which the scattered radiation is emitted. However, if the exciting photon has a greater-than-gap energy, the transition is to a real state and a much stronger excitation results. With nonpenetrating radiation, the scattered photons must be viewed in reflection.

If the lattice of the semiconductor is already in an excited state—i.e., if it already has an appreciable density of phonons—the scattering process can result in the emission of a more energetic photon:

$$h\nu_s = h\nu_i + h\nu_o \tag{12-2}$$

This process is shown in Fig. 12-13(b). These up-converted frequency shifts are the anti-Stokes-shifted scattering modes. Normally, the intensity of the anti-Stokes modes is much weaker than that of the Stokes components, because usually there are few phonons to be absorbed compared to the density of phonons that can be emitted, the probability for absorption being lower than the probability for emission by a factor of exp $(h\nu_o/kT)$. However, the phonons resulting from the Stokes-shifted process (phonon emission) can subsequently participate in the anti-Stokes process (phonon absorption). Hence at strong excitation the Stokes and anti-Stokes components can be nearly equal.

In addition to the conservation of energy treated above, momentum must also be conserved in the photon–phonon interaction. The momentum of a wave is $\hbar k$, where k is the wave-propagation vector. The momentum of the scattered radiation is found by vectorial construction, as shown in Fig. 12-14.

The momentum vector of a photon ($k = 2\pi/\lambda$) is very small compared to the range available to phonons (up to $k = 2\pi/a$, a being the lattice spacing). Because the interaction involves two photons and one phonon, the momentum of the phonon is restricted to small values, extending at most to twice the momentum of a photon.

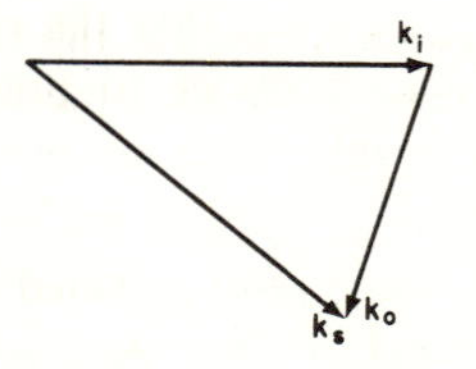

Fig. 12-14 Conservation of momentum in photon–phonon interaction.

Raman scattering and Brillouin scattering are important, accurate tools for studying the phonon spectra of semiconductors at low values of propagation vector. Furthermore, stimulated Raman scattering and Brillouin scattering provide a source of tunable coherent radiation.

12-B-1 RAMAN SCATTERING[30]

As mentioned above, Raman scattering is an interaction of light with the optical modes of lattice vibration. In the scattering process, energy and momentum must be conserved. The optical-phonon dispersion spectrum $h\nu_o(k)$ is practically constant over the range of interest, since the momentum of the incident photon extends only to k_i (Fig. 12-15). The conservation conditions are

$$h\nu_{s_1} = h\nu_i \pm h\nu_o$$
$$k_{s_1} = k_i \pm k_o$$

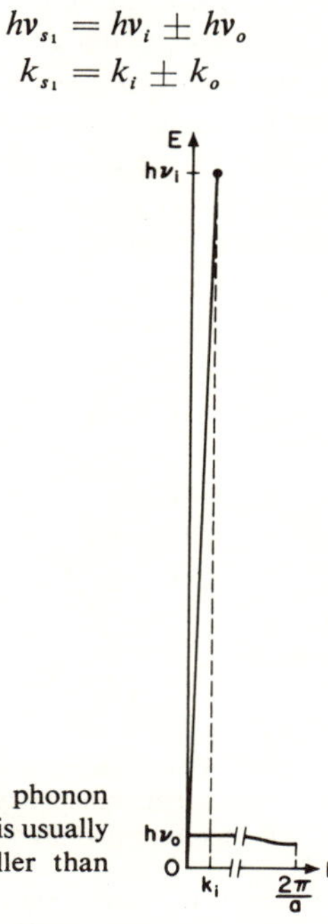

Fig. 12-15 Photon and optical phonon dispersion curves. The value of k_i is usually three orders of magnitude smaller than $2\pi/a$.

[30] R. Loudon, *Advances in Phys.* **13**, 423 (1964).

where the minus sign is used for the Stokes shift and the plus sign applies to the anti-Stokes case. Note that the resulting photon of energy $h\nu_{s_1}$ can in turn interact with the lattice to produce a second photon of still lower energy:

$$h\nu_{s_2} = h\nu_{s_1} - h\nu_o = h\nu_i - 2h\nu_o$$

This cascade process can be repeated m times, emitting

$$h\nu_{sm} = h\nu_i - mh\nu_o$$

However, the intensity of the interaction decreases as the order of the process increases. In CdS, as many as nine orders of Raman scattering have been observed. Figure 12-16 shows the nine peaks shifted to lower energies by multiples of the longitudinal optical phonon. The intensity of the multiphonon shifted radiation is enhanced when its energy coincides with the energy of an exciton.[31,32]

At very high intensities, using coherent light for excitation, stimulated Raman emission can be obtained. In this case the scattered radiation is also coherent and so are the phonons—the semiconductor functions as a parametric amplifier. Momentum conservation determines the direction of propagation of the stimulated Raman emission, which usually occurs at a small angle to the incident beam. In principle it should be possible to obtain stimulated Raman scattering in semiconductors over several orders of interaction, each order being emitted into a separate cone coaxial with the first order and with the incident beam, just as it can be demonstrated with liquids.[33] However, with a semiconductor the cone may be slightly warped due to a possible anisotropy of the optical-phonon dispersion characteristic and to a possible anisotropy of the index of refraction.

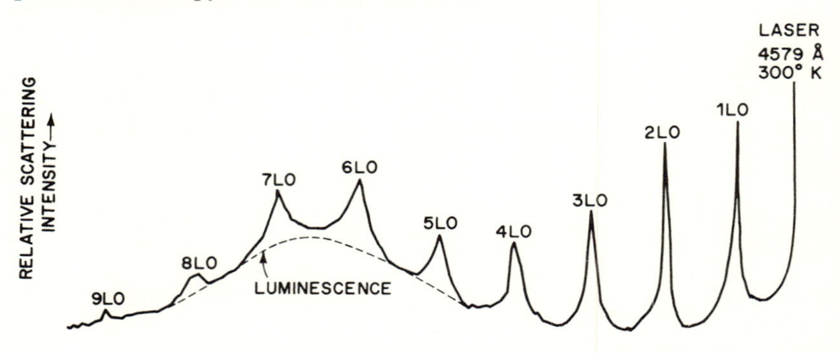

Fig. 12-16 Reflectance spectrum of CdS excited by an argon laser.[31] Note that some of the excitation produces electron–hole pairs and therefore photoluminescence.

[31]R. C. C. Leite, J. F. Scott, and T. C. Damen, *Phys. Rev. Letters* **22**, 780 (1969).
[32]M. V. Klein and S. P. S. Porto, *Phys. Rev. Letters* **22**, 782 (1969).
[33]R. W. Minck, R. W. Terhune, and W. G. Rado, *Appl. Phys. Letters* **3**, 181 (1963).

Recall that at extremely low values of the momentum vector, where the optical phonon is a polariton, a photon-like dispersion curve is obtained. Raman scattering with polaritons occurs only at very small angles with respect to the incident light beam, since k_o must be very small. Then as k_o decreases, the Raman scattering changes from a two-photon–one-phonon to a three-photon scattering process.[34] Stimulated Raman scattering from polaritons is also possible.[35]

12-B-2 BRILLOUIN SCATTERING[36]

In Brillouin scattering, the radiation is scattered from an acoustic wave in the semiconductor. The acoustic phonon can be generated thermally, by the incident photon, or by other means such as the acoustoelectric effect to be discussed in the next section.

A longitudinal acoustic wave consists of a uniformly spaced alternation of regions of higher and lower density. This periodic structure acts as a grating moving at the velocity of sound. Light is scattered by this grating in the same way as X rays are scattered by a lattice (Fig. 12-17). This analogy leads to the expectation that total Bragg reflection will occur when the radiation reflects from the planar wavefronts of the acoustic wave at an angle θ such that

$$2\lambda_A \sin\theta = m\lambda_i \tag{12-3}$$

where λ_A is the acoustic wavelength; m is an integer indicating the order of the scattering; and λ_i is the wavelength of the incident photon inside the semiconductor. Therefore, the scattered radiation emerges at angle 2θ to the incident beam. The dependence of the deflection θ on the wavelength of the acoustic wave can be used to deflect a light beam. The material is driven by a transducer which is energized at the appropriate frequency to induce the desired deflection. A different deflection is obtained by changing the frequency of the acoustic wave. For practical applications it must be remembered that

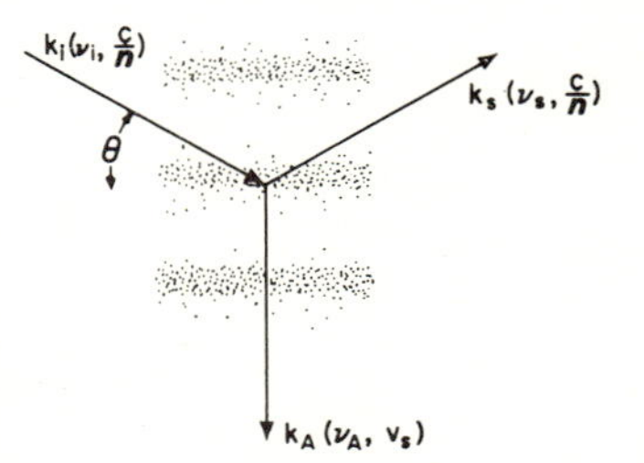

Fig. 12-17 Brillouin scattering. Momentum conservation is expressed by the relationship between the propagation vectors.

[34] C. H. Henry and J. J. Hopfield, *Phys. Rev, Letters* **15**, 964 (1965).
[35] S. K. Kurtz and J. A. Giordmaine, *Phys. Rev. Letters* **22**, 192 (1969).
[36] L. Brillouin, *Ann. de Phys.* **17**, 88 (1922); P. Debye and F. W. Sears, *Proc. Nat. Acad. Sci.* **18**, 409 (1932).

it requires more than 100 nanoseconds for sound to travel across a light beam one millimeter in diameter; this consideration affects the response time of the deflector.

As shown in Fig. 12-17, the motion of the acoustic wave with a velocity v_s ($v_s = 2\pi\nu_A/k_A$) causes a Doppler shift of the optical frequency such that

$$\nu_s = \nu_i - \nu_A \tag{12-4}$$

One could also view Eq. (12-4) as the generation of an acoustic phonon with the emission of the remaining energy $h\nu_s$ at a lower frequency ν_s. Note that when the acoustic wave propagates toward the incoming beam, the Doppler shift increases the frequency of the scattered photon.

Since the dispersion of acoustic phonons, $h\nu_A(k_A)$, allows a continuum of frequencies (Fig. 12-18), Brillouin scattering results in a continuum of Stokes and anti-Stokes components, each being emitted in a different direction. The Stokes components propagate in the general direction of the incoming beam; the anti-Stokes components propagate in the opposite direction.

Stimulated Brillouin scattering can occur at high photon intensities.[37] The electromagnetic energy is then converted into coherent acoustic waves. The photon–phonon coupling is maximum along the optical beam for acoustic waves propagating with or against the photon flux. Because the photon waves and the phonon waves are coherent and phase-locked through the harmonic relation of Eq. (12-3), the semiconductor acts as a parametric amplifier.

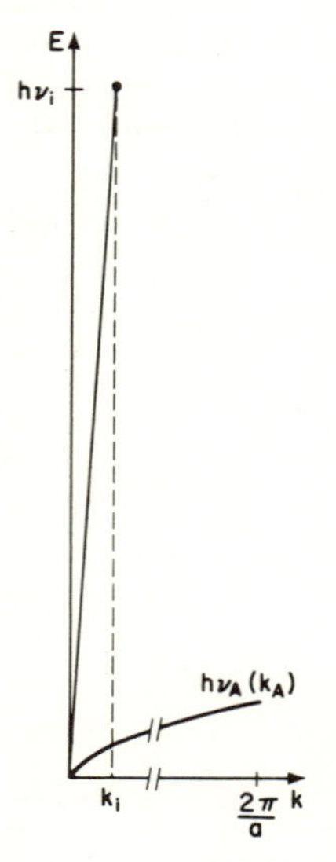

Fig. 12-18 Photon and acoustical phonon dispersion curves. The value of k_i is usually three orders of magnitude smaller than $2\pi/a$.

[37]R. Y. Chiao, C. H. Townes, and B. P. Stoicheff, *Phys. Rev. Letters* **12**, 592 (1964).

The use of lasers has permitted the generation of coherent phonons whose wavelength ranges from the size of the crystal to a few thousand angstroms—i.e., from about 100 kHz to several GHz.

Since phonons are bosons, the number of phonons that can be generated in the coherent acoustic wave—i.e., the phonon intensity—can keep growing along the path of interaction until the mechanical stress fractures the crystal. Stimulated Brillouin emission is believed to account for some of the catastrophic degradation of injection lasser.[38] In semiconductor lasers the cavity

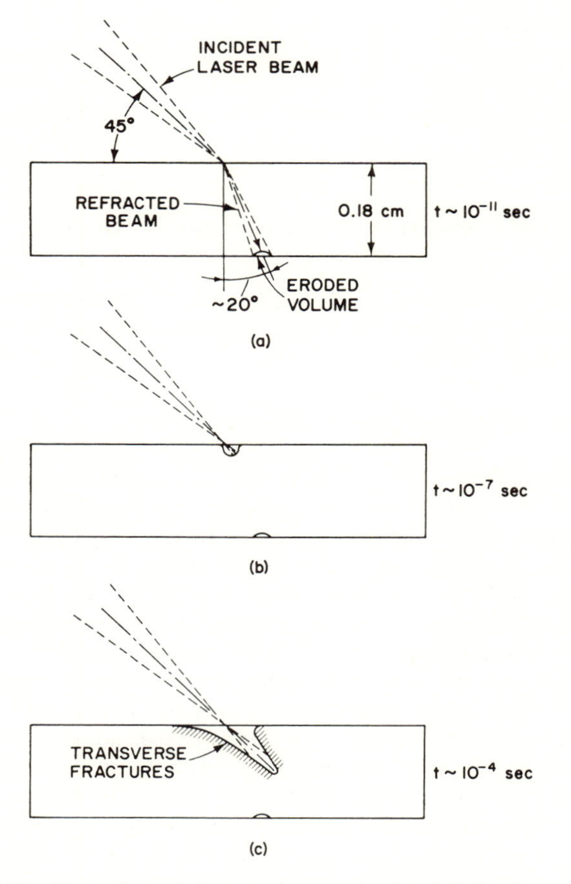

Fig. 12-19 Illustration of damage due to stimulated Brillouin scattering in CdS as a function of time.[39]

[38] H. Kressel and H. Mierop, *J. Appl. Phys.* **38**, 5419 (1967).
[39] D. A. Kramer and R. E. Honig, *Appl. Phys. Letters* **13**, 115 (1968).

facets become pitted at optical-power densities in excess of 10^7 W/cm^2, the surface being the weakest part of the crystal. Mechanical damage due to stimulated Brillouin emission has been demonstrated in CdS traversed by 10^8 W/cm^2 of coherent photons from a ruby laser.[39] As shown in Fig. 12-19, the laser beam, building up the phonon intensity along its path, first blasts out a small pit at the exit facet; then a larger crater develops at the irradiated facet coaxially with the incident beam. This front-facet crater is attributed to stimulated Brillouin emission of coherent phonons propagating in the direction opposite to the laser beam. An examination of the front crater reveals periodic transverse fractures which are spaced 4×10^{-3} cm apart. If this spacing is assumed to equal a phonon wavelength λ_A, a phonon frequency $\nu_A = v_s/\lambda_A = 100$ MHz is obtained ($v_s = 4 \times 10^5$ cm/sec is the longitudinal phonon velocity in CdS). An audible "sonic crack" accompanies the cratering.

12-C Optical Properties of Acoustoelectric Domains

Although all the effects to be described in this section do not necessarily involve coherent radiation, the present topic logically follows the discussion of photon–phonon interactions.

12-C-1 THE ACOUSTOELECTRIC EFFECT[40-42]

The acoustoelectric effect is a cooperative phenomenon between electrons and phonons. Under the influence of an electric field, when the carriers are accelerated to a velocity comparable to the velocity of sound in the material, they transfer some of their kinetic energy to the lattice in the form of phonons. As the electron tends to exceed the sound velocity, it gives up more energy to phonons. Hence after an initial acceleration, the velocity of the electron on the average saturates at the sound velocity v_s. The process of energy transfer continues along the statistical path of the electron, thus building up the intensity of the accompanying phonons. In a uniform n-type semiconductor the phonon build-up which starts at the cathode is the one which can grow the most, since it has the longest path to traverse.

A velocity saturation corresponds to a decrease in mobility and, therefore, to an increase in resistivity. With a constant voltage applied across the specimen, the field distribution along the crystal evolves from an initially

[39] D. A. Kramer and R. E. Honig, *Appl. Phys. Letters* **13**, 115 (1968).

[40] P. O. Sliva and R. Bray, *Phys. Rev. Letters* **14**, 372 (1965).

[41] C. Hervouet, J. Lebailly, P. Leroux-Hugon, and R. Veillex, *Solid State Comm.* **3**, 413 (1965).

[42] H. Hayakawa, M. Kikuchi, and Y. Abe, *Japan J. Appl. Phys.* **5**, 734 and 735 (1966).

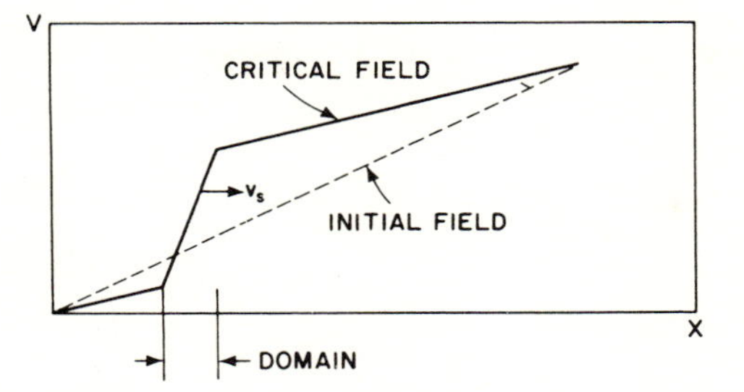

Fig. 12-20 Potential distribution along the semiconductor before and after acoustoelectric domain formation.

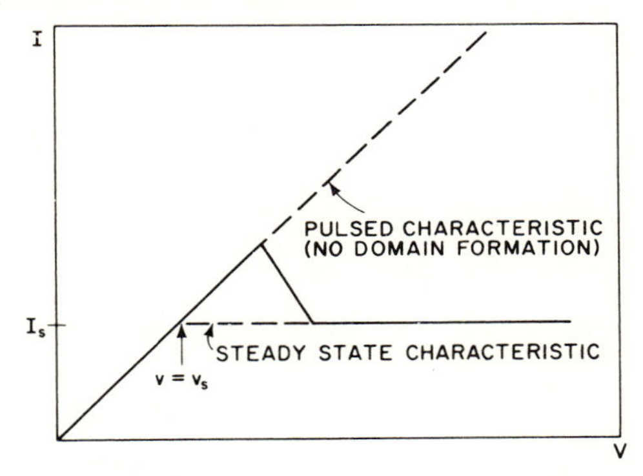

Fig. 12-21 The $I(V)$ characteristics of semiconductors exhibiting the acoustoelectric effect. Under pulsed excitation for a time too short to allow domain formation the ohmic characteristic is obtained. For a long excitation, the steady-state characteristic is obtained, the current saturating at $I_s = qnv_s$ (n is the carrier concentration). For an intermediate duration of excitation the solid curve is obtained.

uniform field to a stepped electric field: as electrons near the cathode reach their saturation velocity, the local resistivity increases, forcing the local field to grow at the expense of the field in adjacent regions (Fig. 12-20). The high-field region, called the "accoustoelectric domain," travels with the velocity of sound. Note that as soon as the domain is formed, an increasing fraction of the applied voltage is developed across the domain, where more electrical energy is transformed into phonons. Hence the domain tends to grow until it eventually saturates when a large portion of the applied voltage appears across the domain, while the field outside the domain drops to the critical value. The critical field is that field at which the electrons travel at the sound velocity.

The initial field being in excess of the critical field before the domain is

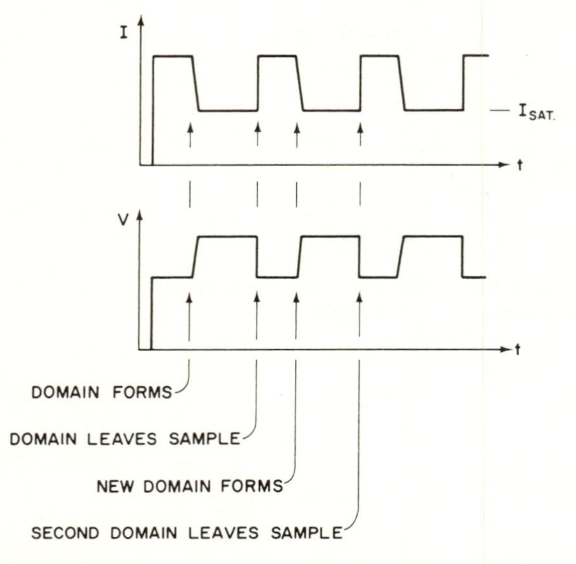

Fig. 12-22 Time dependence of current and voltage resulting from the acoustoelectric effect.

formed, the carriers travel at a velocity greater than v_s and the current, which still obeys Ohm's law, is high. However, while the domain travels along the semiconductor, the current in the external circuit is lower than its initial value, since the carriers now drift at the saturated velocity. The corresponding $I(V)$ characteristic is shown in Fig. 12-21 and time dependences of current and voltage are shown in Fig. 12-22.

When the domain reaches the end of the specimen, the field inside the crystal suddenly increases and the current surges back to its ohmic value. As soon as the domain has been swept out, a new domain forms at the cathode and the current again drops to the saturation value. With a long enough excitation ($t > l/v_s$, $l =$ length of specimen), a number of such cycles can be obtained (Fig. 12-22). If the specimen is not driven at constant voltage (due to a finite circuit resistance), the voltage across the specimen increases when the current decreases; hence the voltage wave shape is a mirror image of the current wave shape.

We must point out that, although energy exchange between electrons and phonons occurs in all semiconductors, the acoustoelectric domain can be formed only in piezoelectric semiconductors. Therefore, the occurrence of acoustoelectric domains depends on crystallographic orientation, just as does the piezoelectric effect (e.g., it is a maximum along the [110] axis in GaAs).

12-C-2 LIGHT TRANSMISSION AT ACOUSTOELECTRIC DOMAIN

Acoustoelectric domains have been observed directly by transmission of photons having a lower-than-gap energy.[43] The use of a light probe has revealed the evolution of the domain: its growing amplitude and simultaneous narrowing as it propagates down the specimen, and eventually its saturation (Fig. 12-23). In the domain, the absorption edge seems shifted to a lower energy. Hence near-gap radiation, to which most of the specimen is transparent, is absorbed at the domain. The increased absorption has been attributed to several mechanisms: enhanced phonon-assisted transition; local thermal narrowing of the gap, since the high phonon density corresponds to a localized hot spot; piezo-optic and Franz–Keldysh effects associated with the high electric field in the domain.

Strain-induced birefringence in the domain changes the polarization of

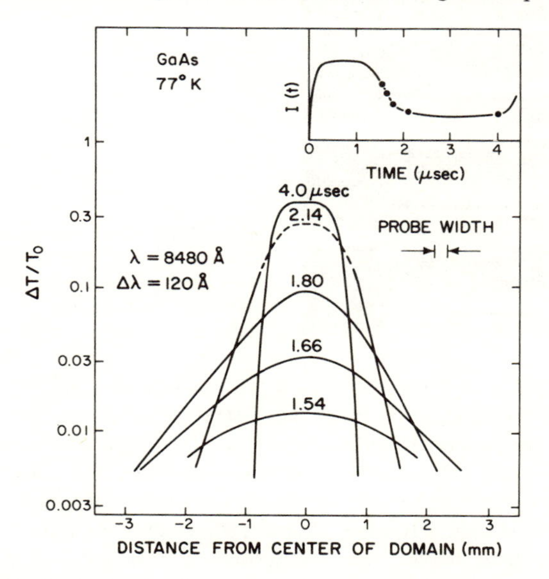

Fig. 12-23 Evolution in shape of the acoustoelectric domain, in terms of the modulation of optical transmission $\Delta T/T_0$. The time at which a given spatial distribution of optical signal was obtained is correlated with the current trace $I(t)$ in the inset. The center of the propagating domain moves with the sound velocity during these stages. For better comparison of the domain shapes, the centers were shifted to a common point.[44]

[43] C. S. Kumar, P. O. Sliva, and R. Bray, *Phys. Rev.* **169**, 680 (1968).
[44] D. L. Spears and R. Bray, *Appl. Phys. Letters* **13**, 268 (1968).

the light traversing the semiconductor. Thus, the motion of an acoustoelectric domain in CdS has been filmed by strobing the specimen with polarized visible light from a $GaAs_{0.6}P_{0.4}$ injection laser.[45] Six frames of this film are shown in Fig. 12-24.

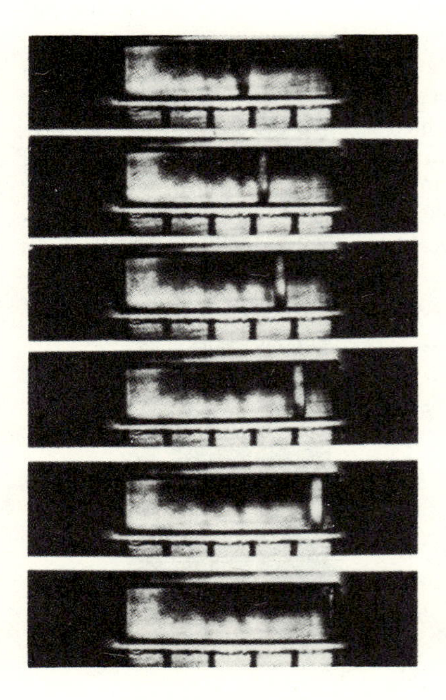

Fig. 12-24 Time evolution of acousto-electric domain. The voltage was applied two microseconds before the first frame. Interval between pictures is 0.25μ sec. Scale below crystal is 1 mm/div.[45]

In principle, two types of phonon can propagate: longitudinal and transverse modes (compressional and shear strains). If these two modes were launched in the specimen and were allowed to propagate in a field-free medium, they would propagate at different velocities, the longitudinal wave propagating faster. Each mode would impart a different rotation to the transmitted polarized light. However, the acoustoelectric domain generates only transverse acoustic waves, which determine the velocity of the domain. If the electric field is turned off when the domain arrives at the end of the sample, upon reflection the energy in the domain is transformed into longitudinal and transverse modes and one can observe both acoustic waves reflected by the anode. Now, in the absence of a driving field, the two modes are free to propagate independently. Their velocities of propagation can be studied by measuring the time of arrival of the two differently polarized transmission signals at various positions along the specimen.[46]

[45]A. R. Moore, *Appl. Phys. Letters* **13**, 126 (1968).
[46]R. Bray, personal communication.

12-C-3 LIGHT EMISSION BY ACOUSTOELECTRIC DOMAIN

A burst of light emission has been observed in the dumbell-shaped specimen of Fig. 12-25 when the acoustoelectric domain reaches the end of the bar.[47] This emission, the spectrum of which corresponds to near-gap recombination, is associated with the sudden collapse of the electric field in the domain when the latter reaches the wider cross-section of the semiconductor near the anode. Since this emission occurs before the domain reaches the electrode, the possibility of injection at a poor (nonohmic) contact is unlikely. The mechanism proposed to explain this emission is as follows: the high electric field in the domain separates electrons and holes into a moving dipole (here, the holes are field-ionized minority carriers)[48-50]; when the electric field collapses at the flared end of the specimen, the electron and hole clouds are allowed to overlap and then to recombine radiatively.

If an array of electrically floating p–n junctions is built on the semiconductor bar, each p–n junction can be caused to emit in turn as the domain passes by.[51-53]

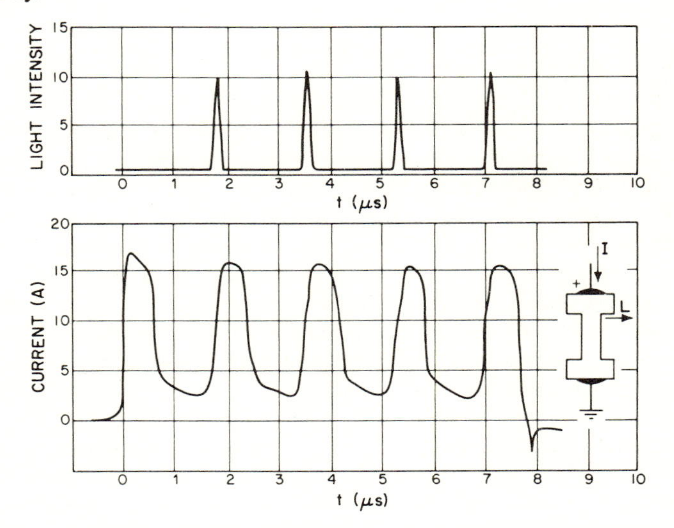

Fig. 12-25 Light emission in n-type GaAs when the acoustoelectric domain reaches the end of the bar in the dumbell-shaped specimen.[47]

[47] A. Bonnot, *Comptes Rendus* **263**, 388 (1966).
[48] S. S. Yee, *Appl. Phys. Letters* **9**, 10 (1966).
[49] N. I. Meyer, N. H. Jørgensen, and I. Balslev, *Solid State Comm.* **3**, 393 (1965).
[50] A. Bonnot, *Phys. Stat. Solidi* **21**, 525 (1967).
[51] B. W. Hakki, *Appl. Phys. Letters* **11**, 153 (1967).
[52] J. I. Pankove and A. R. Moore, *RCA Review* **30**, 53 (1969).
[53] Y. Nannichi and I. Sakuma, *J. Appl. Phys.* **40**, 3063 (1969).

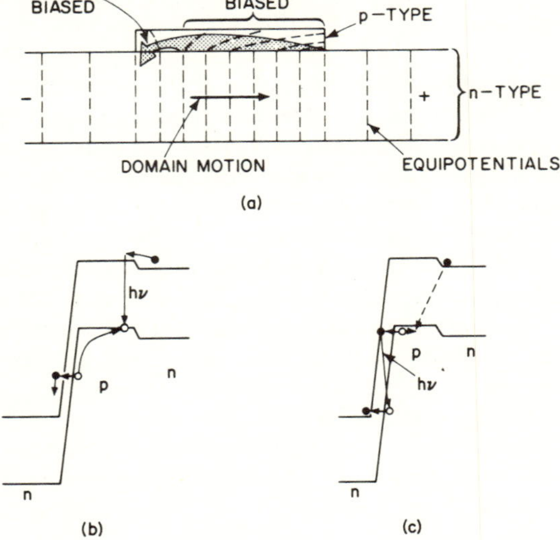

Fig. 12-26 Current flow in a floating *p–n* junction.

A floating *p–n* junction[54] is a structure such as shown in Fig. 12-26, where the net current through the *p–n* junction is zero. In the presence of a potential distribution such that the equipotential planes intersect the *p–n* junction, part of the junction is reverse-biased and the rest of the junction is forward-biased. The ratio of forward-biased–to–reverse-biased areas adjusts itself in such a manner that the forward current is equal to the reverse current. Since the average reverse-current density is much lower than the average forward-current density, the reverse-biased area is much larger than the forward-biased area. The forward-biased portion of the *p–n* junction can emit by injection luminescence, while the reverse-biased portion can generate light in the breakdown mode (Sec. 8-D-3-c).

A floating junction is equivalent to a "*n–p–n* hook" [Fig. 12-26(b)]; one side is reverse-biased and the other is forward-biased. To maintain charge neutrality in the *p*-region, the electrons injected into the *p*-region by the forward-biased junction are compensated by holes collected at the reverse-biased junction (these holes may result from electrons tunneling to the conduction band in the high-field region). The radiative recombination is very efficient in the case of two-carrier flow to the *p*-type region, which is the case for GaAs and $GaAs_{1-x}P_x$ devices.[52] In the case of breakdown luminescence, which occurs with floating Cu_2S–CdS heterojunctions[51] [Fig. 12-26(c)], the two carriers are formed in the high-field region, where their encounter

[54] A. R. Moore and W. M. Webster, *Proc. IRE* **43**, 427 (1955).

time is very short. The holes flow to the *p*-type region, where they are majority carriers. In the forward-biased Cu_2S–CdS junction, charge neutrality must be obtained by nonradiative recombination (or tunneling), since injection luminescence was not found in this Cu_2S–CdS system.

It is conceivable that the principle of the acoustoelectric domain inducing a spacially sweeping injection luminescence along an array of *p–n* junctions may someday result in a practical solid-state flying-spot scanner.

12-C-4 BRILLOUIN-SCATTERING STUDIES OF ACOUSTOELECTRIC DOMAINS

Once it has been established that the acoustoelectric effect converts electrical energy into acoustical energy, one wishes to know the spectrum of the phonons generated in the domain. The use of a transducer near the cathode to launch phonons of a known frequency and one near the anode to detect the received phonons has revealed that the acoustoelectric effect amplifies the introduced phonons[55] and, therefore, also all the phonons already present. A study of the acoustic spectrum in the absence of a coherent acoustic input has shown that the phonons at the anode describe thermal noise amplified by up to 60 dB.[56]

The use of Brillouin scattering (Sec. 12-B-2) is a convenient tool for studying the phonon spectrum in the acoustoelectric domain.[57] Intense light is incident on a spot on the semiconductor. When the acoustic wave passes the illuminated region, the radiation is scattered in a direction which depends on the phonon frequency, and at the same time its frequency is shifted by an amount which is equal to the frequency of the scattering phonon. Hence Brillouin scattering can give the spectrum of thermal phonons.[58] In the case of the acoustoelectric domain, the change in intensity of the scattered light from the zero-field value is a measure of the phonon intensity (Fig. 12-27). Thus a complete spectrum of the phonons in the acoustoelectric domain is obtained by measuring the intensity of the Stokes or anti-Stokes components as a function of the scattering angle. Whether the scattered components increase or decrease is determined by the relative directions of the electric field with respect to the direction of the incident beam.

[55] A. R. Hutson, J. H. McFee, and D. L. White, *Phys. Rev. Letters* **7**, 237 (1965).

[56] A. R. Moore, *J. Appl. Phys.* **38**, 2327 (1967).

[57] J. Zucker, S. Zemon, E. Conwell, and A. Ganguly, "Brillouin Scattering Study of Acoustoelectric Effects in Piezoelectric Semiconductors," *Light Scattering Spectra of Solids*, ed. G. B. Wright, Springer-Verlag (1969), p. 615.

[58] R. W. Smith "Fabry-Perot Analysis of the Acoustoelectric Interation in CdS." *Light Scattering Spectra of Solids*, ed. G. B. Wright, Springer-Verlag (1969), p. 611.

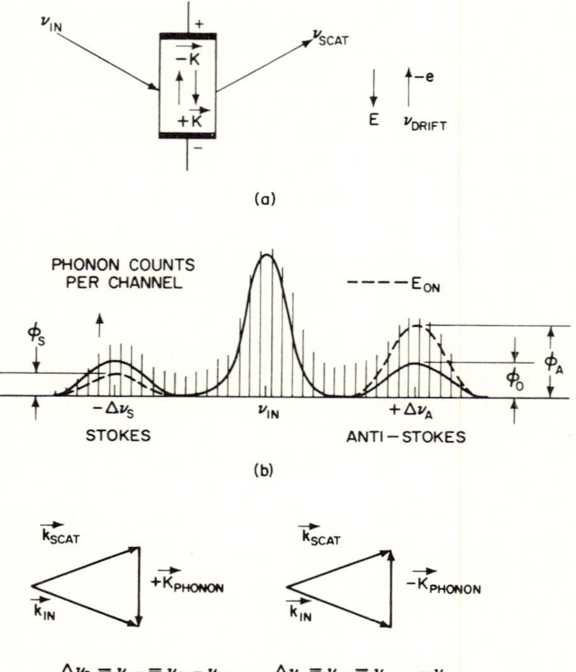

Fig. 12-27 (a) Crystal and light beam orientation for Brillouin scattering experiment. (b) Brillouin scattering spectrum. One Fabry–Perot order divided into 39 channels of a multi-scaler and synchronously scanned. Solid line is the spectrum with the electric field off (thermal scattering). The dashed line is the spectrum with the field on. (c) k-vector diagram.[59]

[59]Courtesy of R. W. Smith and A. R. Moore.

PHOTOELECTRIC EMISSION

13

Here we shall discuss an interaction between a photon and a semiconductor resulting in the emission of an electron out of the semiconductor. This process occurs through the boundary between the semiconductor and the vacuum; therefore, we must be concerned with the behavior of energy levels at the surface.

13-A Threshold for Emission

Figure 13-1 will serve to illustrate the definition of terms. The vacuum level E_{vac} is the energy at which an electron would emerge from the semiconductor's surface and appear in the vacuum with practically zero kinetic energy. The energy χ separating the edge of the conduction band from the

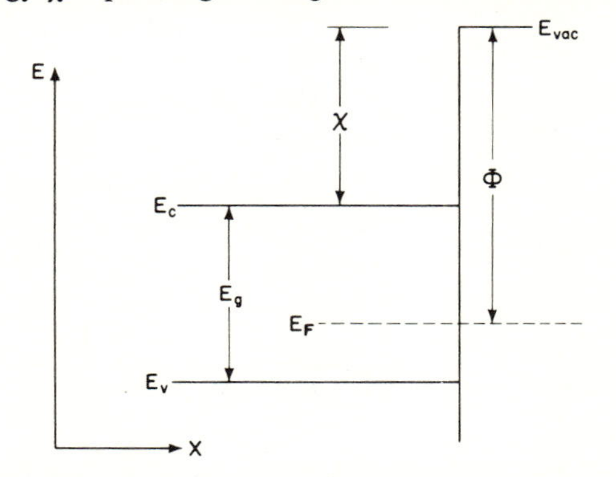

Fig. 13-1 Band structure and definition of symbols.

vacuum level is the "electron affinity." The quantity Φ, which is the energy difference between the Fermi level and the vacuum level, is called the "work function." The work function of most materials ranges between 3 and 5 eV. The threshold for photoelectric emission, E_T, is the lowest photon energy which will excite an electron out of the semiconductor.

We can readily predict what the threshold for emission ought to be in special cases. For intrinsic or nondegenerate semiconductors, as in Fig. 13-1, where the Fermi level lies inside the gap, the highest level occupied by electrons is the top of the valence band. Therefore,

$$E_T = \chi + E_g \tag{13-1}$$

If the semiconductor is heavily doped p-type, as in Fig. 13-2(a), and the Fermi level is ξ_p below the valence-band edge, then

$$E_T = \chi + E_g + \xi_p \tag{13-2}$$

If, on the other hand, the semiconductor is heavily doped n-type, as in Fig. 13-2(b), and the Fermi level is ξ_n above the conduction-band edge, then

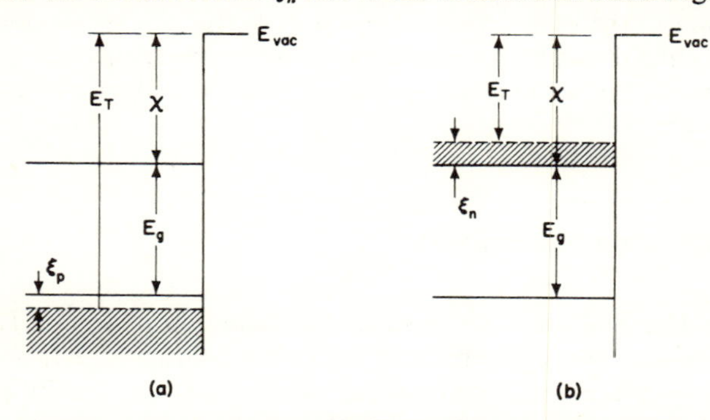

Fig. 13-2 Dependence of threshold energy for emission on doping.

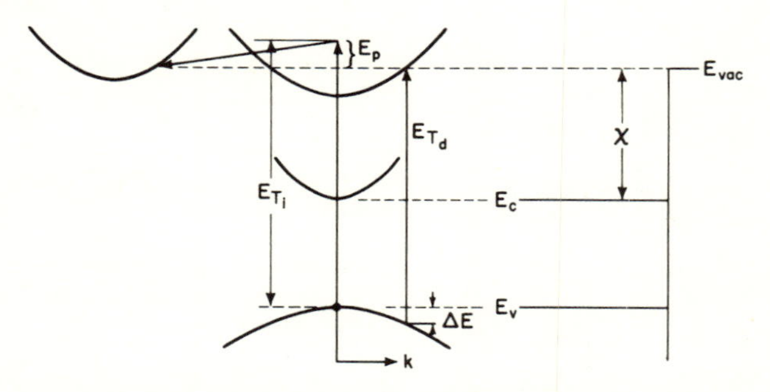

Fig. 13-3 Photoelectric thresholds for direct (E_{T_d}) and indirect (E_{T_i}) excitations.

$$E_T = \chi - \xi_n \tag{13-3}$$

The threshold for emission in Eqs. (13-1) through (13-3) gives the minimum conceivable value for E_T. To determine the actual value of E_T, one must take into account the fact that the optical excitation is a transition between real states (an absorption process) and, therefore, momentum must be conserved as well as energy. Thus for the direct transition of Fig. 13-3, the initial state must be below the valence-band edge by an amount ΔE; for an indirect transition where momentum is conserved by phonon emission,

$$E_T = \chi + E_g + E_p \tag{13-4}$$

13-B Photoelectric Yield

As the photon energy is increased above the threshold value, the number of electrons that can be emitted increases. A typical curve of photoelectric quantum yield as a function of the exciting photon energy is shown in Fig.

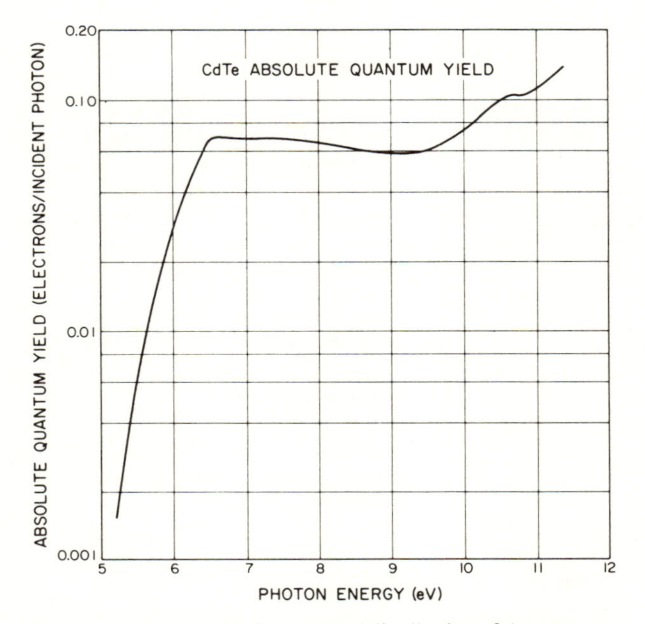

Fig. 13-4 An example of the spectral distribution of the quantum yield (electrons emitted per photon absorbed). The curve is for CdTe. The threshold of emission is about 5 eV.[1]

[1]W. E. Spicer, *Proc. 8th Int. Conf. on Phenomena in Ionized Gases*, Int. Atomic Energy Agency, Vienna, (1968), p. 271.

13-4. It is characterized by a fast initial rise and a plateau about 1 eV above threshold. The plateau is structured, reflecting properties of the semiconductor's band structure.

The initial fast-rising portion of the yield curve $Y(h\nu)$ is characterized by a dependence of the form

$$Y = A(h\nu - E_r)^r \tag{13-5}$$

where A and r are constants. The value of r can be predicted[2] on the basis of the model assumed for the transition generating the energetic electron and for the scattering process considered. The processes and the predicted values of r are listed in Table 13-1. Actual data for several semiconductors is shown in Fig. 13-5. This data suggests that different types of transition occur over different spectral ranges. In particular, near threshold, a cube law seems to characterize the dependence of yield on energy in excess of threshold, while a linear dependence seems to prevail at higher energies. One could generalize Eq. (13-5) for the possibility of different transitions occurring over various spectral ranges:

$$Y = \sum_n A_n(h\nu - E_{T_n})^{r_n} \tag{13-6}$$

Table 13-1[2]

Threshold energies and energy dependence of the photoelectric yield for different excitation and scattering processes

Excitation from	*Transition*	*Scattering process*	*Threshold E_T*	*r*
		BULK PROCESSES		
valence band	indirect	unscattered	$E_T = \chi + E_c - E_v$	$\frac{5}{2}$
		scattered		
	direct	unscattered	$E_T \geq \chi + E_c - E_v$	1
		scattered	$E_T \geq \chi + E_c - E_v$	2
		SURFACE PROCESSES		
valence band		diffuse surface scattering	$E_T = \chi + E_c - E_v$	$\frac{5}{2}$
		specular surface scattering		$\frac{3}{2}$
discrete localized surface states below E_F			$E_T = \chi + E_c - E_t$	1
continuous distribution of surface states at E_F			$E_T = \chi + E_c - E_F$	2
surface-state band below E_F	indirect		$E_T > \chi + E_c - E_F$	2
	direct		$E_T > \chi + E_c - E_F$	1
surface-state band at E_F	indirect		$E_T = \chi + E_c - E_F$	$\frac{5}{2}$
	direct			$\frac{3}{2}$

[2] E. O. Kane, *Phys. Rev.* **127**, 131 (1962).

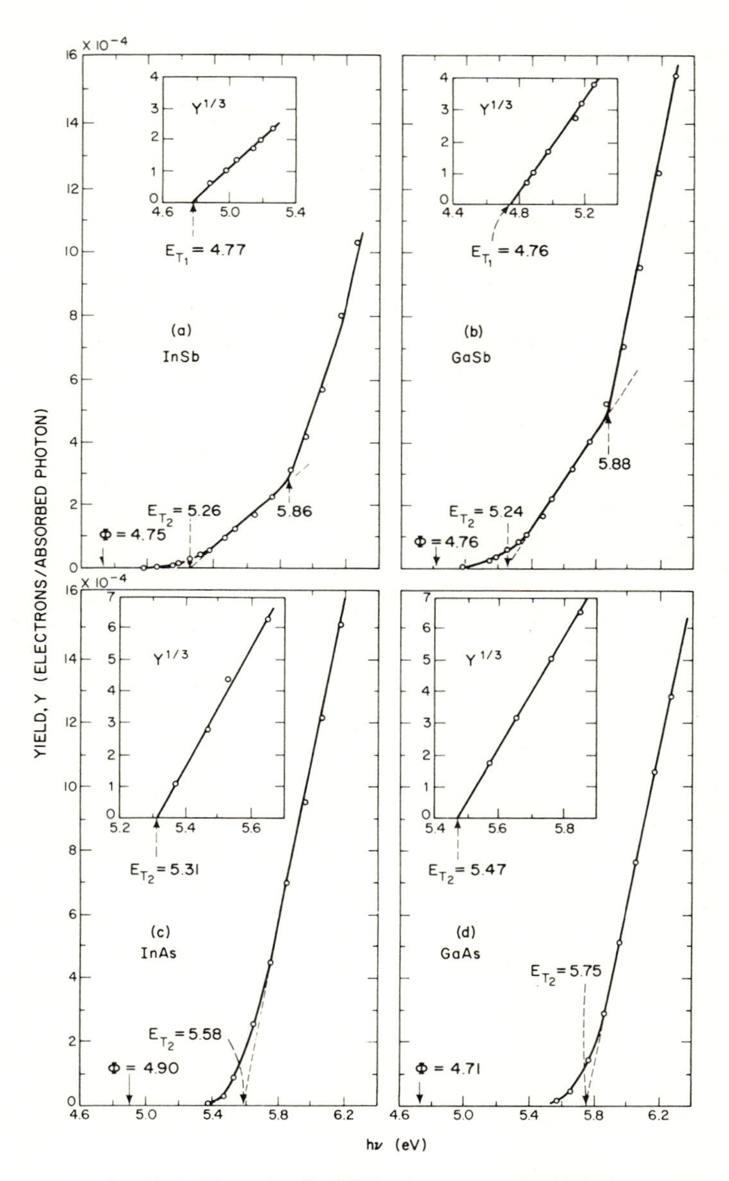

Fig. 13-5 Photoelectric yield in electrons per absorbed photon for cleaved (110) surfaces of (a) InSb, (b) GaSb, (c) InAs, (d) GaAs. Inserts show cube law theshold extrapolations. Φ is the work function; E_{T_1} is the lowest photoelectric threshold, and E_{T_2} is the photoelectric threshold for direct transitions.[3]

[3]G. W. Gobeli and F. G. Allen, "Photoelectric Threshold and Work Function," *Semiconductors and Semimetals*, ed. R. K. Willardson and A. C. Beer, Vol. 2, Academic Press (1966), p. 263.

Note that the photoelectric emission is mostly a bulk process. Excitation from surface states is a small contribution because the total number of surface states is small compared to the number of participating bulk states. Furthermore, there is evidence that most of the excitation occurs in the bulk. If the thickness of the material is gradually increased, the emission yield builds up until a saturation value sets in when the material's thickness exceeds either the excitation depth or the escape depth.[1] We shall overlook the occurrence of interference effects at critical thicknesses: the emission goes through a minimum at thicknesses equal to an odd number of quarter-wavelengths of the light in the semiconductor and through a maximum at thicknesses equal to an integral number of half-wavelengths.[4]

Electron emission (Fig. 13-6) comprises a chain of three processes: (1) the photon interaction exciting an electron to a high-energy state; (2) a scattering of the excited electron, which loses energy in the process; and (3) the escape of the excited electron, carrying away some energy in excess of E_{vac} in the form of kinetic energy.

The excitation process occurs inside the semiconductor within a range of 60 to 300 Å from the surface. The excited electron may interact with the lattice (phonon scattering) and lose energy by emitting a phonon. When the excited electron is several electron-volts above the edge of the conduction band, the phonon emission represents a very small but significant loss in energy. Thus in Si the mean free path between collisions is 60 Å and the mean loss per collision is 0.06 eV.[5] Therefore, it takes 17 collisions to lose 1 eV in silicon. This corresponds to a total travel of $17 \times 60 \approx 1000$ Å. Note that if the electron travels at thermal velocity (10^7 cm/sec), its escape time is $(10^{-5}\text{ cm})/(10^7\text{ cm/sec}) = 10^{-12}$ sec. However, the multiple collisions impose a random walk path for the escaping electron; hence an order-of-magnitude estimate for the escape depth is[6]

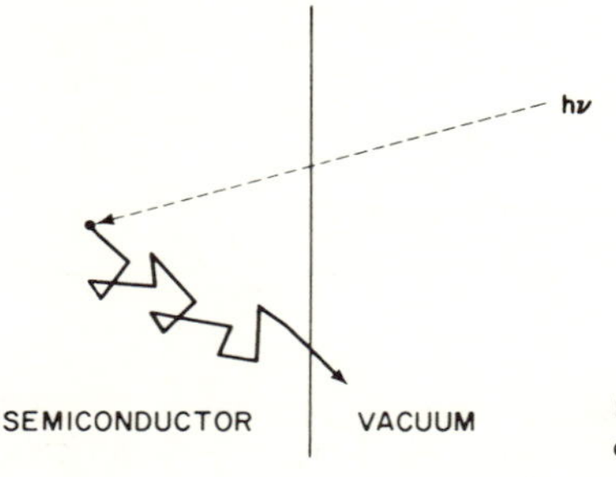

Fig. 13-6 Excitation, scattering and escape of photoelectron.

[1]W. E. Spicer, *Proc. 8th Int. Conf. on Phenomena in Ionized Gases*, Int. Atomic Energy Agency, Vienna, (1968), p. 271.

[4]E. G. Ramberg, *Appl. Optics* **6**, 2163 (1967).

[5]C. A. Lee, R. A. Logan, R. L. Batdorf, J. J. Kleimack, and W. Wiegmann, *Phys. Rev.* **134**, A761 (1964).

[6]R. E. Simon and B. F. Williams, *IEEE Trans. on Nuclear Science* **15**, 167 (1968).

$$d \approx \left(\frac{N_c}{3}\right)^{1/2} l = 140 \text{ Å}$$

where N_c is the number of collisions (17 in this case) and l is the electron mean free path (60 Å, here).

Energy may also be lost by an electron–electron collision. This is equivalent to an Auger effect as shown in Fig. 13-7: the excited electron may drop to a much lower level, experiencing an energy loss greater than gap energy; the energy lost by the first electron is gained by a valence-band electron which is then excited into the conduction band. A consequence of this collision is that the first electron loses too much energy to escape over the vacuum level, while the second electron may not be energetic enough to escape either.

If the absorption $\alpha(\nu)$ and reflectance $R(\nu)$ of a semiconductor are known (for example, from absorption measurements) and if the mean free path l is known (for example, from secondary-emission experiments), one can calculate the quantum yield.[1] Let a radiation of intensity $L_0(\nu)$ be incident on a semiconductor which, for our purpose, is semi-infinite to avoid a consideration of interference effects (Fig. 13-8). The intensity of the radiation at some distance x in the semiconductor is

$$L(\nu, x) = L_0(\nu)[1 - R(\nu)]e^{-\alpha(\nu)x} \tag{13-7}$$

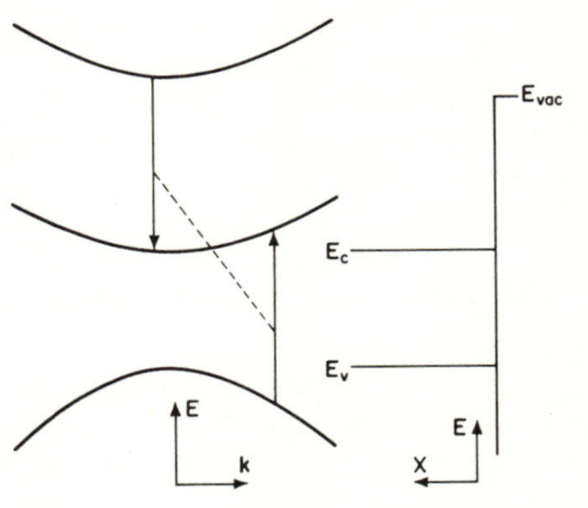

Fig. 13-7 Energy loss by electron-electron interaction.

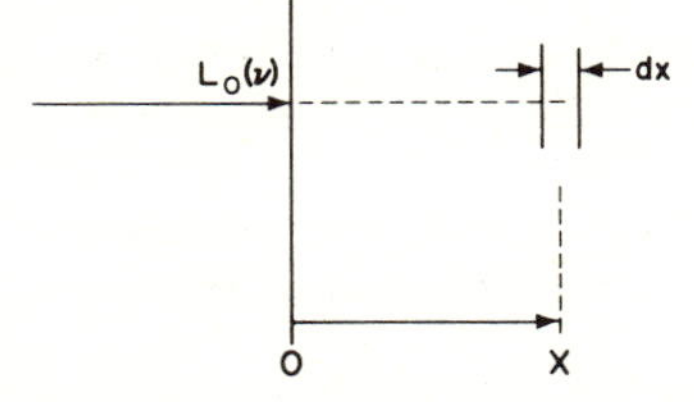

Fig. 13-8

[1] W. E. Spicer, *Proc. 8th Int. Conf. on Phenomena in Ionized Gases*, Int. Atmic Energy Agency, Vienna, (1968), p. 271.

Let $a\alpha(\nu)$ be the efficiency for exciting an electron; then the photocurrent generated at x within a slice dx is

$$di = a\alpha(\nu)L(\nu, x)\,dx \tag{13-8}$$

The escape probability is

$$P_{\text{esc}} = B(\nu)e^{-x/d} \tag{13-9}$$

where $B(\nu)$ is independent of x, and d is the escape depth; both B and d depend on the energy of the electron. Integrating the contributions of such slices from 0 to ∞, the emission current is

$$\begin{aligned} I &= \int_0^\infty a\alpha BL_0(1 - R)e^{-(\alpha+1/d)x}\,dx \\ &= \frac{a\alpha BL_0(1 - R)}{\alpha + 1/d} \end{aligned} \tag{13-10}$$

Therefore, the quantum yield is

$$Y = \frac{I}{L} = \frac{a\alpha B}{\alpha + 1/d} = \frac{aB}{1 + 1/\alpha d} \tag{13-11}$$

where B, α, and d depend on the photon energy and a is constant. Equation (13-11) states that the quantum yield will increase when the photon energy is so large that its penetration depth $1/\alpha(\nu)$ is short compared to the escape depth. Equation (13-11) also states that Y will decrease when the escape depth d becomes very short, as would result from electron–electron collisions. In GaAs the escape depth is approximately equal to the electron diffusion length, i.e., about 1.5 μm.

13-C Effect of Surface Conditions

Until now, we have assumed that the energy bands are flat up to the surface. Since the emission process is a bulk property, it was simple to relate the electron affinity to the vacuum level at the surface, as we did in Eqs. (13-1) through (13-3). However, surface states often induce a space charge which bends the bands either up or down near the surface. (For a discussion of surface states, see Sec. 16-A.) Less than a monolayer of nitrogen or oxygen adsorbed on a surface can result in a 1-eV change in work function; annealing germanium or silicon in vacuum above 450°K can drop the work function by 0.3 eV.[7]

Let us first consider the case of a nearly intrinsic semiconductor. If the bands bend up by ΔE [Fig. 13-9(a)] and the electron affinity at the surface is

[7] F. G. Allen and G. W. Gobeli, *Proc. Int. Conf. Semiconductor Physics, Exeter (1962)*, The Institute of Physics and the Physical Soc., London (1962), p. 818.

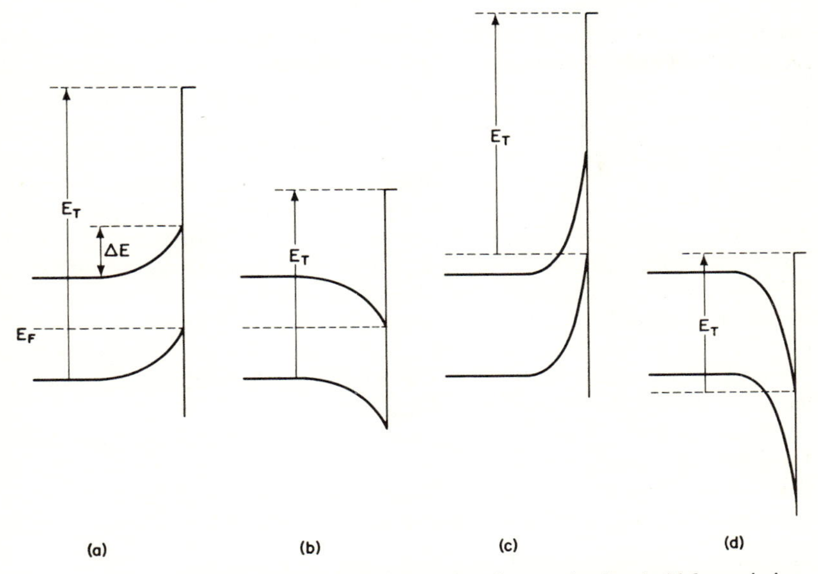

Fig. 13-9 Effect of band bending on the threshold for emission.

not changed, the threshold for excitation from the bulk, beyond the space-charge region, is increased by ΔE. Conversely, if the bands bend down at the surface by ΔE, the threshold for emission from the bulk decreases by ΔE [Fig. 13-9(b)].

As the impurity concentration increases to make the semiconductor degenerately doped, the Fermi level moves into a band. In the case of an *n*-type semiconductor [Fig. 13-9(c)], a surface inversion layer would increase the photoelectric threshold to its maximum value $\chi + \Delta E - \xi_n$; where ΔE can be approximately equal to the energy gap, since the Fermi level at the surface is usually pinned to the appropriate band edge. In a degenerately doped *p*-type material, on the other hand [Fig. 13-9(d)], the Fermi level in the bulk is below the edge of the valence band, but the surface-inversion layer can lower the threshold to its minimum value, $\chi - \Delta E + \xi_p$. Evidently, the lower photoelectric threshold (and, therefore the higher yield) will be obtained with an *n*-type surface on a *p*-type bulk.[8] The higher the doping, the narrower the depletion layer at the surface inversion, thereby allowing more of the bulk to be effective for photoelectric emission. Thus in *p*-type silicon doped with 10^{20} acceptors/cm^3, the bending of the bands due to the surface inversion layer extends only to a depth of about 20 Å.[9]

If an extremely thin layer of a low–work-function material (thickness $\ll$

[8]W. E. Spicer, *RCA Review* **19**, 555 (1958).
[9]W. E. Spicer and R. E. Simon, *Phys. Rev. Letters* **9**, 385 (1962).

mean free path) is placed on the surface of the semiconductor, the photoelectrons will be able to traverse the surface material without much loss in energy, and the lower work function of the new surface will reduce the threshold. Thus cesium, which has a work function of 1.9 eV, can be deposited as a monolayer on the semiconductor surface and effect a considerable reduction in work function, electron affinity and threshold. Actually, the work function of a cesiated surface is even lower than the work function of cesium. A monolayer of cesium adsorbed on the surface of heavily doped *p*-type GaAs lowers the work function to less than 1.5 eV. Since the energy gap of GaAs is about 1.5 eV, the electron affinity becomes nearly zero and, therefore, the threshold for emission is approximately the energy gap of GaAs.[10] Such a photoelectric emitter has a very high quantum yield (Fig. 13-10). The presence of oxygen during cesiation seems to depress further the work function, so that values of about 0.7 eV have been obtained.[11]

When the energy gap of a *p*-type semiconductor is larger than the surface work function, a negative electron affinity is obtained [Fig. 13-11(a)]. This is the case for GaP ($E_g = 2.2$ eV), the cesiated surface of which has a work function of 1.2 eV, leaving a negative electron affinity of 1.0 eV.[12] Conse-

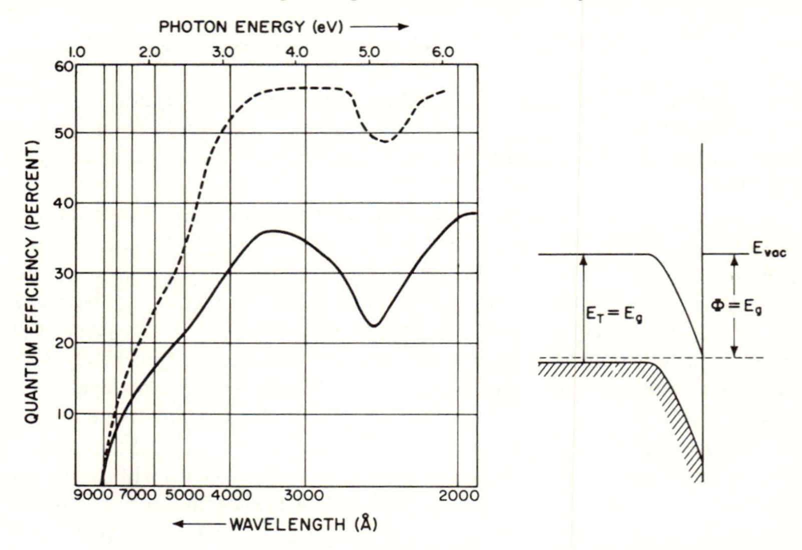

Fig. 13-10 Spectral distribution of the photoelectric yield for GaAs + Cs. The drawn curve shows the efficiency in electrons per incident quantum and the dashed curve the efficiency per absorbed quantum.[10]

[10]J. J. Scheer and J. Van Laar, *Solid State Comm.* **3**, 189 (1965).
[11]B. F. Williams, *Appl. Phys. Letters* **14**, 273 (1969).
[12]B. F. Williams and R. E. Simon, *Phys. Rev. Letters* **18**, 485 (1967).

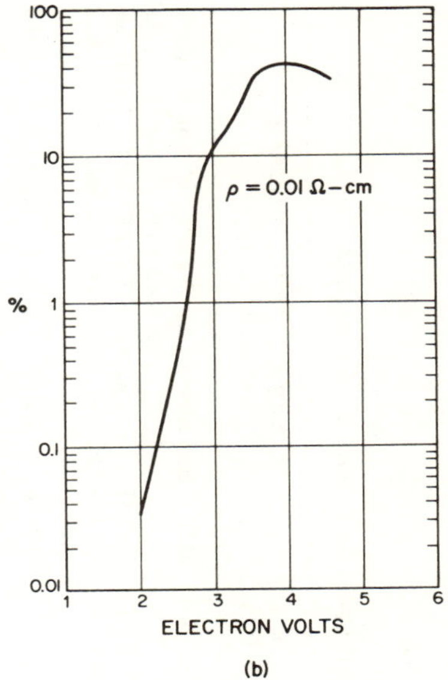

Fig. 13-11 Photoelectric emission form cesiated p-type GaP: (a) band bending at the surface; (b) the quantum efficiency in percent as a function of incident-photon energy. The high yield where the absorption constant is low indicates an escape depth of 2000–2500 Å.[12]

quently, almost every photoelectron excited within the escape depth, even though it may just lie at the bottom of the conduction band, can escape—actually, only 50% of the electrons travel toward the surface—therefore, the quantum yield is very high [Fig. 13-11(b)]. A measurement of the energy distribution of the emitted electrons (Fig. 13-12) (we shall discuss such measurements in the next section) provides a means for studying the scattering process which degrades the electron's energy. An analysis of the quantum-yield data indicates that the escape depth is at least 2000 Å, so that most of the photo-excited electrons have thermalized to the bottom of the conduction band by the time they enter the bent-band region, which extends about 100 Å into the semiconductor. Since 0.05 eV is lost per phonon-emitting collision, it is possible to calculate the mean free path in the bent-band region resulting in the observed energy distribution of emitted electrons. The mean free path turns out to be between 22 and 25 Å.

The experimentally determined work function Φ, the photoelectric threshold E_T, and the threshold for direct transition E_{Td} are listed in Table 13-2 for several noncesiated semiconductors in which the band bending has been taken into account.

Table 13-2[3]

Work functions and photoelectric threshold values for several semiconductors

Compound and plane	E_g (eV)	Φ (eV)	E_T (eV)	E_{Td} (eV)	$\chi = E_T - E_g$	$\delta = (E_F - E_v)_S$†	E_F‡ (eV)	*Sample resistivity, band bending, surface-state density#*
AlSb(110)	1.5	4.86	5.15	5.75	3.65	0.29	—	0.10 Ω-cm *n*-type, $N_D = 1.8 \times 10^{17}$ cm^{-3}, surface strongly *p*-type; bands bend up to surface by ~1.0 eV
GaAs(110)	1.40	4.71	5.47	5.75	4.07	0.76	1.35	0.08 Ω-cm *n*-type, $N_D = 2 \times 10^{16}$ cm^{-3}; bands bend up to surface by ~0.6 eV, surface near intrinsic $N_{ss} \geq 10^{12}$ cm^{-2}
GaSb(110)	0.70	4.76	4.76	5.24	4.06	~0	0.08	0.07 Ω-cm *p*-type, $N_A = 1.2 \times 10^{17}$ cm^{-3}; surface degenerately *p*-type
InAs(110)	0.41	4.90	5.31	5.58	4.90	~0§	0.31	0.01 Ω-cm *n*-type, $N_D = 2.4 \times 10^{16}$ cm^{-3}; surface degenerately *n*-type
InSb(110)	0.18	4.75	4.77	5.26	4.59	~0	0.12	0.019 Ω-cm *n*-type, $N_D = 5.5 \times 10^{14}$ cm^{-3} at 77°K; surface probably strongly *p*-type
InP(110)	1.3	4.45	5.68	5.94	4.38	1.23	—	0.03 to 3 Ω-cm, $N_D = 1.3$ to 80×10^{15} cm^{-3}; surface strongly *n*-type
Si(111)	1.09	4.83	5.10	5.45	4.01	0.27	—	200 Ω-cm *p*-type (flat bands), $N_{ss} \geq 2 \times 10^{14}$ cm^{-2}
Ge(111)	0.67	4.80	4.80	5.22	4.13	0	—	0.2 Ω-cm *p*-type (nearly flat bands), $N_{ss} \geq 2 \times 10^{13}$ cm^{-2}

†$\delta = (E_F - E_V)_S$ indicates where the Fermi level is pinned at the surface.
‡E_F is the energy of the Fermi level referred to the valence-band-edge in the bulk.
#N_{ss} is the density of surface states.
§T. E. Fisher, F. G. Allen, and G. W. Gobeli, *Phys. Rev.* **163**, 703 (1967).

[3] G. W. Gobeli and F. G. Allen, "Photoelectric Threshold and Work Function," *Semiconductors and Semimetals*, ed. R. K. Willardson and A. C. Beer, Vol. 2, Academic Press (1966), p. 263.

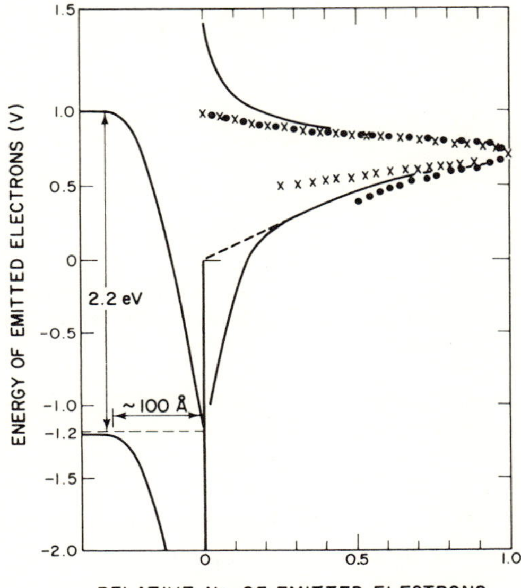

Fig. 13-12 The energy distribution of emitted electrons superimposed on a band diagram showing the band bending at the surface of cesiated p-type GaP. The dots and the crosses represent calculated results.[12]

13-D Energy Distribution of Emitted Electrons

The energy distribution of the emitted electrons can be measured by a variety of techniques. Thus magnetic deflection provides a means for velocity selection. The most commonly used method employs a retarding potential to collect only those electrons having a kinetic energy greater than the retarding potential (Fig. 13-13). A small ac voltage modulates the retarding potential; the resulting ac-current signal describes the slope of the distribution curve at the retarding potential, and by sweeping the latter over a range of bias the complete distribution curve is obtained.

We saw at the end of the previous section that the energy-distribution spectrum permitted a good evaluation of the electron mean free path. If it were not for the degradation of energy by scattering, the energy distribution of the emitted electrons would be a replica of the band structure in the semiconductor.[13] If a stationary emission peak is obtained while the energy of

[13]W. E. Spicer and R. E. Simon, *J. Phys. Chem. Solids* **23**, 1817 (1962).

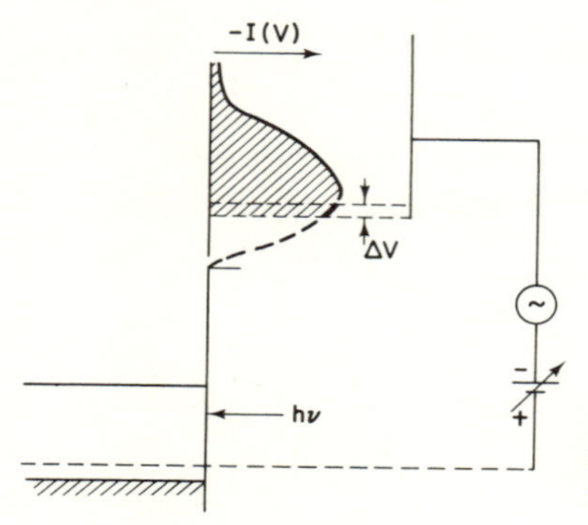

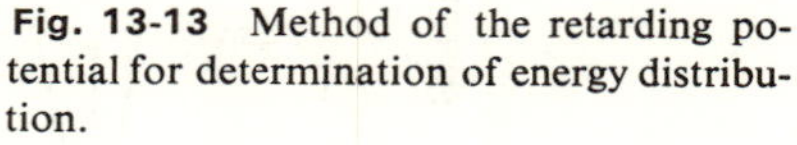
Fig. 13-13 Method of the retarding potential for determination of energy distribution.

the exciting photons is increased, one can conclude that the final states in the optical transition consist of set of bunched states in the conduction band. As the photon energy is increased, the initial states are deeper in the valence band.

It is interesting to expose successively higher-lying levels in the conduction band by increasing the energy of the photo-excitation.[14] The maximum energy that a photoelectron can have is given by Einstein's photoelectric equation:

$$E_{\max} = h\nu - \Phi$$

where Φ is the work function. Exciting cesiated GaAs with monochromatic radiation will first allow the escape of electrons which have thermalized to the direct Γ-valley at about 1.4 eV (Fig. 13-14). At photon energies greater than 1.7 eV, the indirect X-valleys can receive electrons. If the emission time is comparable to the intervalley relaxation time, the emission-velocity distribution spectrum will show a two-peak structure—the 1.4-eV peak for the electrons in the direct Γ-valley, and the 1.7-eV peak for the indirect X-, or $\langle 100 \rangle$-, valleys.

Problem 1. Use the information of Fig. 13-10 to design an integrated light-source–electron-emitter device. Assume a very thin p^+ layer (~ 1000 Å thick). For a light source, chose some semiconductor p–n junction in which a high luminescence efficiency is expected.

Describe the difficulties you can anticipate in the realization and operation of this cold cathode.

[14]R. C. Eden, J. L. Moll, and W. E. Spicer, *Phys. Rev. Letters* **18**, 597 (1967).

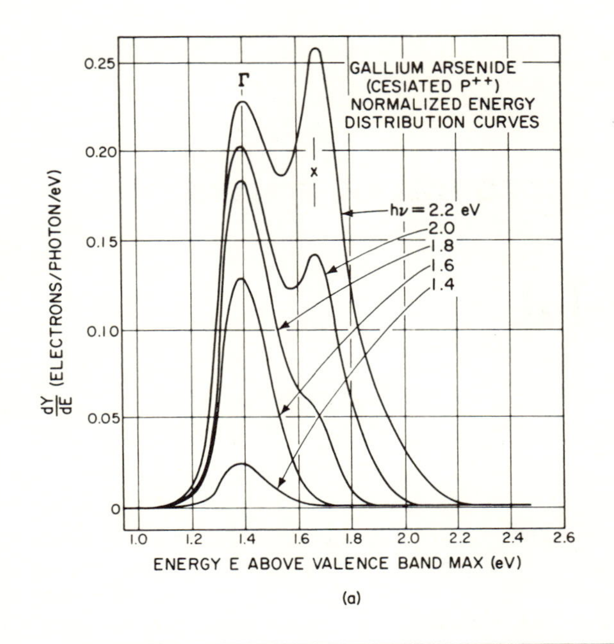

(a)

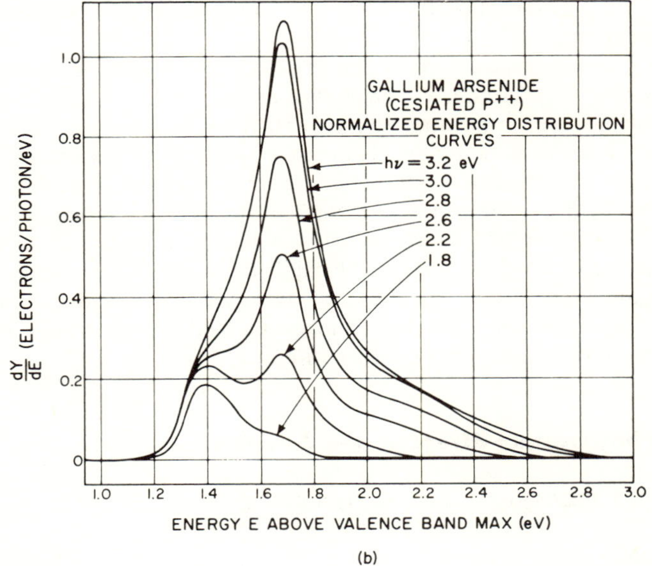

(b)

Fig. 13-14 Energy distributions for photoemitted electrons for photon energies between 1.4 and 3.2 eV. The distributions have been normalized to quantum yield.[14]

PHOTOVOLTAIC EFFECTS

14

Photovoltaic effects constitute a class of phenomena in which light generates a voltage across a portion of the semiconductor. Actually, the light produces only an excess of free carriers. The free carriers move in response to local fields and accumulate in regions where they produce a net space charge. It is this deviation from thermal equilibrium that is responsible for the photovoltage.

The photo-generated carriers can be driven by a variety of fields to establish a potential difference. We shall dismiss the externally applied electric field, which is used to measure photoconductivity. Internal electric fields can be due to a variation in doping (as in a *p–n* junction), to a variation in composition (as in heterotransitions), or to both (as in heterojunctions). A discontinuity of the crystal, such as a surface, can generate a Schottky barrier and the consequent field. Local fluctuations due to a nonuniform distribution of impurities, strains, etc. engender local fields; most of these fields occur in random directions, so that the net emf is zero. We shall also discuss diffusion as a driving force and show how a magnetic field can generate a potential difference in response to optical excitation.

14-A Photovoltaic Effect at p–n Junctions

14-A-1 ELECTRICAL CHARACTERISTICS

Consider the *p–n* junction of Fig. 14-1(a) illuminated with photons of greater-than-gap energy. The absorption of the photon results in the creation of a hole–electron pair. The "built-in" field at the junction drives the two carriers in opposite directions—the electron to the *n*-type region, the hole to the *p*-type region. The charge separation results in a potential difference V

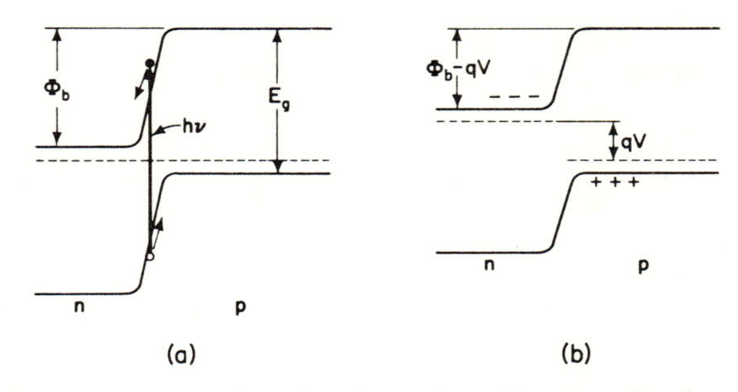

Fig. 14-1 Generation of a photovoltage V at a p–n junction.

across the junction [Fig. 14-1(b)]. Note that in this process the carriers have become majority carriers and, therefore, are endowed with an infinite lifetime. However, the potential difference set up by the photo-generated, field-separated pair biases the p–n junction in the forward direction. Hence some of the carriers, overcoming the lowered barrier $\Phi_b - qV$, will be injected into the opposite region, where they become minority carriers and recombine. If an external circuit is connected across the p–n junction, one can measure the photovoltage V or, if the load resistance is low, a photocurrent, the illuminated junction acting as a battery.

The $I(V)$ characteristic of the junction is given by the diode equation:

$$I = I_0\left(\exp \frac{qV}{kT} - 1\right) \tag{14-1}$$

The current I is the injection current which would flow through the junction under the influence of a forward bias V. Here I_0 is the "saturation current" representing free carriers which can flow through the junction, overcoming the barrier Φ_b under the influence of thermal activation.

Light generates additional electrons and holes at a rate G. If the diode is short-circuited, the current in the circuit is essentially the current flowing through the junction under the influence of the built-in field—i.e., it comprises all the minority carriers generated within a diffusion length of the junction (we assume that the diffusion lengths L_e and L_h are large compared to the widths of the depletion layers). In other words, not only are the pairs generated inside the junction separated by the built-in field, but also most of the pairs within a diffusion length from the region containing a field can diffuse into the built-in field and be separated—i.e., the junction extracts the minority carriers optically generated within a diffusion length. Then the short-circuit current is

$$I_{sc} = Aq(L_e + L_h)G \tag{14-2}$$

where A is the area of the p–n-junction. Note that this short-circuit current flows in a direction opposite to that obtained during forward bias.

If, instead of a short circuit, a finite load resistance is used, allowing the photovoltage to build up to some value V, the net current through the load would be lower than the short-circuit value because of the charge leakage in the opposite direction in the form of an injection current. Hence the photovoltaic current for an arbitrary photovoltage V is

$$I = I_{sc} - I_0\left[\exp\frac{qV}{kT} - 1\right] \tag{14-3}$$

From this expression we can deduce the open-circuit photovoltage V_{oc} by setting $I = 0$ in Eq. (14-3). Then,

$$V_{oc} = \frac{kT}{q}\ln\left(\frac{I_{sc}}{I_o} + 1\right) \tag{14-4}$$

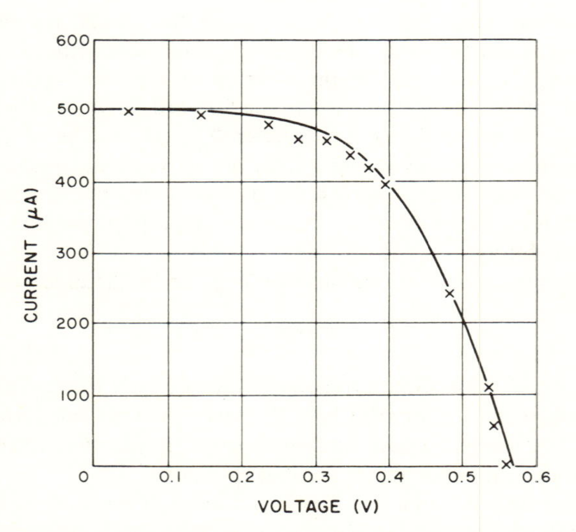

Fig. 14-2 Comparison of the theoretical current–voltage characteristic with experimental results on a GaAs photodiode.[1]

The behavior of the relation in Eq. (14-3) is illustrated in Fig. 14-2. As the light intensity is increased, both the short-circuit current and the open-circuit voltage increase. The current I_{sc} increases linearly with light intensity, as expressed by Eq. (14-2), but the open-circuit photovoltage increases only logarithmically [Eq. (14-4)]. This behavior is illustrated by the data plotted in Fig. 14-3.

As the light intensity is increased, the photovoltage rises until the barrier which opposes internal charge leakage disappears. The barrier height represents the maximum photovoltage that can ever be achieved. The maximum

[1] D. A. Jenny, J. J. Loferski, and P. Rappaport, *Phys. Review* **101**, 1208 (1956).

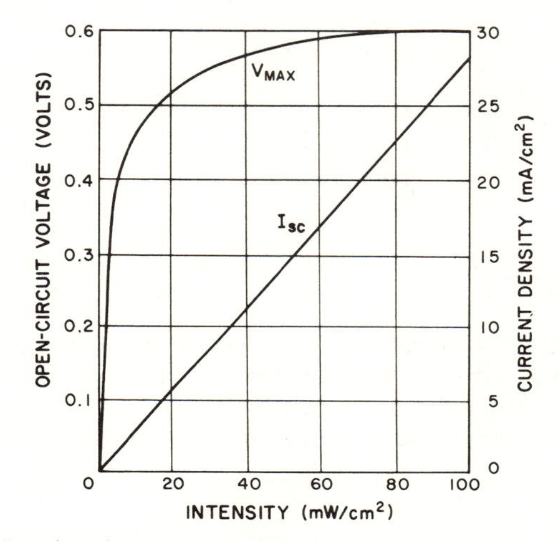

Fig. 14-3 Junction photocurrent and photovoltage as a function of light intensity.[2]

photovoltage, then, depends on the doping; in practical cases this limit would correspond to the energy gap. A photovoltage approaching the value expected from the energy gap has been measured[3] in a GaAs *p–n* junction excited with a He–Ne laser at a level of about 10^3 W/cm^2.

14-A-2 SPECTRAL CHARACTERISTICS

The charges which establish a voltage across the *p–n* junction are generated by an absorption process. All those absorption transitions which result in at least one free carrier (they are described in Secs. 3-A through 3-F) can result in a photovoltage. The dominant transition is expected to be the fundamental absorption which occurs at photons whose energy is nearly equal to the gap energy.

Because of the field present in the junction, electron–hole pairs can be generated at photon energies substantially lower than E_g by a tunneling-assisted transition (Franz–Keldysh effect) (Fig. 14-4). In fact, the low-energy edge of the photovoltaic spectrum is usually an exponential tail (Fig. 14-5) the slope of which correlates with the abruptness of the *p–n* junction: the higher the built-in field, the farther the tail of photosensitivity extends into the infrared.[4] If a reverse bias is applied to the junction and the photoconduc-

[2]P. Rappaport, *RCA Review* **20**, 373 (1959).
[3]N. Holonyak, Jr., J. A. Rossi, R. D. Burnham, B. G. Streetman, and M. R. Johnson, *J. Appl. Phys.* **38**, 5422 (1967).
[4]J. I. Pankove, unpublished observations.

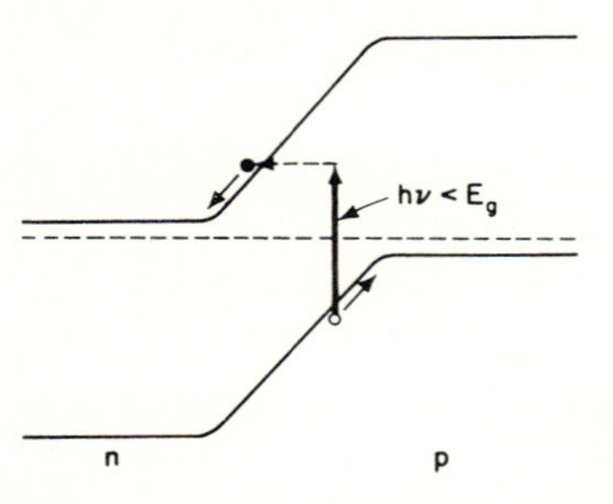

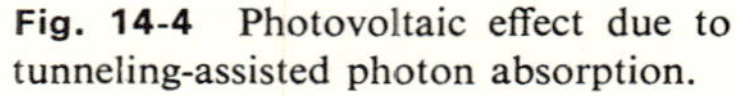

Fig. 14-4 Photovoltaic effect due to tunneling-assisted photon absorption.

tivity spectrum is measured, the low-energy edge of the spectrum shifts to lower energies as the electric field in the junction is increased.[5]

The more energetic photons (such that $E_g < h\nu \leq \frac{3}{2}E_g$) generate hot carriers. Their quantum yield (number of pairs created per incident photon)

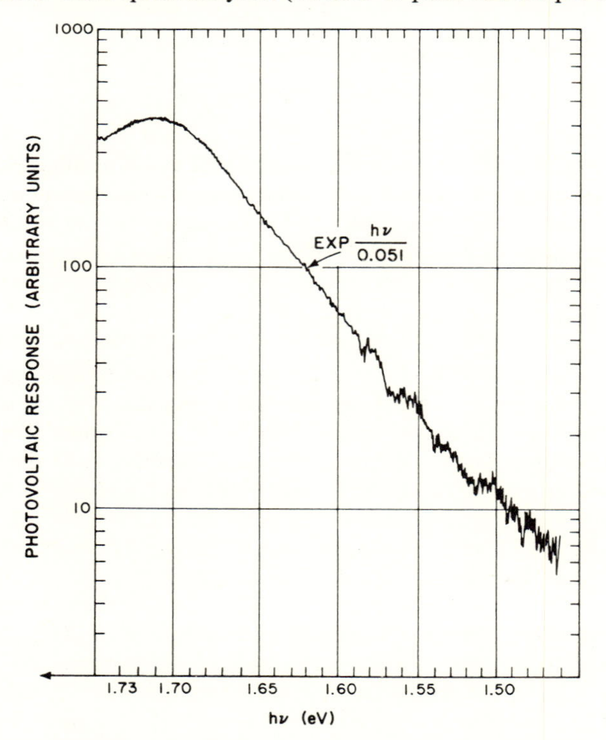

Fig. 14-5 The photovoltaic spectrum of a *p–n* junction in $GaAs_{0.85}P_{0.15}$.[4]

[5]A. A. Gutkin and D. N. Nasledov, *Sov. Phys.—Solid State* **4**, 999 (1962).

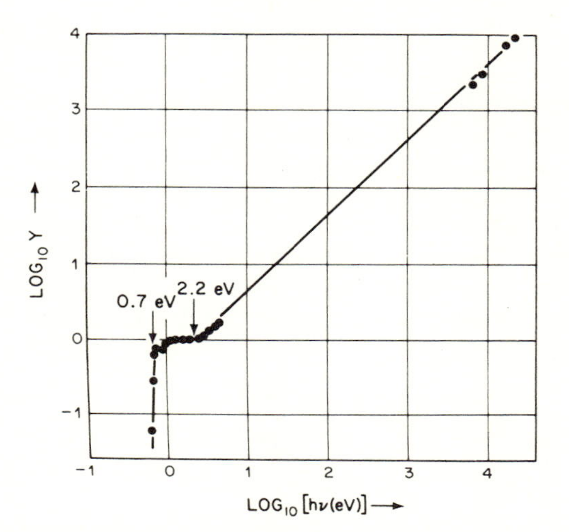

Fig. 14-6 Dependence of photovoltaic quantum yield on the photon energy in germanium.[6]

is at most unity. However, if the hot carriers are sufficiently energetic to excite secondary pairs by impact ionization, then a quantum yield greater than one can be obtained. This is illustrated in Fig. 14-6, which shows that the quantum yield increases rapidly to unit at $h\nu = E_g$; then it stays constant up to some critical photon energy E_s beyond which the quantum yield increases linearly with $h\nu - E_s$. In germanium, $E_g = 0.7$ eV and $E_s = 2.2$ eV.

In measuring photovoltaic spectra, one must take into account absorption of radiation in field-free regions. The spectrum reaching the junction can be different from the spectrum incident on the diode, the higher-energy photons being absorbed at the surface layer. This consideration is crucial when the *p–n* junction is parallel to the illuminated surface and further than one diffusion length away from this surface. Carriers generated within one diffusion length from the surface may recombine rapidly through surface states before they have a chance to reach the junction and contribute to the photovoltage. This reduces the quantum yield.

14-A-3 THE SOLAR CELL[7]

The most important application of the photovoltaic effect is the conversion of solar radiant energy into electrical energy. For this application the dominant parameters are conversion efficiency and power output. The solar

[6] J. Tauc, *Rev. Modern Phys.* **29**, 308 (1957).

[7] For a more detailed discussion, see P. Rappaport and J. J. Wysocki, "The Photovoltaic Effect," *Photoelectronic Materials and Devices*, ed. S. Larach, D. Van Nostrand (1965), p. 239.

spectrum dictates the range of materials that can be used for the generation of photoelectricity: the smaller the energy gap of the semiconductor, the larger the portion of the solar spectrum which is utilized; but the maximum photovoltage obtainable is correspondingly smaller. On the other hand, a larger energy gap can give a higher photovoltage and a lower leakage across the junction (or saturation current I_0). Therefore, an important factor in the design of the solar cell is an optimization of its spectral response. For space applications in which atmospheric absorption does not enter the problem, the optimum energy gap is about 1.6 eV. Atmospheric absorption would move the optimum gap to lower energies. The calculated efficiency is plotted as a

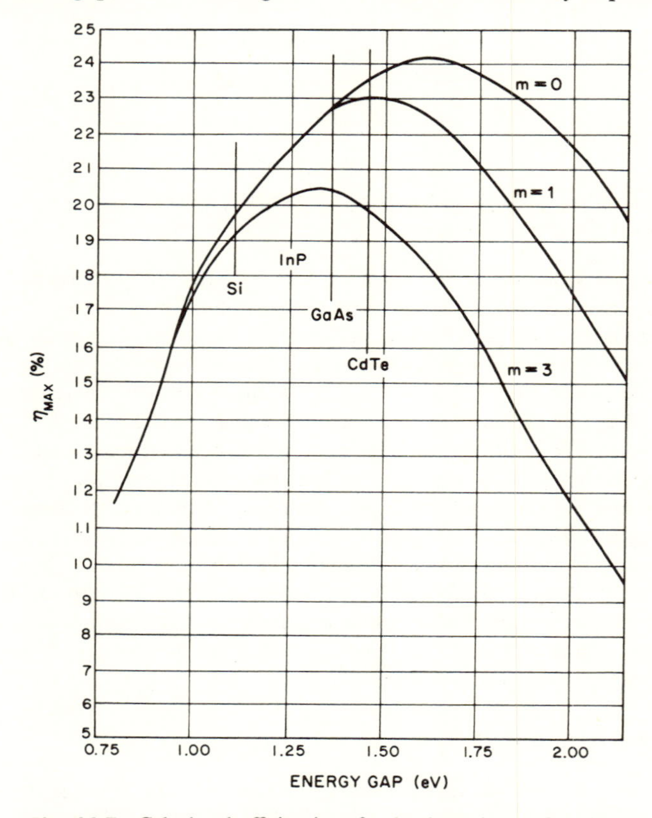

Fig. 14-7 Calculated efficiencies of solar batteries as functions of the energy gap E_g of the semiconductor. The absorption of the sunlight in the atmosphere is taken into account ($m = 1/\cos\theta$, where θ is the angle the sun makes with the zenith).[8]

[8]J. J. Loferski, *J. Appl. Phys.* **27**, 777 (1956).

function of energy gap in Fig. 14-7 for different thicknesses of atmosphere. Although CdTe promises the highest efficiency, the best overall performance thus far has been obtained with silicon, the material which has benefited from the most intensive technological effort. An efficiency of 14% has been reported for silicon solar cells.[9] The higher-gap material, GaAs, has yielded an efficiency of 11%.[10] In order to extend the spectral sensitivity of solar cells, heterojunctions have been made between semiconductors of different energy gaps, the radiation entering the junction through the material with the largest E_g.[11]

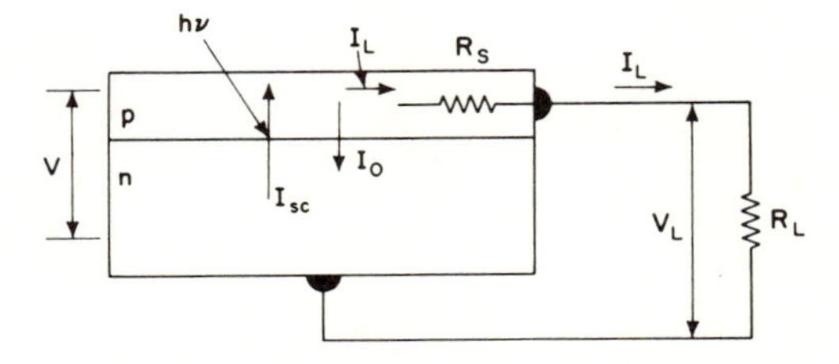

Fig. 14-8 Equivalent circuit of a solar cell.

The solar cell can be represented by the approximate equivalent circuit of Fig. 14-8. Here, R_L is the load resistance and R_S is the internal series resistance of the diode. The radiation absorbed within one diffusion length of the junction generates a current I_{sc}. This current splits into two components—a part I_L which flows through the load and a portion I_0 which is injected across the junction. Because of the internal series resistance R_S, the voltage V_L appearing across the load is lower than the photovoltage generated across the junction:

$$V_L = V - I_L R_S \tag{14-5}$$

If we combine Eqs. (14-5) and (14-3), where $I \equiv I_L$, we get

$$V_L = \frac{kT}{q} \ln\left(1 + \frac{I_{sc} - I_L}{I_0}\right) - I_L R_S \tag{14-6}$$

Figure 14-9 shows the influence of the internal series resistance on the performance of a solar cell. Evidently, the losses across R_S reduce the maximum power output and, therefore, the overall efficiency.

Figure 14-10 shows a comparison of the electrical characteristics of solar cells made of different gap materials under identical test conditions. It is

[9] M. Wolf, *Proc. IRE* **48**, 1246 (1960).
[10] P. Rappaport and J. J. Wysocki, *Acta Electronica* **5**, 364 (1961).
[11] D. A. Cusano, *Solid State Elec.* **6**, 217 (1963).

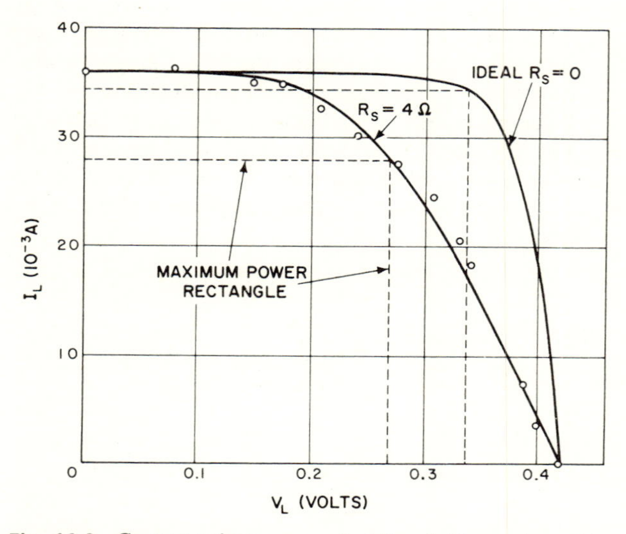

Fig. 14-9 Current–voltage curves of a 1.7 cm² silicon solar cell in bright sunlight. The circles represent experimental data, while the solid lines represent Eq. (14-6).[12]

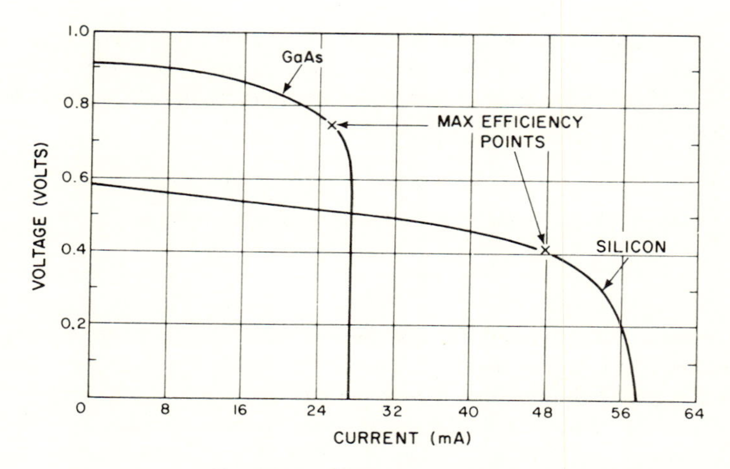

Fig. 14-10 $I(V)$ curves of GaAs- and Si-solar cells.[7]

[12]L. V. Azaroff and J. J. Brophy, *Electronic Processes in Materials*, McGraw-Hill (1963), p. 282.

evident that the larger-gap material produces the largest open-circuit voltage, while the lowest-gap material, which absorbs a greater portion of the incident spectrum, generates the largest short-circuit current. Also, Si, having a larger carrier-diffusion length than GaAs, collects more photo-generated carriers. The point of maximum power output and of maximum efficiency occurs at the "knee" of the curve. For given values of V_{oc} and I_{sc}, the more rectangular the curve, the higher the efficiency.

In order to collect the carriers generated at the illuminated surface by nonpenetrating radiation ($h\nu \gg E_g$), the junction is placed very close to the surface (within one diffusion length). The thinness of the diffused layer results in a high R_S. Hence an important design problem is how to make a judicious compromise between a high collection efficiency and a low series resistance.

The loss of radiation by reflection is usually overcome to a great extent by covering the surface with an antireflection coating.

When solar cells are used in outer space, they are subjected to bombardment by energetic particles which induce a degradation called "radiation damage." The damage consists in the generation of Frenkel pairs. (Atoms are dislodged from a lattice site to an interstitial position, leaving behind a vacancy; the interstitial can migrate away until it finds a stable site, such as a surface, while the vacancy may migrate so some imperfection or impurity with which it forms a complex—e.g., an "*A*" center, which is the combination of a vacancy with an oxygen atom in Si.) The consequence of radiation damage is the formation of centers where efficient nonradiative recombina-

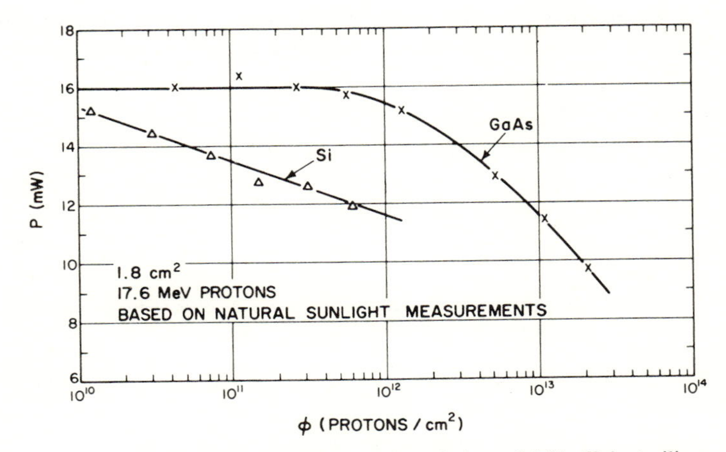

Fig. 14-11 Comparison of degradation of 9% efficient silicon and GaAs solar cells.[13]

[13] J. J. Wysocki, *IEEE Trans. Nuclear Sci.* **NS-10**, 60 (1963).

tion can occur. Hence minority carriers have a shortened lifetime when the material is damaged. This in effect shortens the diffusion length and, therefore, reduces the volume of material which responds to optical excitation.

Various semiconductors differ in their sensitivity to radiation damage. The threshold for damage in GaAs is twice as high as it is in Si. This difference in sensitivity is manifest in Fig. 14-11.

Radiation damage can be annealed by heating, which allows the interstitial atoms to migrate back into vacant substitutional sites. Radiation damage can also be reduced by doping silicon with lithium.[14] Lithium has a high diffusivity in Si at room temperature. By some still unexplained mechanism, lithium can restore the photosensitivity of the solar cell.

14-B Photovoltaic Effects at Schottky Barriers

14-B-1 THE SCHOTTKY BARRIER

When a metal is brought into contact with a semiconductor, there is usually a redistribution of charges which results in the formation of a depletion layer in the semiconductor. This deformation of the band edge at the interface is called a "Schottky barrier."

Let us illustrate this effect in the case of a metal with a work function Φ_M, larger than Φ_S, that of the n-type semiconductor (Fig. 14-12). After contact, electrons must leave the semiconductor surface to equilibrate the Fermi levels. If $\Phi_M < \Phi_S$, electrons would flow to the semiconductor and the opposite band bending would result, forming a Schottky barrier to holes.

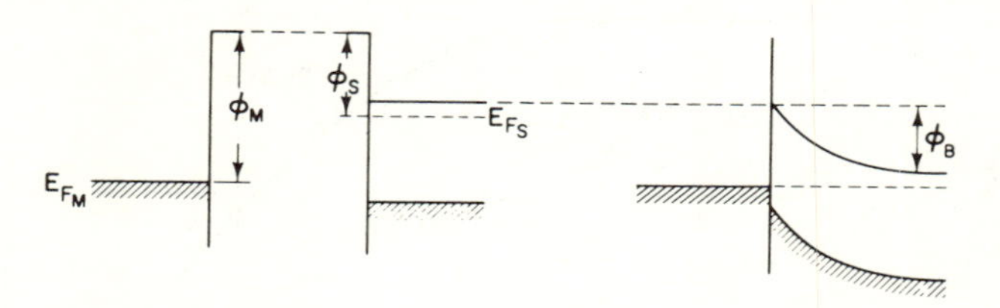

Fig. 14-12 Formation of a Schottky barrier in a semiconductor in contact with a metal.

The height Φ_B of the Schottky barrier depends on the work functions at the surfaces of both materials—i.e.,

$$\Phi_B = \Phi_M - \Phi_S \tag{14-7}$$

The current density through the metal–semiconductor contact is given by[15]

[14] J. J. Wysocki, P. Rappaport, E. Davidson, R. Hand, and J. J. Loferski, *Appl. Phys. Letters* **7**, 44 (1966).

[15] C. A. Mead, *Solid State Elec.* **9**, 1023 (1966).

$$J = J_0 \exp\left(-\frac{\Phi_B}{kT}\right)\left[\exp\left(\frac{qV}{kT}\right) - 1\right] \tag{14-8}$$

where

$$J_0 \approx 120T^2 \quad \text{A/cm}^2$$

when the effective mass of the electron is approximately equal to the free-electron mass.[16] Equation (14-8) implies that Φ_B can be determined from a judicious plot of the experimental data $J(V)$ and $J(T)$ as in Figs. 14-13 and 14-14.

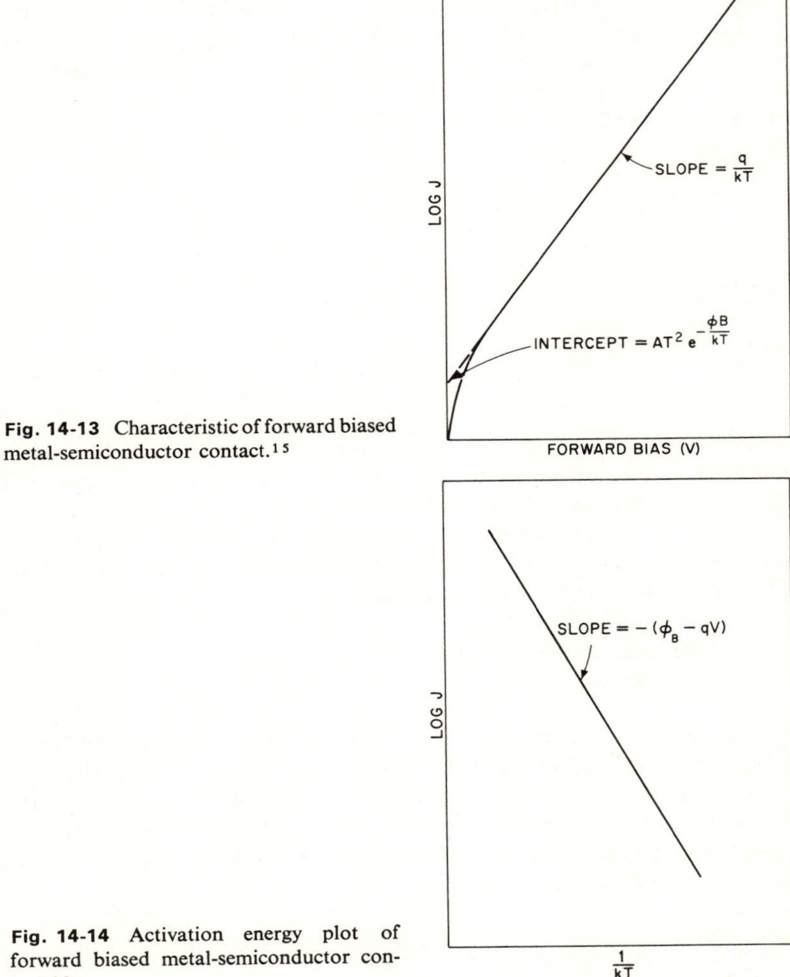

Fig. 14-13 Characteristic of forward biased metal-semiconductor contact.[15]

Fig. 14-14 Activation energy plot of forward biased metal-semiconductor contact.[15]

[16]C. R. Crowell, *Solid State Elec.* **8**, 395 (1965).

14-B-2 PHOTO-EFFECTS

14-B-2-a Internal Photoelectric Emission

Now let us illuminate the structure of Fig. 14-12 with photons such that $h\nu > \Phi_B$. Optical excitation will raise electrons in the metal to a total energy sufficient to overcome the barrier. Although hot electrons move in all directions, some move toward the interface and, if they have not suffered too many collisions, have enough energy to enter the semiconductor, which thus acquires negative charges, generating a photovoltage across the barrier. This is similar to the photoelectric emission discussed in Chapter 13, except that here the photoelectrons are emitted into the semiconductor instead of into the vacuum (process 1 in Fig. 14-15).

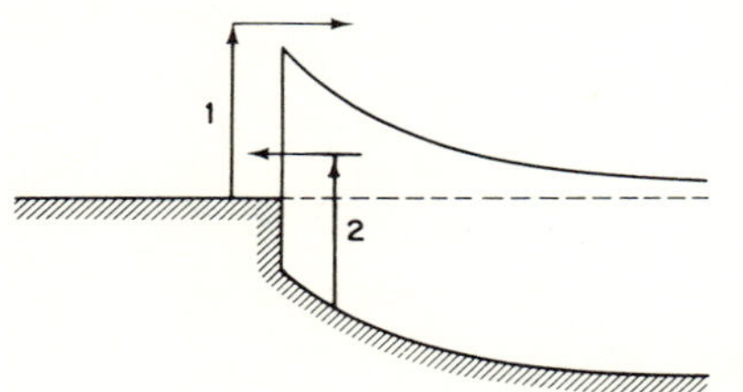

Fig. 14-15 Two possible internal photo-electric emission processes.

Note that it is also possible to photo-emit electrons from the valence band of the semiconductor into the metal above the Fermi level (process 2 in Fig. 14-15). However, because this process requires tunneling assistance, its probability is much lower than for the above transition along the two conduction bands. Furthermore, this process 2 would not generate a photo-voltage since the hole would drift to the metal.

14-B-2-b Photovoltaic Effect at a Schottky Barrier

With $h\nu > E_g$, electron–hole pairs generated inside the Schottky barrier are separated by the local field to generate a photovoltage between the metal electrode and the bulk of the semiconductor (Fig. 14-16). Schottky-barrier photovoltaic diodes can be made into large-area photocells.[17] The uniformity of their sensitivity can be probed with a flying-spot scanner and presented as a topographic display on a kinescope.[18]

[17] J. I. Pantchechnikoff, *Rev. Scientific Instruments* **23**, 135 (1952).
[18] J. I. Pantchechnikoff, S. Lasof, J. Kurshan and A. R. Moore, *Rev. of Scientific Instr.* **23**, 465 (1952).

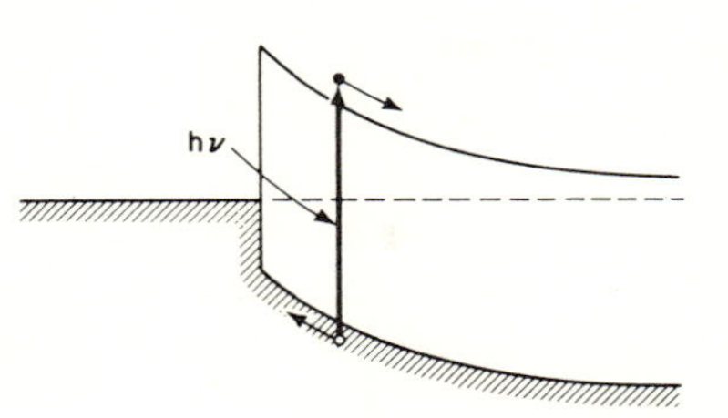

Fig. 14-16 Photovoltaic effect at Schottky barrier.

14-B-2-c Information Learned from Internal Photoelectric Emission

The barrier height can be determined from the spectral dependence of the photovoltaic or photoconductive response. Figure 14-17 shows two regions: first, a large response corresponding to the across-the-gap pair generation (the process of Fig. 14-16); and second, a tail extending to lower energies due

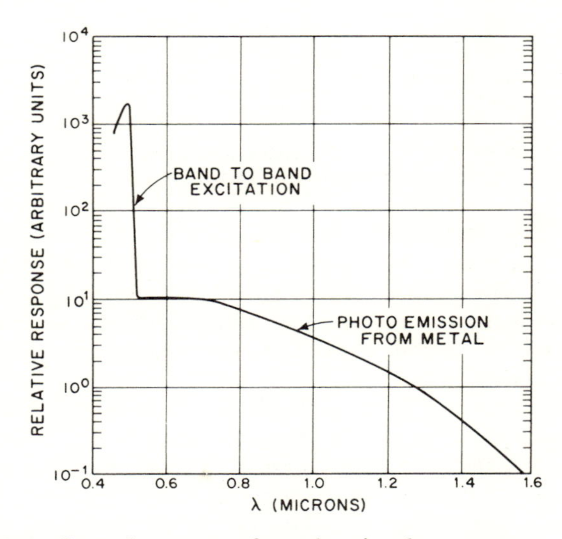

Fig. 14-17 Spectral response of metal-semiconductor contact showing regions of different excitation processes.[15]

to the migration of hot electrons from the metal over the barrier into the semiconductor (process 1 of Fig. 14-15). The low-energy threshold is then equal to the barrier height. The theoretical prediction that the rate of photoelectric emission should be proportioned to $(h\nu - \Phi_B)^2$ is demonstrated in the plot of Fig. 14-18, where the intercept with the $h\nu$-axis determines Φ_B.

[15]C. A. Mead, *Solid State Elec.* **9**, 1023 (1966).

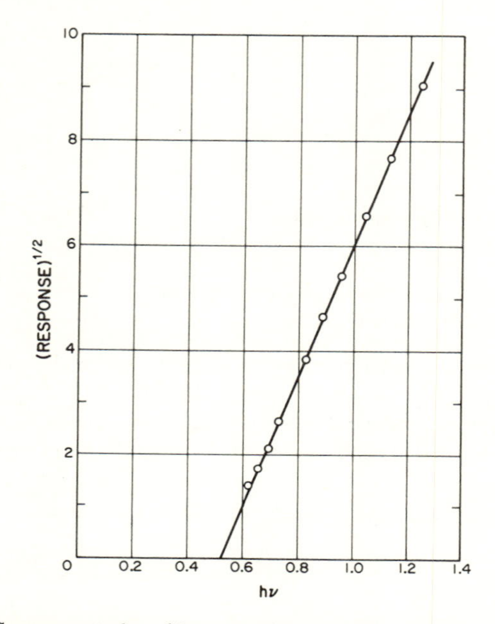

Fig. 14-18 Square root plot of long wavelength photo-emission data showing extrapolation to obtain barrier energy.[15]

It is possible to verify that the barrier height depends on the work function of the metal as per Eq. (14-7) (assuming a constant work function for the semiconductor). The metal work function (electronegativity) is known from experiments with electron emission into vacuum, whereas the barrier height can be measured by electron emission into the semiconductor. This comparison is shown in Fig. 14-19.

In general, however, the barrier height does not change as much as the metal's work function, as shown by the GaAs data of Fig. 14-19. This disagreement is attributed to surface states.[15,19] The presence of a high density of surface states on the semiconductor usually "pins" the Fermi level at the surface at some energy lower than the center of the energy gap. This results in the formation of a Schottky barrier even before the semiconductor contacts a metal, and fixes the height of the barrier. When the metal contacts the semiconductor, any charge needed to equilibrate the Fermi levels can come mostly from the surface states without appreciably shifting the pinning of the Fermi level at the interface with respect to the band edges.

Figure 14-20 shows that in n-type semiconductors the electrons of the metal see the same barrier height Φ_B regardless of the applied bias V. The

[15] C. A. Mead, *Solid State Elec.* **9**, 1023 (1966).
[19] D. V. Geppert, A. M. Cowley, and B. V. Dore, *J. Appl. Phys.* **37**, 2458 (1966).

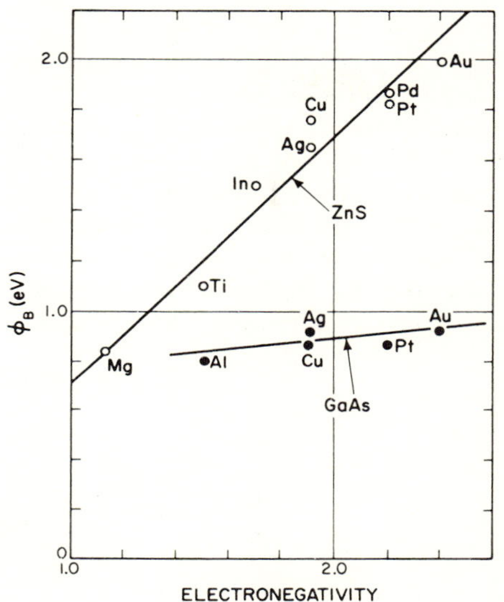

Fig. 14-19 Barrier energies of various metals on ZnS (electronegativity controlled) and GaAs (surface state controlled).[15]

semiconductor's electrons see a barrier Φ_b, the height of which depends on the bias, just as in a p–n junction. If the Fermi level is pinned at an energy E_0 at the surface and the Fermi level in the bulk is located at the conduction-band edge, then $E_c - E_0 = \Phi_B = \Phi_b$ is the barrier height. An allowance

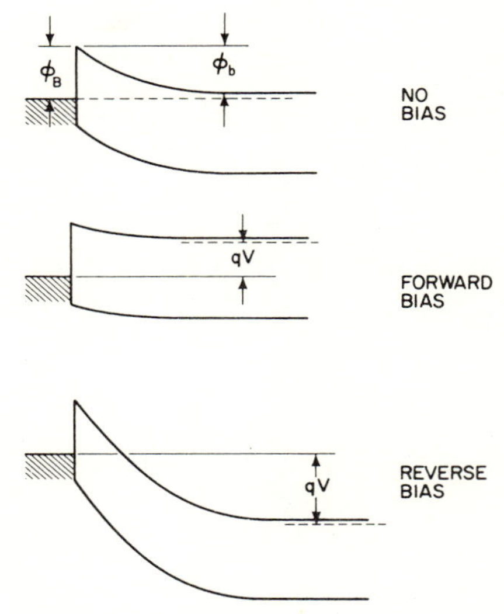

Fig. 14-20 Energy band diagrams for metal contacts on n-type semiconductor for various bias conditions.

[15] C. A. Mead, *Solid State Elect.* **9**, 1023 (1966).

for the position of the Fermi level can be made for any doping. A plot of $E_c - E_0$ vs. E_g (Fig. 14-21) reveals that $E_c - E_0 = \frac{2}{3}E_g$ in many semiconductors. However, some materials (InAs, InP, GaSb, CdTe, and CdSe) do not follow this dependence.

If the semiconductor is *p*-type and $\Phi_S > \Phi_M$, the Schottky barrier blocks the flow of holes. Optical excitation of an electron from a depth greater than Φ_{B_p} to the Fermi level permits the injection of the resulting hole into the semiconductor (Fig. 14-22). The threshold of spectral sensitivity for this process is Φ_{B_p}. Measurements of Φ_{B_p} in *p*-type materials and of Φ_{B_n} in *n*-type materials for differently doped samples of GaAs and of Si revealed that[20,21]

$$\Phi_{B_n} + \Phi_{B_p} = E_g \tag{14-9}$$

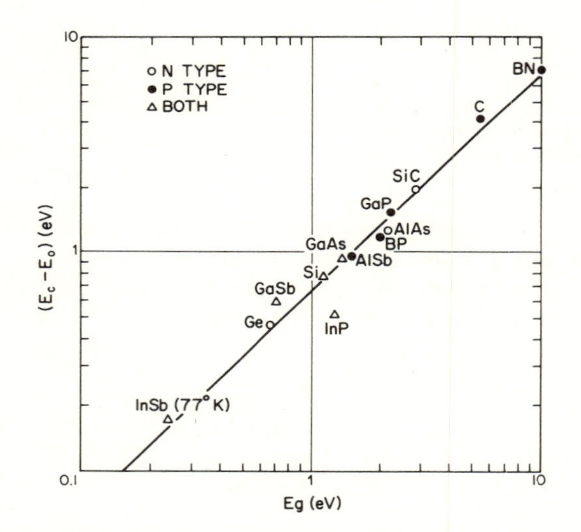

Fig. 14-21 Location of interface Fermi level relative to conduction band edge for gold contacts on various surface-state-controlled materials. The line is $E_c - E_o = 2/3E_g$.[15]

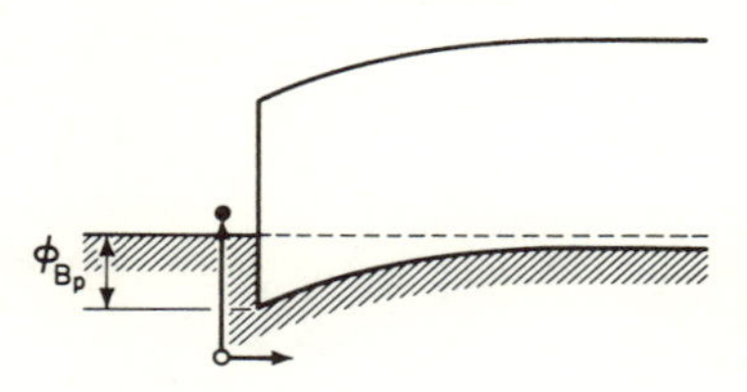

Fig. 14-22 Photoemission of hole into *p*-type semiconductor.

[20]C. A. Mead and W. G. Spitzer, *Phys. Rev. Letters* **10**, 471 (1963).
[21]W. G. Spitzer and C. A. Mead, *J. Appl. Phys.* **34**, 3061 (1963).

This result suggests that the surface states have the same distribution in both *n*- and *p*-type materials, so that the Fermi level is always pinned at the same energy, regardless of the doping.

Further photoelectric-threshold experiments have shown that, at least in Si, the pinning potential at the surface is fixed with respect to the valence-band edge; hence the distribution of surface states would also be fixed with respect to the valence-band edge.[22] The temperature dependence of the photoelectric threshold in diodes made of Au on *n*-type Si is identical to the temperature dependence of the energy gap. This means that the shift of the barrier height, $E_c - E_0$, is due only to the shift of the conduction-band edge.

14-B-3 PARTICLE DETECTORS

Photovoltaic-radiation detectors are suitable for the detection of particles other than photons. Thus α-particles which have a short range in semiconductors can be detected in a Schottky barrier at the contact between a thin gold film and either Ge or Si,[23] and also between gold and GaP.[24] In Ge and Si, the number of electron–hole pairs generated is proportional to the energy of the α particle up to 7.5 MeV. Since pairs are generated inside the depletion layer, an α-voltaic emf can be measured.

X rays can excite electrons from inner atomic shells, producing hot carriers which are energetic enough to ionize secondary pairs—a chain of events which results in a high quantum yield of electron–hole pairs per incident photon.[25]

In general, the incident particle dissipates its energy by forming a cloud of pairs along its track, the number of pairs being proportional to the energy of the incident particle. In bulk-particle detectors,[26] the particle is sensed as a sudden increase in conductivity; the energy of the particle is measured in terms of the change in the conductance. Bulk-particle detectors are used in conjunction with multiple-channel analyzers and counters which integrate in separate channels the number of ionizing events. The display of the counter outputs vs. the channel number forms a "bar graph" of the energy spectrum for the incident particles.

Clearly, if all the ionization is produced in a junction or in a Schottky barrier, an efficient particle-voltaic effect can be obtained. However, since most particles are very penetrating (e.g., γ rays), only the pairs generated within a diffusion length would contribute to the signal and those generated in the rest of the material would be wasted. The collecting volume can be

[22] C. R. Crowell, S. M. Sze, and W. G. Spitzer, *Appl. Phys. Letters* **4**, 91 (1964).
[23] J. W. Mayer, *J. Appl. Phys.* **30**, 1937 (1959).
[24] B. Goldstein, *J. Appl. Phys.* **36**, 3853 (1965).
[25] H. Pfizter, *Z. angew, Phys.* **37**, 1191 (1959).
[26] J. M. Taylor, *Semiconductor Particle Detectors*, Butterworths (1963).

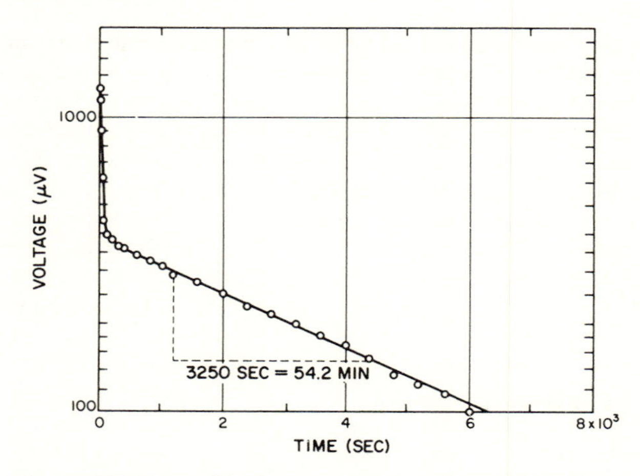

Fig. 14-23 Decay of β-voltage in InP after neutron irradiation.[28]

increased by grading the junction over a long distance—this can be achieved by Li-drifting techniques.[27]

Thermal neutrons are nonionizing particles, yet they can indirectly induce a photovoltage by a curious effect: in InP, thermal neutrons cause the transmutation of In^{115} into the radioactive isotope In^{116}; In^{116} decays into Sn^{116} with the emission of energetic electrons (β particles), which in turn can generate electron–hole pairs. A p–n junction inside the InP can detect this ionization and signal it with an emf.[28] This "β voltage" decays in time with the same characteristic time constants as In^{116}—i.e., 13 sec and 54.3 min, as shown in Fig. 14-23.

14-C Bulk Photovoltaic Effects

14-C-1 DEMBER EFFECT

Strongly absorbed radiation can generate a high density of electron–hole pairs at the surface of a semiconductor. The carriers diffuse away from the illuminated region. The electrons, having a higher mobility than the holes, will extend the electron cloud somewhat farther into the crystal than the cloud of holes. In the absence of any other effect, the differential diffusion will tend to make the surface more positive than the bulk. The direction of the resulting electric field is such as to accelerate the lower-mobility carrier and to slow down the more mobile carrier, so that the net current is nil.

[27]E. M. Pell, *J. Appl. Phys.* **31**, 291 (1960).
[28]R. Gremmelmaier and H. Welker, *Z. Naturforsch.* **11a**, 420 (1956).

If no other electric field is present, the Dember voltage can be readily evaluated by solving the diffusion equations

$$\left.\begin{aligned} j_e &= q(n_0 + \Delta n)\mu_e \mathscr{E}_D + qD_e \frac{d\Delta n}{dx} \\ j_h &= q(p_0 + \Delta p)\mu_h \mathscr{E}_D - qD_h \frac{d\Delta p}{dx} \end{aligned}\right\} \tag{14-10}$$

where $\mathscr{E}_D$ is the Dember field and $\Delta n = \Delta p$ is the concentration of excess carriers. Under steady-state conditions, the total current density $j = j_e + j_h$ is zero. Then one can extract the Dember field $\mathscr{E}_D$ and, integrating it from the surface to deep in the bulk (several diffusion lengths), where the concentration of excess carriers vanishes, and using Einstein's relation $D = kT\mu/q$, one obtains the Dember voltage:

$$V_D = \frac{kT}{q}\frac{b-1}{b+1}\ln\left[1 + \frac{(b+1)\Delta n}{bn_0 + p_0}\right] \tag{14-11}$$

where b is the ratio of the highest to the lowest mobilities.

The Dember voltage is usually very small, of the order of millivolts. Equation (14-11) shows that if $\mu_e = \mu_h$ (i.e., $b = 1$), then $V_D = 0$. In the usual case ($b > 1$), the Dember voltage is not much greater than kT/q.

Note that although the Dember effect does not require the presence of a potential barrier, the semiconductor must still be nonhomogeneous in some other way in order to develop a photovoltage. In this instance, the geometry of a semi-infinite semiconductor is asymmetric.

The Dember voltage occurs near a semiconductor surface, or near a metal–semiconductor contact, or along a semiconductor between illuminated and shadowed regions (which is where Dember discovered this photovoltage).[29]

The Dember voltage occurs independently of other effects at a surface barrier; it is an additional component which may add or subtract from the photovoltage generated at the barrier. Because ever-present surface states always induce a surface barrier, special precautions are needed when one wants to observe the Dember effect.

14-C-2 PHOTOMAGNETOELECTRIC EFFECT

For a detailed theoretical treatment, see references 30 and 31.

Illumination with strongly absorbed light causes a concentration gradient of electron–hole pairs diffusing away from the illuminated surface in the x-direction (Fig. 14-24). If a magnetic field H_z is applied transversely to this diffusion current, the electrons and holes will be deflected in opposite direc-

[29]H. Dember, *Physik. Zeits.* **32**, 554 and 856 (1931); **33**, 209 (1932).
[30]O. Garreta and J. Grosvalet, *Progress in Semiconductors* **1**, 165 (1956).
[31]W. Van Roosbroeck, *Phys. Rev.* **101**, 1713 (1956).

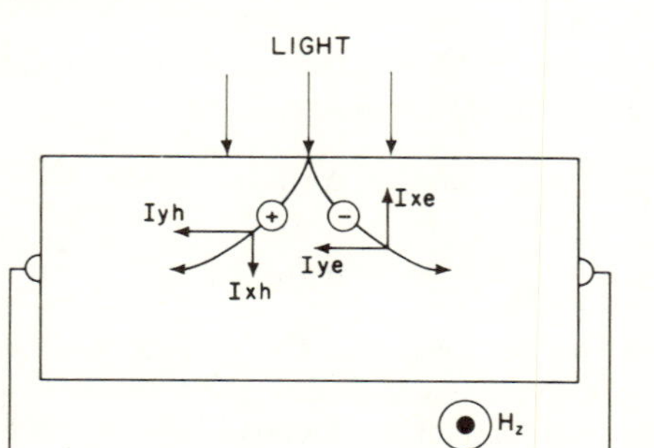

Fig. 14-24 Generation of photomagnetoelectric voltage V_y.

tions (as in the Hall effect), with the result that a voltage V_y will appear across the ends of the specimen.

Unlike the Dember effect, which required the carriers to have different mobilities, in the photomagnetoelectric (PME) effect the mobilities need not be different. For a given illumination, the PME voltage depends on the surface recombination s and on the lifetime τ of carriers in the volume. The PME effect has been used to find these quantities s and τ which are so important in the performance of semiconductor devices.[32-34]

The PME drift of carriers at the illuminated surface forms a high-density current (to the left in Fig. 14-24) which develops a measurable voltage V_y. The voltage V_y in turn drives a lower current density from right to left in the rest of the crystal (Fig. 14-25). Suppose the magnetic field is tilted in the y–z plane; then the H_y component of the magnetic field interacts with the circulating current to produce a torque. This is the "photomechanical" effect.[30]

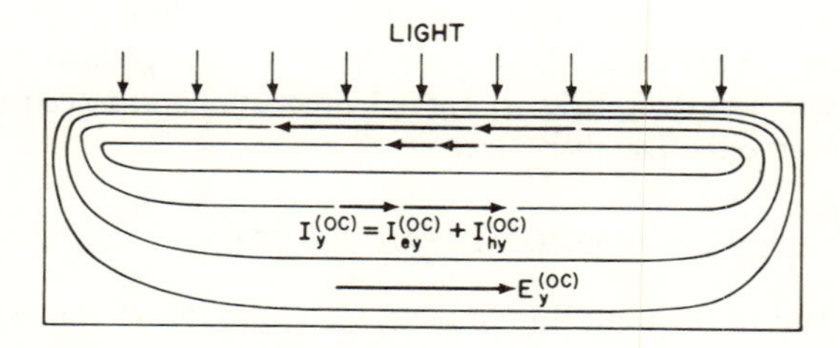

Fig. 14-25 The PME effect in a semiconductor slab of finite length: Open-circuit current, without electrodes.[31]

[30] O. Garreta and J. Grosvalet, *Progress in Semiconductors* **1**, 165 (1956).
[31] W. Van Roosbroeck, *Phys. Rev.* **101**, 1713 (1956).
[32] T. S. Moss, L. Pincherle, and A. M. Woodward, *Proc. Phys. Soc.* **66B**, 743 (1953).
[33] P. Aigrain and H. Bulliard, *Compt. Rend.* **236**, 595 and 672 (1953).
[34] H. Bulliard, *Phys. Rev.* **94**, 1564 (1954).

A measurement of the torque can then be substituted for the electrical measurement, with the advantage that electrodes on the crystal are not needed.

14-D Anomalous Photovoltaic Effect

Some semiconductors in the form of a thin layer exhibit a high-voltage photovoltaic effect.[35] The voltage can be many times larger than the potential drop across the energy gap. Although this phenomenon is spectacular and very intriguing, little progress has been made in explaining it in a detailed and definitive way. The main requirement for the occurrence of this effect is that when the film is grown, the insulating substrate must be inclined with respect to the gradient of the incident vapor as shown in Fig. 14-26. The anomalous photovoltaic effect has been observed in a large number of semiconductor films (Table 14-1) and even in single crystals of ZnS[36,37] and of sulfur.[38]

Table 14-1

Materials in which the anomalous photovoltaic effect has been observed

Material	*Reference*	*Material*	*Reference*
PbS	39	HgTe	45
CdTe	35	Sb_2Se_3	46
Si	40	Sb_2S_3	46
Ge	40	$Sb_{2x}Bi_{2(1-x)}S_3$	46
GaAs	41, 42	ZnSe ("single" crystal)	35
InP	43	ZnS ("single" crystal)	36
GaP	44		

[35] B. Goldstein and L. Pensak, *J. Appl. Phys.* **30**, 155 (1959).
[36] S. G. Ellis, E. E. Loebner, W. J. Merz, C. W. Struck, and J. G. White, *Phys. Rev.* **109**, 180 (1958).
[37] W. J. Mertz, *Helvetica Physica Acta* **31**, 625 (1958).
[38] W. Ruppel and P. M. Grant, *Solid State Comm.* **4**, 649 (1966).
[39] J. Starkiewicz, L. Sosnowski, and O. Simpson, *Nature* **158**, 26 (1946).
[40] H. Kallman, B. Kramer, E. Heidemenakis, W. J. McAller, H. Barkemeyer, and P. I. Pollark, *J. Electrochem. Soc.* **108**, 247 (1961).
[41] E. I. Adirovich, V. M. Rubinov, and Yu. M. Yuabov, *Sov. Phys.—Solid State* **6**, 2540 (1965).
[42] S. Martinuzzi, *Compt. Rend.* **258**, 1769 (1964).
[43] M. D. Uspenskii, N. G. Ivanova, and I. E. Malkis, *Sov. Phys.—Semiconductors* **1**, 1059 (1968).
[44] E. I. Adirovich, V. F. Roslyakova, and Yu. M. Yuabov, *Sov. Phys.—Semiconductors* **2**, 848 (1969).
[45] S. Martinuzzi and J. Fourny, *Bull. d'Inform. du COMPLES* **5**, (1963).
[46] V. M. Lyubin and G. A. Fedorova, *Sov. Phys.—Dokl.* **5**, 1343 (1960).

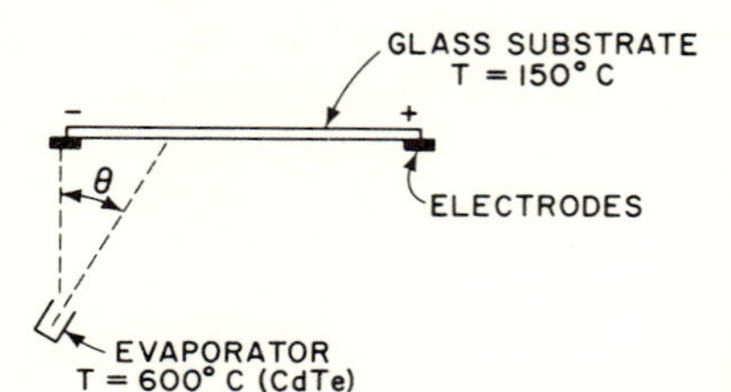

Fig. 14-26 Schematic arrangement for depositing film material at an angle.[35]

14-D-1 CHARACTERISTICS OF ANOMALOUS PHOTOVOLTAIC CELLS

There is considerable variability in the performance of semiconductor films prepared to generate a high photovoltage. A collection of photovoltage–vs.–light-intensity curves for different materials is shown in Fig. 14-27.

Photovoltages as high as 5000 V have been obtained.[47] Experiments have shown that the voltage is generated continuously along the length of the film,[35] and that it is not a contact effect. The end of the film nearest the evaporator is almost always negative. Yet the tapering off by the film thickness away from the evaporator is not a relevant factor, as shown in an experiment where a moving shutter assured a constant film thickness during oblique deposition.[48]

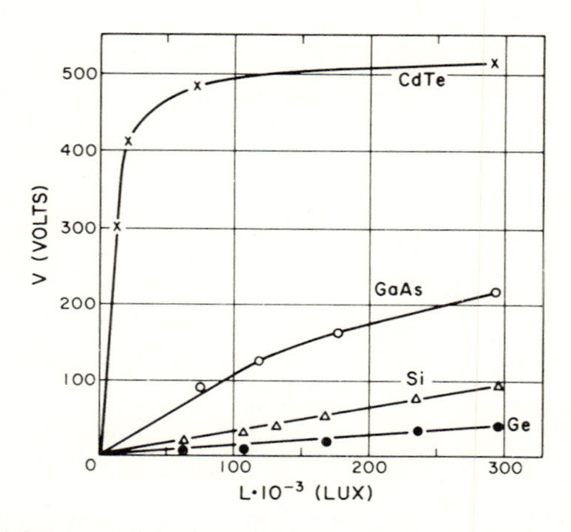

Fig. 14-27 Voltage vs. light intensity characteristics of films made of different semiconductors.[41]

[47]E. I. Adirovich, T. Mirzamakhmudov, V. M. Rubinov, and Yu. M. Yuabov, *Sov. Phys.—Solid State* **7**, 2946 (1966).

[48]E. I. Adirovich, V. M. Rubinov, and Yu. M. Yuabov, *Sov. Phys.—Dokl.* **10**, 844 (1966).

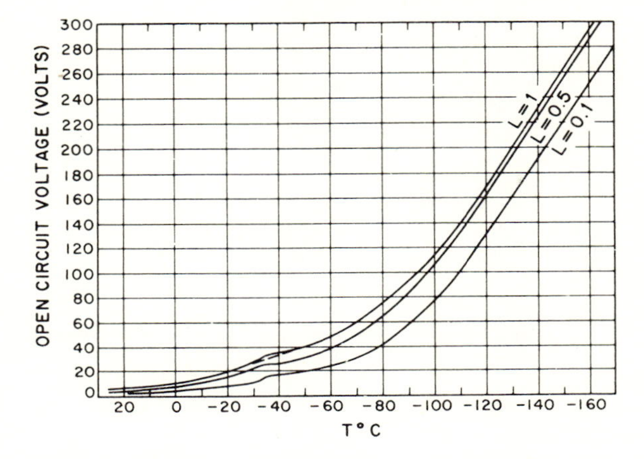

Fig. 14-28 Open circuit photovoltage vs temperature at varying light intensities L.[35]

The $I(V)$ characteristics of the film are usually linear and exhibit a very high resistance of the order of 10^{10} to 10^{14} ohms. In some cases the resistance of the film is not affected by illumination (no detectable photoconductivity),[47] but sometimes the cell exhibits both the photovoltaic effect and a photoconductive drop in resistance.[35]

The amplitude of the photovoltage increases with decreasing cell temperature as shown in Fig. 14-28. However, in some samples the open-circuit voltage goes through a minimum at 430°K, increasing at both lower and higher temperatures.[49]

The spectral sensitivity of the anomalous photovoltaic effect extends to photon energies lower than the gap energy of the single-crystal semiconductor.[47] This is not surprising in view of the fact that films are a somewhat disordered system with consequent strong local perturbations of the band structure. But what is surprising is that sometimes the polarity of the photovoltage reverses over part of the spectral range.[46]

Response times (rise and fall) of the photovoltage have been reported from 10^{-4} sec to as long as 40 sec in references 46 and 50 respectively.

14-D-2 CONDITIONS FOR OBTAINING THE ANOMALOUS PHOTOVOLTAIC EFFECT

The conditions under which the semiconducting film must be prepared are more critical for some materials than for others. In general, the angle of

[35] B. Goldstein and L. Pensak, *J. Appl. Phys.* **30**, 155 (1959).
[46] V. M. Lyubin and G. A. Fedorova, *Sov. Phys.—Dokl.* **5**, 1343 (1960).
[49] G. Cheroff, *Bull. Am. Phys. Soc.* **6**, 110 (1961).
[50] H. W. Brandhorst and A. E. Potter *J. Appl. Phys.* **35**, 1997 (1964).

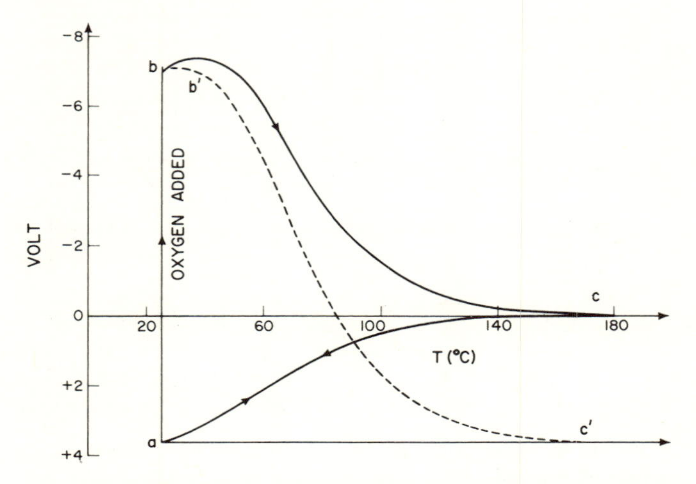

Fig. 14-29 Temperature dependence of the anomalous photovoltaic effect in a high vacuum.[52] The polarity is that of the thick end with respect to the thin end.

evaporation with respect to the substrate must be in the range of 30–60°. The substrate temperature should be in the range of 50–100°C. The film thickness has an optimum value. The pressure and composition of the residual gases in the vacuum chamber have a pronounced influence; and often a treatment is needed after the deposition of the film on the substrate. The choice of substrate material and its cleanliness are also of importance.

The adsorption and desorption of oxygen affect the amplitude of the photovoltaic effect and can result in polarity reversal.[51,52] A study of the influence of various gases adsorbed on CdTe films shows that the end of the film which is closest to the vapor source is positive soon after the film has been formed, and remains positive even at a low argon pressure.[52] However, this same end becomes negative when the film is exposed to oxygen. The polarity can be reversed back to positive only after heating the sample in vacuum ($\sim 10^{-7}$ Torr). This oxygen adsorption–desorption cycle is reproducible. Figure 14-29 shows the photovoltaic data taken in a high vacuum. In the absence of oxygen, curve *ac* is obtained, which gives a positive photovoltage with a reproducible reversible temperature dependence. After oxygen has been adsorbed, the polarity is reversed and curve *bc* is obtained. Note that *bc*, attributed to the desorption of oxygen upon warm-up, does not have a reversible temperature dependence. The curve *b'c'*, the difference between the negative and positive curves, represents the contribution of the adsorbed

[51] L. Pensak, "Effect of Gases on the High Voltage Photovoltaic Effect," *Structure and Properties of Thin Films*, ed. C. A. Neugebauer et al., Wiley (1959), p. 503.

[52] G. Brincourt and S. Martinuzzi, *Comptes Rendus* **266**, B-1283 (1968).

oxygen. Hence the temperature dependence is due to two competitive processes: that of the reversible oxygen-free condition, which is proportional to $\exp(0.70/kT)$, and that due to the presence of oxygen, which is proportional to $\exp(0.85/kT)$ and represents the activation of the desorption process.

14-D-3 MODELS FOR THE ANOMALOUS PHOTOVOLTAIC EFFECT

The still-tentative models which have survived critical scrutiny are based on the summation of elementary photovoltages generated in an array of small photocells. These photovoltages arise either from the Dember effect in microcells or from photovoltaic effect at *p–n* junctions or at grain boundaries. However, neither of these models fits all the observations, although one of them may be adequate for a given material. Further work is still needed to resolve present ambiguities.

All the models have in common the concept of an array of asymmetric cells. The asymmetry may be a difference in area between the illuminated and the shadowed surfaces, or it may be an asymmetry between the *p–n* and the *n–p* junctions in a *p–n–p–n–p–n* sequence, or it may be an asymmetry in a similar sequence of Schottky barriers; or else strains[53] at grain boundaries induce asymmetric deformation potentials (these strains may be due to stacking faults or to the differential thermal expansion of the semiconductor film and the substrate). Most probably, a combination of these factors plays a part in the anomalous photovoltaic effect.

In the case of CdTe films, optical-reflection and optical-transmission as well as electron-diffraction data indicate that the film consists of (111) planes inclined at a slight angle to the substrate; this angle depends on the obliqueness of the deposition.[54] A measurement of the photocell capacitance suggests that the cell is made up of a large number of thin insulated microlayers the area of which is larger than the transverse cross-section of the film.[55] Hence this measurement also indicates that the microlayers must make a small angle with the substrate.

14-D-3-a Stacking-fault Model[37]

The stacking-fault model is the model advanced for the case of ZnS needles or platelets where the anomalous photovoltaic effect is found. Microscopic examination reveals that the crystal is made up of an alternation of cubic and hexagonal layers. The photovoltage develops transversely to the striations.

Cubic ZnS and hexagonal ZnS have approximately the same energy gap.

[37]W. J. Merz, *Helvetica Physica Acta* **31**, 625 (1958).
[53]A. R. Hutson, *Bull. Am. Phys. Soc.* **6**, 110 (1961).
[54]M. Kamiyama, M. Haradome, and H. Kukimoto, *Japanese J. of Appl. Phys.* **1**, 202 (1962).
[55]M. I. Kosunskii and V. M. Fridman, *Sov. Phys.—Solid State* **8**, 213 (1966).

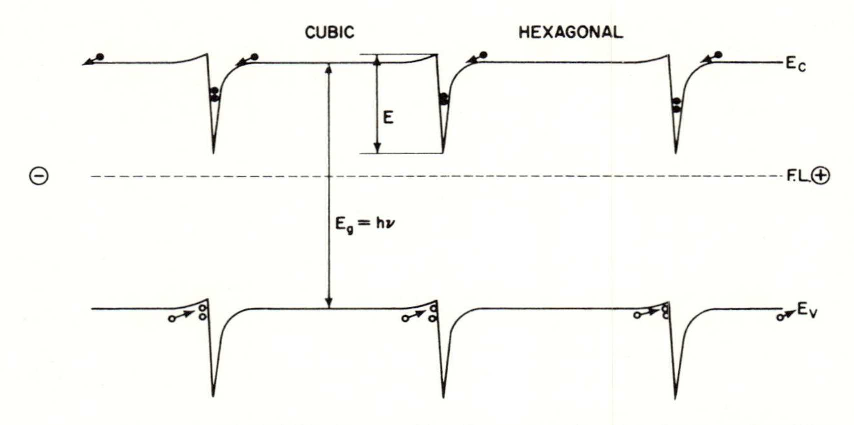

Fig. 14-30 Proposed band structure for cubic–hexagonal ZnS.[37]

However, one goes from one structure to the other by rotating the ZnS tetrahedron around one bond along the hexagonal axis by 60°. The resulting change in short-range periodicity would induce the perturbation of the band structure shown in Fig. 14-30. This perturbation has the asymmetry needed to generate additive potentials of the order of 0.1 V per boundary.

14-D-3-b Model Based on p–n Junctions

A chain of *p–n–p–n–p–n* transitions requires that only junctions of the same type (e.g., *p–n*) be illuminated or photovoltaic, while those of the opposite polarity (*n–p*) be inactive (shadowed or very leaky). Figure 14-31(a) shows a schematic exaggeration of this model. The voltage to be expected from ***m***

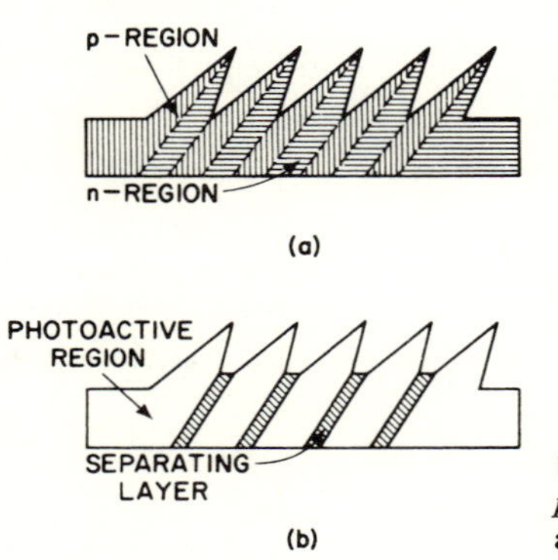

Fig. 14-31 Models of the film: (a) micro *p-n* junctions; (b) photodiffusion microareas.[56]

[56] E. I. Adirovich, V. M. Rubinov, and Yu. M. Yuabov, *Sov. Phys.—Dokl.* **11**, 512 (1966).

junctions is given by[56]

$$V_{oc} = \frac{kT}{q} \sum_{i=1}^{m} \ln\left(1 + \frac{J_{sc_i}}{J_{s_i}}\right) \tag{14-12}$$

where J_{sc_i} is the short-circuit photocurrent density of the ith junction and J_{si} is the saturation-current density of the ith junction.

14-D-3-c Model Based on the Dember Effect.[57]

The film is assumed to consist of elementary microcells separated by insulating layers [Fig. 14-31(b)]. A Dember voltage is generated at the illuminated facet of each cell. The smaller shadowed boundary makes a negligible bucking contribution.

For the Dember model, the open-circuit voltage to be expected is given by[56]

$$V_{oc} = \frac{kT}{q} \frac{b-1}{b+1} \sum_{i=1}^{m} \ln \frac{\sigma_{2_i}}{\sigma_{1_i}} \tag{14-13}$$

where σ_{1i} and σ_{2i} are the total conductivities (dark conductivity plus photoconductivity) in, respectively, the regions at the left and right of the ith boundary; and b is the ratio of the electron and hole mobilities.

Equations (14-12) and (14-13) show that unless one knows the number of cells $\boldsymbol{m}$ and the ratio of the effect of illumination on either side of each boundary, one cannot distinguish between the last two models solely from the dependence of V_{oc} on light intensity.

14-D-4 ANGULAR DEPENDENCE OF PHOTOVOLTAIC EFFECTS

14-D-4-a Anomalous Photovoltaic Effect

Obliquely deposited films often show the same polarity of photovoltage regardless of the angle of incidence of the light beam with respect to the film,

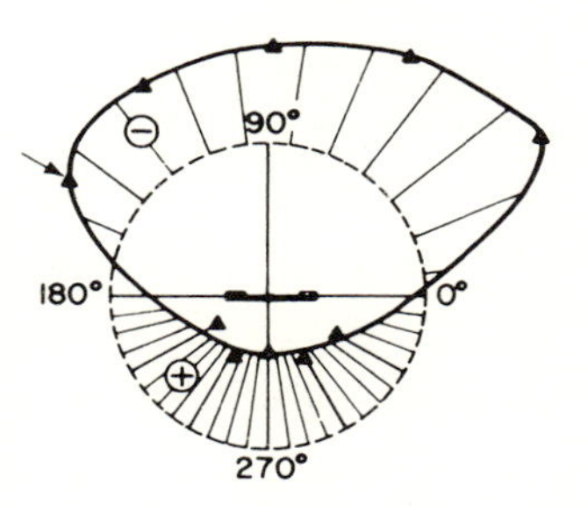

Fig. 14-32 Dependence of the anomalous photovoltage on the angle of illumination for a Ge film. The arrow indicates the direction of the vapor stream during deposition. The dashed circle is the level of zero voltage, V being measured along a radial vector. The polarity sign refers to the left contact.[56]

[56]E. I. Adirovich, V. M. Rubinov, and Yu. M. Yuabov, *Sov. Phys.—Dokl.* **11**, 512 (1966).
[57]I. P. Zhadko and V. A. Romanov, *Phys. Stat. Solidi* **28**, 797 (1968).

even when illuminated through the substrate. There is a tendency for the photovoltage to be maximum for illumination at about 90° to the direction of deposition.[54] In some films the polarity of the photovoltage reverses when the film is illuminated through the substrate (as if the *n–p* instead of the *p–n* junctions of Fig. 14-31 were now activated). An example of such a dependence is shown in Fig. 14-32.

14-D-4-b "Photoangular" Effect[58]

Names for anomalous effects becoming scarce, the present designation was applied to observations made on relatively thick films (2–50 μm) of polycrystalline GaAs and Si. These films were deposited by vapor-phase transport across a small gap between the source crystal and the substrate.[59] Here the deposition is on the average normal to the substrate and, therefore, the structure of this film should be very different from the thin angularly grown films in which the anomalous photovoltaic effect is observed.

When the film is illuminated at normal incidence, no photovoltage appears. As the light beam is tilted with respect to a normal to the film, in the azimuthal plane passing through the two small electrodes (Fig. 14-33), a photovoltage develops and grows with further inclination of the beam. The photovoltage has opposite polarities on either side of normal incidence. Sometimes the potential difference developed is larger than the energy gap

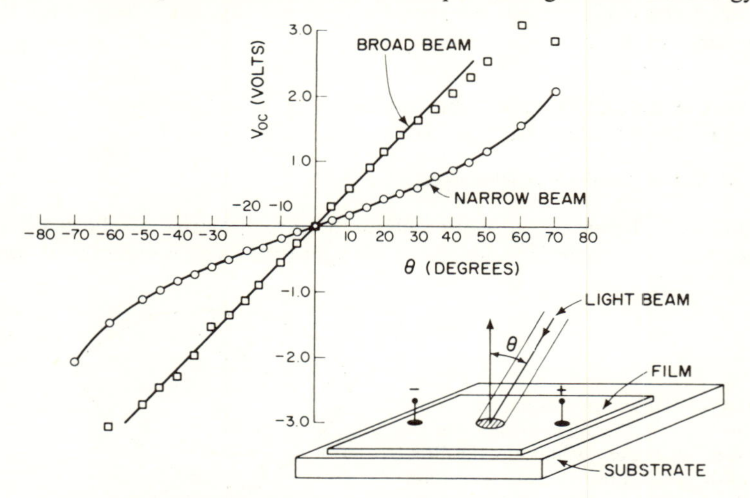

Fig. 14-33 Variation of open-circuit voltage with angle of incidence.[58]

[54] M. Kamiyama, M. Haradome, and H. Kukimoto, *Japanese J. of Appl. Phys.* **1**, 202 (1962).

[58] D. Perkins and E. F. Pasierb, Paper II-A-2, *Proc. 4th Photovoltaic Specialists Conference* Cleveland, Power Information Center (1964).

[59] F. H. Nicoll, *J. Electrochem. Soc.* **110**, 1165 (1963).

of the semiconductor. A linear dependence of open-circuit photovoltage on angle of illumination is obtained over a large angle of scan (Fig. 14-33). A signal of 1 mV/min of arc was measured with a 9-mW-radiation input.

The photoangular effect is strikingly reminiscent of Dember's experiment with an angularly shaped crystal,[29] except that in the present case the surface may consist of an array of angular ridges.

No photovoltage is obtained when the light is incident through the substrate. The photovoltage develops only on the free surface of the film and ceases if the free surface is either polished or etched. The spectral response cuts off abruptly at photon energies lower than the gap energy, and extends undiminished to 4 eV. This confirms that the photoangular effect occurs at the free surface of the polycrystalline layer.

14-E Other Photovoltaic Effects

14-E-1 LATERAL PHOTOEFFECT[60]

Thus far we have considered photovoltages generated transversely to assumed or real boundaries, such as a *p–n* junction. For the lateral photoeffect the photovoltage is parallel to the *p–n* junction.

The device consists of two ohmic connections on either side of a floating *p–n* junction, as shown at the top of Fig. 14-34. Because the p^+ region has a high conductance, it forms an equipotential. The "normal" photovoltage due to a small light spot forward-biases the *p–n* junction. The excess holes can leak out of the p^+ region over the whole junction area. If the light spot illuminates the center of the junction, the hole-current flows symmetrically out of the junction and no net voltage appears across the ohmic connections. If, on the other hand, the light spot illuminates the left side of the junction, most of the holes leak out of the right side of the junction, making the right side more positive than the left side. Electrons in the *n*-type region instantly redistribute themselves to preserve local space-charge neutrality. This charge redistribution generates a voltage across the two ohmic connections. The polarity of the lateral photovoltage reverses when the light spot illuminates the right side of the junction.

14-E-2 OPTICALLY INDUCED BARRIERS

Here we shall consider how light can form an electrostatic barrier in a uniformly doped semiconductor. This effect contrasts with the normal photovoltaic effect, which tends to eradicate an already present barrier.

[29]H. Dember, *Physik. Zeits.* **32**, 554 and 856 (1931); **33**, 209 (1932).
[60]J. T. Wallmark, *Proc. IRE* **45**, 474 (1957); J. I. Alferov, V. M. Andreev, E. L. Portnoi and I. I. Protasov, *Sov. Phys.—Semiconductors* **3**, 1103 (1970).

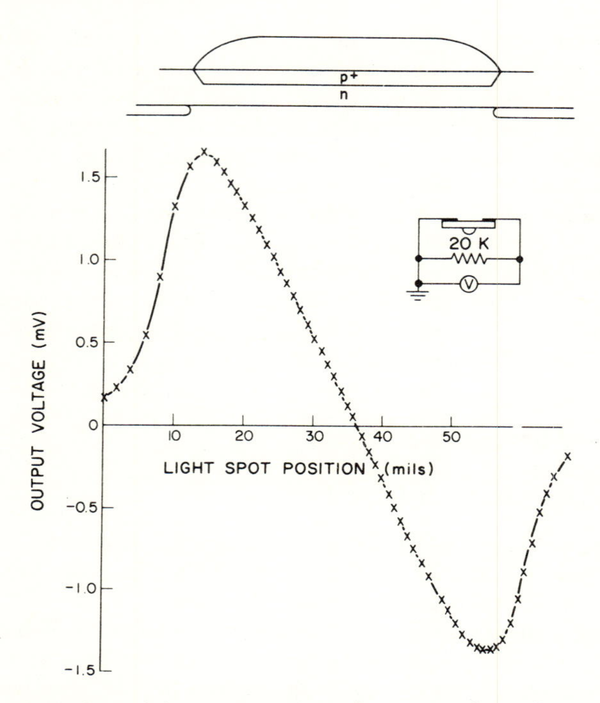

Fig. 14-34 Lateral photo-response curves for p^+ on n photocell.[60]

Consider an n-type semiconductor. The Fermi level is located between the edge of the conduction band and the donor levels. If one-half of the crystal is illuminated to produce more pairs than the number n_D of majority carriers present in the dark half of the crystal, the quasi-Fermi level for electrons in the illuminated portion of the crystal will move nearer to the edge of the conduction band by an amount ΔE:

$$\Delta E = E_{F_n} - E_F = kT \ln\left(\frac{n_L}{n_D}\right)$$

where n_L is the electron concentration in the illuminated region. The values of E_{F_n} and E_F are measured from the valence band edge in the light and dark regions, respectively. The Fermi level in the dark region remains fixed. Hence a barrier to electrons develops between the light and dark regions[61] (Fig. 14-35). A barrier height of several tenths of an eV can be obtained in CdS. Note that the depletion layer extends mostly into the dark region. This

[60] J. T. Wallmark, *Proc. IRE* **45**, 474 (1957); J. I. Alferov, V. M. Andreev, E. L. Portnoi and I. I. Protasov, *Sov. Phys.—Semiconductors* **3**, 1103 (1970).

[61] G. Wlérick, *Annales de Phys. Ser.* **13**, **1**, 623 (1956).

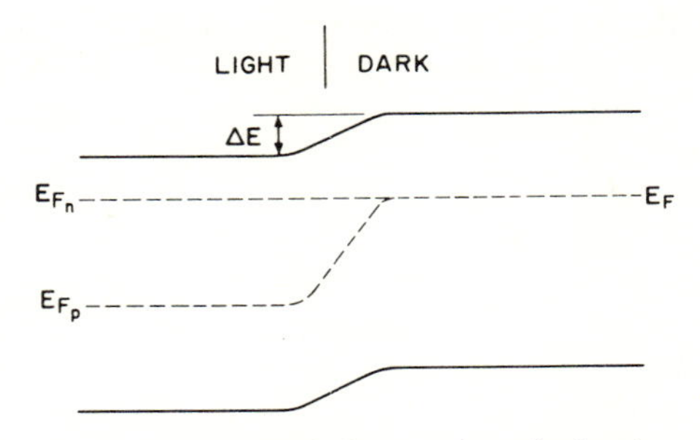

Fig. 14-35 Optical generation of a barrier.

is of practical importance in image tubes, where the extent of the depletion layer ($\sim 10^{-3}$ cm) limits the picture resolution. A slight rectification might be expected at such a transition, but in practice it is the barrier layer at the electrodes which dominates the $I(V)$ characteristics.

14-E-3 PHOTOVOLTAIC EFFECT AT A GRADED ENERGY GAP

The energy gap of a semiconductor can be modified by changing the material's composition, as in the alloy between two miscible semiconductors. A gradual transition between two semiconductors with different energy gaps results in a local field.[62] In the example of Fig. 14-36, the local field drives the minority carriers faster than the majority carriers. In contrast to another photovoltaic effect occurring in a uniformly doped semiconductor—the Dember voltage which results from different diffusion velocities—here, the photovoltage is due to different drift fields acting on each type of carrier.

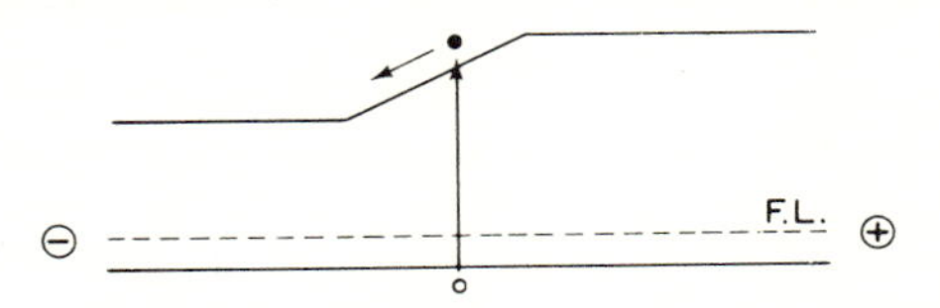

Fig. 14-36 Generation of photovoltage at a graded gap.

Another means of changing the energy gap of semiconductors is the deformation resulting from pressure. In the "photopiezoelectric" effect,[63] a uniaxial strain is impressed at a point on a crystal; the energy gap is deformed symmetrically with respect to the pressure point; hence a light spot scanning

[62] H. Kroemer, *RCA Review* **18**, 332 (1957).
[63] J. Tauc and M. Zavetova, *Czechosl. J. Phys.* **9**, 95 (1959).

the crystal near the pressure point obtains a polarity reversal of the photovoltage when the light spot passes the pressure point (Fig. 14-37).

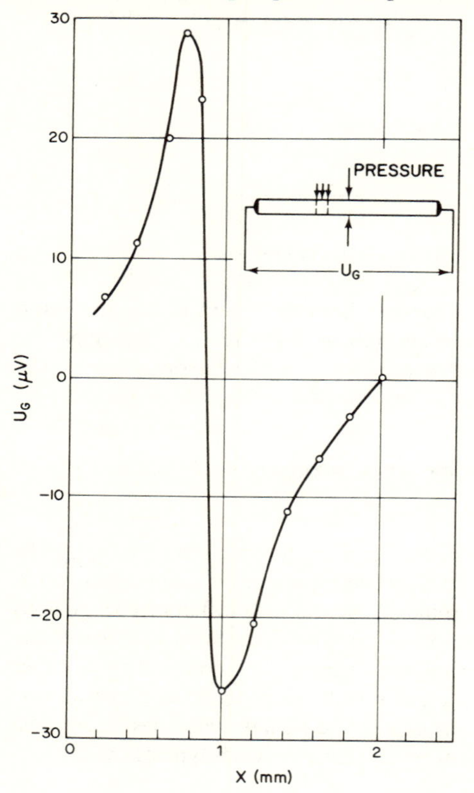

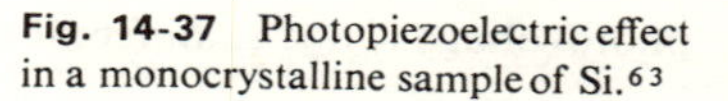
Fig. 14-37 Photopiezoelectric effect in a monocrystalline sample of Si.[63]

Problem 1. Consider a *p–n* junction as shown in Fig. 14P-1. Upon shining light of energy $h\nu$ on the *p*-side, a short-circuit current will flow which is proportional

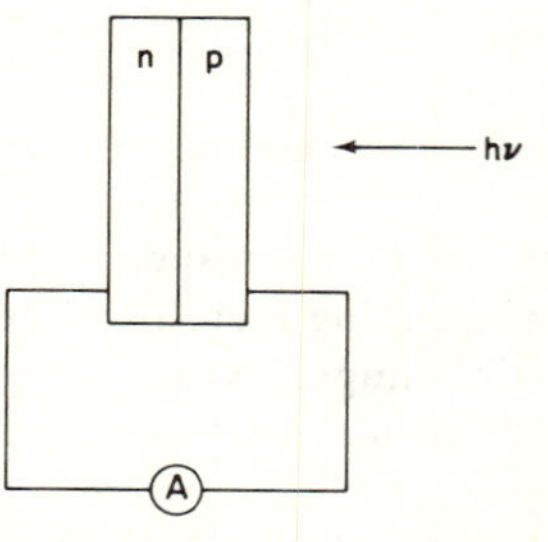

Fig. 14P-1

[63] J. Tauc and M. Zavetova, *Czechosl. J. Phys.* **9**, 95 (1959).

to the incident flux of radiation. Draw a qualitative graph of the short-circuit current vs. $h\nu$ and indicate how to find the position of the maximum. Neglect recombination of carriers in the p-type region.

Problem 2. Consider the configuration given in the diagram. Assume the light to be incident on a facet perpendicular to the junction and the width of the beam to be less than $d + L_h + L_e$, where L_h and L_e are the diffusion lengths of holes and electrons, respectively, and d is the junction thickness. Neglect tunneling effects. Assume that all the carriers generated are collected so that there is no variation along the x-direction perpendicular to the beam. Draw a qualitative $i_{sc}(h\nu)$ spectrum where i_{sc} is the short-circuit current.

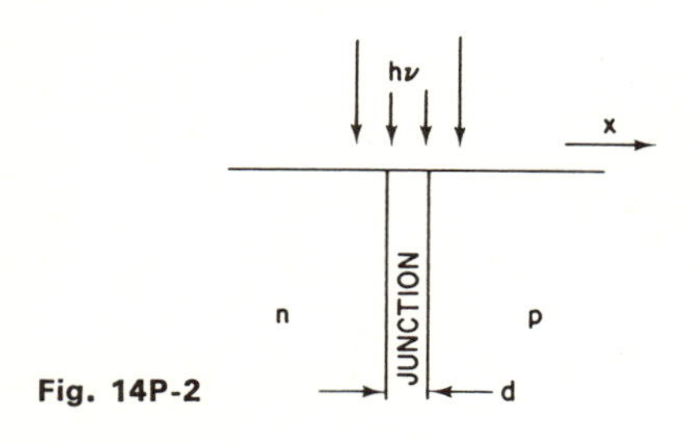

Fig. 14P-2

Problem 3. In a practical p–n-junction photocell, the maximum value of the open-circuit photovoltage is given by the barrier height as shown in Fig. 14-1(a). In a junction between degenerately doped semiconductors, the barrier height can be larger than the energy gap. Discuss the photovoltaic characteristics you would expect in such a diode.

Problem 4. A metal with a work function $\Phi_M = 4.8$ eV is evaporated onto a semiconductor having a work function $\Phi_S = 3.6$ eV. Assume that the semiconductor is n-type with a Fermi level 0.2 eV below the conduction-band edge and that it has an energy gap of 1.5 eV. Furthermore, assume that the semiconductor has no surface states for pinning the Fermi level at the surface.

(a) Find the height of the Schottky barrier.
(b) What is the maximum open-circuit photovoltage that could be obtained with such a photocell?

Problem 5. Speculate on the possibility that the "lateral" photoeffect plays a part in the "photoangular" effect. What sort of surface structure would be needed to account for this possibility?

Problem 6. In the accompanying figure, the practical application of a semiconductor photoconductor is illustrated. The light beam incident on the device, which is merely a resistive bar, creates extra electron–hole pairs, thus increasing the conductivity of the detector. This changes the current through the circuit and, there-

POLARIZATION EFFECTS

15

In cubic semiconductors, radiation propagates with the same velocity c/n in any direction. In noncubic semiconductors, the optical properties are not isotropic; the dielectric constant is a tensor the elements of which are direction-dependent. The optical anisotropy affects especially the polarization characteristics of the propagating wave. The presence of a uniaxial strain or of an electric or a magnetic field can induce optical anisotropy in a normally isotropic crystal. We shall briefly review the various optical properties of nonisotropic materials in order to define the terminology used in dealing with polarization effects and then we shall discuss how these can be induced or modulated by externally applied perturbations.

15-A Birefringence

Birefringence is that property of a material which propagates two different polarizations at different velocities. Double refraction, a consequence of birefringence, is characterized by the fact that, in general, for an arbitrary orientation, a light beam incident on a parallelepiped emerges as two parallel beams (Fig. 15-1). The two beams are differently polarized and each obeys a Snell's law with a different refractive index. However, one of these beams, the ordinary ray, obeys the same Snell's law in all directions, while the other beam, the extraordinary ray, does not. When the light has a normal incidence to the birefringent crystal, the ordinary ray goes right through, undisplaced, while the extraordinary ray, after traversing the crystal, comes out parallel to the ordinary ray, but, in general, displaced from it.

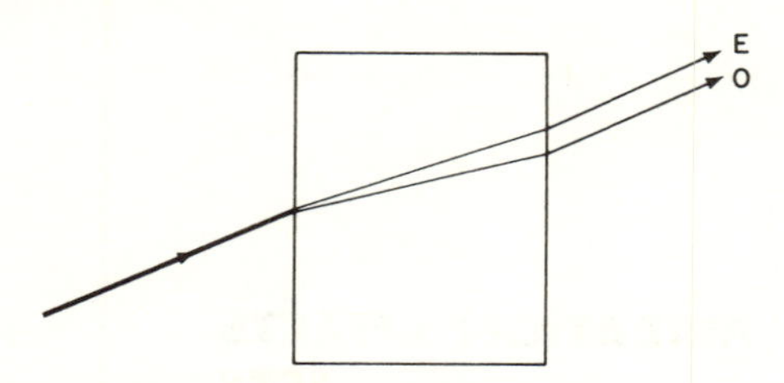

Fig. 15-1 Beam transmission by a birefringent crystal. E: extraordinary ray; O: ordinary ray.

15-A-1 BIREFRINGENCE IN UNIAXIAL CRYSTALS

In uniaxial crystals, the indices of refraction for the ordinary and extraordinary rays are equal along a unique direction called the "optic axis." Therefore, along the optic axis both rays have the same propagation velocity, whereas at an angle to the optical axis the two rays propagate at different velocities. For negative uniaxial crystals, $n_E < n_O$; while in positive uniaxial crystals, $n_O < n_E$ (n_E and n_O being the indices of refraction for the extraordinary and ordinary rays, respectively). These properties are illustrated in Fig. 15-2, where the circle is the equiphase front for a spherical ordinary wave generated at the central point, and the oval is the equiphase front for the extraordinary wave emanating from the same central point S. The two waves have mutually perpendicular polarizations (orientation of the displacement vector $\vec{D}$); the ordinary ray always has a zero component of $\vec{D}$ along the optic axis.

The ordinary ray, obeying Snell's law, is always in a plane defined by the incident ray and the normal to the crystal facet at the point of incidence. The extraordinary ray, in general, is not in this plane of incidence. Note by exam-

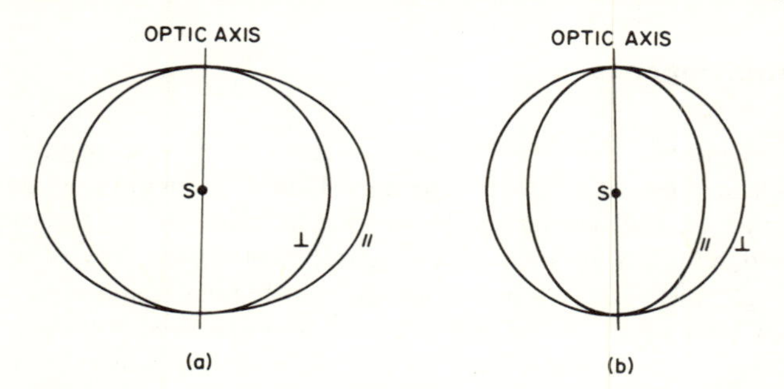

Fig. 15-2 Locus of equal phase for ordinary and extraordinary rays for negative (a) and positive (b) uniaxial crystals. The eccentricity of the ovals is exaggerated. The symbols $\perp$ and $/\!/$ refer to the polarization of the wave with respect to the plane of the paper.

ining Fig. 15-2 that the two beams coincide and are in phase only along the optic axis, where the two indices of refraction are identical.

The plane containing the optic axis and the ordinary ray is called the principal plane of the ordinary ray. The plane containing the optic axis and the extraordinary ray is called the principal plane of the extraordinary ray. These two principal planes intersect at the optic axis but, in general, do not coincide. The extraordinary ray is polarized in the principal plane of the extraordinary ray, while the ordinary ray is polarized perpendicularly to the principal plane of the ordinary ray. Hence the ordinary ray is always polarized transversely to the optic axis.

By way of example, selenium is a uniaxial birefringent semiconductor with $\boldsymbol{n}_O = 2.78$ and $\boldsymbol{n}_E = 3.58$ between 9 and 23 μm.[1]

15-A-2 ELLIPTICAL POLARIZATION

Consider a radiation propagating along a direction transverse to the optic axis of a uniaxial crystal. As evident from Fig. 15-3, the ordinary and extraordinary rays travel with different velocities. Hence there will be a gradual separation in the phases of the two waves. After the waves have traveled a distance d, the phase difference is

$$\Delta\phi = \frac{2\pi}{\lambda} d(\boldsymbol{n}_O - \boldsymbol{n}_E). \qquad (15\text{-}1)$$

If the incident light is linearly polarized at 45° to the optic axis, the wave can

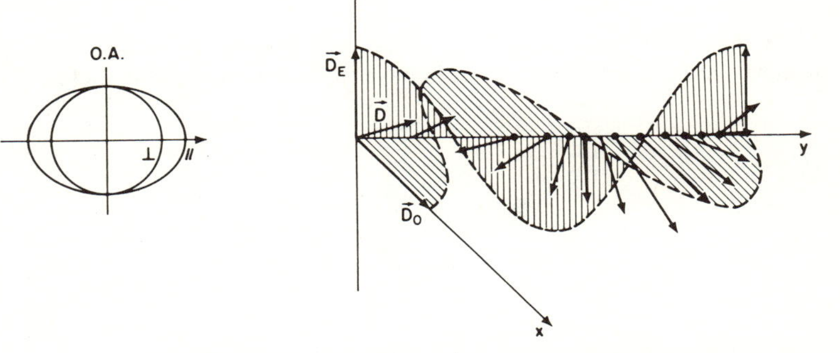

Fig. 15-3 Propagation of $\vec{D}$ vector of light wave along y direction in a negative uniaxial birefringent crystal. The figure shows the initial orientation of the $\vec{D}$ vector and its orientation at subsequent times when a total phase change somewhat smaller than $\pi/2$ has developed.

[1] R. S. Caldwell and H. Y. Fan, *Phys. Rev.* **114**, 664 (1959).

be conceptually decomposed into an ordinary component, with the $\vec{D}$-vector perpendicular to the optic axis and an extraordinary component whose $\vec{D}$-vector is parallel to the optic axis. Both vectors oscillate at the optical frequency. Because $\vec{D}_E$ and $\vec{D}_O$ propagate with different velocities, their sum along the direction of propagation adds up to a fluctuating $\vec{D}$-vector which rotates about the direction of propagation. In the present case (incident radiation linearly polarized at 45° to the optic axis), the $\vec{D}$-vector, in general, traces an ellipse on the exit facet of the crystal. There are, however, two special cases in which the phase difference between the two rays results in either linear or circular polarization—two extreme cases of elliptical polarization.

When the thickness of the crystal is such that the ordinary and extraordinary rays are out of phase by an integral number of half-wavelengths, the emerging light is again linearly polarized. When the integer is odd, the emergent linear polarization is perpendicular to the incident-polarization direction.

When the thickness of the crystal is such that the two rays are out of phase by an odd number of quarter-wavelengths, the emerging light is circularly polarized. Right and left circular polarizations are obtained on alternate odd numbers of quarter-wavelengths. A quarter-wave plate is a thin birefringent crystal the thickness of which has been adjusted to produce a $\pi/4$ phase difference between the ordinary and extraordinary rays at the operating wavelength. A quarter-wave plate transforms a linear polarization into a circular polarization and vice versa.

15-A-3 BIREFRINGENCE IN BIAXIAL CRYSTALS

In biaxial crystals there are two optical axes along which one quasi-ordinary and the extraordinary rays have the same propagation velocity and,

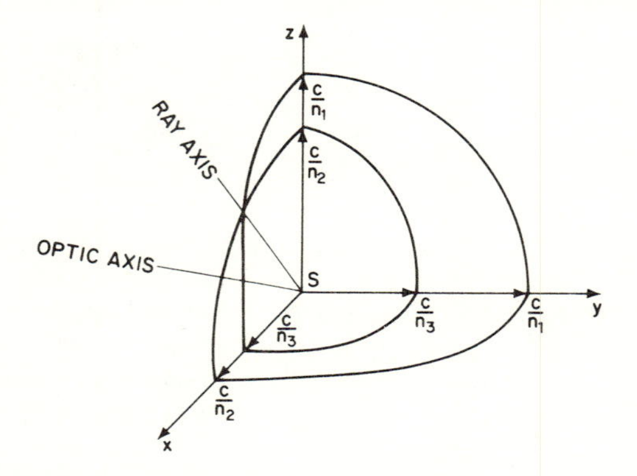

Fig. 15-4 Constant phase wavefront emanating from a point source at S in a biaxial crystal.

therefore, experience a common value of n. It will soon become clear that in biaxial crystals the incident beam always splits into two rays and that both are extraordinary—with the exception, however, that in each of the three coordinate planes, xy, yz, and zx, one of the two rays has the properties of an ordinary ray; we shall call this the quasi-ordinary ray, since its occurrence is restricted to three specific planes (Fig. 15-4).

The "ray axis" is defined as the direction along which there is only one ray velocity. There are two such axes. The optic axis is the direction along which the wave normals propagate with the same velocity. The wave normal is found by drawing a tangent to the two wave surfaces. The radius vector to the point of tangency on the quasi-ordinary wave is the optic axis. Optic axes also occur in pairs. The two optic axes are at an angle to each other; this angle usually depends on the wavelength. A similar statement can be made about the two ray axes. Note that uniaxial birefringence is a special case of biaxial birefringence in which the two optic axes coincide.

The occurrence of two optic axes in biaxial crystals corresponds to the presence of three differently valued components of the index of refraction. The three principal values of the refraction index are $n_1 < n_2 < n_3$. A constant-phase wavefront acquires the complex configuration of Fig. 15-4. Each equiphase component has a circular intersection in one plane and an elliptical intercept in the other two planes perpendicular to the first plane. For the case illustrated in Fig. 15-4, the optic axes and the ray axes are located in the xz-plane. (For clarity, the xz-plane is reproduced in Fig. 15-5.) In the xz-plane, the quasi-ordinary ray propagates along any radius vector with the same velocity c/n_2, while the extraordinary ray has the propagation characteristic shown by the oval. Note from Fig. 15-4 that in the other two planes xy and yz, the oval is either entirely inside or entirely outside the circle. The propagation velocity for the extraordinary ray is correspondingly faster or

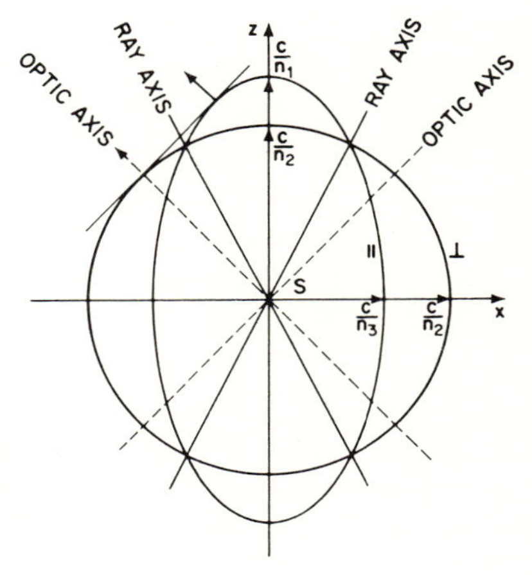

Fig. 15-5 Constant phase wavefronts in the xz-plane containing the ray axes and the optic axes for radiation emitted at the central point S in a biaxial crystal.

slower than for the quasi-ordinary ray in all directions within the plane considered.

The polarization of the waves can be visualized from Fig. 15-6, which shows that the quasi-ordinary ray is polarized with the $\vec{D}$-vector perpendicular to the plane in which the quasi-ordinary ray propagates, while the extraordinary ray is polarized in its own plane of propagation.

In an anisotropic crystal the electrical displacement $\vec{D} = \epsilon\vec{\mathscr{E}}$ in general is not parallel to the electric field $\vec{\mathscr{E}}$, because the dielectric susceptibility ϵ is a tensor. The electrical polarization induced by a driving electric field, the $\vec{\mathscr{E}}$-vector of the radiation field, results from a charge displacement oriented like $\vec{D}$. We can choose a Cartesian system such that the displacement is parallel to the electric field along the three axes:

$$\vec{D}_x = \epsilon_x \vec{\mathscr{E}}_x, \qquad \vec{D}_y = \epsilon_y \vec{\mathscr{E}}_y, \qquad \vec{D}_z = \epsilon_z \vec{\mathscr{E}}_z \tag{15-2}$$

Recall the relationship between the dielectric susceptibility and the index of refraction, $\epsilon = \boldsymbol{n}^2$; then, let

$$\epsilon_x = (\boldsymbol{n}_1)^2, \qquad \epsilon_y = (\boldsymbol{n}_2)^2, \qquad \epsilon_z = (\boldsymbol{n}_3)^2 \tag{15-3}$$

One can construct a dielectric ellipsoid (Fig. 15-7) defined by

$$\frac{x^2}{(\boldsymbol{n}_1)^2} + \frac{y^2}{(\boldsymbol{n}_2)^2} + \frac{z^2}{(\boldsymbol{n}_3)^2} = 1 \tag{15-4}$$

If one places through the origin of the dielectric ellipsoid a plane perpendicular to the common direction of propagation of two independent plane waves, one obtains an elliptic intercept with the dielectric ellipsoid. The minor and major axes of the resulting ellipse are the refractive indices $\boldsymbol{n}_a$ and $\boldsymbol{n}_b$ for the two waves. These two plane waves propagate with respective velocities $c/\boldsymbol{n}_a$ and $c/\boldsymbol{n}_b$ in a direction perpendicular to the plane of the ellipse. The polarization of one wave is oriented parallel to the minor axis of the el-

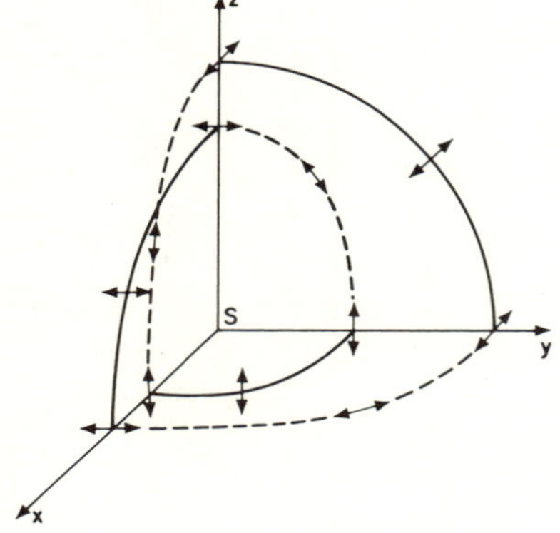

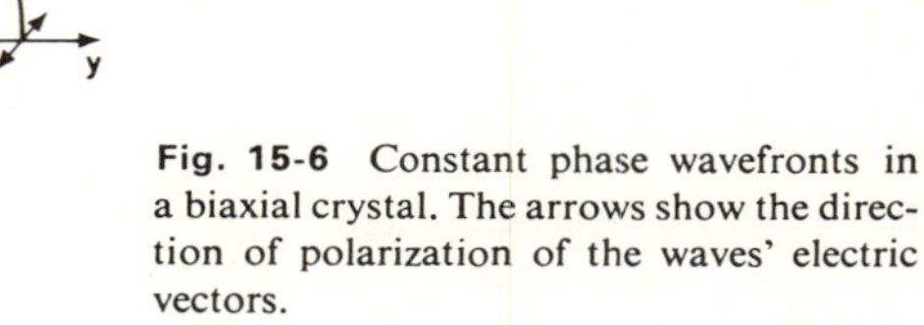
Fig. 15-6 Constant phase wavefronts in a biaxial crystal. The arrows show the direction of polarization of the waves' electric vectors.

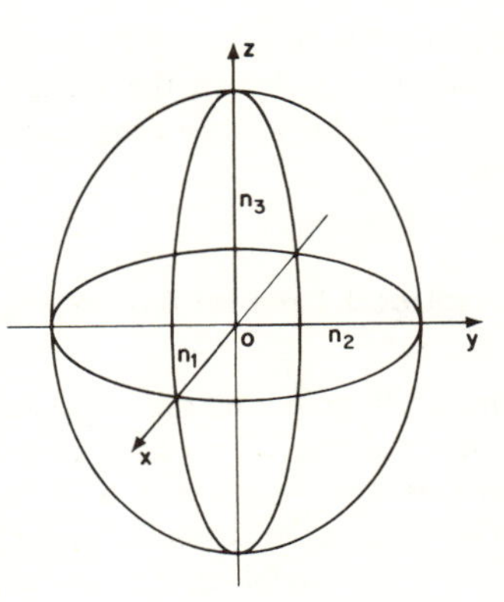

Fig. 15-7 Dielectric ellipsoid for a biaxial crystal.

lipse and the polarization of the other wave is parallel to the major axis of the ellipse.

More generally, the propagation direction can be defined by a unit vector $\vec{s}$ which is described by the three Cartesian components $\vec{s}_x$, $\vec{s}_y$, and $\vec{s}_z$. The values of $\boldsymbol{n}_a$ and $\boldsymbol{n}_b$ can be obtained by solving the Fresnel equation:

$$\frac{(\vec{s}_x)^2}{\boldsymbol{n}^2 - (\boldsymbol{n}_1)^2} + \frac{(\vec{s}_y)^2}{\boldsymbol{n}^2 - (\boldsymbol{n}_2)^2} + \frac{(\vec{s}_z)^2}{\boldsymbol{n}^2 - (\boldsymbol{n}_3)^2} = \frac{1}{\boldsymbol{n}^2} \tag{15-5}$$

which is a quadratic equation in $\boldsymbol{n}$. Its two solutions are $\pm\boldsymbol{n}_a$ and $\pm\boldsymbol{n}_b$. The $\pm$ sign indicates that the direction of propagation for the two waves can be reversed.

Thus far we have assumed a special case in which the two extraordinary rays coincide in the direction $\vec{s}$ and are distinguishable only by their mutually perpendicular polarizations and different propagation velocities along $\vec{s}$. In the general case, the two extraordinary rays do not coincide: one ray extends in a direction $\vec{s}$ and the other in a direction $\vec{s}\,'$. A solution of Fresnel's equation [Eq. (15-5)] for $\vec{s}$ and $\vec{s}\,'$ yields two pairs of refractive indices—$\boldsymbol{n}_a$, and $\boldsymbol{n}_b$, and $\boldsymbol{n}'_a$ and $\boldsymbol{n}'_b$—and the corresponding two pairs of polarization directions. From each pair only one value of $\boldsymbol{n}$ and one polarization can be assigned to each ray. This choice is guided by considering the boundary conditions at the surface where the light enters the crystal.

Note that if the crystal is uniaxial, the dielectric ellipsoid has a high degree of rotational symmetry which is described by

$$\frac{x^2}{(\boldsymbol{n}_O)^2} + \frac{y^2}{(\boldsymbol{n}_O)^2} + \frac{z^2}{(\boldsymbol{n}_E)^2} = 1 \tag{15-6}$$

and z is the optic axis.

Thus far we have assumed the crystal perfectly transparent to the various rays. In fact, the semiconductor has a finite absorption which may be different along several axes. This property is called "pleochroism." Dichroic crystals transmit more readily one of the two rectilinear components of the radiation and absorb the other; hence the transmitted light is linearly polarized. A trichroic crystal has different absorption coefficients along each of the three axes.

15-B Induced Optical Anisotropy

A circularly polarized radiation consists of a wave whose $\vec{D}$-vector rotates about the direction of propagation. Clockwise and counterclockwise rotations are possible. A crystal is said to be "optically active" when the clockwise and counterclockwise circular polarizations propagate at different velocities. Some birefringent materials exhibit this gyration property in their natural state, while other materials exhibit optical activity when they are acted upon by a magnetic field.

Note that a linearly polarized radiation can be decomposed into synchronous clockwise and counterclockwise circularly polarized components which travel at the same velocity. Hence when a linearly polarized light passes through an optically active material, the plane of polarization gradually rotates.

Many semiconductors which are normally isotropic in their optical properties become anisotropic when subjected to externally applied electric, magnetic, or mechanical stress.

15-B-1 ELECTRO-OPTIC KERR EFFECT

In the electro-optic Kerr effect, an electric field applied across the crystal in a direction transverse to the light beam induces birefringence; depending on the crystallographic orientation, the crystal becomes uniaxial, and its optic axis is aligned with the electric field (Fig. 15-8). In the general case, the semiconductor becomes biaxial.[2] If the electric vector of the linearly polar-

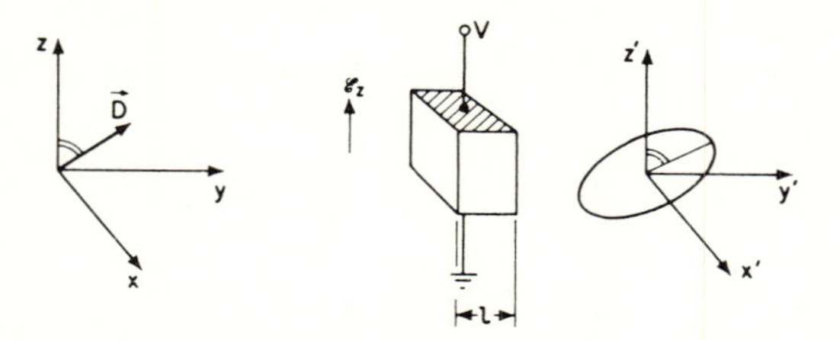

Fig. 15-8 Kerr effect on a linearly polarized wave. After transmission through the electro-optic crystal light becomes elliptically polarized.

[2]For a detailed treatment of the relationship between optical anisotropy and crystal symmetry, see. J. F. Nye, *Physical Properties of Crystals*, Part 4, Clarendon (1957).

ized light wave is inclined at 45° to the applied electric field, its components parallel and perpendicular to the applied electric field travel with different velocities, so that the light emerges elliptically polarized. The optical-path difference for the two components is given by

$$\Delta = c_K l (\mathscr{E}_z)^2 \lambda \tag{15-7}$$

where c_K is the Kerr constant and l is the path length under the electrodes.

15-B-2 POCKELS EFFECT OR LINEAR ELECTRO-OPTIC EFFECT

In some semiconductors which are normally optically isotropic, the presence of an electric field induces biaxial birefringence. To visualize the effect of the electric field on the index of refraction, consider the three-dimensional dielectric ellipsoid (Fig. 15-7). When the electric field is zero, the index of refraction is isotropic and the dielectric ellipsoid is a sphere. When an electric field is applied along some direction which we shall call z, the index of refraction $\boldsymbol{n}_z$ does not change; but in the transverse plane the refractive index is altered by an amount $\Delta\boldsymbol{n}$, proportional to the electric field $\mathscr{E}_z$. The modification of the dielectric sphere occurs in such a way that the refractive index changes by $\pm\Delta\boldsymbol{n}$ in mutually perpendicular directions. Then we can select two directions x and y such that

$$\left.\begin{aligned} \boldsymbol{n}_x &= \boldsymbol{n} + \Delta\boldsymbol{n} \\ \boldsymbol{n}_y &= \boldsymbol{n} - \Delta\boldsymbol{n} \end{aligned}\right\} \tag{15-8}$$

The actual orientation of the crystal for a maximum effect is guided by a choice of crystallographic directions. The optimum choice depends on the crystal-group symmetry to which the material belongs.

The magnitude of the linear electro-optic effect is

$$\Delta\boldsymbol{n} = \frac{\boldsymbol{n}^3}{2} r \mathscr{E}_z \tag{15-9}$$

where r is the electro-optic coefficient.

In the arrangement of Fig. 15-9, the two principal components of the

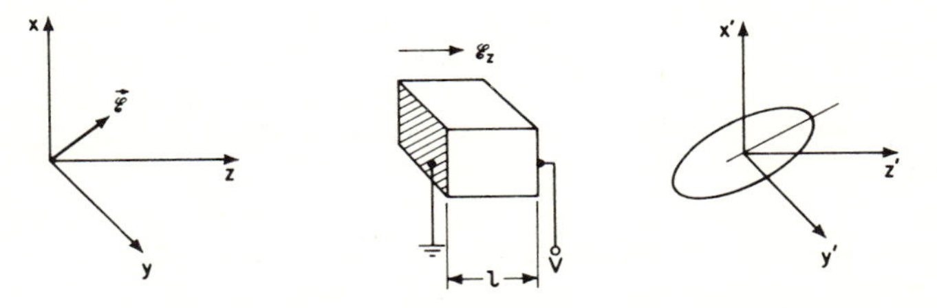

Fig. 15-9 Transformation of linear polarization into elliptical polarization by the Pockels effect.

wave's electric vector propagate at different velocities, which results in a phase retardation:

$$\Delta\phi = \frac{\pi}{\lambda} n^3 r \mathscr{E}_z l \tag{15-10}$$

Note that $\mathscr{E}_z l$ is the applied voltage. Hence with light propagating in the direction of the applied field, a linear electro-optic effect is obtained which, for a given material, depends only on the applied voltage.

Values of the electro-optic coefficient r for several semiconductors are listed in Table 15-1.

Table 15-1

Electro-optic properties of semiconductors

Material	*Electro-optic coefficient at room temperature* r(cm/V $\times 10^{-10}$)	*Refractive index at zero field* n	$n^3 r$ (cm/V $\times 10^{-10}$)	*References*
GaAs	1.6	3.34	59	3
ZnTe	1.4	2.79	34	3
ZnS	2.0	2.37	27	4
CdTe	6.8	2.6	120	5
CdS	5.5	2.3	66	3
Se	2.5	2.8	55	6

15-B-3 FARADAY EFFECT[7]

Some materials become optically active in a magnetic field. When linearly polarized radiation is transmitted by the material in the direction of the magnetic field, the plane of polarization is rotated. The amount of rotation θ is proportional to the magnetic field H and to the path length l:

$$\theta = C_v H l$$

where C_v is the so-called Verdet constant (rotation per unit length per unit magnetic field). The Verdet constant varies with the wavelength. The direction of rotation of the plane of polarization depends on the polarity of the magnetic field. The convention is that positive rotation is obtained for a right-hand screw advancing in the direction of the field. Note that the direction of rotation is then the same for a wave propagating in either direction along the magnetic field.

[3] I. P. Kaminov, *IEEE J. Quantum Electronics* **4**, 23 (1968).
[4] S. Namba, *J. Opt. Soc. Am.* **51**, 76 (1961).
[5] J. E. Kiefer and A. Yariv, *Appl. Phys. Letters* **15**, 26 (1969).
[6] M. C. Teich and T. Kaplan, *IEEE J. Quantum Electronics* **2**, 702 (1966).
[7] Treatments of Faraday rotation in semiconductors can be found in B. Lax and S. Zwerdling, *Progress in Semiconductors* **5**, 251 (1960), and L. M. Roth, *Phys. Rev.* **133**, A542 (1964).

Free carriers exhibit a Faraday effect which can be readily observed in heavily doped semiconductors. The value of the Faraday rotation θ by free carriers is given by[8]

$$\theta = \frac{q^3 N \lambda^2 H l}{2\pi c^4 \mathbf{n}(m^*)^2} \tag{15-11}$$

where N is the free-carrier concentration; λ is the wavelength of the radiation; $\mathbf{n}$ is the index of refraction in the absence of a magnetic field; and m^* is the free-carrier effective mass. Although a uniform effective mass is assumed in Eq. (15-11), corrections can be made for the ellipsoidal shape of the valleys[9] and for the presence of two types of carriers.

The quadratic dependence of θ on λ suggests that the free-carrier Faraday rotation should be measured at long wavelengths. However, the angular frequency at which the electrons are driven must be large compared to the cyclotron frequency $\omega_c = qH/m^*$ and furthermore the corresponding period must be short compared to the carrier-relaxation time.

If the carrier concentration N is known from other measurements, the Faraday effect can be used to evaluate the effective mass.

15-B-4 VOIGT EFFECT

A magnetic field induces birefringence for radiation propagation transversely to the magnetic field. This is the magneto-optic equivalent of the Kerr effect. For linearly polarized radiation, the component of the wave's $\vec{\mathscr{E}}$-vector which is parallel to the magnetic field travels at a different velocity than the transverse component, causing the radiation to become elliptically polarized.

The Voigt effect due to free carriers in ellipsoidal valleys exhibits the following phase shift:[10]

$$\Delta\phi = \frac{q^4 N \lambda^3 H^2 l}{4\pi^2 c^6 \mathbf{n}(m_v^*)^3}$$

where N is the carrier concentration; l is the sample thickness; $\mathbf{n}$ is the refractive index at zero field; and m_v^* is a Voigt effective mass, which is a complicated average of the effective masses in the ellipsoidal valley. The value of m_v^* exhibits a strong crystallographic dependence.

15-B-5 STRAIN-INDUCED BIREFRINGENCE

A material which normally has isotropic optical properties becomes birefringent when a uniaxial strain is present. Strain-induced birefringence

[8] T. S. Moss, *Optical Properties of Semiconductors*, Butterworths (1959), p. 85.
[9] B. Lax and S. Zwerdling, *Progress in Semiconductors* **5**, 251 (1960).
[10] E. D. Palik, *J. Phys. Chem. Solids* **25**, 767 (1964).

is often called the "photoelastic" effect. In semiconductors the elements of the stress tensor are strongly dependent on crystallographic directions; therefore, the direction of maximum birefringence is tied to crystallographic considerations.

We saw in Chapter 2 that, in *n*-type silicon, a uniaxial strain in the [100] direction redistributes the electrons so as to preferentially populate those indirect valleys which lie in the direction of maximum strain. An infrared or a microwave radiation will interact with the free electrons differently when the electric vector of the radiation is aligned with the stress than when it is transverse to it. This anisotropy, due to the fact that the electrons are confined to an elongated ellipsoidal valley, results in birefringence.[11] This anisotropic free-carrier effect can also lead to dichroism,[12] an optical property which has been proposed as a convenient tool for studying deformation potentials and intervalley scattering.

15-B-6 DEFLECTION AND MODULATION OF A LIGHT BEAM

Any phenomenon which changes the index of refraction of a semiconductor can be used to deflect a light beam to which the material is transparent. The electro-optic effect is a convenient means of modulating the index of refraction for this purpose.[13] The crystal is shaped into an optimally oriented prism as shown in Fig. 15-10. The transmitted ray emerges at an angle to the incident beam. The application of a voltage V across the prism induces a change in the index which is either proportional to V (Pockels effect) or to V^2 (Kerr effect), and the emerging beam is deflected accordingly. High voltages are needed to obtain a useful though very small deflection (care must be taken not to exceed the breakdown field of the material). Another problem may be the formation of two beams due to the induced birefringence.

Here we shall mention another possible method for deflecting light which is based on the Brillouin scattering discussed in Sec. 12-B-2, but without

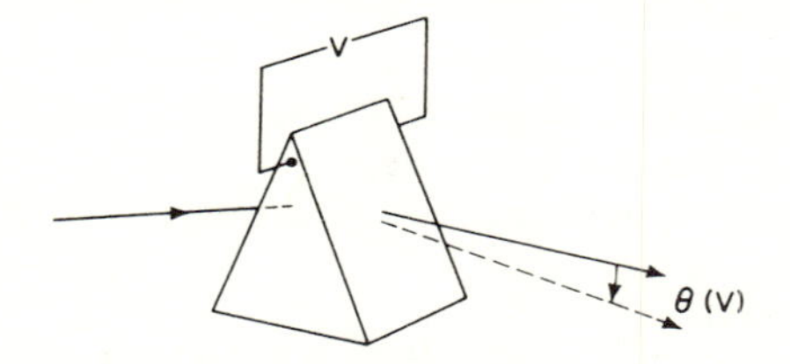

Fig. 15-10 Electro-optic beam deflection.

[11] K. J. Schmidt-Tiedemann, *Phys. Rev. Letters* **7**, 372 (1961).
[12] J. K. Furdyna and G. P. Soardo, *Proc. Int. Conf. Physics of Semiconductors*, Paris, Dunod (1964), p. 171.
[13] R. A. Soref and D. H. McMahon, *Electronics* **56**, (Nov. 29, 1965).

considering polarization effects. Light is scattered by regions of nonhomogeneous index of refraction, such as those resulting from changes in density or from local stress. A periodic fluctuation in the density can be induced by a longitudinal acoustic wave, or a periodic distribution of sheer stress can be generated by a transverse acoustic wave. The acoustic wave forms a grating which deflects the incident radiation by an angle θ such that, for the first-order mode,

$$\sin\frac{\theta}{2} = \frac{\lambda}{2nL_A}$$

where λ/n is the wavelength of light in the semiconductor and $L_A = v_s/f$ is the phonon wavelength (v_s is the sound velocity and f is the phonon frequency). Hence deflection of the light beam can be controlled by varying the frequency of the acoustic wave.

Another important application of electro-optic and magneto-optic effects is the modulation of light. Since the electro-optic and magneto-optic effects alter the plane of polarization of the transmitted light, the intensity of the radiation can be modulated by inserting an analyzer beyond the semiconductor. The analyzer is a filter which transmits only light linearly polarized in a certain direction. The analyzer is then oriented to transmit the radiation only when the external field is zero. When the applied field has a finite, nonzero value, the plane of polarization is rotated by an angle θ and the intensity transmitted by the analyzer is reduced to

$$L = L_0 \cos\theta$$

where L_0 is the intensity of the radiation emerging from the semiconductor.

The compound GaAs, which has a large electro-optic coefficient, has been used successfully to modulate the 1.06-μ output of a cw laser over a 5-MHz bandwidth.[14] A 100% modulation was obtained with 300 V. This experiment demonstrated the practicability of transmitting a television signal over a laser beam.

The light-guiding ability of p–n junctions and their small volume make them well suited for use as electro-optic modulators.[15] The smallness of the active region minimizes the power requirement.

Problem 1. Consider a uniaxial crystal with the ordinary and extraordinary refractive indices n_O, n_E; $n_E > n_O$. Denote the respective indices by n_a and n_b when the propagation direction makes an angle θ with the optic axis. Describe n_a and n_b in terms of n_O, n_E, and θ. If the ordinary and extraordinary waves travel a distance d through the material, draw an approximate plot of the phase change $\Delta\phi = (2\pi d/\lambda)|n_a - n_b|$ vs. θ. [*Hint:* Read the paragraph before Eq. (15-5).]

[14] W. J. Hannan, J. Bordogna, T. E. Penn, and T. E. Walsh, *Proc. IEEE* **53**, 171 (1965).
[15] D. F. Nelson and F. K. Reinhart, *Appl. Phys. Letters* **5**, 148 (1964); W. L. Walters, *J. Appl. Phys.* **37**, 916 (1966); W. G. Oldham and A. Bahraman, *IEEE J. Quantum Elec.* **3**, 278 (1967).

Problem 2. Show that in materials with inversion symmetry the linear electro-optic effect cannot be observed. Compare Si and GaAs, draw a lattice for each and discuss if you expect either of these to exhibit the linear electro-optic effect.

Problem 3. In cubic zincblende crystals application of a large electric field in the [111] direction makes the crystal uniaxial with the optic axis along [111]. The ordinary and the extraordinary refractive indices are in this case given by

$$n_O = n + \frac{1}{2\sqrt{3}} n^3 r \mathscr{E} \qquad n_E = n - \frac{1}{\sqrt{3}} n^3 r \mathscr{E}$$

Consider the sample shown in Fig. 15P-3. Linearly polarized light is incident perpendicular to the #1 facet of the sample. Describe the phase and polarization of the ray reaching the end of the sample in each of the following cases: light incident with its electric field (1) parallel to [111] direction; (2) perpendicular to [111] direction; (3) making a 45° angle with [111] direction. Assume $d = 2\mathbf{m}\,\lambda/n$ where λ is the wavelength, and $\mathbf{m}$ is any integer.

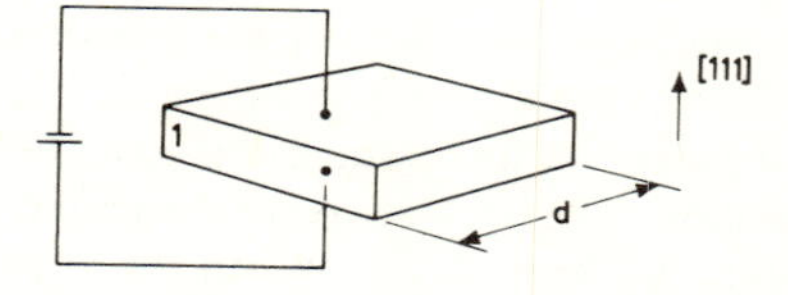

Fig. 15P-3

Problem 4. In the electro-optic KDP crystals (chemical formula KH_2PO_4) the modulating field $\vec{\mathscr{E}}$ and the propagation direction of the light beam are parallel, as shown in Fig. 15-9. In the cubic zincblende structure of Prob. 3, however, the two directions were perpendicular. Discuss the relative merits of the two phase modulators. Assume $n^3 r$ is the same for both.

Problem 5. In Prob. 3 the use of an electro-optic crystal for phase modulation and polarization modulation was investigated. The present problem deals with intensity modulation. An actual experimental setup is shown in Fig. 15P-5 using a class $\bar{4}3$m cubic crystal, with the modulating field $\mathscr{E}$ parallel to the [111] direction which is also taken as the z-axis. The refractive indices, n_O and n_E, for this

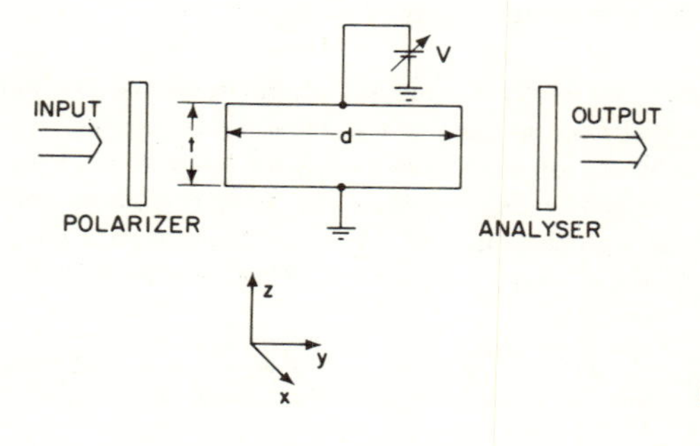

Fig. 15P-5

geometry are given in Prob. 3. The polarizer is parallel to the x-z plane, with its axis at 45° to the z-axis. This makes linearly polarized light from the input unpolarized light, with the electric field $\vec{\mathscr{E}}$ at 45° to the z-axis. The analyzer is set also parallel to the x-z plane, but with its axis at $-45°$ to the z-axis. Thus the analyzer has its axis at 90° to that of the polarizer, and will transmit light rays with $\vec{\mathscr{E}}$ vector parallel to its axis. Find the ratio of the output intensity to the input intensity, and find the voltage that makes this ratio a maximum in terms of the device parameters $\boldsymbol{n}$, r, d, t, and wavelength λ. Describe how the output intensity varies as V changes from zero to this value.

PHOTOCHEMICAL EFFECTS

16

The optical excitation of an electron out of a valence-bonding state, transforming it into a free carrier, represents the breaking of a bond between two atoms and, therefore, constitutes a chemical reaction. In solids the atoms are, in general, stationary (ignoring thermal effects); when the electron returns to the ground state, the bond is re-established and, therefore, it is irrelevant to talk of chemical reaction. However, there are cases in which, with weakened bonds, atoms become mobile under the influence of light; then it is appropriate to consider the event a photochemical reaction. Photochemical effects occur at the surface of a semiconductor in contact with a gaseous or liquid ambient transparent to the radiation. In the bulk, atomic motion is possible at the site of dislocations. Finally, the reorganization of orbitals about an impurity allows a different type of complexing with adjacent atoms (a chemical change) without either the impurity or an adjacent atom leaving its site.

With the advent of lasers it is possible to induce photochemical processes with a greater yield than ever before. Surely, the field of photochemistry in semiconductors, which has barely been explored with incoherent light, must still have a wealth of phenomena yet to be uncovered with the use of coherent radiation. Powerful laser beams have been used to evaporate and ionize semiconductors for mass spectrometry; but since this is a conversion of a high-power density from radiation into heat, rather than a spectrally dependent operation, this type of heating cannot be classified as a photochemical process, even though optical excitation may occur concurrently to a less significant degree.

16-A Photochemistry with a Gaseous Ambient

Atoms and molecules can be adsorbed or desorbed from the semiconductor surface, the kinetics (and sometimes the direction) of the process being controlled by the light incident on the surface. Molecules can be dissociated on the surface of the crystal, which then acts as a catalyst the effectiveness of which is controlled by the radiation. Adsorbed molecules have a characteristic optical-absorption spectrum which can be brought out by techniques of internal-reflection spectroscopy (which is a standard technique for physicochemical and chemical analysis). When a crystal is grown from the vapor phase, the crystallographic properties can be improved by illumination which promotes the mobility of atoms along the substrate's surface until a stronger bonding site is found.

16-A-1 SURFACE STATES

When a semiconductor is cleaved, a set of unsaturated bonds appears at the exposed surface. These bonds are available for strong interactions (chemisorption) with atoms and molecules of the ambient. Surface states appear which can either give or receive electrons, thus connecting the ambient to the crystal. On theoretical grounds, one should expect a very high density of surface states, because the discontinuity of the crystal lattice at the surface would consist of at least one "dangling bond" per atom. The unsatisfied bond is equivalent to an acceptor state. These surface states are known as "Tamm states."[1] Their maximum density should be of the order of 10^{15} cm^{-2}. However, an actual surface usually has only about 10^{11} detectable states/cm^2.

Before trying to account for this apparent decrease in the density of surface states, it must be pointed out that an understanding of surface states is just beginning to emerge.[2] The number of detectable surface states depends on how the surface atoms are arrayed. The perturbation introduced by the discontinuity in the lattice where the surface is formed may result in a rearrangement of surface atoms compared to their disposition along the same plane within the crystal, with possibly a resulting stretching and twisting of bonds.

The periodic distribution of atoms on the surface forms a two-dimensional lattice which determines an $E(k)$ diagram for surface states. As shown in Fig. 16-1, the surface states may be grouped deep in the gap with a low $E(k)$ dispersion—this results in a high density of surface states clustered in

[1] I. Tamm, *Phys. Z. Sowjetunion* **1**, 733 (1932).
[2] S. G. Davison and J. D. Levine, "Surface States," *Solid State Physics*, **25**, ed. F. Seitz and D. Turnbull, Academic Press (1970).

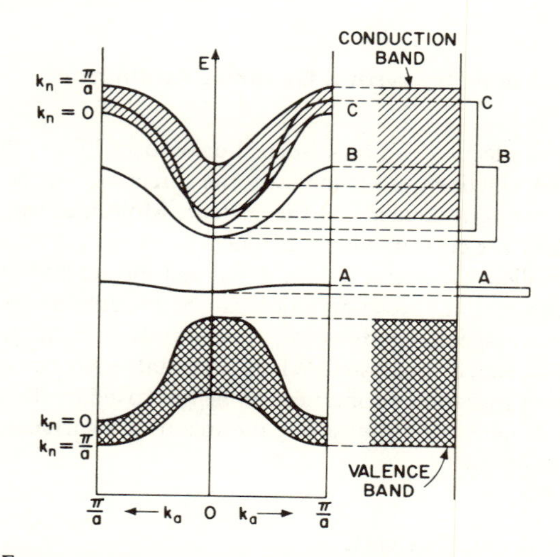

Fig. 16-1 Energy versus wave vector and density-of-states for some possible surface state models.[2]

a narrow distribution which pins the Fermi level at the surface (we saw a manifestation of this in Sec. 13-C); or the surface states may have a large $E(k)$ dispersion and partly overlap an intrinsic band. In this case, the overlapping states are indistinguishable from intrinsic bulk states. Furthermore, in ionic or in partly ionic crystals, the surface cations and anions function respectively as donors and acceptors[3,4] and may act as compensated pairs.

16-A-2 ADSORPTION AND DESORPTION

As a result of the surface states, the energy bands are usually bent near the surface. If the surface state can trap an electron, it acts as an acceptor; therefore, as shown in Fig. 16-2(a), the bands bend upward at the surface. Conversely, if the donor-like trap can lose an electron, forming a positive charge at the surface, it will cause the bands to bend down [Fig. 16-2(b)].

The bending of the bands can be followed experimentally by measuring the change in contact potential or the change in conductivity parallel to the surface. A more extensive depletion layer corresponds to a lower surface conductance.

If the approaching gas atom or molecule has a greater electron affinity than the work function of the semiconductor, it may capture an electron from the surface and, thus behaving as an acceptor, will tend to bend the

[3] J. D. Levine and P. Mark, *Phys. Rev.* **144**, 751 (1966).
[4] J. D. Levine, *Bull. Am. Phys. Soc.* **14**, 787 (1969).

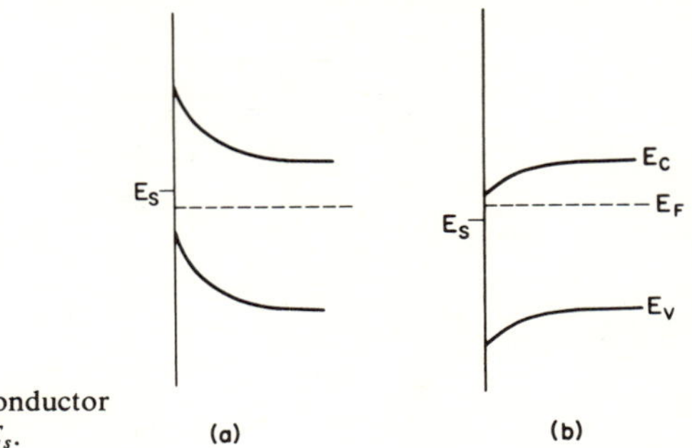

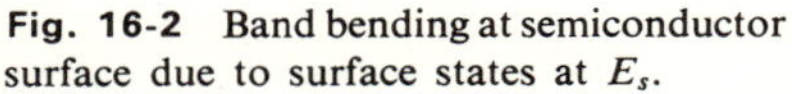
Fig. 16-2 Band bending at semiconductor surface due to surface states at E_s.

bands upward. Oxygen, which is very electronegative, causes an upward bending of the bands (Fig. 16-3). When oxygen is adsorbed on an *n*-type semiconductor, the depletion layer increases and the surface conductance decreases. On the other hand, oxygen adsorbed on a *p*-type material causes the surface to become more negative, increasing the hole concentration at the surface, building up an accumulation layer and, therefore, increasing the surface conductance (Fig. 16-3).

The adsorption process may be helped by optical stimulation, which makes more electrons available to the strongly electronegative gas, and more holes available to the donor-like species. Figure 16-4 shows that illumination with white light accelerates the adsorption of oxygen on CdSe (an *n*-type semi-

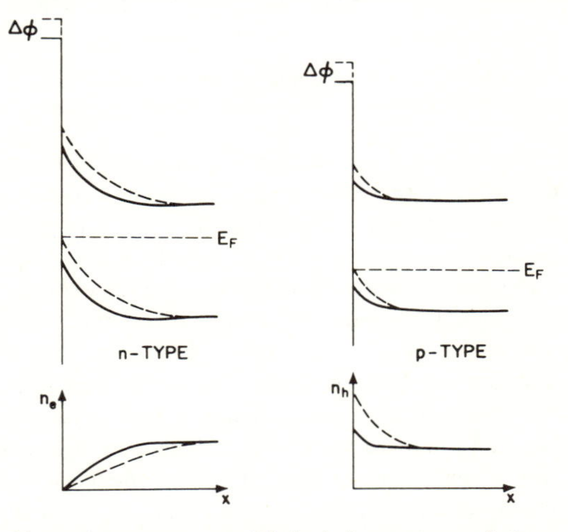

Fig. 16-3 Change in contact potential, depletion or accumulation layer and carrier concentration in *n*-type and *p*-type semiconductor due to surface adsorption of an electronegative ion. Solid and dashed lines show the status before and after adsorption, respectively.

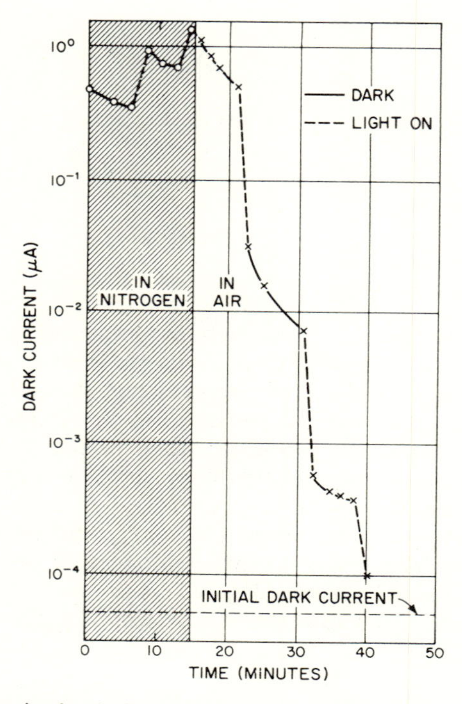

Fig. 16-4 Photostimulated adsorption of oxygen on CdSe crystal at room temperature. Measurements were made only in the dark.[5]

conductor). In this case, the adsorption was monitored by measuring, in the dark, the conductance of the sample parallel to the surface.

The adsorption of oxygen was also studied on ZnS.[6] It was found that the rate of adsorption increases with increased illumination ($h\nu > E_g$) and with increased oxygen pressure. In this case, the adsorption was monitored by the change in contact potential, which is a direct measure of the change in surface potential.

The change in surface potential was also monitored in experiments with germanium, where the ambient was alternately wet oxygen and dry oxygen; the largest effect was obtained while the surface was illuminated (Fig. 16-5). Water vapor and oxygen seem to have opposing effects, with water dominating. The surface potential changes in opposite directions for *n*- and *p*-type germanium.

Actually, illumination can cause either increased adsorption or the op-

[5]R. H. Bube, *J. Chemical Physics* **27**, 496 (1957).
[6]A. Kobayashi and S. Kawaji, *J. Phys. Soc.* (Japan) **10**, 270 (1955).

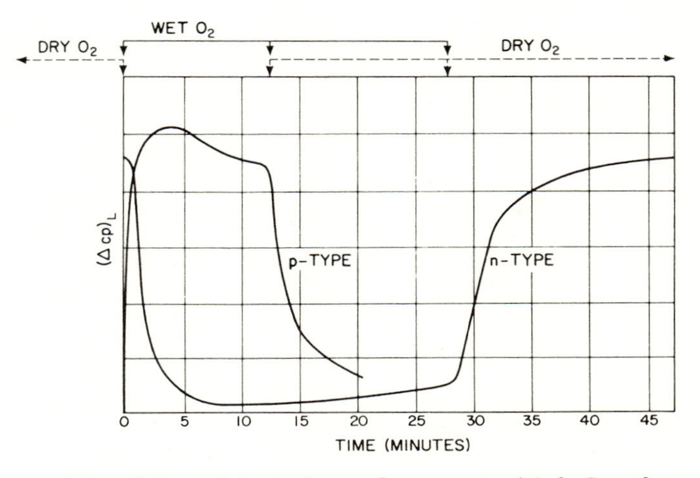

Fig. 16-5 Variation in change of contact potential of a Ge surface under illumination. Change from dry oxygen to wet oxygen ambient at $t = 0$. Return to dry oxygen ambient at end of 12 min for *p*-type sample, and at end of 28 min for *n*-type sample.[7]

posite effect, increased desorption.[8] The direction of the effect depends on the nature of the gas, the semiconductor, its doping, the bending of the bands at the surface, and such experimental conditions as temperature and pressure. In fact, so many parameters affect the photoadsorptive behavior of the crystal that it is not surprising that different experimenters have reported conflicting results.

Theoretical predictions have been made as to whether photoadsorption or photodesorption should be the net result of illumination depending on the position of the Fermi level (i.e., doping and temperature) and the band bending at the surface (via gas pressure, which affects the surface coverage by the adsorbed gas). These predictions are summarized in Fig. 16-6. Thus the adsorption of oxygen (an acceptor) on an intrinsic semiconductor (Fermi level at midgap) makes the bands curve upward at the surface. When the oxygen coverage is low (low pressure), the band bending is small, corresponding to the region between points *a* and *b* of Fig. 16-6. In this range of positive photo-effect, photoadsorption occurs. At higher pressure more oxygen is adsorbed, the system moves to the right of point *a*, and illumination causes photodesorption.

In the case of ZnO, oxidation and diffusion of oxygen into the crystal cause a gradual deviation from stoichiometry at the surface. This may result in band bending and may counteract the effect of adsorption on the surface.

[7] S. R. Morrison, *J. Phys. Chem.* **57**, 860 (1953).
[8] Th. Wolkenstein and I. V. Karpenko, *J. Appl. Phys.* **33**, 460 (1962).

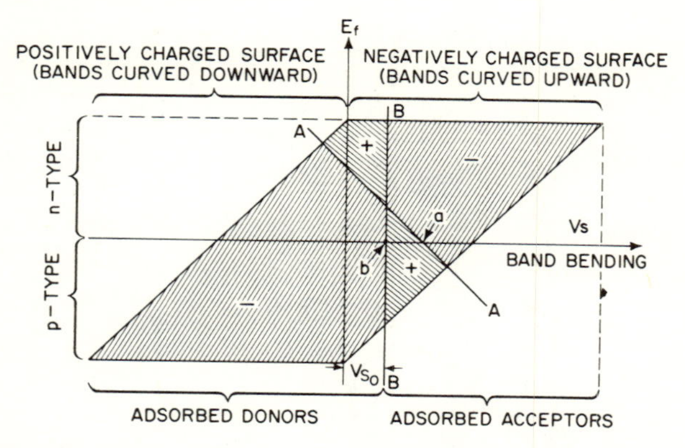

Fig. 16-6 Permissible values of Fermi level, E_F, and of band bending, V_s, as a function of adsorbed particles. The lines AA and BB are deduced from theory. The offset of BB with respect to the zero line represents the initial values of the band bending. The areas labeled "+" give photoadsorption, the areas labeled "−" give photodesorption.[8]

The adsorption of oxygen on CdTe can cause a very large polarity reversal in the anomalous photovoltaic effect of CdTe[9] (see Sec. 14-D for a discussion of the anomalous photovoltaic effect). Oxygen is very tightly bound to the CdTe surface, for it requires a heating in vacuum to cause its desorption and a return to the initial polarity of photovoltage. However, it was noted that the adsorption of oxygen was helped by illumination.

16-A-3 PHOTOCATALYSIS

When molecules are adsorbed on the surface of a semiconductor, the binding energy between the constituent atoms is altered by virtue of the interaction with the substrate. The bonding can be sufficiently weakened to allow incident photons to complete the decomposition of the adsorbed molecules. Thus ZnO decomposes hydrogen peroxide under illumination.[10] Note that this is a poor example in this section, since H_2O_2 is a liquid; however, it is conceivable that photocatalysis could also operate on a gaseous species.

16-A-4 SPECTROSCOPIC ANALYSIS OF ADSORBED SPECIES

Many adsorbed molecules can be identified by their characteristic *IR*-absorption spectra. The technique of internal-reflection spectroscopy[11] is

[9] G. Brincourt and S. Martinuzzi, *Comptes Rendus* **266**, B1283 (1968).
[10] W. Hnojevig Ref. 71 by J. T. Law, *Semiconductors* ed. Hannay, Reinhold (1959), p. 702.
[11] N. V. Harrick, *Internal Reflection Spectroscopy*, Wiley (1967), p. 260.

particularly suitable for this study. When light propagating inside the semiconductor is incident on the surface at an angle such that total internal reflection results, the radiation, upon touching the surface, probes the surface bonds with its fringing field (or "evanescent" wave).

The semiconductor can be shaped in a number of configurations which allow the radiation to make many reflections along the surface of the crystal, thus enhancing the absorption spectrum due to the adsorbate. Figure 16-7 shows such a structure.

Examples of spectra obtained by internal-reflection spectroscopy are shown in Fig. 16-8. The various germanium oxides as well as the hydride are clearly distinguishable. In silicon, the O–H bond gives an absorption peak at 2.9 μ and CO_2 a band at 4.27 μ; SiH_4 produces an absorption band in the range 4.7 to 4.8 μ; the exact location of the absorption peak depends on the energy of the hydrogen ion used in bombarding the Si-crystal.[14]

As we saw in Sec. 3-F (Fig. 3-33), this technique of internal-reflection

Fig. 16-7 Trapezoidal plate for multiple internal reflections.[12]

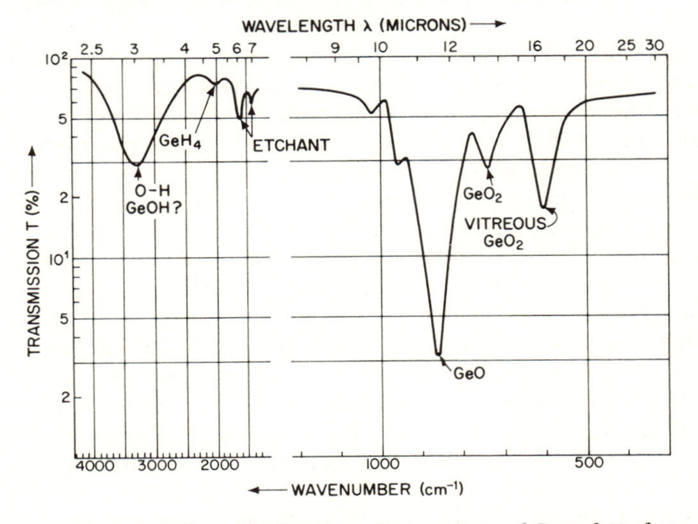

Fig. 16-8 Internal reflection surface spectrum of Ge surface after etching in HF/HNO_3. The complexes formed on the surface are clearly detectable.[13]

[12] N. V. Harrick, *Internal Reflection Spectroscopy*, Wiley (1967), p. 101.
[13] K. H. Beckmann, *Surface Sci.* **5**, 187 (1966).
[14] G. E. Becker and G. W. Gobeli, *J. Chem. Phys.* **38**, 2942 (1963).

spectroscopy has been used to study the oxidation of a germanium surface. On a clean surface the absorption spectrum showed a transition from a surface state near the top of the valence band to a set of states 0.16 eV below the conduction band. Oxidation caused a disappearance of the absorption band.[15]

16-A-5 EPITAXIAL GROWTH

The probability of external atoms sticking on a semiconductor surface assumes particular importance in the case of crystal growth, when the atom to be adsorbed is similar to those forming the crystal. This is the case of crystal growth from the vapor phase either by pyrolysis or by evaporation onto a heated substrate. In order to promote a well-ordered crystal growth propagating the lattice of the substrate, the incident atoms must be able to migrate along the surface until the appropriate location (a lattice site) has been found. Another possible mechanism of crystal growth is that the incident atom is rejected unless it arrives at the appropriate lattice site. The rejection of the atom is the result of thermal activation. This is why the substrate must be heated to an elevated temperature in order to obtain a more nearly perfect crystal structure. Since optical excitation can cause desorption of atoms from a gaseous ambient, optical "treatment" of the surface during crystal growth has often been attempted. With the advent of optical heating with high-intensity "sun-gun" lamps—a process which supplies heat for pyrolysis (thermal decomposition of a gaseous compound) and for substrate heating—the specific action of light on adsorption becomes an integral and indistinguishable part of the crystal-growing process.

The effect of light on crystal growth can be studied when heating is supplied by a resistance furnace or from an rf-source. However, at these high temperatures, infrared emission is very abundant and may well include the spectral region where photoadsorption and photodesorption are important mechanisms.

Some influence of light on the epitaxial growth of silicon has been found.[16] The crystal was grown by pyrolysis (thermal decomposition) of $SiCl_4$ at 730 to 820°C. It was found that irradiating the silicon substrate with 10 W/cm^2 of light from a high-pressure mercury lamp (mostly UV) increased the growth rate. Figure 16-9 shows a plot of the growth rate vs. $1/T$, from which an activation energy for the crystal-growth process can be derived. It is evident that illumination causes a drop of 2.1 kcal/mole in the heat of activation. Hence the light provides part of the energy normally supplied by the thermal or chemical reaction. Therefore, although there is no definite threshold

[15] G. Chiarotti, G. Del Signore, and S. Nannarone, *Phys. Rev. Letters* **21**, 1170 (1968).

[16] M. Kumagawa, H. Sunami, T. Terasaki, and J. Nishizawa, *Japanese J. Appl. Phys.* **7**, 1332 (1968).

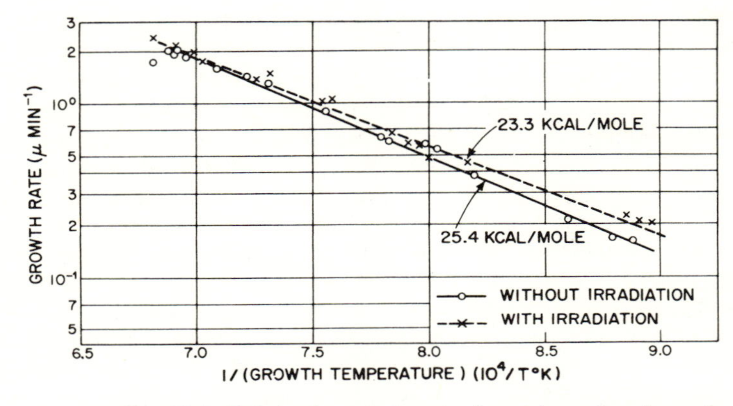

Fig. 16-9 Relation between the rate of crystal growth and growth temperature with and without light irradiation of the substrate surface in the rf furnace.[16]

temperature for growth, it was possible to lower the substrate temperature by 40° and still obtain good crystalline growth at a temperature where no appreciable growth rate is obtained without illumination.

The fact that even under illumination no growth occurs on quartz substituted for silicon at the specimen temperature demonstrates the epitaxial nature of the deposition process. Since illumination increases the growth rate, it may be possible to selectively build up the crystal layer according to a predetermined pattern generated by optical imaging. Such a patterning technique would have an obvious advantage over standard masking techniques, which can introduce contaminants.

16-B Photochemistry with a Liquid Ambient

16-B-1 CHEMICAL ETCHING[17]

When a semiconductor is immersed in an etching solution, some components of the liquid are adsorbed on the surface via a chemical reaction. Subsequently, the reaction product is desorbed with the result that some of the semiconductor is removed. The etchant normally operates by continuously oxidizing the surface, while the oxide is removed by dissolution in another constituent of the etch—hydrofluoric acid, for example. Since oxidation can be enhanced by illumination, the etching rate can be acceler-

[17] B. A. Irving, "Chemical Etching of Semiconductors," *The Electrochemistry of Semiconductors*, ed. P. J. Holmes, Academic Press (1962), p. 256.

ated by light. If the light forms a pattern, such as by optical imaging, photoengraving is obtained.

The etching rate depends on crystal orientation. This could be expected from the fact that the density of atoms on a surface should vary with the crystallographic plane. However, the bonding properties in different directions vary also; hence it may be difficult to predict the directional dependence of the etching rate. In fact, Fig. 16-10 shows that even in the [111] direction, successive layers require breaking alternatively three bonds per atom and then one bond per atom. In III–V compounds, a (111) double layer consists of type-III atoms on one side and type-V atoms on the opposite facet. The reactivities of the two facets can be extremely different.[18] Usually, the (111) or *A*-facet, consisting of type-III atoms, is more inert than the $(\bar{1}\bar{1}\bar{1})$ or *B*-facet which consists of type-V atoms. It is believed that due to electron redistribution the B^{V}-facet possesses more electrons and therefore is more readily oxidized than the A^{III}-facet. The impurity concentration in the semiconductor should have an effect on the etching rate, since it controls the availability of carriers participating in the bonding reaction. A nonhomogeneous impurity distribution causes local potential fluctuations which modulate the etching rate and results in either hollows (faster etch) or hillocks (slower rate). Crystal defects, with their associated strains, usually etch faster, forming characteristic pits whose shape can be controlled by an artful formulation of the etching solution. Note that in A^{III}–B^{V} crystals, the inertness of the *A*-facet enhances the contrast of etch pits and gives this surface a very rough appearance. Since illumination tends to flatten the energy bands near the

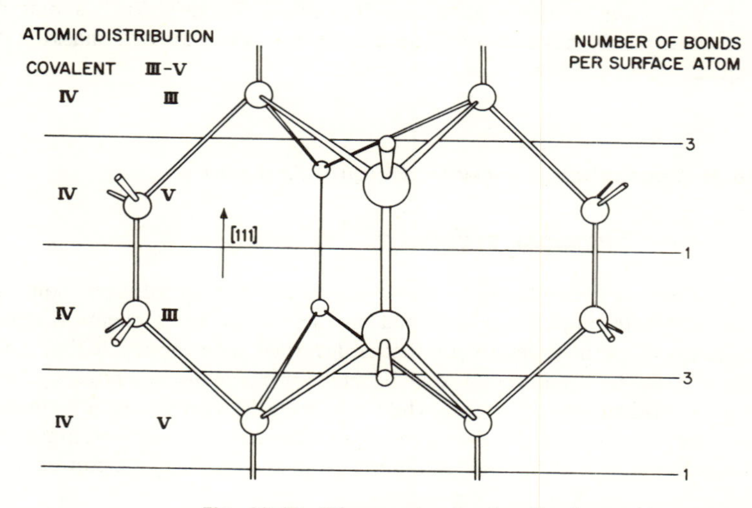

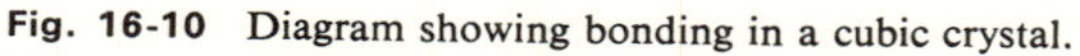
Fig. 16-10 Diagram showing bonding in a cubic crystal.

[18] H. C. Gatos and M. C. Lavine, *J. Electrochem. Soc.* **107**, 427 (1960).

surface, the local potential differences due to nonhomogeneities can be eliminated. Then, under illumination, the uniformly high rate of attack results in a smoother surface.[19]

16-B-2 ELECTROLYTIC ETCHING

If the semiconductor is immersed in an electrolyte and biased positively with respect to the solution, anodic reaction occurs whereby the surface atoms either are oxidized or form other complexes with molecules of the solution and subsequently become desorbed.[20] The composition of the electrolyte is usually not critical. Although the etching rate increases with increasing current, it is the hole component of the current which most affects the dissolution rate.[21] Hence in *p*-type semiconductors, where the current flow is predominently carried by holes, the etching rate is proportional to the current. On the other hand, *n*-type semiconductors dissolve faster if holes are injected into the surface either from a nearby forward-biased *p–n* junction or optically with photons having greater-than-gap energy. Note that an inversion layer at the surface of an *n*-type semiconductor forms a surface barrier. Since the electrolyte biases this barrier in the reverse direction, the current through the interface is the saturation current through the surface barrier. This etching current can be increased by illumination.

Photoengraving can be achieved readily with electrolytic etching, since holes can be generated optically where desired by projecting the appropriate light pattern. A parallel beam of light can be used to drill holes or wells in the semiconductor. To obtain a good contrast in photoengraving it is desirable to eliminate the ambient light and to cool the crystal and the electrolyte, thus minimizing the dark current through the unexposed regions of the surface.

When this technique is used on InSb, however, the surface becomes oxidized.[22] After an initial oxide layer has been built up, further oxidation becomes light-dependent (provided the wavelength is less than 5200 Å). By this photoanodization process a visible image can be formed on a polished surface of InSb.

16-B-3 PLATING

In plating, the semiconductor is made negative with respect to the solution. This allows metallic ions to deposit on the semiconductor surface. However, unlike electrolytic etching, plating requires that the composition of

[19]M. V. Sullivan, D. L. Klein, R. M. Finne, L. A. Pompliano, and G. A. Kolb, *J. Electrochem. Soc.* **110**, 412 (1963).

[20]Details of such reactions are discussed in D. R. Turner "Experimental Information on Electrochemical Reactions at Germanium and Silicon Surfaces," *The Electrochemistry of Semiconductors*, ed. P. J. Holmes, Academic Press (1962), p. 171.

[21]W. H. Brattain and C. G. B. Garrett, *Bell Syst. Tech. J.* **34**, 129 (1955).

[22]J. D. Venables and R. M. Broudy *J. Appl. Phys.* **30**, 1110 (1959).

the solution be carefully selected. Since plating depends on conduction, it is possible to control the rate of plating by illumination and, therefore, the geometry of the plating deposit can be controlled at the same time.

If the semiconductor contains a *p–n* junction, the *p*-type region tends to plate faster than the *n*-type region because the *p*-type region tends to be more negative, as shown in Fig. 16-11. However, it is not always the *p*-type side of the *p–n* junction which gets plated:[23] when a *p–n* junction is illuminated by light of energy greater than the gap energy, the resulting photovoltaic effect biases the junction in the forward direction. With the junction immersed in an electrolyte, the space charge can leak out through the electrolyte, which effectively short-circuits the *p–n* junction. In this process, the charges must cross two surface barriers Y_n and Y_p (Fig. 16-12); Y_n is the surface barrier of the *n*-type region and is equal to the change in energy of the bottom of the conduction band as it reaches the surface. This quantity is positive if the conduction-band edge moves away from the Fermi level. An analogous definition applies to Y_p, the surface barrier of the *p*-type region. The photogeneration of carriers at the surface tends to flatten the bands and hence to reduce Y_n and Y_p.

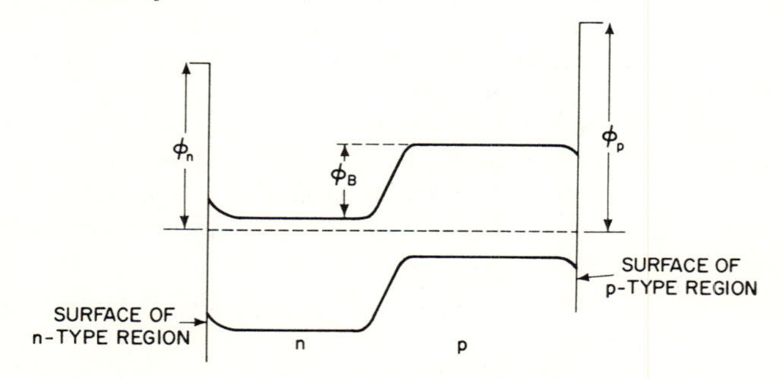

Fig. 16-11 Energy level diagram illustrating how the internal contact potential can render the *p*-type surface more electronegative than the *n*-type surface.

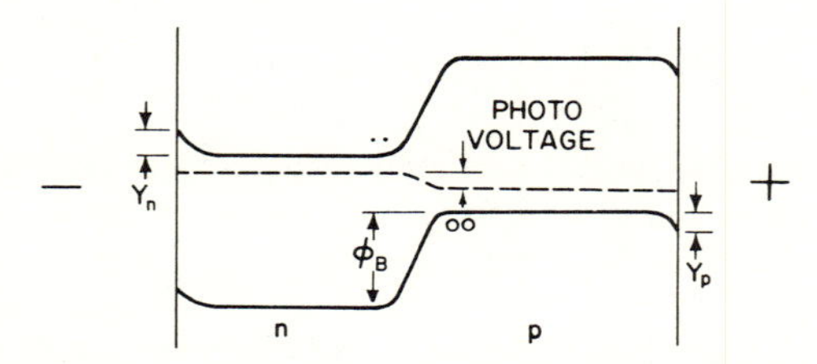

Fig. 16-12 Energy level diagram illustrating how a photovoltaic effect causes preferential plating on the *n*-type region.

[23]J. I. Pankove, "Practical Applications of Electrolytic Treatments to Semiconductors," *The Electrochemistry of Semiconductors*, ed. P. J. Holmes, Academic Press (1962), p. 299.

Clearly, since $Y_n + Y_p < \Phi_b$ (Φ_b being the barrier height of the p–n junction), most of the photocurrent will flow through the surface. The conventional current flow through the electrolyte is from the p-type to the n-type regions. In this case, preferential etching occurs at the p-type region and, depending on the electrolyte used, plating can take place at the n-type region.

16-C Photochemical Reactions Inside the Crystal

Light is capable of helping the displacement of atoms inside semiconductors and of changing the bonding state of impurities, thus affecting their chemical nature.

16-C-1 PHOTO-INDUCED ANNEALING

When a semiconductor is irradiated with high-energy particles, neutrons, or gamma rays, atoms can be displaced from their lattice sites to an interstitial position. They thus form Frenkel pairs, which are point defects con-

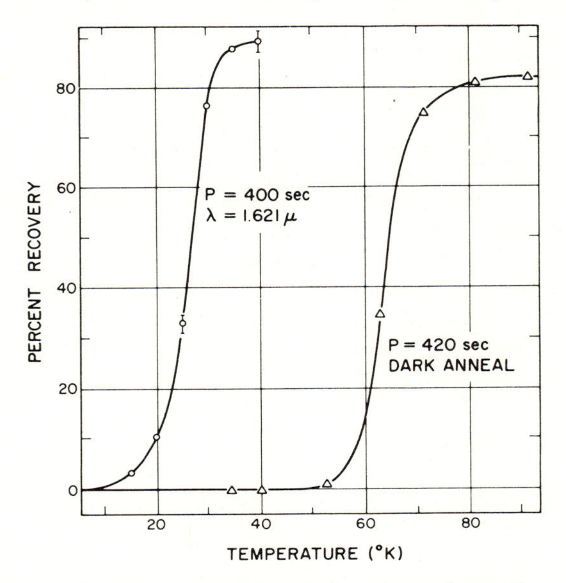

Fig. 16-13 Temperature dependence of photo-induced annealing in presence of 1.621 micron light (compared with isochronal anneal in dark). Each point represents 400 sec of illumination.[25]

[25] I. Arimura and J. W. MacKay, "Photo-induced Annealing in n-type Germanium," *Radiation Effects in Semiconductors*, ed. F. L. Vook, Plenum (1968), p. 204.

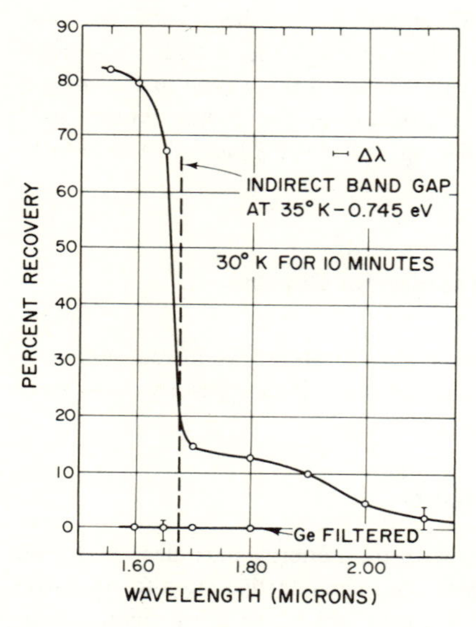

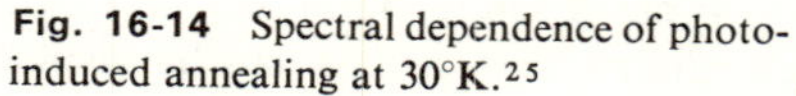
Fig. 16-14 Spectral dependence of photo-induced annealing at 30°K.[25]

sisting of an interstitial and a vacancy.[24] Such a defect can heal itself by letting the displaced atom diffuse back to a substitutional site. The energy needed for this motion can be supplied by heat or by light. In lightly doped n-type germanium, thermal recovery can be obtained at 65°K.[25] The recovery is monitored by measuring the Hall effect and the conductivity after various stages of optical and thermal treatments.

Figure 16-13 shows that light allows recovery at a much lower temperature than would be possible in the dark. Figure 16-14 shows that annealing occurs mostly when the photon energy is larger than the gap energy. However, it was also found that recovery from damage is obtained if, instead of monochromatic radiation $h\nu > E_g$, a broad band of infrared is used with all the photons in the range $h\nu < E_g$. In this case of broad-band infrared illumination recovery can be obtained at temperatures as low as 4.5°K.

16-C-2 PHOTOCHROMISM

A photochromic material becomes colored when exposed to UV light and remains colored for some time until it is bleached either thermally or by exposure to visible light. This effect has been observed in wide-gap materials such as alkali halides,[26] rare-earth-doped CaF_2,[27] or transition-metal-doped

[24] V. S. Vavilov, *Effects of Radiation on Semiconductors*, Consultants Bureau (1965), Chapter 5.
[26] J. Markham, *F Centers in Alkali Halides*, Academic Press (1966).
[27] D. L. Staebler and Z. J. Kiss, *Appl. Phys. Letters* **14**, 93 (1969).

$SrTiO_3$, $CaTiO_3$, or TiO_2.[28] Although these materials are classed as insulators, similar effects could occur in semiconductors. However, since deep levels are involved in the present process, transitions which occur in the visible region of the spectrum in an insulator would be scaled down to the infrared in a semiconductor and, therefore, the word "chromism" would not be used.

The photochromic process consists of the transfer of an electron from an impurity ion to some trapping center. As a result of this charge transfer, the valences of the metallic ion and of the trapping center are changed. Note that this is just an optical impurity-ionization process. However, since the change in valence can last for a long time, this process can be classed as a photochemical reaction:

$$M^{n+} \longrightarrow M^{(n+1)+} + 1 \text{ electron}$$

where M is the metal ion. Bleaching is the reverse reaction, which can be induced either optically or thermally. The excitation to the colored state can also be induced by electron bombardment (cathodochromism).[29]

Three processes are available for the creation of the color center, as shown in Fig. 16-15:

1. The electron can be transferred from the metallic ion (a shallow acceptor) to the trap via the conduction band;
2. A valence-band electron can be excited to the trap and the acceptor loses an electron to the resultant hole (in other words, the hole is trapped at the impurity);
3. The valence-band electron can reach the trap via the conduction band, and the hole is trapped by the acceptor.

As a result of this charge transfer, the impurity becomes a deeper acceptor and the trap a deeper donor. Now, because optical transitions between the deep impurity and the associated band are possible, the material becomes absorbing at $h\nu < E_g$.

In order to convert the deep centers into their original shallow states, the

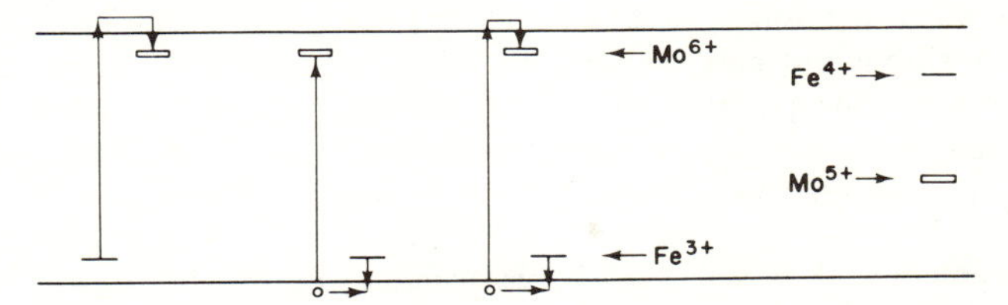

Fig. 16-15 Three possible mechanisms for exciting a photochromic material into the colored state. ($SrTiO_3$: $Fe^{3+} + Mo^{6+}$ shown in this example).

[28] B. W. Faughnan and Z. J. Kiss, *Phys. Rev. Letters* **21**, 1331 (1968).
[29] W. Phillips and Z. J. Kiss, *Proc. IEEE* **56**, 2072 (1968).

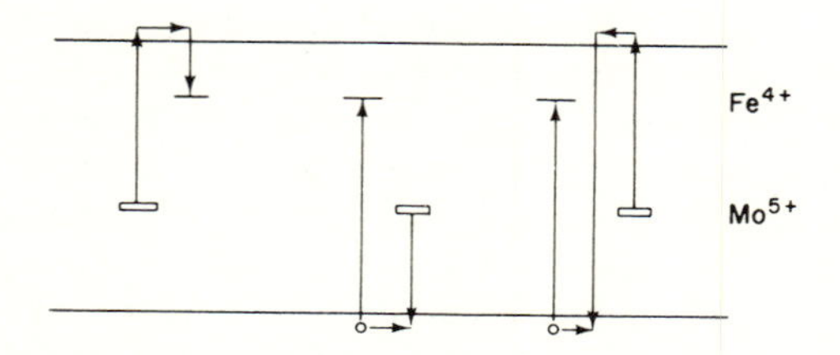

Fig. 16-16 Three possible mechanisms for bleaching the photochromic material of Fig. 16-15.

electron must be transferred back from the trap to the metallic ion. This can occur via either of the three reverse processes shown in Fig. 16-16:

1. Excitation of electron from the trap to the conduction band and its return to the acceptor;
2. Excitation of electron from valence band to the acceptor followed by recombination of electron in the deep donor (the trap);
3. Excitation of electrons from valence band to the deep acceptor and from deep donor to conduction band followed by an across-the-gap recombination.

Note that lower-energy photons are needed to bleach the material than to excite it into the colored state.

By way of example, $SrTiO_3$ doped with Fe^{3+} and Mo^{6+}, which is transparent over the visible range, becomes opaque when irradiated with UV ($h\nu \approx 3.1$ eV, $\lambda \approx 0.4\ \mu$). A change in absorption coefficient of the order of $100\ cm^{-1}$ can be obtained. Presumably, after excitation the two impurities become respectively Fe^{4+} and Mo^{5+}. The bleaching mechanism might be a transition from the valence band to the Fe^{4+}-level (giving the photochemical reaction Fe^{4+} + electron $\rightarrow Fe^{3+}$), which requires a 2.5-eV photon ($\lambda \approx 0.5\ \mu$), while the transition from the Mo^{5+} level to the conduction band (reaction: Mo^{5+} − electron $\rightarrow Mo^{6+}$) requires a 2.0-eV photon. Note that the observed absorption bands at 2.0 and 2.5 eV may not be due entirely to bleaching transitions but may include other processes as well. Figure 16-17 shows the absorption spectrum of $SrTiO_3$ doped with nickel. Here the trap is believed to be an oxygen–vacancy–nickel complex, while the metallic ion switches between the Ni^{2+}- and the Ni^{3+}-states.

Problem 1. A crystal of n-type germanium with 2×10^{15} donors/cm^3 has acceptor-like surface states the density of which is $10^{11}\ cm^{-2}$. The dielectric constant is 16. Use Poisson's equation to describe the bending of bands at the surface.

Problem 2. The crystal of Problem 1 is immersed in a solution of KOH and biased positively with respect to the electrolyte. Neglect band bending and assume it takes four holes per atom to etch the crystal. How long should it be etched at

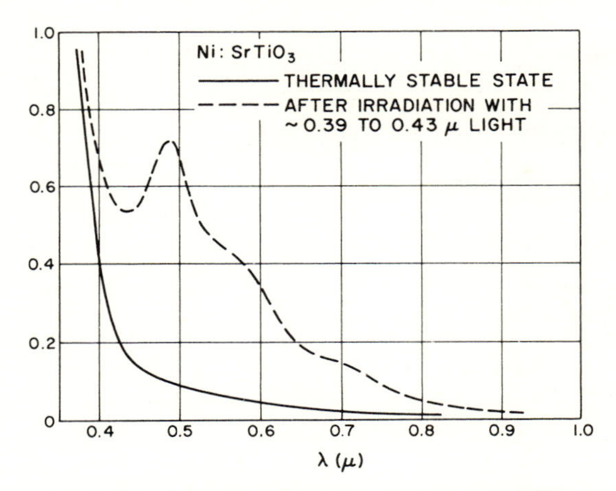

Fig. 16-17 Absorption spectrum of Ni-doped $SrTiO_3$.[30]

1 A/cm² to remove 1 μm of germanium from the surface? What field is necessary in the crystal to provide the needed current? Discuss how illumination with a flux of 1 mW/cm² at 1.0 eV can help the etching process.

There are about 5×10^{22} atoms of Ge per cm³. The intrinsic carrier concentration at room temperature is 2.4×10^{13} cm^{-3}. Other pertinent data can be taken from Appendix II.

[30] B. W. Faughnan and Z. J. Kiss, *IEEE J. Quantum Elec.* **5**, 17 (1969).

EFFECT OF TRAPS ON LUMINESCENCE

17

Traps are metastable states which capture an electron from a higher-energy state and retain it for a considerable time at the end of which the electron is released back to the higher-energy state, from which it can make a transition to a much-lower-energy state—for example, a hole in the valence band. This last transition may or may not be radiative. In the case of a final radiative recombination, the trapping process is responsible for the persistence of phosphorescence, or afterglow, which occurs after the excitation is stopped. The traps are also responsible for the gradual growth of luminescence and even for delays observed in lasers.

17-A Growth and Decay of Luminescence

As the excitation begins, the carriers start filling the traps and, therefore, are not available for luminescence. After the traps are filled, the luminescence intensity reaches a steady state (Fig. 17-1). Hence at very low temperatures,

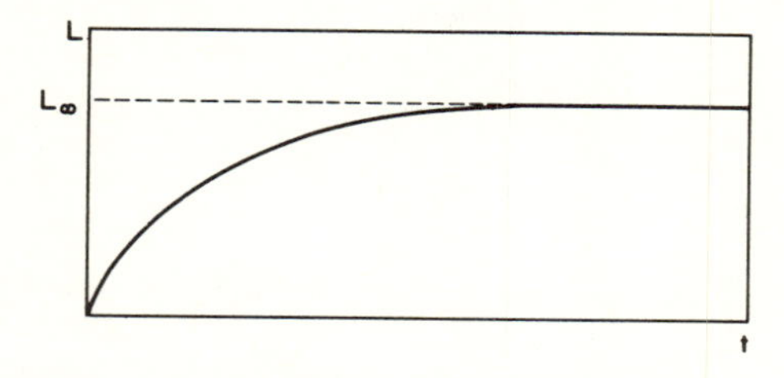

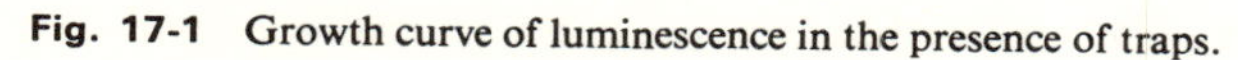
Fig. 17-1 Growth curve of luminescence in the presence of traps.

where the carriers are not thermally excited out of the traps, the area between the growth curve and the steady-state luminescence can be correlated with the trap density. Consider a step excitation which pumps carriers at a rate I into a volume V of active region in the semiconductor the traps of which are initially empty. If the luminescence-growth curve is represented by a function $L(t)$, the trap density is

$$N_t = \frac{1}{V\eta}\int_0^\infty [L_\infty - L(t)]\,dt \tag{17-1}$$

where η is the quantum efficiency for the radiative recombination; $L(t)$ is expressed in photons/sec; and L_∞ is the steady-state value of $L(t)$.

Of course, if the temperature is elevated, the traps may empty by thermal activation. In that instance, the kinetics of trap filling become more complicated. For the simplest case, in principle, the area between the luminescence-growth curve and the steady-state or maximum value of luminescence intensity extrapolated to $t = 0$ is proportional to the trap density.

In p-type materials, if the traps are shallower than the upper state for radiative recombination, the trapped electrons will gradually leak out from the trapping centers into the radiating level. Then, when the excitation is stopped, as the traps release gradually their stored electrons, the luminescence intensity decays. The rate of luminescence decay is, therefore, a measure of the carrier lifetime in the trap. If, on the other hand, the trap is deeper than the upper state for the radiative-recombination process and the temperature is sufficiently low to keep the electron in the trap, the luminescence stops when the excitation is terminated. In n-type materials, the above arguments are valid for the trapping of holes, and the depth of the states is measured with respect to the top of the valence band.

17-B Thermoluminescence

When a crystal is excited optically (with $h\nu > E_g$) or with an electron beam, deep traps can be filled. Those carriers which have been captured by the deep levels remain trapped after the excitation is terminated. The trapped carriers subsequently can be excited out of the traps either optically ($E_t < h\nu < E_g$) or thermally (E_t is the depth of the trap). The most frequently used technique is the thermal activation. The temperature at which the carrier is excited out of the trap correlates with the depth of the trapping level. When the carriers are released from the traps, they can participate in measurable events such as drifting in an electric field (thermally stimulated current[1]) or recombining radiatively (thermally stimulated luminescence,[2] or "thermoluminescence" for short).

[1] R. H. Bube, *Photoconductivity in Solids*, Wiley (1960), p. 292.
[2] H. W. Leverenz, *An Introduction to Luminescence of Solids*, Wiley (1950), p. 299.

The probability per unit time for an electron to escape from the trap is given by

$$P = s \exp\left(-\frac{E_t}{kT}\right) \tag{17-2}$$

The quantity s is given by[3]

$$s = N_B v \sigma_t$$

where N_B is the density of states in the band into which the carriers escape; v is the carrier's thermal velocity; and σ_t is the trap's capture cross-section.

Typical values of s are 10^8 sec^{-1} for CdS and 10^9 for ZnS.[4] The escape probability P is related to the time τ spent in the trap: $P = 1/\tau$. Equation (17-2) shows that τ has a strong dependence on the depth of the trap and on the temperature. Thus at room temperature, τ may vary from seconds to days for values of E_T ranging from 0.5 to 0.8 eV.

If n_t is the density of trapped electrons, its rate of change due to thermal excitation is

$$\frac{dn_t}{dt} = -n_t P \tag{17-3}$$

If the freed electrons are not recaptured by the traps, the solution of Eq. (17-3) is

$$n_t = n_{t0} \exp\left(-\frac{t}{\tau}\right) \tag{17-4}$$

where n_{t0} is the initial density of trapped electrons. If the radiative lifetime τ_r is much shorter than the escape time τ, light will be emitted at a rate $-\eta(dn_t/dt)$. Hence from Eqs. (17-3), (17-4), and (17-2), the light intensity will be given by

$$L(t, T) = n_{t0}\eta s \exp\left(-\frac{t}{\tau}\right) \exp\left(-\frac{E_t}{kT}\right) \tag{17-5}$$

One must take into account the fact that the quantum efficiency η is also a function of the temperature. In Sec. 7-C [Eq. (7-5)] we saw that the efficiency is given by

$$\eta = \frac{\eta_0}{1 + C \exp\left(-\frac{E^*}{kT}\right)} \tag{17-6}$$

where C is a constant, E^* is an activation energy, and η_0 is the efficiency at $T = 0$ [in Eq. (7-5) η_0 was taken as unity].

Equation (17-5) suggests two experimental techniques for determining the depth of the trap: (1) measuring the decay time of luminescence at various constant temperatures, and (2) measuring the temperature dependence of thermoluminescence at various known times. Note that these two techniques

[3]R. H. Bube, *Photoconductivity in Solids*, Wiley (1960), p. 291.
[4]D. Curie, *Luminescence in Crystals*, Wiley (1953), p. 170.

are equivalent, since they both require a determination of t and T. The temperature dependence of radiative efficiency can be determined in a separate experiment, such as measuring the efficiency during steady-state excitation.

Figure 17-2 shows an example of the determination of the temperature dependence of the time constant for decay of thermoluminescence. In this figure the data follows Eq. (17-2) rewritten as

$$\tau = \frac{1}{s} \exp\left(\frac{E_t}{kT}\right)$$

The slope of the straight line determines E_t, which, in this case of Cu-doped ZnS, is equal to 1.0 eV. Presumably, this is the energy needed to excite an electron from the valence band to the Cu acceptor, which permits a conduction-band electron to make a radiative transition to the hole in the valence band. Hence the Cu impurity acts as a hole trap. On the other hand, if an electron trap were operant, thermal excitation would take the electron from the trap to the conduction band or to a donor from which the electron would make a radiative transition (Fig. 17-3). If traps were present at different depths, a plot of log τ vs. $1/T$ would exhibit a set of straight segments having

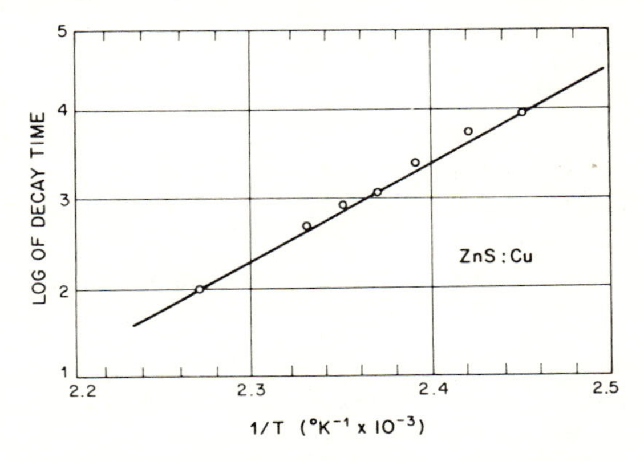

Fig. 17-2 Logarithm of the decay time of luminescence as a function of temperature for Cu-doped ZnS.[5]

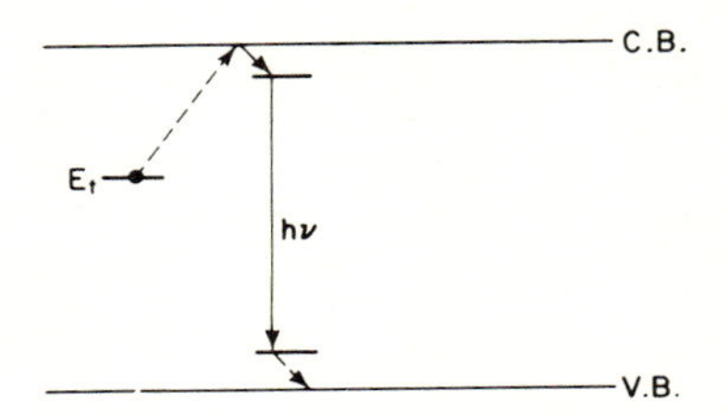

Fig. 17-3 Thermoluminescence where the electron released by a deep trap radiates by a donor-to-acceptor transition.

[5] R. H. Bube, *Phys. Rev.* **80**, 655 (1950).

steeper slopes toward higher temperatures, where more traps can be emptied.

The second technique for measuring thermoluminescence consists in raising the temperature of the semiconductor at a constant rate. As the temperature increases, the shallow traps empty first. If only one trapping level is present, the thermoluminescence goes through a maximum and then decreases as the traps are nearly exhausted. Figure 17-4 shows thermoluminescence curves for four samples of ZnS, each doped with a different donor in addition to copper. The peaked nature of $L(T = at)$ is evident from the relation in Eq. (17-5); $\exp -(E_t/kT)$ is responsible for the rising portion of the glow curve while $\exp -(t/\tau)$ causes the drop in luminescence.

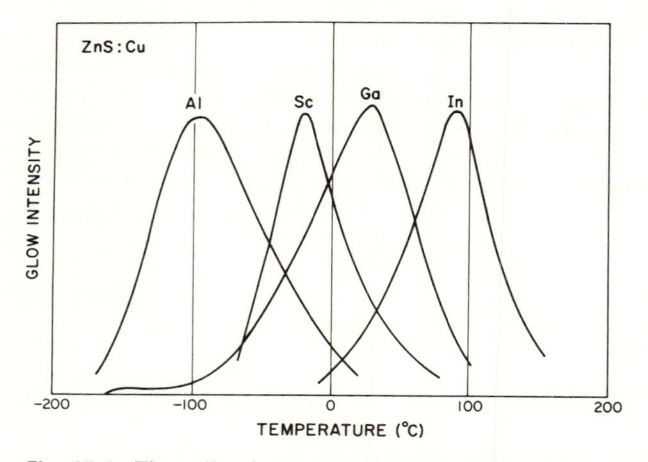

Fig. 17-4 Thermally stimulated luminescence emission for ZnS doped with Cu and aluminum, scandium, gallium, or indium.[6]

When many trapping levels are present, each level contributes a thermoluminescent peak at a different temperature, as in the case of Fig. 17-5. The temperature at which each peak occurs has been used to determine the depths of the trapping levels. This determination is done by integrating Eq. (17-5) up to every value T^* where $dL/dT = 0$.[7] By repeating the measurements at different heating rates, one can obtain more accurately the trap depth E_t and the capture cross-section σ_t. Slower heating rates result in greater structure. If the peaks overlap, the substructure can be brought out by stopping the heating and keeping the temperature constant for some time in order to let one set of traps empty completely before activating the next set of traps.

In principle, by starting at the lowest practical temperature—say, 4.2°K—it should be possible to reveal even shallow traps. But in practice it is found

[6] H. A. Klasens, *J. Electrochem. Soc.* **100**, 72 (1953).
[7] D. Curie, *Luminescence in Crystals*, Wiley (1963), p. 167.

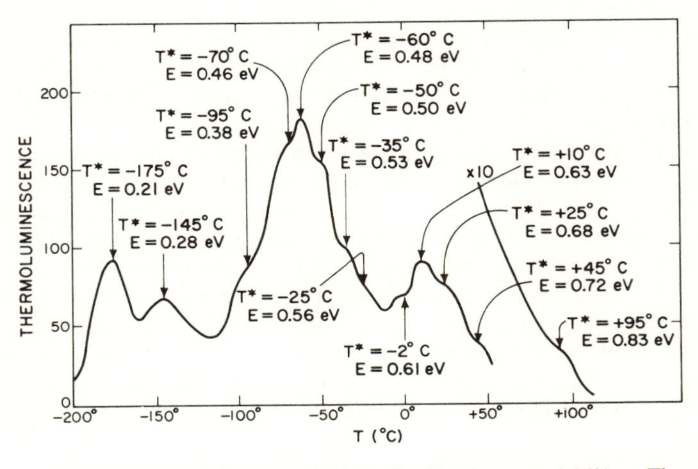

Fig. 17-5 Glow curve of ZnS: Cu. Heating rate: 0.06°/sec. The trap depths are computed by the formula:[7]

$$E(\text{eV}) = \frac{T^*(°\text{K}) - 7}{433}$$

that traps which are 0.1 eV deep lose their charge slowly and spontaneously while the material is maintained at near–liquid-helium temperature. This leakage out of the traps is attributed to tunneling, as evidenced by the time dependence of the afterglow spectrum: the high-energy edge of the spectrum

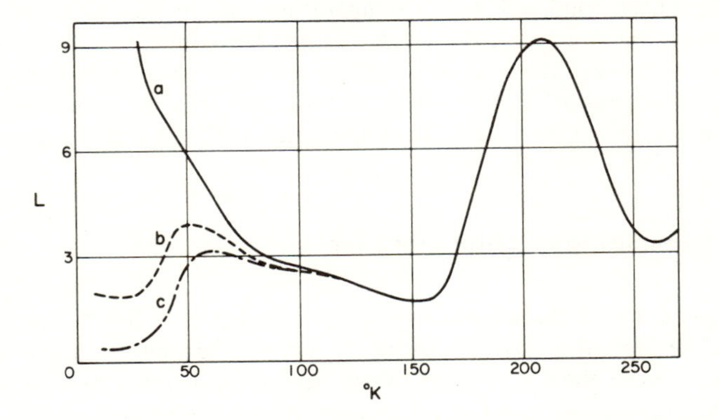

Fig. 17-6 Glow curves of ZnS: Cu containing traps with a quasi-continuous distribution in energy.[8] [Time for starting warmup after end of excitation: a, 0 min; b, 8.5 min; c, 34 min.]

[7] D. Curie, *Luminescence in Crystals*, Wiley (1963), p. 167.

[8] N. Riehl, G. Baur, L. Mader, and P. Thoma, "Afterglow in ZnS due to Tunneling," *II–VI Semiconducting Compounds*, ed. D. G. Thomas, Benjamin (1967). p. 724.

decays faster than the low-energy edge. The thermally stimulated luminescence curves taken after different postexcitation times spent at 6°K before the material is heated up show a decrease in low-temperature thermoluminescence (Fig. 17-6).

Note that heating releases electrons from a trap to donor or conduction-band states from which radiative transitions initiate; since different sets of levels can be coupled by the radiative transitions, a structured or a broad emission spectrum can be obtained from a single set of traps (which in a glow curve shows only one thermoluminescent peak). On the other hand, when different trapping levels are emptied (structured thermoluminescence), the free carriers may recombine via the same transition, resulting in a single emission peak. Hence, since release from traps and recombination are different processes, no correlation should be expected between the structures of the two types of curves—the glow curve $L(T)$ and the emission spectrum $L(h\nu)$.

17-C Infrared Stimulated Luminescence

Carriers can be excited out of traps by optical excitation. Since in most semiconductors the depth of the trap is less than 1.5 eV, the excitation can be obtained by infrared radiation. After the carrier is excited out of the trap, it makes a radiative transition, emitting a photon $h\nu > E_t$. Hence after the crystal has been "pumped" at low temperature, the luminescence which occurs when the traps are emptied by *IR* is called "*IR*-stimulated luminescence." The depth of the traps can be readily obtained from a spectrum of optical stimulation. Thus in ZnS, visible luminescence can be stimulated at low temperatures when the material is illuminated with 1.2-μ radiation.[9]

17-D Quenching of Luminescence

In some materials, luminescence can be quenched either during the afterglow of phosphorescence or during excitation. Quenching of luminescence can result from the application of an electric field, from heating, or from illumination with infrared. One may choose from several models for the quenching mechanism, depending on circumstances. The lower state for the radiative transition may become filled, or a nonradiative recombination path may become available.

[9] S. Shionoya, H. P. Kallmann, and B. Kramer, *Phys. Rev.* **121**, 1607 (1961).

Thermal quenching permits the onset of nonradiative recombination—several models were considered in Chapter 7. In either case, a barrier required that an activation energy E^* be overcome, which would give the drop in luminescence efficiency expressed by Eq. (17-6). The same barrier E^* could be overcome by optical excitation with photons of energy $h\nu_Q \geq E^*$.

Figure 17-7 illustrates the excitation of electrons from the valence band or from a shallow acceptor to a deeper acceptor which forms the lower state for the radiative transition, thereby quenching the latter. This *IR*-quenching mechanism requires that the probability for a nonradiative transition from near the conduction band to the valence band be more probable than for the radiative transition $h\nu$. This model for optical filling of the lower state of the luminescent transition is consistent with the observation of the reverse process—namely, a radiative transition (emission $h\nu'$ in Fig. 17-7). An emission at about 1.2 eV has been observed in ZnS[10] which corresponds to the range of photon energies responsible for quenching the visible luminescence.

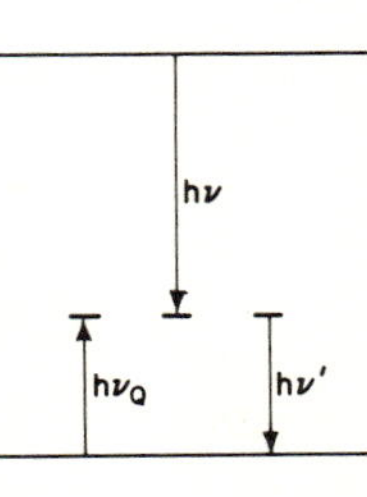

Fig. 17-7 Quenching of luminescence $h\nu$ by filling the recombination center with *IR* excitation $h\nu_Q$. Note the possibility of *IR* emission $h\nu' = h\nu_Q$ from the recombination center.

It must be pointed out that quenching and stimulation of luminescence are complex and sometimes competing processes. In ZnS a broad band of radiation at about 0.75 μ always quenches the fluorescence and the phosphorescence; but the band at 1.2 μ, which always quenches the luminescence at room temperature, sometimes stimulates the luminescence at 77°K. If the visible excitation and the infrared radiations are both incident on ZnS, the fluorescence is stimulated for a short time after the *IR* is turned on and subsequently it is lower than the luminescence without *IR*.[9]

17-E Trapping Effects in Lasers

17-E-1 TIME DELAY IN LASERS

In some semiconductor lasers the coherent output is delayed with respect to the beginning of a step excitation. Lasing delays have been observed in

[10] G. Meijer, *J. Phys. Chem. Solids* 7, 153 (1958).

GaAs[11-18] and in $GaAs_{1-x}P_x$[19] injection lasers. In the latter case, delays as large as 1.2 μs have been obtained. These effects have been attributed to the presence of traps in the active region of the semiconductor.

Increasing the current causes a faster rise of the spontaneous emission and a shorter delay (Fig. 17-8). A plot of the inverse of the delay time ($1/t_d$) vs. the current through the junction gives a straight line the extrapolation of which intercepts the current axis (Fig. 17-9). The intercept occurs at I_∞, which is the threshold current for infinite delay. If we call the current in excess of I_∞ the "overdrive," we find that the delay is inversely proportional to the overdrive and the constant of proportionality has the dimension of a charge:

$$t_d = \frac{Q}{I - I_\infty}$$

A typical value of Q is 5×10^{-8} coulombs.

If one assumes that the delay is due to trap filling in the active region at the rate of one electron per trap, one can estimate the trap density as

$$N_t = \frac{Q}{(1.6 \times 10^{-19})V}$$

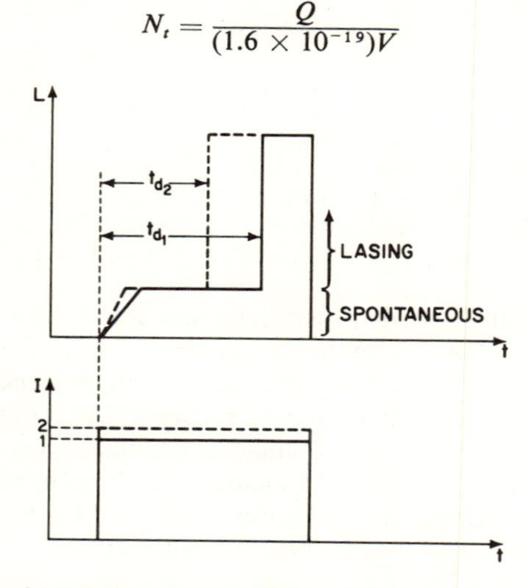

Fig. 17-8 Lasing delays in injection laser for two different excitation levels.

[11] G. E. Fenner, *Solid State Elec.* **10**, 753 (1967).
[12] K. Konnerth and C. Lanza, *Appl. Phys. Letters* **4**, 120 (1964).
[13] R. O. Carlson, *J. Appl. Phys.* **38**, 661 (1967).
[14] C. D. Dobson, J. Franks, and F. S. Keeble, *IEEE J. Quantum Elec.* **4**, 151 (1968).
[15] G. Guekos and M. J. O. Strutt, *Elec. Letters* **3**, 276 (1967).
[16] J. C. Dyment and J. E. Ripper, *IEEE J. Quantum Elec.* **4**, 155 (1968).
[17] K. Konnerth, *IEEE Trans. Elec. Devices* **10**, 506 (1965).
[18] N. N. Winogradoff and H. K. Kessler, *Solid State Comm.* **2**, 119 (1964).
[19] J. I. Pankove, *IEEE J. Quantum Elec.* **4**, 161 (1968); ibid., **4**, 427 (1968); *Proc. Int. Conf. Physics of Semiconductors* (Moscow) 559 (1968); *Bull. Am. Phys. Soc.* **13**, 1657 (1968).

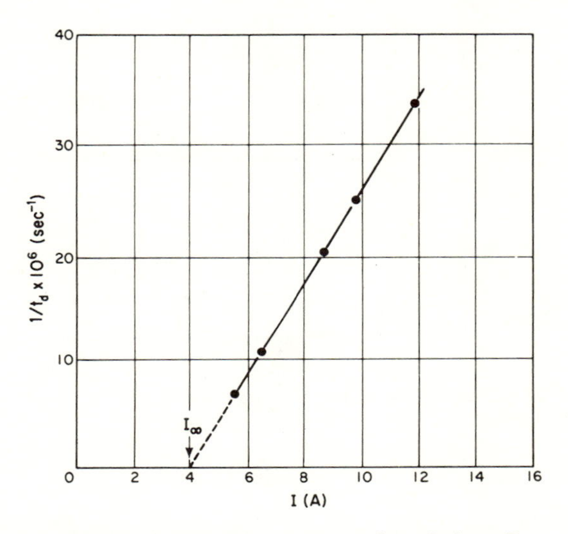

Fig. 17-9 Plot of reciprocal delay vs. current for a $GaAs_{0.6}P_{0.4}$ laser at 119°K.

where the numerical coefficient in the denominator is the charge of one electron, and V is the volume of the active region. For a typical laser, the area of the p–n junction is 3×10^{-4} cm^2 and the thickness of the active region is about one micron. Hence N_t is about 10^{19} cm^{-3}. Similar trap densities have been estimated in the case of GaAs lasers.[12,13]

17-E-2 TRAP STORAGE TIME

Once a trap is filled, how long does it retain the charge? This question can be answered by using a double-pulse technique: the current is adjusted to barely fill the traps during the first pulse; at the second pulse, the traps being full, the diode lases (Fig. 17-10). As the spacing between pulses increases beyond the decay time of the trapped charge, the lasing signal begins to decrease. The decay of the lasing signal (Fig. 17-11) shows that the charge leaks out of the trap in a time of the order of milliseconds.

17-E-3 TRAPS AS SATURABLE ABSORBERS

An attempt was made at emptying the traps optically. This was done by irradiating the $GaAs_{1-x}P_x$ laser with the output from a GaAs laser. The two lasers were closely spaced (about 25 μm apart) with the two junctions coplanar (Fig. 17-12). The GaAs laser radiates a power density of up to several megawatts/cm^2 emitting 1.48-eV photons, which are not energetic enough to

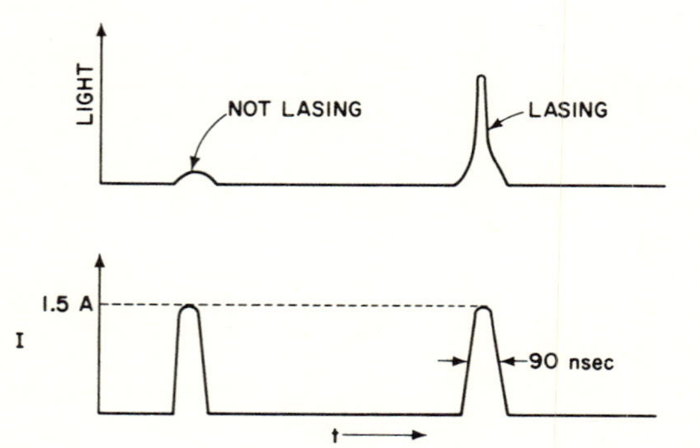

Fig. 17-10 Double pulse technique for measuring the decay of trapped charges.

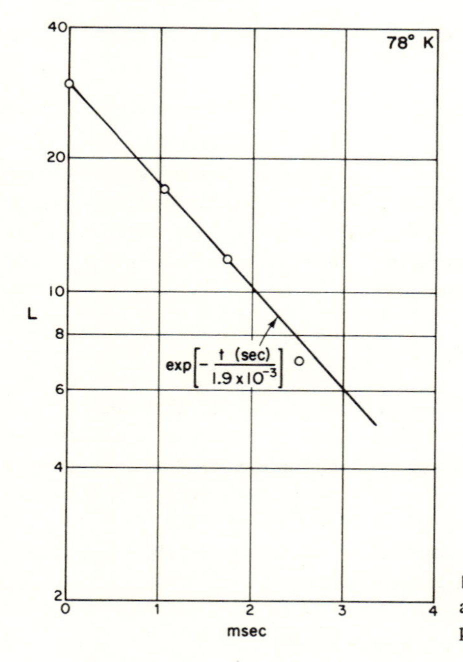

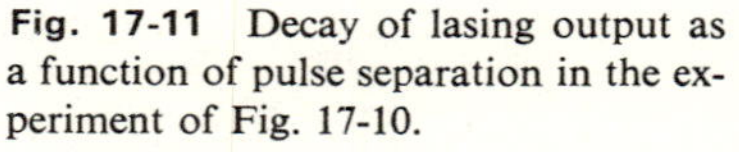
Fig. 17-11 Decay of lasing output as a function of pulse separation in the experiment of Fig. 17-10.

be absorbed in a fundamental transition (across the 1.9-eV gap of $GaAs_{0.6}P_{0.4}$) but probably energetic enough to excite electrons out of traps into the conduction band. After the traps were filled by the first pulse, the GaAs laser was turned on briefly. If the traps had been emptied, the second excitation of the $GaAs_{1-x}P_x$ laser would have exhibited either a nonlasing or a reduced output. However, the opposite effect was found: the output at the second pulse was strengthened, indicating that the 1.48-eV irradiation replenished the traps by optical excitation of electrons from the valence band.

The results of another trap-filling experiment are shown in Fig. 17-13. The alloy diode is irradiated for 400 ns (nanoseconds) with the 8400-Å emission from the GaAs laser, then, 400 ns later, a pulse of current is passed through the alloy diode. With a small excitation of the GaAs laser (0.5 A), a lasing delay of 400 ns is obtained in the alloy diode. When the GaAs current is raised to 1.5 A, the delay is reduced to about 240 ns. It was possible to adjust the current through the alloy diode so that the latter would lase only if the GaAs laser had been "on" less than one millisecond earlier—i.e., less than the trap-leakage time. If the duration of the optical pumping by the GaAs laser is lengthened, the traps remain filled for a longer time. For a given level of optical pumping with the GaAs laser, the change in the lasing delay inside the alloy laser is proportional to the duration of the optical

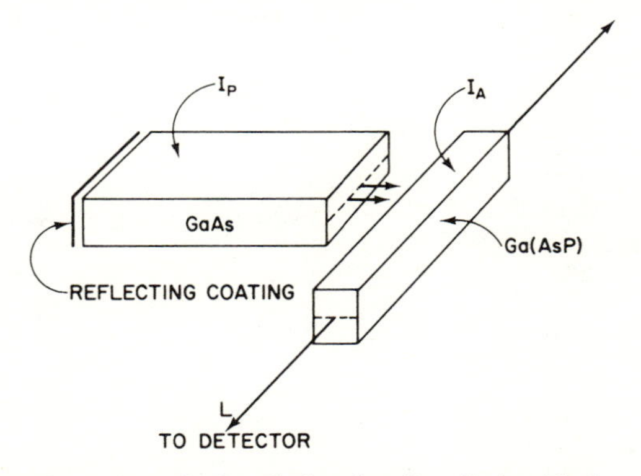

Fig. 17-12 Arrangement for irradiating the alloy diode with the output from a GaAs laser.

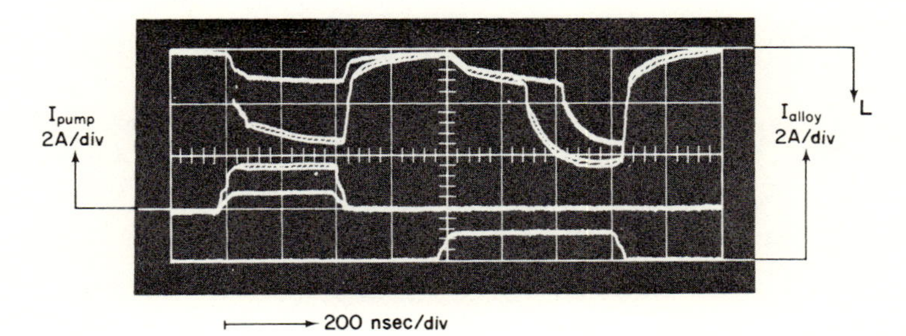

Fig. 17-13 Demonstration of optical trap-filling using the arrangement of Fig. 17-12. Increasing the GaAs laser output (via I_{pump}) decreases the lasing delay in the alloy diode.

pumping. These definitive experiments demonstrate that the traps responsible for the delay can be filled optically.

From the above set of optical trap-filling experiments, one can conclude that the traps act as saturable absorbers. Some of the light generated in the alloy diode is absorbed by a trap-filling transition. Once the trap is filled, it remains filled for about 10^{-3} sec and during this time cannot absorb another photon. Thus the material is less lossy and lasing can occur. Since the traps can also be filled by lower-energy photons (1.48 eV), it should be possible to observe a saturation of absorption at 8400-Å corresponding to the radiation emitted by a GaAs diode. This was done in the next experiment (Fig. 17-14), which detects the 8400-Å radiation transmitted by the alloy diode. Figure 17-14 shows that increasing the excitation of the alloy diode increases its transmission at 8400 Å. If the transmission of the 8400-Å radiation is measured at various times after the traps have been filled by a pulse of current through the alloy diode, one can monitor the decay of charges from the traps. One can make sure that the traps are initially filled by observing lasing at 6440 Å at the end of the current pulse. Figure 17-15 illustrates the transmission profile obtained, indicating the return to the unfilled-trap condition in a time of the order of one millisecond.

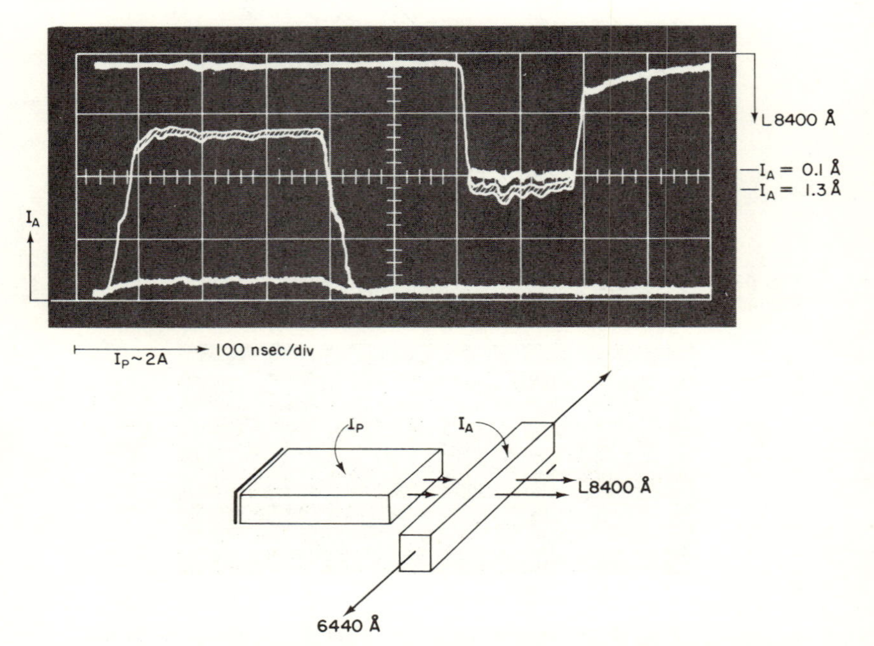

Fig. 17-14 Demonstration that traps in the alloy diode can act as saturable absorbers at 8400 Å.

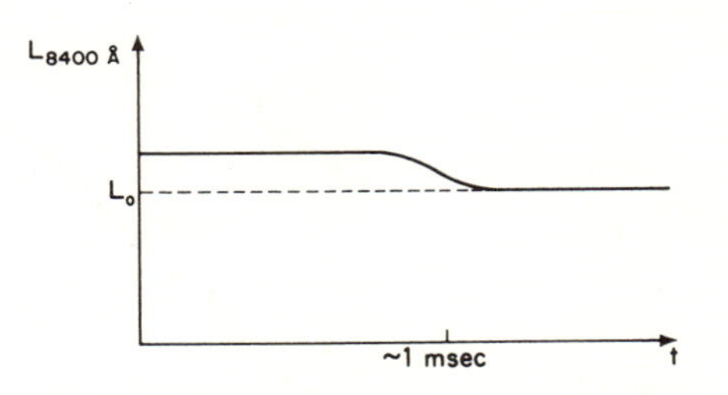

Fig. 17-15 Decay of transmitted GaAs radiation as a function of the time elapsed after pulsing the $GaAs_{1-x}P_x$ laser. L_0 is the transmission level when the traps in the alloy diode are empty.

The optical filling of the traps by the 1.48-eV radiation clearly indicates that most of the trapping levels are deep, at least 0.4 eV below the conduction band (which is 1.9 eV above the valence band). However, this does not negate the possibility that there may be a distribution of trapping states extending to and beyond the conduction band.

In summary, $GaAs_{1-x}P_x$ injection lasers contain about 10^{19} deep traps/cm^3 in the active region. These traps, the nature of which is still unknown, act as saturable absorbers which are filled by absorbing the spontaneous emission when the laser is first turned on. The electron-storage time in the trap is long (10^{-3} sec), so that the trap-filling process can be integrated over a prolonged time, which is the delay time. The delay is reduced if the emission rate is increased.

17-E-4 TEMPERATURE DEPENDENCE OF TRAPPING IN GaAs INJECTION LASERS

In contrast to the long delays described above for $GaAs_{1-x}P_x$ injection lasers, where they were observed over a large range of temperatures extending to at least 2°K, in GaAs injection lasers long delays have not been found at low temperatures. Below 100°K, the delays are at most of the order of nanoseconds, but at higher temperatures delays of hundreds of nanoseconds can be obtained.[13-15,17] The transition temperature T_t beyond which long delays can be observed correlates with a kink in the curve representing the temperature dependence of the threshold current (Fig. 17-16).[16] The transition temperature depends on the doping of the crystal and on thermal treatments. In correlating the kink in the $I_{th}(T)$ curve with the dependence of delay on

[13] R. O. Carlson, *J. Appl. Phys.* **38**, 661 (1967).
[14] C. D. Dobson, J. Franks, and F. S. Keeble, *IEEE J. Quantum Elec.* **4**, 151 (1968).
[15] G. Guekos and M. J. O. Strutt, *Elec. Letters* **3**, 276 (1967).
[16] J. C. Dyment and J. E. Ripper, *IEEE J. Quantum Elec.* **4**, 155 (1968).
[17] K. Konnerth, *IEEE Trans. Elec. Devices* **10**, 506 (1965).

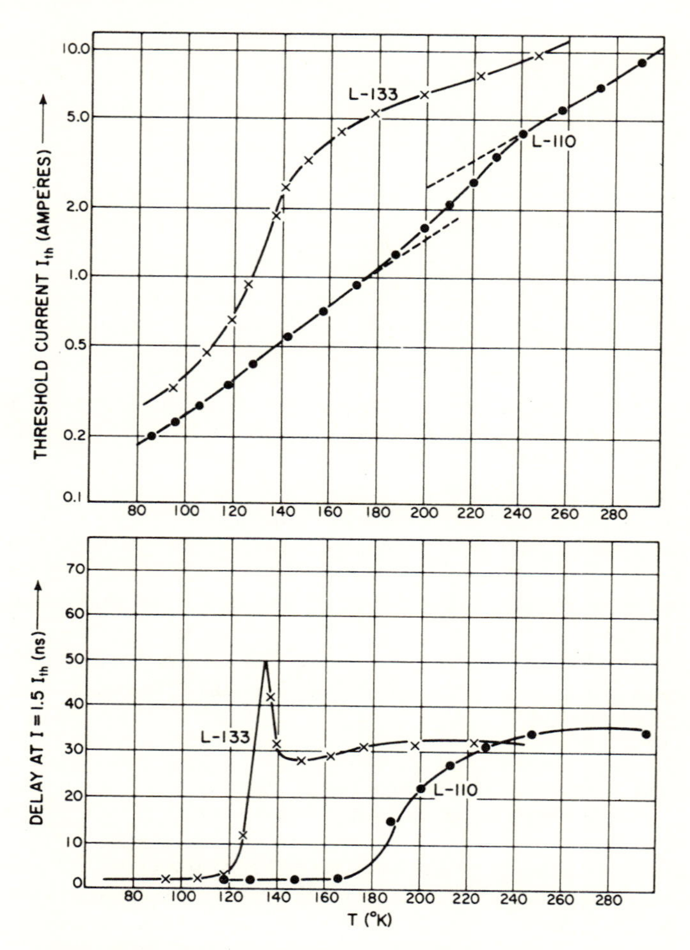

Fig. 17-16 Correlation of I_{th} and the delay t_d at $I = 1.5I_{th}$ with temperature for diodes L-110 (1-hour diffusion time) and L-133 (4-hour diffusion time). Both diodes are made from the same substrate.[16]

temperature, it is found that the steeper kink in $I_{th}(T)$ corresponds to a sharper transition between short and long delays (Fig. 17-16).

If the traps act as saturable absorbers, some aspects of the lasing delay can be explained: the cavity is lossy when the traps are empty; as these fill up, the losses become smaller than the gain, whereupon the radiation be-

[16] J. C. Dyment and J. E. Ripper, *IEEE J. Quantum Elec.* **4**, 155 (1968).

comes coherent.[11] These trapping centers could be filled either by trapping electrons injected into the conduction band or by trapping valence-band electrons excited by the spontaneous emission.

17-E-5 DOUBLE-ACCEPTOR MODEL

A double acceptor with a coulomb barrier for trapping the second electron has been proposed to account for the temperature dependence of the delay.[11] But the model which has met with the most success thus far is a double acceptor which acts as a saturable absorber only if the lower level is filled.[16]

Figure 17-17 shows the three possible states of the trap. In its first state Tr_1, the trap can capture an electron at a level E_1 near the valence band, whereupon it goes to a second state Tr_2. The first state Tr_1 is in thermal equilibrium with the valence band; hence its occupation probability is determined by the quasi-Fermi level for holes E_F. If E_F is above E_1, the traps are in the Tr_2 state. This would be the case of lightly doped p-type GaAs or it might occur in compensated materials. In heavily doped GaAs the quasi-Fermi level is inside the valence band. Therefore, E_1 is occupied only at higher temperatures, whereupon the traps become saturable absorbers Tr_2. When the electron returns from E_1 to the valence band, the trap switches from the Tr_2 to the Tr_1 state in a time of the order of 10^{-11} sec.[20] From its Tr_2 state the trap can go to a Tr_3 state by capturing a second electron at the level E_2. The level E_2 can be filled by an electron from the conduction band or by one excited optically from the valence band. Thus Tr_2 is an absorbing lossy condition of the trap which prevents lasing until after the delay, when the level

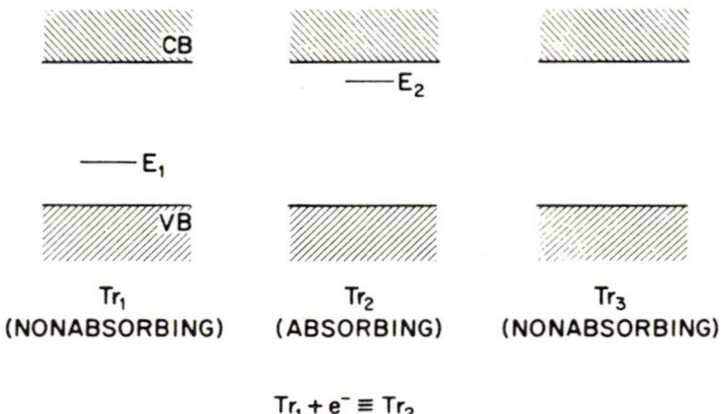

Fig. 17-17 Representation of the energy levels available for the capture of electrons in the states Tr_1, Tr_2, and Tr_3 of a single trapping center.[16]

[11] G. E. Fenner, *Solid State Elec.* **10**, 753 (1967).
[20] J. E. Ripper, *IEEE J. Quantum Elec.* **5**, 391 (1969).

E_2 is sufficiently filled to make the losses lower than the gain. At low temperatures, if E_1 is above the quasi-Fermi level E_F, the trap remains in the Tr_1 state, which is not a lossy state; hence no appreciable delay is observed. The position of the quasi-Fermi level depends very much on the impurity profile at the *p–n* junction and, therefore, on the fabrication technique and on any subsequent thermal treatment.

17-E-6 INTERNAL Q-SWITCHING

In some GaAs injection lasers, over a narrow range of temperatures and over a limited range of currents a short burst of stimulated emission is obtained at the end of the current pulse (Fig. 17-18).[21] The duration of the spike is of the order of 300 ps (picoseconds) or less. The intensity of the spike is independent of the duration of the exciting current pulse. The drive can be varied from 2 ns to several microseconds without affecting the intensity or duration of the spike. This behavior is also independent of the pulse-repetition rate, provided heating effects are avoided. The only parameter which affects the intensity of the lasing spike at a given temperature is the amplitude of the driving current; the intensity of the spike increases with increasing currents.

When the occurrence of this end-of-pulse spiking is charted over the current–temperature space (Fig. 17-19), it is found that internal *Q*-switching

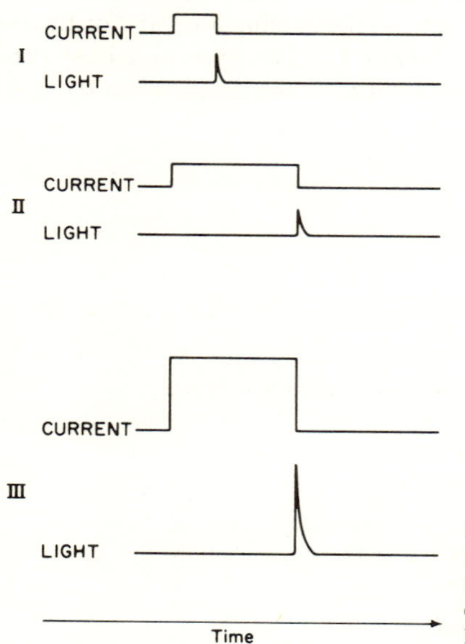

Fig. 17-18 Diagram showing the behavior of *Q*-switching in GaAs lasers; the narrow light pulse occurs at the end of the current pulse in spite of increases in pulse length (II) or amplitude (III).[21]

[21] J. E. Ripper and J. C. Dyment, *Appl. Phys. Letters* **12**, 365 (1968).

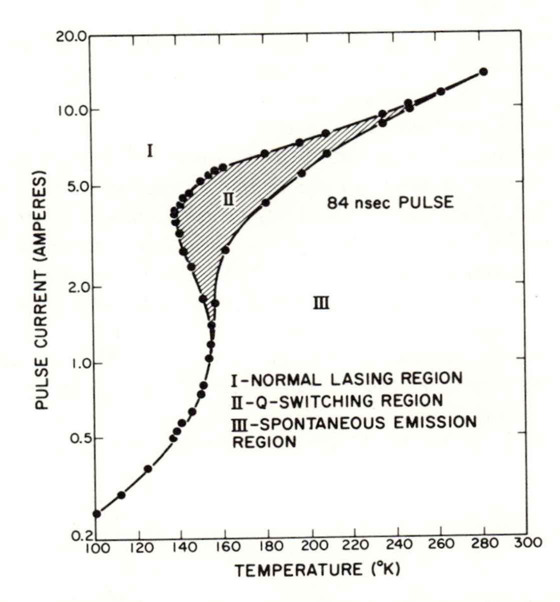

Fig. 17-19 Characteristics of a typical diode showing the Q-switching region (II) as a function of injection current and heat-sink temperature. In this region stimulated emission only occurs after the end of the injection pulse in contrast with the normal lasing region (I) where it occurs during the current pulse, and the spontaneous emission region (III) where no lasing is observed.[21] The transition temperature T_t is about 140°K.

happens only in the shaded region II, which is above the transition temperature beyond which long lasing delays are obtained. In region I, lasing occurs during excitation (with a longer delay when $T > T_t$). In region III, only spontaneous emission is obtained.

The internal Q-switching has been interpreted in terms of the double-acceptor-trap model, the same trapping model used to explain delay effects (both effects are observed in the same device). This model requires that the level E_1 be emptied at the end of the current pulse, causing the disappearance of state Tr_2 and thus allowing the lasing spike to occur.

17-F Triboluminescence

17-F-1 STRAIN-EXCITED LUMINESCENCE

Triboluminescence is the emission of light in response to mechanical excitation. The stress can generate a local field by the piezoelectric effect.

With local fields of 10^6 V/cm, which can be readily achieved, the deformation energy is transformed into radiation. Zener breakdown can set in to generate electron–hole pairs which will then recombine. Some of these carriers will be trapped and will participate in a radiative process at a later time.

The triboluminescent properties of Mn-doped ZnS have been extensively studied.[22] In this case, triboluminescence has been generated by impact from a solenoid-operated hammer which allows a reproducible control of the stress. After the blow, the luminescence decays with a time constant ranging from 0.1 to 1 ms. Triboluminescence generated by friction between moving particles of ZnS gives a nearly continuous emission the spectrum of which can be measured. The triboluminescent spectrum was found to be identical to the photoluminescent spectrum as shown in Fig. 17-20. This demonstrates that in triboluminescence the radiative-recombination process is similar to that of free carriers.

In the impact experiment,[22] the luminescence intensity increased with

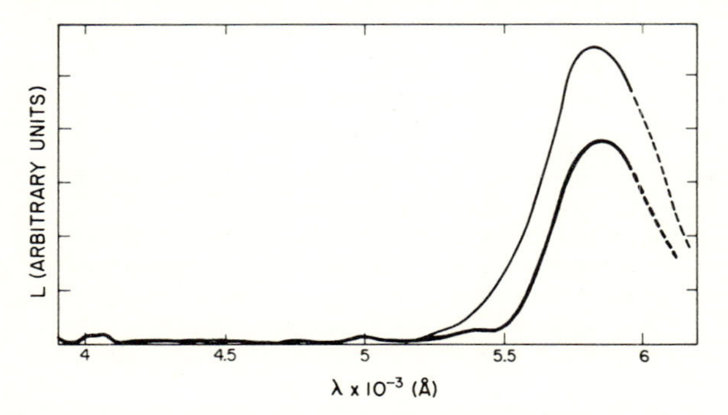

Fig. 17-20 Triboluminescence spectrum of ZnS: Mn (thick line), photoluminescence spectrum of ZnS: MN (thin line).[22]

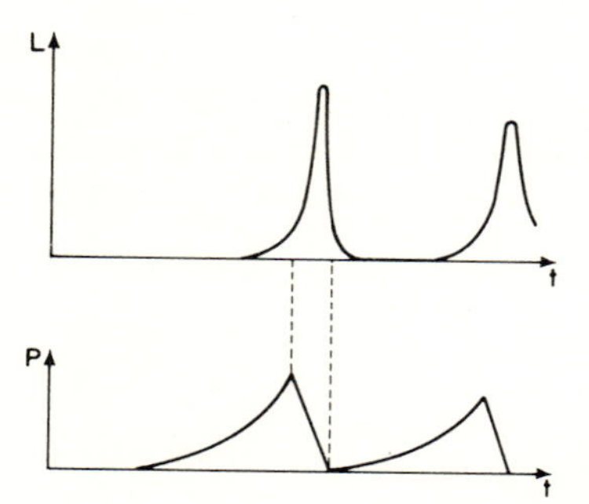

Fig. 17-21 Relationship between triboluminescence L and excitation P in ZnS.[23]

[22] I. Chudacek, *Czechoslovak J. Phys.* **15**, 359 (1965) and **17**, 34 (1967).
[23] L. Sadomka, *Czechoslovak J. Phys.* **14**, 800 (1964).

pressure up to a maximum and, at higher pressures, decreased as some quenching process set in. A time-dependent study of impact triboluminescence showed that emission occurs only while this pressure decreases (Fig. 17-21).[23] This observation would agree with the above hypothesis that the local field produced by the piezoelectric strain generates electron–hole pairs which are spatially separated by the local field. As the strain decreases, the electric field drops and the electrons and holes can diffuse toward each other for a radiative recombination.

17-F-2 STRAIN-STIMULATED LUMINESCENCE

If the ZnS is cooled while it is mechanically activated, such as by grinding, the carriers can go to deep traps where they remain frozen for a long time. If now the material is heated, the carriers are released from the traps and emit light. This process is called "tribothermoluminescence."

We shall also include in the triboluminescent category the mechanical activation of carriers out of traps in the case where the electron–hole pairs have been generated by other means of excitation, such as light. A strain operates as described in the configuration diagram, moving the carrier along a metastable state to higher energies and thus enabling it to escape to a radiating state (Fig. 17-22). Crystals of CdS pumped at low temperature with radiation having a wavelength shorter than 6900 Å retain carriers in traps for a very long time. If, subsequently, the crystal is struck with a hammer in the direction of the *C*-axis, it emits a flash of blue-green light.[24] Hundreds of flashes can be obtained by repeated blows after a single optical pumping.

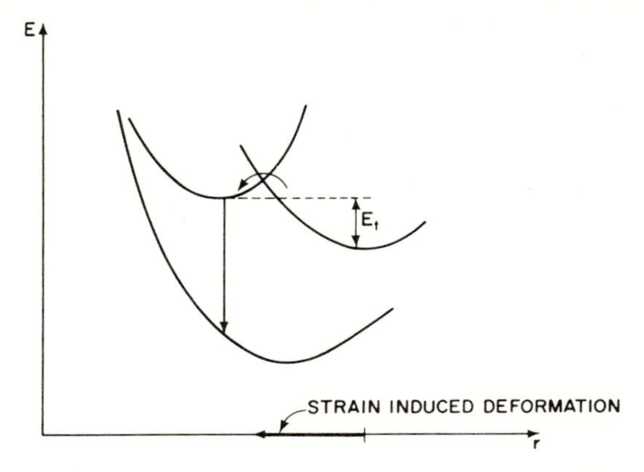

Fig. 17-22 Configurational model for triboluminescent stimulation of luminescence.

[24]D. M. Warschauer and D. C. Reynolds, *J. Phys. Chem. Solids* **13**, 251 (1960), and C. W. Litton and D. C. Reynolds, *Phys. Rev.* **125**, 516 (1962).

The emission spectrum exhibits four or five peaks separated by LO-phonons. This spectrum resembles that obtained by photoluminescence in the same material, with the exception that the second peak dominates in impact luminescence, whereas the first peak dominates in photoluminescence.

17-F-3 FRACTURE LUMINESCENCE

Another form of triboluminescence is obtained when a semiconductor is broken. When a crystal is fractured in the dark, one can sometimes see a blue flash. This emission is believed to be due to the formation of surface states on the newly generated surfaces and to the rearrangement of carriers at the surface. Presumably, the initial redistribution of electrons to surface states having the most positive energies is radiative (in contrast to any subsequent surface recombination which is not radiative). The triboluminescence by fracture has been studied in silicon.[25,26] To measure the emission spectrum, a series of closely spaced flashes was obtained by sand-blasting the Si-crystal. This process causes small fractures at a fast rate. To keep the light source stationary, the crystal was slowly rotated in front of the jet of sand so that the point of impact was always at the focus of the optical system.[25] The emission spectrum is very broad, extending over the visible range.

It is possible that the light emitted during fracture is excited by the strain which, incidentally, fractures the crystal as well. In this case, surface states would not be involved in the luminescence process. It is also possible that fracture luminescence was nothing else but air glow resulting from an arc between charged freshly cleaved surfaces.

[25] O. Parodi, private communication (1957).
[26] D. A. Jenny, *J. Appl. Phys.* **28**, 1515 (1957).

REFLECTANCE MODULATION

18

Although reflectance modulation offers little device interest, its greatest value is as a tool for studying the band structure of semiconductors.[1] The most obvious practical (but rather ineffective) application of reflectance modulation would be in a device for controlling the intensity of reflected radiation. A more important practical application is to determine effects of radiation damage in a nondestructive test[2] where the reflectance measurement probes changes in the concentration of carriers and in their relaxation time.

Let us distinguish between true surface-reflectance modulation (which occurs at $h\nu > E_g$) and the modulation of light reflected by some internal surface, occurring at $h\nu < E_g$. In the latter case, the light may travel twice (in and out) through a region where the absorbance can be modulated so that, although the reflected beam appears modulated, the active process is not reflectance but absorption. An example of absorption modulation of a reflected beam occurs in the case of a *p–n* junction below an illuminated *n*-type surface. When the junction is forward-biased, the hole concentration in front of the junction increases, intensifying free-carrier absorption by transition between valence subbands. These transitions have a characteristic spectral dependence which appears in the modulation of the radiation reflected by the *p–n* junction. Thus during hole injection, an 80% decrease in reflected intensity has been obtained at about 9 μ in germanium.[3]

In contrast to the above modulation by absorption occurring at photon

[1]D. L. Greenaway and G. Harbeke, *Optical Properties and Band Structure of Semiconductors*, Pergamon (1968); and D. E. Aspnes, P. Handler, and D. F. Blossey, *Phys. Rev.* **166**, 921 (1968); M. Cardona, *Modulation Spectroscopy*, Solid State Physics Supplement 11, ed. F. Seitz, D. Turnbull and H. Ehrenreich, Academic Press (1969). A number of review papers on various aspects of reflectance modulation are to appear in *Semiconductors and Semimetals*, Vol. 8, Academic Press (1971).

[2]A. Kahan, L. Bouthillette, and W. G. Spitzer, *Bull. Am. Phys. Soc.* **14**, 326 (1969).

[3]J. I. Pankove, *Annales de Physique* **6**, 331 (1961).

energies much lower than the gap energy, the earliest observations[4] of reflectance modulation in germanium were made near or slightly below the gap energy. These were small reflectance changes (in the order of 1% or less) in response to injection near the reflecting outer surface. This modulation of the reflectance was interpreted in terms of a change in the complex index of refraction due to the effect of free carriers on the electrical susceptibility.[5] The surface layer with a dielectric constant somewhat different from that of the bulk acts as a reflecting or antireflecting dielectric layer the index and thickness of which are modulated by the injected carriers. A very high density of electron–hole pairs can be generated by the intense light of a Q-switched laser, imparting a metallic-type reflectance to the semiconductor.[6]

Here we shall consider how reflectance is affected by periodically perturbing the band structure at and near the surface of the semiconductor. Although in some cases the modulating process may also change the band structure in the bulk, the reflectance measurement primarily probes the surface. Parameters which can control reflectance are pressure, temperature, and electric field.

18-A Dependence of Reflectance on the Band Structure

The reflectance R is given by the following relationship:

$$R = \frac{(\boldsymbol{n} - 1)^2 + \boldsymbol{k}^2}{(\boldsymbol{n} + 1)^2 + \boldsymbol{k}^2} \tag{18-1}$$

where $\boldsymbol{n}$ is the real part of the index of refraction and $\boldsymbol{k}$ is the extinction coefficient. In Sec. 4-A (Eq. 4-7) we found that

$$\boldsymbol{k} = \frac{c\alpha}{4\pi\nu} \tag{18-2}$$

where α is the absorption coefficient. Near and below the energy gap, $\boldsymbol{k}$ is usually small compared to $\boldsymbol{n} - 1$; therefore, R is dominated by $\boldsymbol{n}$:

$$R = \left(\frac{\boldsymbol{n} - 1}{\boldsymbol{n} + 1}\right)^2 \tag{18-3}$$

In Sec. 4-C we saw that the Kramers–Kronig relation for the index of refraction is

$$\boldsymbol{n}(E) - 1 = \frac{ch}{2\pi^2} P \int_0^\infty \frac{\alpha(E')}{(E')^2 - E^2}\, dE'$$

where P means the Cauchy principal value of the integral. Integration by parts leads to

[4]I. Filinski, *Phys. Rev.* **107**, 1193 (1957).
[5]L. Sosnowski, *Phys. Rev.* **107**, 1193 (1957).
[6]M. Birnbaum and T. L. Stocker, *Bull. Am. Phys. Soc.* **9**, 729 (1964).

$$n(E) - 1 = \frac{ch}{4\pi^2} P \int_0^\infty \ln \frac{1}{(E')^2 - E^2} \frac{d\alpha(E')}{dE'} dE' \qquad (18\text{-}4)$$

When the logarithmic factor becomes large, the integral is governed by the derivative term. Now, $\alpha(E')$ depends on the optical joint density of states and becomes large for transitions in the neighborhood of critical points. Critical points are defined as regions where $\nabla_k[E_c(k) - E_v(k)] = 0$. The simplest critical points are those for which

$$\frac{dE_c(k)}{dk} = \frac{dE_v(k)}{dk} = 0$$

There are four general types of such critical points in the $E(k)$ distributions that can be linked by optical transitions. These are tabulated in Fig. 18-1, which shows the shape of the $E(k)$ dependence along each direction in three-dimensional momentum space. The corresponding optical joint density of states are illustrated in Fig. 18-2.

Since $\alpha(E')$ is proportional to the joint density of states, $d\alpha(E')/dE'$ in Eq. (18-4) will be large and positive above E_0 and below E_1, and it will be large

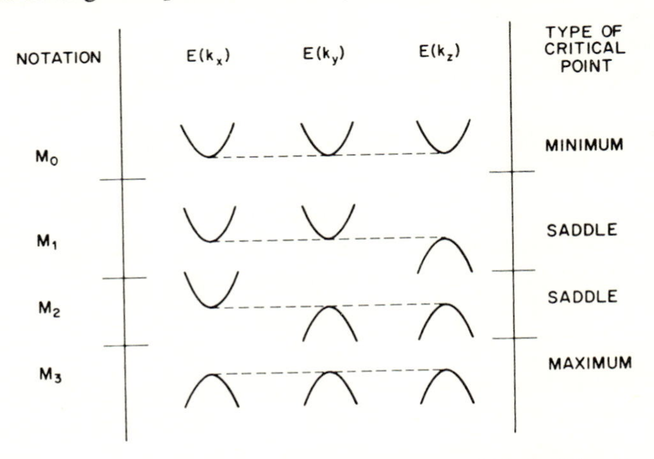

Fig. 18-1 Critical points for which $\nabla_k E(k) = 0$.

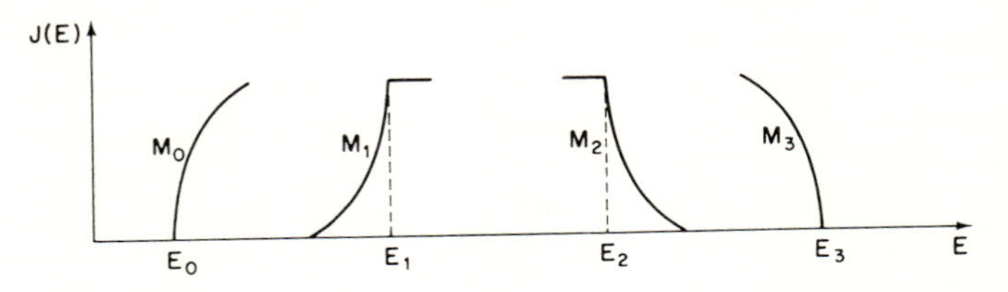

Fig. 18-2 Optical joint density of states distributions for the four main types of critical points.

and negative above E_2 and below E_3. Then, the index of refraction in Eq. (18-4) will go through a maximum at E_0 and E_1 and a minimum at E_2 and E_3. We can obtain further insight into the significance of the maxima and minima in terms of the distribution of states: direct transitions between parabolic bands across an energy gap E_g give an $\alpha(E)$ proportional to $(E - E_g)^{1/2}$. At this M_0 critical point, $d\alpha(E)/dE$ is proportional to $(E - E_g)^{-1/2}$, which has a square-root singularity at E_g. On the other hand, for forbidden transitions where $\alpha(E)$ is proportional to $(E - E_g)^{3/2}$ and for indirect transitions, where $\alpha(E)$ is proportional to $(E - E_g)^2$, the derivative $d\alpha(E)/dE$ goes gradually to zero at the critical point. When excitons occur, their joint distribution of states is narrow; therefore, excitons are associated with a large $d\alpha(E)/dE$.

Hence the index of refraction exhibits a structure whenever $d\alpha(E)/dE$ goes through a large maximum or minimum. The reflectance spectrum replicates the structure of the index of refraction.

It is possible to enhance the structure in the refractive index and, therefore, in the reflectance spectrum by periodically modulating a parameter which affects the joint density of states. A synchronous detection of the modulated reflection extracts the time-dependent portion of the reflection from the steady-state portion. By way of example, let us consider the effect of modulating the electric field.

The index of refraction at some field $\mathscr{E}$ is given by[7]

$$\boldsymbol{n}(E, \mathscr{E}) - 1 = \frac{c\hbar}{\pi} \int_0^\infty \frac{\alpha(E', 0)}{(E')^2 - E^2}\, dE' + \frac{c\hbar}{\pi} \int_0^\infty \frac{\Delta\alpha(E', \mathscr{E})}{(E')^2 - E^2}\, dE' \qquad (18\text{-}5)$$

where the first integral gives $\boldsymbol{n}(E, 0) - 1$, which determines $\boldsymbol{n}(E)$ at zero field. Hence the change in $\boldsymbol{n}(E)$ due to the applied field is

$$\Delta n(E, \mathscr{E}) = \frac{c\hbar}{\pi} \int_0^\infty \frac{\Delta\alpha(E', \mathscr{E})}{(E')^2 - E^2}\, dE' \qquad (18\text{-}6)$$

The variations of the real and imaginary parts of the dielectric constant, ϵ_1 and ϵ_2 respectively, as a function of photon energy $E = \hbar\omega$ are shown in Fig. 18-3 for zero field and for a finite field. The field-induced change in the dielectric constant can be visualized readily for the different types of critical points. The variations of $\Delta\boldsymbol{n}$ and $\Delta\alpha$ of Eq. (18-6) are similar to those of $\Delta\epsilon_1$ and $\Delta\epsilon_2$, respectively, of Fig. 18-3.

Recall that in Sec. 2-C-2 we saw the Franz–Keldysh broadening of a discrete level. In the case of a band of states in a homogeneous field, the probability of finding a particle in a field-broadened system is described by an Airy function.[9] The Airy function is characterized by an exponential tailing on one side of the zero-field band edge and, on the other side, an oscil-

[7] B. O. Seraphin and N. Bottka, *Phys. Rev.* **139**, A560 (1965).
[9] L. D. Landau and E. M. Lifshitz, *Quantum Mechanics*, Pergamon (1965), p. 73.

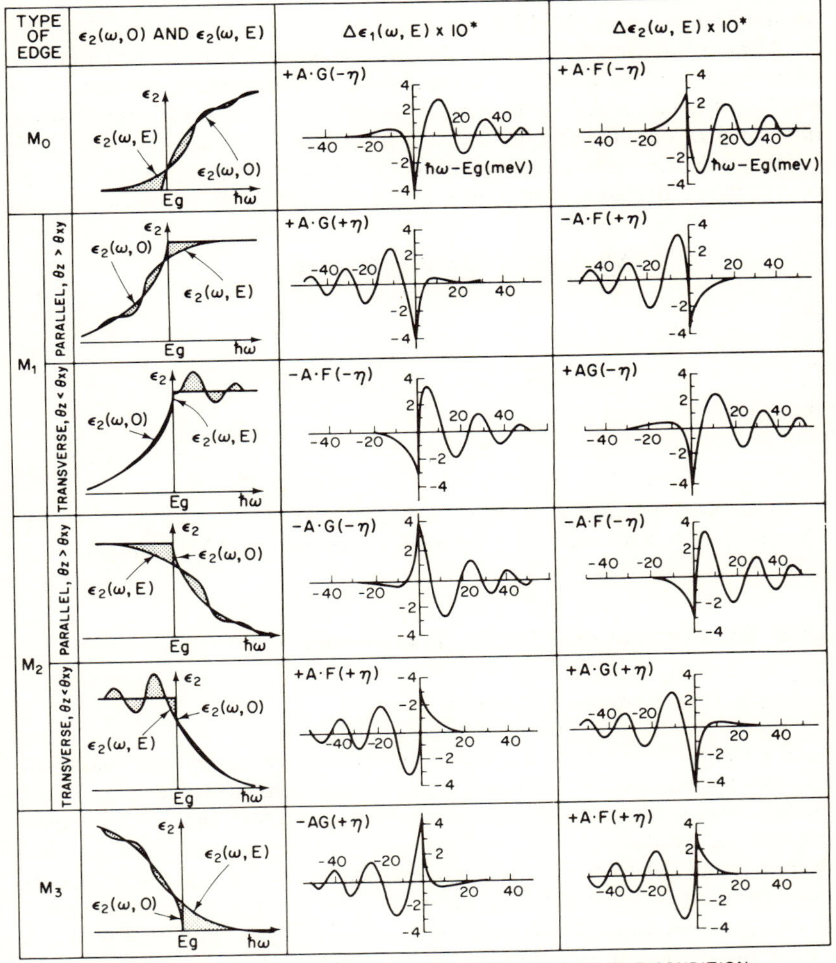

*LINE SHAPES OF $\Delta\epsilon_1(\omega, E)$ AND $\Delta\epsilon_2(\omega, E)$ CALCULATED AT THE CONDITION: $\hbar\theta = 10$ meV, $E_g = 0.8$ eV AND $B = 1$, HERE $\eta = (\hbar\omega - E_g)/\hbar\theta$, $A = (B - \theta^{1/2})/\omega^2$

Fig. 18-3 A summary of the field-induced change in the dielectric constant at the various types of edges.[8]

[8] Y. Hamakawa, P. Handler, and F. A. Germano, *Phys. Rev.* **167**, 709 (1968).

lation the fluctuations of which decrease in amplitude and period. These fluctuations can be seen in Fig. 18-3. However, when the electric field is not homogeneous, the interpretation of the reflectance data is more difficult, as we shall point out in Sec. 18-B-1.

Although all the energies contribute to the integral of Eq. (18-6), only the changes in α in the vicinity of critical points are important. Transitions at much higher and much lower energies make negligible contributions to the integral at E, since the absolute value of the denominator becomes very large.

When the Kramers–Kronig relation is used to calculate the relative change in reflectance due to a change in n and α, the following expression is obtained:

$$\frac{\Delta R}{R} = A\frac{c\hbar}{\pi}\int_0^\infty \frac{\Delta\alpha(E')\,dE'}{(E')^2 - E^2} + \frac{c\hbar}{2E}B\Delta\alpha(E) \tag{18-7}$$

where

$$A = \frac{n^2 - k^2 - 1}{[(n+1)^2 + k^2][(n-1)^2 + k^2]}$$

and

$$B = \frac{8nk}{[(n+1)^2 + k^2][(n-1)^2 + k^2]}$$

In germanium,[7] $\Delta R/R = 0.267\Delta n + 0.0022\Delta k$. In GaAs also,[10] $A \gg B$. Therefore, $\Delta R/R$ is much more sensitive to Δn than to $\Delta k = \lambda\Delta\alpha/4\pi$.

18-B Reflectance-modulation Techniques

18-B-1 ELECTROREFLECTANCE

An electric field can be applied along the surface of the semiconductor. For this purpose, an ac voltage is applied across two electrodes evaporated on the front surface so that an electric field appears in the uniform space between the parallel edges of the electrodes.[11] If the thickness of the material is thin compared to the interelectrode spacing, the field is approximately uniform through the crystal. In this case, the reflectance data can be compared to the transmission data under identical field conditions. Note that with this configuration, it is possible to vary the polarization of the incident radiation with respect to the direction of the modulating electric field.

Usually, however, the electric field is oriented transversely to the surface —i.e. nearly parallel to the direction of light propagation. For example, a

[10] W. E. Engeler, H. Fritzsche, M. Garfinkel, and J. J. Tiemann, *Phys. Rev. Letters* **14**, 1069 (1965).

[11] R. A. Forman and M. Cardona, "Exciton Electroreflectance in II–VI Compounds," *II–VI Semiconducting Compounds*, ed. D. G. Thomas, Benjamin (1967), p. 100.

transparent electrode spaced from the surface by a 10-μm thick layer of mylar can be used to apply the alternating electric field.[12] The opposite facet is made rough to reduce reflections from the rear surface of the specimen. This technique is capable of detecting relative changes in reflectance, $\Delta R/R$, as low as 5×10^{-6}.

Considerable care must be applied to the interpretation of electroreflectance data. When the field is applied transversely to the surface, the field inside the semiconductor is not uniform. This nonuniformity has a profound effect on the spectral shape to be expected due to averaging effects and to mixing of the real and imaginary parts of the dielectric function.[13] Figure 18-4 shows how the oscillatory portion of the dielectric functions is strongly

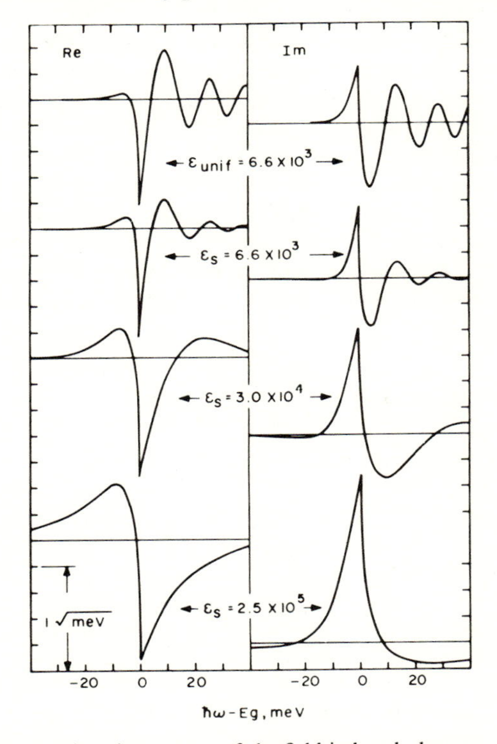

Fig. 18-4 Real and imaginary parts of the field-induced change in the dielectric function for various fields at the fundamental direct gap in Ge. The top pair of curves represent the uniform field approximation. Fields in V/cm; $\langle \Delta\epsilon_1 \rangle$ proportional to $\Delta R/R$ on the left side and $\langle \Delta\epsilon_2 \rangle \simeq (nc/\omega) \langle \Delta\alpha \rangle$ on the right side.[13]

[12] B. O. Seraphin, *Proc. Int. Conf. on Semiconductor Phys.* (Paris), Dunod (1964), p. 165.
[13] D. E. Aspnes and A. Frova, *Solid State Comm.* 7, 155 (1969).

affected by the nonuniformity of the electric field (the two upper curves correspond to a uniform field). An experimental verification of the effect of a nonuniform field on the electroreflectance spectrum has confirmed the need to be concerned about field nonuniformity.[14]

In view of the averaging process inherent to synchronous-detection techniques, it is important that the electric field be modulated as a square wave varying between two fixed values so that this differential technique results in a spectrum corresponding to a fixed and preferably uniform field.

A simpler technique for applying an electric field transversely to the surface consists in immersing the semiconductor in an electrolyte such as a weak solution of KCl in water.[15] An ac voltage is applied between the specimen and a platinum electrode immersed in the electrolyte. Radiation is incident on the sample through a window in the cell. The short-wavelength cutoff of water is 1700 Å; at the other end of the spectrum, a wavelength of 2.6 μ has been used in a system with a thin aqueous solution or with an organic electrolyte free of OH radicals.[16] This simple technique allows the sample to be cleaved in the electrolyte without being exposed to air. Stresses can be readily applied for the purpose of studying deformation effects on the band structure.[17] The use of electrolytes is restricted, however, to temperatures greater than 200°K.

Note that surface states and the nature of adsorbed impurities determine the surface potential and, consequently, the electric field at and near the sur-

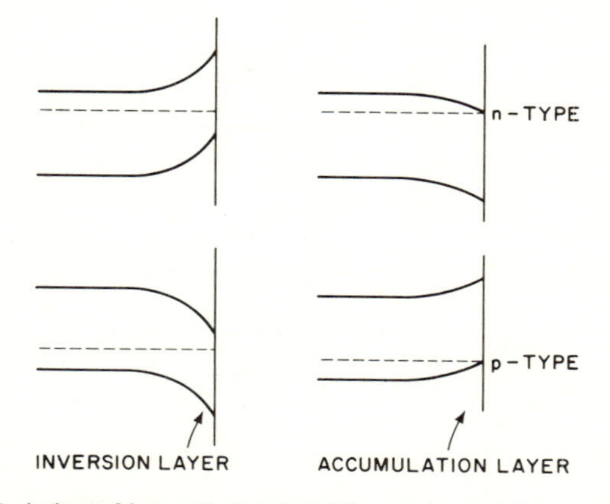

Fig. 18-5 Variation of internal electric field at various surfaces. The electric field is the slope of the band edges.

[14] A. Frova and D. E. Aspnes, *Phys. Rev.* **182**, 795 (1969).
[15] K. L. Shaklee, F. H. Pollak, and M. Cardona, *Phys. Rev. Letters* **15**, 883 (1965).
[16] M. Cardona, F. H. Pollak, and K. L. Shaklee, *J. Phys. Soc. Japan* **21**, Supplement, 89 (1966).
[17] F. H. Pollak, M. Cardona, and K. L. Shaklee, *Phys. Rev. Letters* **16**, 942 (1966).

face, as illustrated in Fig. 18-5. Applying an external potential induces a surface charge which can accentuate or reduce the band bending or even reverse the polarity of the surface field. Because *n*- and *p*-type semiconductors can have opposite band bendings at the surface, the peaks in the differential-reflectance spectrum can have opposite polarities in the two types of materials.

The ideal case of measurement from the flat-bands condition can be nearly achieved by applying a dc bias so adjusted as to make the surface potential zero. The electroreflectance signal goes through a minimum when the bands have been flattened at the surface.[8]

18-B-2 OPTICAL MODULATION OF REFLECTANCE[18,19]

When the bands are bent, the surface potential can be varied by a high concentration of carriers generated at the surface. The carriers can be generated by optical excitation with a modulated secondary beam of strongly absorbed radiation. Optical injection flattens the bands, thus reducing the surface potential. Hence the maximum magnitude of the modulation depends on the initial band bending which, in turn, depends strongly on the ambient. This optical technique, sometimes called "photoreflectance," is somewhat less sensitive than electroreflectance ($\Delta R/R \approx 10^{-4}$ minimum), but has the advantage of not requiring any electrode. This method also permits an independent control of the temperature and of the ambient.

In order to avoid interference from the modulating light in the reflectance spectrum, the primary beam is produced by a monochromator and the reflected beam is received by a spectrometer which is driven in synchronism with the monochromator. Thus, both instruments being tuned to the same wavelength, the background of modulating secondary light is rejected.

18-B-3 CATHODOREFLECTANCE MODULATION

Note that since a high density of electron–hole pairs can be generated by electron bombardment, the surface potential can be modulated also by an electron beam. Hence electron bombardment can also be used to modulate the reflectance of a semiconductor;[20] we could call this measurement "cathodoreflectance." However, as we saw in Sec. 11-C, electron-beam excitation is not an efficient pair-generation method. Therefore, localized heating may result. The advantage of using an electron beam for modulating the reflectance

[8] Y. Hamakawa, P. Handler, and F. A. Germano, *Phys. Rev.* **167**, 709 (1968).

[18] E. Y. Wang, W. A. Albers, and C. E. Bleil, "Light-modulated Reflectance of Semiconductors," *II–VI Semiconducting Compounds*, ed. D. G. Thomas, Benjamin (1967), p. 136.

[19] R. E. Nahory and J. L. Shay, *Phys. Rev. Letters* **21**, 1569 (1968).

[20] J. H. McCoy and D. B. Wittry, *Appl. Phys. Letters* **13**, 272 (1968).

seems more promising for the wide-gap materials, where optical excitation is not convenient. Another advantage of the electron beam is the relatively greater flexibility of combining electron optics with light optics compared to the simultaneous use of two separate optical systems required by the photoreflectance method discussed in the previous section. One limitation of the cathodoreflectance technique is the need to have vacuum as the ambient.

18-B-4 PIEZOREFLECTANCE MODULATION[10,21]

An alternating strain is induced in the crystal while it is probed with polarized light. The strain is a tensor which has different components along the various crystallographic directions. Thus it is possible to bring out the symmetry properties of the crystal at various critical points.

Figure 18-6 shows a single-crystal silicon bar driven by a quartz trans-

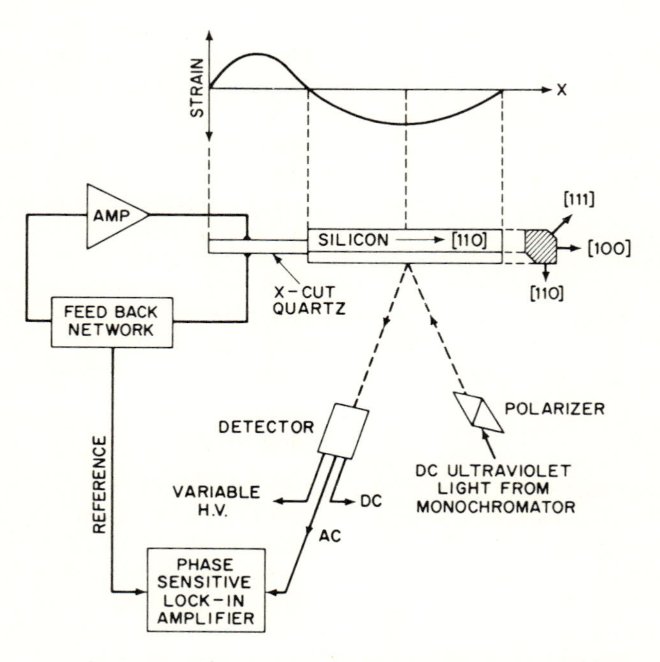

Fig. 18-6 Diagram of the ac piezoreflectance experiment.[21]

[10]W. E. Engeler, H. Fritzsche, M. Garfinkel and J. J. Tiemann, *Phys. Rev. Letters* **14**, 1069 (1965).

[21]G. W. Gobeli and E. O. Kane, *Phys. Rev. Letters*, **15**, 142 (1965).

ducer. The bar is cut along the [110] direction because this axis runs parallel to the three major crystallographic planes—(111), (110), and (100). Thus the relative change in reflectance from the different facets can be studied.

Although electroreflectance is a simpler technique to give a sharp structure, piezoreflectance provides a significantly different type of result. Piezoreflectance yields an oriented derivative of the reflectance spectrum with respect to the photon energy; the stress shifts the energy gap without appreciably changing the distribution of states in the bands. Electroreflectance, on the other hand, probes a field-induced change in the shape of the density-of-states distribution (Franz–Keldysh effect).

18-B-5 THERMOREFLECTANCE MODULATION[22]

A thin sample of a semiconductor presents a small thermal inertia. Therefore, its temperature can be periodically modulated at a low frequency ($\sim$100 Hz). Rapid heating can be obtained by passing current through the sample or by mounting the thin specimen on a low-inertia heater.

The bandgap shifts with the temperature, causing a change in the energy of critical points. However, because the temperature coefficients of various critical points are different, the structure in the reflectivity spectrum will include the differences in the temperature coefficients as well as the joint distribution of states of each critical point.

In contrast to the previously described techniques for modulating the reflectance in which the properties of the semiconductor were changed along a specific direction—that of the field or of the strain—in thermoreflectance the change is isotropic.

18-B-6 WAVELENGTH MODULATION

Wavelength modulation is the simplest modulation scheme. It can be obtained by vibrating the exit slit of a monochromator across the dispersed spectrum[23] or by oscillating a mirror in front of the exit slit to achieve the same result but without moving the image of the slit.[24] However, wavelength modulation describes only the derivative of a static-reflection spectrum, whereas in reflectance modulation some internal parameter, such as the joint density of states, is modulated. Therefore, it appears that the use of several modulation schemes on the same material will yield complementary bits of information.

[22] B. Batz, *Solid State Comm.* **4**, 241 (1966); A. Balzarotti and M. Grandolfo, *Solid State Comm.* **6**, 815 (1968).
[23] I. Balslev, *Phys. Rev.* **143**, 636 (1966).
[24] R. Braunstein, P. Schreiber, and M. Welkowsky, *Solid State Comm.* **6**, 627 (1968).

18-C Some Results

An example of the structure obtainable by electroreflectance in germanium is shown in Fig. 18-7. The 0.798-eV peak occurs slightly below the edge of the direct-energy gap. The much lower peak at 1.09 eV corresponds to the direct transition from the split-off valence subband V_3. The 2.109-eV and 2.322-eV peaks are attributed to transitions at the Λ point along the $\langle 111 \rangle$ direction from the V_1 and V_3 subbands, respectively, which are split by only 0.2 eV at this position in the Brillouin zone. The structure between 3 and 4 eV has been tentatively attributed to a quadruplet transition at the $\Gamma_{25'}$–Γ_{15} point of crystal symmetry. In a separate study, the first electroreflectance peak of germanium was found to depend on the ambient and to be affected by a dc bias superimposed on the modulating fields.[25] Note that most of the early data (obtained before 1969) may need reinterpretation to take into account the nonuniformity of the electric field. The effect of the ambient on the reflectivity of CdS at the fundamental edge is illustrated in Fig. 18-8. Here, the modulation is generated by optical excitation (photoreflectance), which drives the surface potential between a value fixed by the ambient and a lower-field condition determined by the photovoltage developed across the surface barrier.

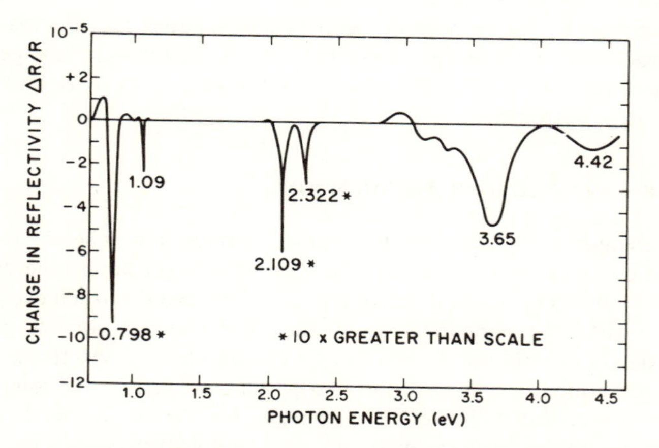

Fig. 18-7 Relative change of the reflectivity, $\Delta R/R$, as a function of photon energy. The recorder trace going negative indicates a decrease of the reflectivity caused by the positive half-wave of the modulating ac field. For three peaks marked *, multiply the scale by 10.[25]

[25] B. O. Seraphin and R. B. Hess, *Phys. Rev. Letters* **14**, 138 (1965).

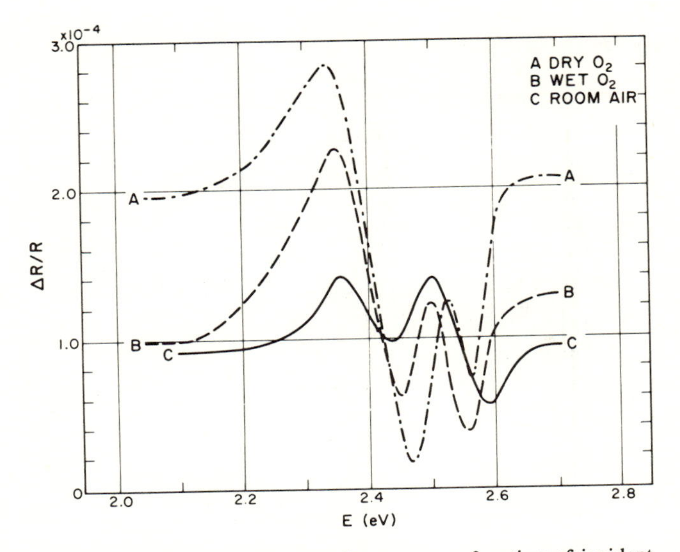

Fig. 18-8 Light modulated reflectance as a function of incident photon energy for hexagonal CdS at room temperature and various ambient gases; light direction parallel to c-axis.[26]

The structure in the electroreflectance spectrum[27] for GaAs in Fig. 18-9 has been labeled to show the corresponding transitions in the band diagram of Fig. 18-10. All these peaks shift with temperature; their intensities have different nonlinear dependences on the amplitude of the driving modulation.[28] The lowest peak at 1.38 eV (30 meV lower than the energy gap at room temperature) is attributed to an impurity level, probably an acceptor. The 1.42-eV peak corresponds to transitions at the direct gap. This peak is replicated at 1.77 eV corresponding to a transition from the split-off valence sub-band ($E_0 + \Delta_0$). Another set of peaks occurs at E_1 and $E_1 + \Delta_1$, which correspond to the $\Lambda_{3v} - \Lambda_{1c}$ transitions from the V_1 and V_2 subbands, respectively. The temperature dependence of all these peaks can be followed in Fig. 18-11 and can be compared with the known temperature dependence of the energy gap obtained by absorption. In addition to the relative temperature dependences of the various edges, the data of Fig. 18-11 yields values for the spin–orbit splitting of the valence band at several crystal symmetry points. Thus Δ_0 (at Γ) $= 0.348 \pm 0.002$ eV and Δ_1 (at Λ) $= 0.232 \pm 0.002$ eV.[28]

The main difference in the band structures of GaAs and GaP is that in GaP the Γ_1-minimum of the conduction band lies above the X_1-minimum,

[26] E. Y. Wang, W. A. Albers, Jr., and C. E. Bleil, "Light-modulated Reflectance of Semiconductors," *II-VI Semiconducting Compounds*, ed. D. G. Thomas, Benjamin (1967) p. 136.

[27] A. G. Thompson, M. Cardona, K. L. Shaklee, and J. C. Woolley, *Phys. Rev.* **146**, 601 (1966).

[28] B. O. Seraphin, *J. Appl. Phys.* **37**, 721 (1966).

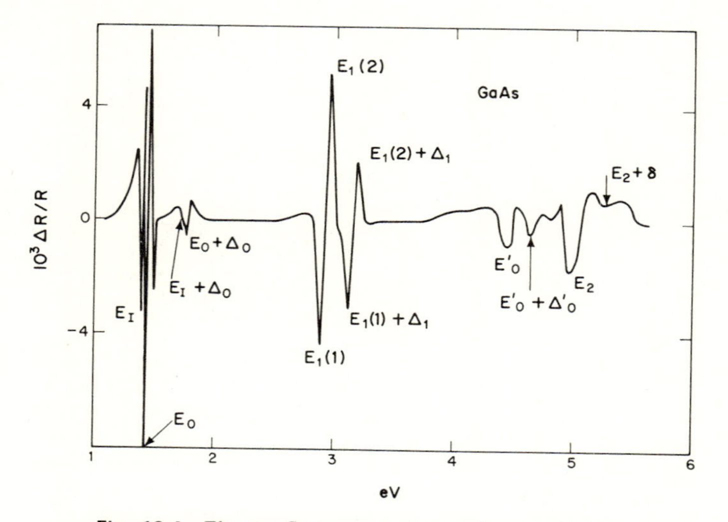

Fig. 18-9 Electroreflectance spectrum of n-type GaAs at room temperature. The dc bias was 1.5 V; the ac modulating voltage, 2.4 V peak-to-peak.[27]

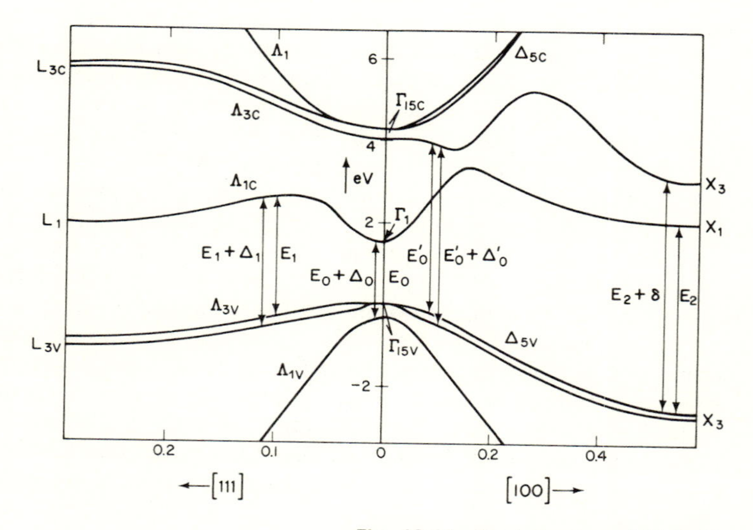

Fig. 18-10 Band structure of GaAs.[28]

[27] A. G. Thompson, M. Cardona, K. L. Shaklee, and J. C. Woolley, *Phys. Rev.* **146**, 601 (1966).

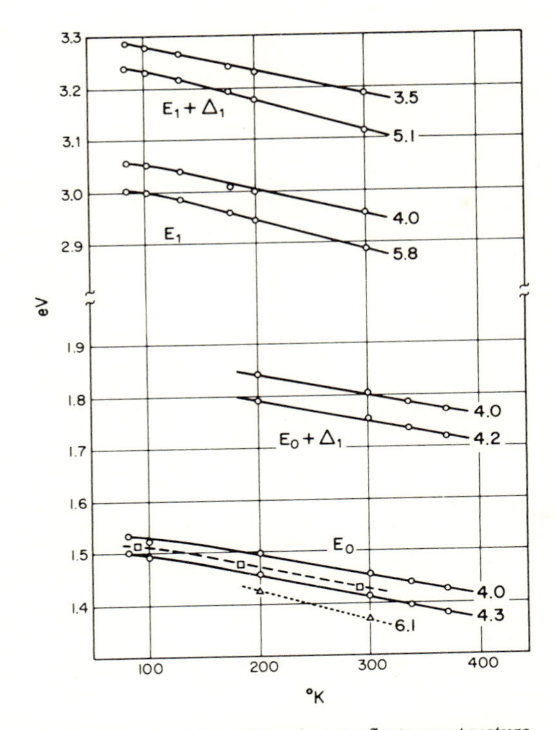

Fig. 18-11 The energy position of the electroreflectance structure as a function of temperature. The three squares connected by a dotted line give the position of the edge according to absorption data.[29] The negative temperature coefficients are given on the right, in units of 1×10^{-4} eV/deg.[28]

making the semiconductor an indirect-gap material. Hence except for the different energies separating the conduction and valence bands at points of symmetry, the two materials, GaAs and GaP, have very similar electroreflectance spectra in which the structure can be identified (Fig. 18-12).

In order to verify the assignment of the peaks, alloys of $GaAs_{1-x}P_x$ with various compositions x were examined by electroreflectance.[27] This allows one to follow the various peaks as the band structure is altered gradually from that of GaAs to that of GaP. This experiment has revealed that, at room temperature, the direct-gap E_0 varies with a small quadratic component as a

[28]B. O. Seraphin, *J. Appl. Phys.* **37**, 721 (1966).
[29]M. D. Sturge, *Phys. Rev.* **127**, 768 (1962).

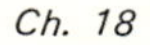

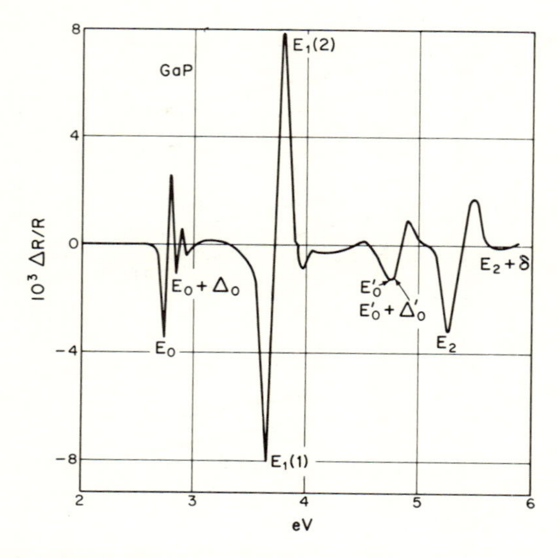

Fig. 18-12 Electroreflectance spectrum of n-type GaP at room temperature. The dc bias was 1.5 V; ac modulation 3 V peak-to-peak.[27]

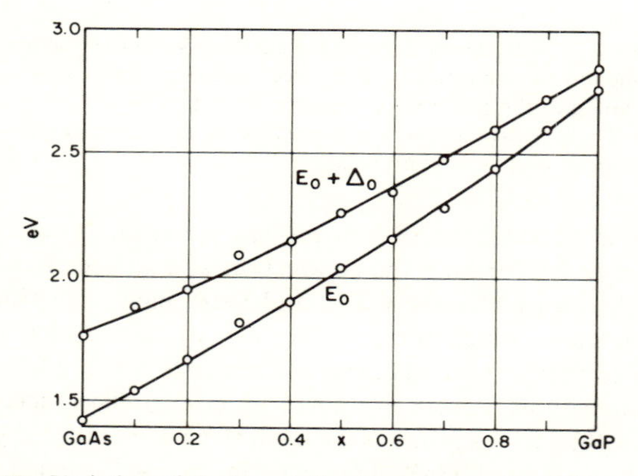

Fig. 18-13 Variation of the lowest direct energy-gap E_0 and its spin-orbit split component $E_0 + \Delta_0$ with concentration x for $GaAs_{1-x}P_x$ obtained from photoreflectance measurements. The points are experimental. The curves represent the expressions $E_0 = 0.210x^2 + 1.09x + 1.441$ eV and $E_0 + \Delta_0 = 0.182x^2 + 0.884x + 1.776$ eV.[27]

function of composition instead of linearly as previously assumed. Thus

$$E_0 = 0.210x^2 + 1.091x + 1.441 \text{ eV}$$

as shown in Fig. 18-13. The spin–orbit splitting varies linearly with composition from 0.33 eV in GaAs to 0.10 eV in GaP.

Although much new information has been gleaned from reflectance-modulation measurements, the interpretation of the experimental data is complicated by many difficulties which have not yet been completely resolved. Thus exciton effects should be taken into account.[30] The electric field shifts the excitons to lower energies (Franz–Keldysh effect) and reduces their contribution (exciton ionization). The bands are usually assumed to be perfectly parabolic; if the exact shape were known, corrections could be attempted. Furthermore, measurements involving surface potentials such as photoreflectance and transverse electroreflectance deal with nonuniform fields. It is difficult to assign weights to the contributions from regions of different fields. Also, the electro-optical effect being nonlinear, the waveform of the optical signal is a distorted version of the driving electrical force. Since the carriers are generated by photons having energies higher than the gap energy, the carriers decay to lower energies in a very short time; hence by uncertainty principle, the energy is less accurately known (lifetime broadening).[31]

[30] J. D. Dow and D. Redfield, *Phys. Rev.* **B4** (1970); also H. I. Ralph, *J. Phys. C.* (Proc. Phys. Soc.) **1**, 378 (1968).

[31] P. T. Landsberg, *Proc. Phys. Soc.* **A62**, 806 (1949).

APPENDIX

I TABLE OF SYMBOLS

II PROPERTIES OF SEMICONDUCTORS

III NOMOGRAPH OF THE TEMPERATURE DEPENDENCE OF THE FERMI LEVEL IN A DEGENERATE PARABOLIC BAND

IV PHYSICAL CONSTANTS

TABLE OF SYMBOLS

a	Lattice constant; Bohr orbit
c	Velocity of light in vacuum
D	Diffusion coefficient
E	Energy
E_c, E_v	Energies of conduction- and valence-band edges
E_g	Gap energy
E_i	Ionization energy; energy of initial state
E_A, E_D, E_X	Binding energies of acceptor, donor, and exciton
E_F	Fermi level
E_f	Energy of final state
E_{F_n}, E_{F_p}	Quasi-Fermi levels for electrons and for holes
E_0	Energy needed to create an electron-hole pair
E_p	Phonon energy
$\mathbf{E}_p$	Energy of primary electron
E_T	Threshold for photoelectric emission
E_t	Energy of trapping level
$\mathcal{E}$	Electric field
F	Force
f	Fermi function
g	Gain
G	Generation rate
h	Planck's constant
$\hbar$	Dirac's constant
i, I	Current
I_0	Saturation current
I_{sc}	Short circuit current
j, J	Current density
k	Boltzmann's constant; wave vector; momentum vector
$\mathbf{k}$	Extinction coefficient
K	Momentum vector

l	Length
L	Light intensity
L_e	Diffusion length for electrons
L_h	Diffusion length for holes
m	Mass of free electron
m_e^*	Effective mass of electron
m_h^*	Effective mass of hole
$\mathbf{n}_c$	Complex index of refraction
$\mathbf{n}$	Real part of index of refraction
n	Electron concentration
N	Concentration
N_c, N_v	Densities of states in conduction and valence bands
N_A, N_a; N_D, N_d	Concentrations of acceptors and donors
N_i	Concentration of impurities
N_p	Phonon density
N_T, N_t	Concentration of traps
P	Probability, pressure, power
q	Electron charge
Q	Total charge
r	Radius
R	Reflection coefficient; recombination rate; resistance
t	Time
T	Temperature, transmission
v	Velocity
V	Voltage, volume
V_D	Dember voltage
V_{oc}	Open circuit voltage
α	Absorption coefficient
ϵ	Dielectric constant
Φ	Work function
Φ_B, Φ_b	Barrier height
η	Efficiency
κ	Specific heat
μ	Mobility
ν	Frequency
ω	Angular frequency
ρ	Resistivity; density of states
σ	Conductivity; capture cross-section
τ	Time constant
θ	Angle
χ	Susceptibility; electron affinity
ξ	Energy of the Fermi level with respect to band edge

PROPERTIES OF SEMICONDUCTORS[1]

II

	Energy gap		Lowest conduction-band minimum, direct or indirect	$\left(\frac{dE_g}{dT}\right) \times 10^4$ (300°K) eV/°K	$\left(\frac{dE_g}{dP}\right)_T \times 10^6$ eV/bar	Effective mass		Refractive index n	Static dielectric constant ϵ	Lattice constant a Å	Mobility	
	E_g (0°K) eV	E_g (300°K) eV				m_e^*	m_h^*				μ_e cm²/V·sec	μ_h cm²/V·sec
IV Si	1.166	1.11	ind 100	−2.3	−1.5	m_l 0.98 m_t 0.19	0.52	3.44	11.7	5.43	1,350	480
IV Ge	0.74	0.67	ind 111	−3.7	5.0	m_l 1.58 m_t 0.08	0.3	4.00	16.3	5.66	3,900	1,900
IV α-Sn	−0.2‡		dir 000		5.0	0.02				6.489	2,000	1,000
IV–IV SiC α	3.0 (6H)	2.8–3.2†	ind	−3.3				2.69 ∥ c 2.65 ⊥ c	10.2	a 3.0817 c 15.1123	400	
IV–IV SiC β	2.68	2.2	ind							4.359		
VI Se	1.95	1.74	dir 0001	−14	−20		0.12	5.56 ∥ c 3.72 ⊥ c	8.5		1	
VI Te	0.334	0.32	dir 0001	−0.3	−19	0.038 ⊥	0.26 ∥ 0.10 ⊥	3.07 ∥ c 2.68 ⊥ c	5.0 ∥ c 2.2 ⊥ c		1,100	

III–V	BP		2	ind					2.6	6.9	4.538		
	AlP	2.5	2.43	ind 100	−3.5		0.13‡		3.0	9.8	5.462	80	
	AlAs	2.24	2.16	ind			0.5	m_l 1.06 m_t 0.49		12	5.66	1,000	~100
	AlSb	1.6	1.6	ind 100	−4	−1.6	0.11	0.39	3.4	11	6.135	50	400
	GaN	3.5	3.4	dir 0000	−4.8	4.2	0.2	0.8	2.4	12	*a* 3.18 *c* 5.16	300	
	GaP	2.4	2.25	ind 100	−5.4	−1.7	0.13	0.67§	3.37	10	5.450	120	120
	GaAs	1.520	1.43	dir 000	−5.0	11	0.07	0.5	3.4	12	5.653	8,600	400
	GaSb	0.81	0.69	dir 000	−4.1	12	0.045	0.39	3.9	15	6.095	4,000	650
	InP	1.42	1.28	dir 000	−4.6	4.6	0.07	0.40	3.37	12.1	5.8687	4,000	650
	InAs	0.43	0.36	dir 000	−3.3	5	0.028	0.33	3.42	12.5	6.058	30,000	240
	InSb	0.235	0.17	dir 000	−2.9	15	0.0133	0.18	3.75	18	6.4787	76,000	5,000 (78°K)
II–VI	ZnO		3.2	dir 0000	−9.5	0.6	0.32	0.27	2.02	7.9	*a* 3.2496 *c* 5.2065	180	
	ZnS α		3.8	dir 0000	−3.8	9	0.28	>1 // 0.5⊥	2.4	8.3	*a* 3.814 *c* 6.257		
	ZnS β		3.6	dir 000	−5.3	5.7	0.39		2.4	8.3	5.406		
	ZnSe	2.80	2.58	dir 000	−7.2	6	0.17		2.89	8.1	5.667	100	
	ZnTe	2.39	2.28	dir 000	−5	6	0.15		3.56	9.7	6.101		7
	CdS	2.58	2.53	dir 0000	−5	3.3	0.20	0.7⊥*c* 5 //*c*	2.5	8.9	*a* 4.136 *c* 6.713	210	
	CdSe	1.85	1.74	dir 0000	−4.6		0.13	2.5 //‡ 0.4⊥		10.6	*a* 4.299 *c* 7.010	500	
	CdTe	1.60	1.50	dir 000	−4.1	1.5	0.11	0.35	2.75	10.9	6.477	600	
	HgS		2.5										
	HgSe	−0.24	−0.15	dir 000			0.045			25	6.085	5,500	
	HgTe	−0.28	−0.15 0.14	dir 000	+5.6		0.029	~0.3	3.7	20	6.42	22,000	100 (20°K)
IV–VI	PbS	0.29	0.37	dir 111	+4	−7	0.1	0.1	3.7	170	5.936	550	600
	PbSe	0.15	0.26	dir 111	+4	−8	m_l 0.07 m_t 0.039	m_l 0.06 m_t 0.03		250	6.124	1,020	930
	PbTe	0.19	0.29	dir 111	+4	−9	m_l 0.24 m_t 0.02	m_l 0.3 m_t 0.02	3.8	412	6.460	1,620	750
	SnTe	0.3	0.18	dir 111							6.328		

[1]The data was gathered from many sources, primarily the following: C. Benoit-à-la-Guillaume et al., *Selected Constants Relative to Semiconductors*, Pergamon (1961); *Solids Under Pressure*, ed. W. Paul and D. M. Warschauer, McGraw-Hill (1963), p. 226; J. Tauc, *Progress in Semiconductors*, **9**, 120 (1965); D. Long, *Energy Bands in Semiconductors*, Wiley (1968); S. Shionoya, "Luminescence of Lattices of the ZnS Type," *Luminescence of Organic Solids*, ed. P. Goldberg, Academic Press (1966), p. 206; G. Giesecke, "Lattice Constants," *Semiconductors and Semimetals*, Academic Press **2**, 73 (1966); and a number of more recent papers. Where conflicting values appeared, the most recent data was used. Many parameters which depend on the purity of the semiconductor are subject to change as material technology improves.

†Depends on polytype.

‡Calculated value.

§C. F. Schwerdtfeger, *Solid State Commun.* **11**, 779 (1972).

NOMOGRAPH OF THE TEMPERATURE DEPENDENCE OF THE FERMI LEVEL IN A DEGENERATE PARABOLIC BAND[1]

III

In a simple parabolic band characterized by an effective mass m^* containing a concentration of n carriers/cm^3 at any temperature, the position of the Fermi level ξ with respect to the band edge is related to the temperature T by the expression

$$n = 4\pi\left(\frac{2m^*}{h^2}\right)^{3/2}\int_0^\infty \frac{E^{1/2}}{1 + \exp\dfrac{E-\xi}{kT}}\,dE \qquad \text{(III-1)}$$

where E is the carrier's energy, E and ξ both being measured with respect to the bottom of the parabolic band; k and h are Boltzmann's and Planck's constants, respectively. Equation (III-1) was solved by computer for various values of n, m^*, and T. For this purpose, Eq. (III-1) was expressed in the intermediate form

$$n = a(m^*)^{3/2}T^{3/2}f(\eta)$$

where

$$\eta = \frac{\xi}{kT}$$

The nomograph (Fig. A3–1) permits a graphical determination of any parameter n, m^*, ξ, or T when any three of them are known. For example. if n and m^* are known, they are joined by a straight line. From the intercept of this straight line with the X- axis one draws a straight line to the desired T to find a value η on the left-hand η-scale. Transferring this value to the right-hand η-scale, one strikes another straight line to T to find ξ. If η is negative, ξ lies outside the parabolic band (i.e., inside the energy gap).

[1]J. I. Pankove and E. K. Annavedder, *J. Appl. Phys.* **36**, 3948 (1965).

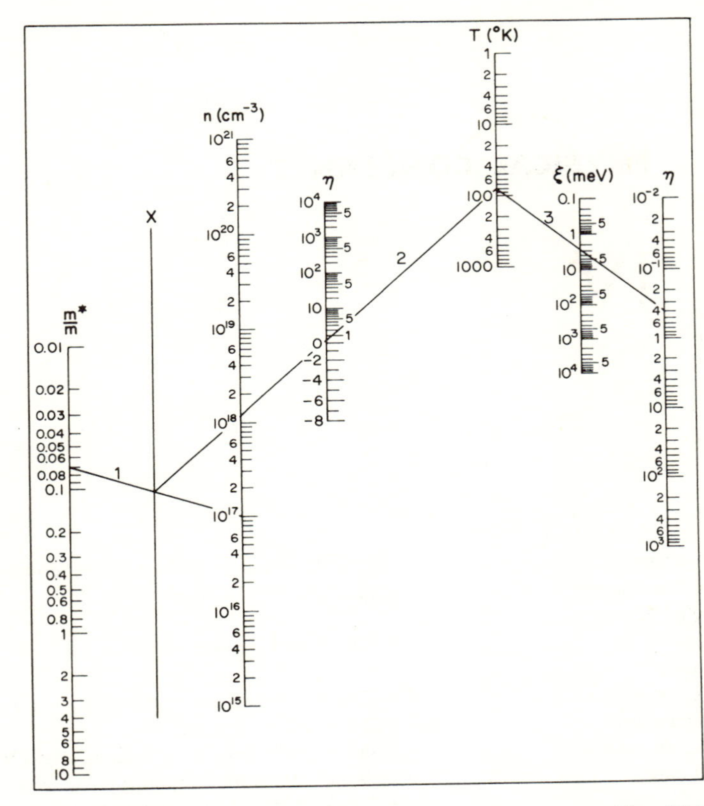

Fig. A3-1

By way of example, the diagram illustrates how one finds the position of the Fermi level ξ at $T = 80°K$ in n-type GaAs ($m_e^* = 0.07m$) having a carrier concentration $n = 1 \times 10^{17}$ cm^{-3}. The three steps labeled 1, 2, 3, lead to $\xi \approx 3$ meV.

Of course, this nomograph is valid only for the case of a temperature-independent free-carrier density and, therefore, it does not apply in the temperature range where carrier freeze-out may occur. Furthermore, a gross approximation is made by assuming that the band is parabolic over the entire range of energies over which the distribution function is appreciable. The neglect of band tailing and of nonparabolicity makes the ξ value determined by this nomograph a maximum value.

PHYSICAL CONSTANTS

IV

Electron charge	$q = 4.8 \times 10^{-10}$ esu $= 1.602 \times 10^{-19}$ coulomb
Mass of free electron	$m = 9.11 \times 10^{-28}$ g
Velocity of light in vacuum	$c = 2.998 \times 10^{10}$ cm/sec
Bohr radius	$a_0 = 5.29 \times 10^{-9}$ cm
Planck's constant	$h = 6.62 \times 10^{-27}$ erg·sec $= 4.5 \times 10^{-15}$ eV·sec
Dirac's constant	$\hbar = 1.054 \times 10^{-27}$ erg·sec
Boltzmann's constant	$k = 1.380 \times 10^{-16}$ erg/°K $= 8.62 \times 10^{-5}$ eV/°
Thermal energy	$kT = 25.9$ meV at room temperature
	$= 6.7$ meV at liquid-nitrogen temperature
	$= 0.36$ meV at liquid-helium temperature
Energy associated with 1 eV	$1 \text{ eV} = 1.602 \times 10^{-12}$ erg
Wavelength in vacuum associated with 1 eV	λ_0 (1 eV) $= 1.239 \times 10^{-4}$ cm
Wave number associated with 1 eV	ν_0 (1 eV) $= 8.06 \times 10^{4}$ cm^{-1}

INDEX

"*A*" center, 311
Abrupt junction, 172
Absorption, 34
  avalanche, 83
  band-to-band, 34
  coefficient, 34, 87
  exciton, 57
  free carrier, 74
  fundamental, 34
  hot electron assisted, 80
  impurity, 62
  intervalley, 71
  intraband, 71
  intravalley, 67
  lattice, 76
  modulation, 391
  self, 127
  tunneling assisted, 48
  two-photon, 268
  vibrational, 80
Acceptor, 8
Acoustoelectric effect, 278
Activation energy, 141
Active region, refractive index, 225
Active region thickness, 218
Adsorption, 354
Amplification of light, 261
Annealing, photo-induced, 365
Anomalous photovoltaic effect, 323
Anti-Stokes shift, 272
Auger effect, 161, 186
Avalanche, 247
  breakdown, 203
  luminescence, 207

Band-filling, 189
Band structure of GaAs, 85, 404
  band structure of GaP, 65
  band structure of Ge, 53
Band tails, 10, 43
Band-to-band transition, 34, 124
Barrier height determination, 315
Barrier induced optically, 331
Beadiness of lasers, 227
Bending of bands, 294, 354
β-Voltage, 320
Biaxial crystals, 340
Birefringence, 337
Bistable operation, 260
Bleaching, 367
Bose–Einstein statistics, 38
Bound exciton, 15, 61, 116
Bragg reflection, 275
Breakdown Luminescence, 207
Brillouin scattering, 271, 275, 286
Brillouin zone, 3
Bulk Excited lasers, 247
Burstein–Moss shift, 39

Capacitance of junction, 173
Capture cross-section, 82, 132, 372
Carrier concentration, 7
Cathodoluminescence, 135, 147, 252
Cathodo-reflectance, 399
Cesiated surface, 296
Circular polarization, 340
Close confinement laser, 226
Coherence, 217
Compensated semiconductor, 149
Conduction band, 2
Configuration diagram, 113, 166
Confinement of radiation, 226
Contrast, 95
Coulomb interaction, 17, 66, 143
Critical points, 13, 393
Cyclotron frequency, 30

Damage, 278
Damage, catastrophic in lasers, 234
Decay of luminescence, 370
Deep levels, 139
Deep transitions, 132
Defects, 80, 165
Deflection of light, 348
Deformation potential, 11
Degradation of lasers, 234, 277
Delay in lasers, 377
Dember effect, 320
Density of states, 6
Depletion layer, 170
Desorption, 354
Detectors of fast particles, 319
Detectors of radiation, 100
Deviation from stoichiometry, 8
Dichroism, 343
Dielectric constant, 81, 91
Dielectric constant, field induced change, 394
Dielectric ellipsoid, 342
Differential external quantum efficiency, 233
Diffusion equation, 321
Diode equation, 180, 303
Dipole layer, 170
Directionality, 220
Direct transitions allowed, 34, 124
Direct transitions forbidden, 36
Donor, 8, 63
Donor-acceptor pairs, 17, 66, 143
Donor-to-acceptor transitions, 143, 184
Doppler shift, 276
Double acceptor trap model, 385
Double peak laser, 244
Double refraction, 337

Effective mass, 3, 70, 91
Efficiency, 111, 166, 231
Electric field effects, 28, 60
Electron affinity, 288
Electron beam excitation, 252
Electro-optic coefficient, 346
Electroreflectance, 396
Elliptical polarization, 339
Emission:
  band-to-impurity, 131
  donor-to-acceptor, 143
  efficiency, 166
  impurity-to-band, 131
  intraband, 154
  tunneling assisted, 151
Energy gap, 2
Entropy of carriers, 193
Epitaxial growth, 360
Escape depth, 292
Etching, chemical, 361
Etching, electrolytic, 363
Excess current, 177, 182
Excitons, 12
  absorption, 57
  bound, 15, 61, 116
  complexes, 14
  free, 12, 114
  molecule, 122
  recombination, 114
  transport, 115
Extinction coefficient, 88
Extraordinary rays, 337

Fabry-Perot cavity, 217, 222
Faraday effect, 346
Far-field pattern, 222, 227
Fermi-Dirac function, 7
Field effects, 60, 83
Field ionization, 29
Field in junction, 174
Filaments, 227
Fluorescence, 108
Fracture luminescence, 390

Franck-Condon shift, 113, 140
Franz-Keldysh effect, 29, 46, 268
Free carrier absorption, 74
Free excitons, 12, 114
Frenkel pairs, 311, 365
Frequency mixing, 266, 270
Fresnel equation, 343
Fundamental transitions, 34, 114

$GaAs_{1-x}P_x$, 19, 156, 210, 405
GaAs band structure, 404
Gain, 217, 218
$Ge_{1-x}Si_x$, 19
Generation rate for pairs, 214
Geometrics for lasers, 234
Glow curve, 375
Graded gap, 197, 333
Graded junction, 173, 320
Growth of luminescence, 370
Gunn domains, 248

Harmonic generation, 265
Heavily doped semiconductors, 128, 134
Heterodyning, 266, 270
Heterojunctions, 197
High-voltage photovoltaic effect, 323
Hook junction, 284
Hot electron assisted absorption, 80
Hot electron effect, 129
Hot light hole emission, 154
Hydrostatic pressure, 22

Impact ionization, 247
Impact-luminescence, 388
Impurity absorption, 62
Impurity gradient, 187
Impurity levels, transitions to and from, 131
Index of refraction, 87, 88
Indirect transitions, 37, 126
Infrared quenching of luminescence, 377
Infrared stimulated luminescence, 376
Injection, 180, 242
Injection laser, 218, 222
Interference, 94
Internal reflection spectroscopy, 66, 358
Interstitial, 8
Inter-valley absorption, 71
Intraband transitions, 67, 71, 154
Inversion layer, 245
Inverter, 260
Ionization energy, 9, 120, 140
Ionization of impurities, 60
Isoelectronic traps, 61, 123

Junction:
  abrupt, 172
  capacitance, 173
  depletion layer, 170
  electric field, 174
  excess current, 177, 182
  floating, 284
  graded, 173, 320
  hetero-, 197
  hook, 284
  leakage current, 202

Kerr effect, electro-optic, 344
Kinetic energy, 3
Kramers–Kronig relations, 89, 392

Landau splitting, 30
Laser:
  delay effects, 377
  electron beam excited, 252
  gain, 217
  injection, 218, 222
  loss, 217
  optical pumping, 249
  temperature effects, 229
Laser-oscillator amplifier pair, 262
Laser-quencher pair, 260
Lasing criteria, 215
Lateral photoeffect, 331
Lattice absorption, 76
Leakage current, 202
Lifetime of carrier, 133
Lifetime of electron-hole pair, 162
Lifetime of photon, 108
Luminescence, 107
  decay, 370
  electron beam excited, 252
  fracture excited, 390
  growth, 370
  impact excited, 388
  IR quenching, 377

Luminescence (contd.):
IR stimulated, 376
optically excited, 249
quenching, 376
thermally stimulated, 371

Magnetic effects, 30, 238
Maxwell's equations, 88
Microplasmas, 204
Microwave generation, 271
MIS structure, 247
Mixing, 266
Mode spacing, 222
Modulation of light, 349
Modulation of reflectance, 391
Momentum vector, 3
Monochromator, 96
Moss rule, 89
Multiple internal reflections, 93
Multiple phonon emission, 167
Multiplying photo-detector, 206

Negative electron affinity, 296
Non-linear optics, 265
Non-radiative recombination, 160

Optical activity, 344
Optical excitation, 249
Optical joint density of states, 393
Optic axis, 338, 341
Ordinary rays, 337

Paired impurities, 143
Pair generation rate, 214
Pair interaction, 61
Particle-voltaic effect, 319
PbTe, 8
Penetration depth of electrons, 255
Performance of injection laser, 234
Phonon absorption, 37, 126
Phonon-assisted transitions, 136
Phonon emission, 37, 126, 132
Phonon-less transition, 139
Phonons, 17, 57
Phosphorescence, 107
Photoadsorption, 354
"Photoangular" effect, 330
Photoanodization, 363
Photocatalysis, 358
Photochemical effects, 352
Photochromism, 366
Photoconductivity, 202
Photodesorption, 354
Photoelastic effect, 348
Photoelectric emission, external, 287
Photoelectric emission, internal, 314
Photoelectric yield, 289
Photoengraving, 362, 363
Photoluminescence, 249
Photomagnetoelectric effect, 321
Photomechanical effect, 322
Photomultiplier, solid state, 206
Photon assisted tunneling, 48, 51, 177, 183
Photon doubling, 252, 265
Photon-phonon interactions, 271
Photon-photon interactions, 258
Photopiezoelectric effect, 333
Photo-reflectance, 399
Photovoltaic current, 304
Photovoltaic effect, anomalous, 323
Photovoltaic quantum yield, 307
Photovoltaic spectrum, 305
Piezo-reflectance, 400
Planck's distribution, 214
Plasma oscillations, 57
Plasma resonance, 92
Plasmons, 17
Plating, 363
Pleochroism, 343
p-n junction, 170
p-n junction, floating, 284
Pockels effect, 345
Poisson's equation, 171
Polariton, 16, 116, 275
Polarizability, 91
Polaron, 17
Population inversion, 216
Power dissipation in lasers, 233
Pressure effects, 22, 239
Principal plane, 339

Q-switching, internal, 386
Quantum efficiency for coherent emission, 235
Quarter-wave plate, 340
Quasi-homojunction, 197

Quaternary alloys, 20
Quenching of laser, 259
Quenching of luminescence, 376

Radiation damage, 182, 311
Radiation-ionization energy, 256
Radiation pattern, 222
Radiative transition, 107
Raman scattering, 271, 273
Ray axis, 341
Recombination rate, 112, 214
Reduced mass, 12
Reflectance, 90
Reflectance modulation, 391
Reflectance spectrum, 57
Reflectivity, 90, 92, 392
Refractive index, 87
Refrigeration by radiation, 193
Relaxation time, 92
Resonant absorption, 161
Reverse bias current, 201

Saddle point, 393
Saturable absorber, 261, 382
Saturation current, 201, 303
Scanner, 285
Scattering, 40
Schottky barrier, 198, 312
Self-absorption, 127
Semiconducting alloys, 18
Semimetal, 20
Shallow transitions, 131
Shifting emission peak, 178, 184
Shrinkage of energy gap, 40
Solar cell, 307
Spectrograph, 96
Spectrometers, 96
Spectroscopy, 96
Spin-orbit interaction, 52, 67, 70, 406
Spontaneous recombination, 213
Stacking fault, 327
Stark effects, 28, 49
Stationary peak, 186
Stimulated Brillouin scattering, 276
Stimulated Raman scattering, 274
Stimulated recombination, 213
Stoichiometry, 8
Stokes shift, 113, 272
Strain-excited luminescence, 387
Strain-induced birefringence, 347
Strain-stimulated luminescence, 388
Suhl effect, 155
Superradiance, 217
Surface barrier, 245
Surface effect, 154
Surface recombination, 164
Surface states, 66, 353

Tamm states, 353
Temperature dependence of lasers, 229
Temperature dependence of lasing delay, 383
Temperature effects, 27
Temperature rise in lasers, 235
Ternary alloys, 19
Thermal activation energy, 140, 167
Thermal quenching of luminescence, 377
Thermodynamics of emission, 193
Thermoluminescence, 371
Thermoreflectance, 401
Threshold for lasing, 218
Threshold for photoelectric emission, 288
Time dependence of emission, 152
Transition probability, 133, 147
Transmission, 93
Trap filling, optical, 382
Traps, 370
Trapping kinetics, 371
Triboluminescence, 387
Tribothermoluminescence, 389
Trichroism, 344
Tunneling, 46, 175
Tunneling-assisted absorption, 48
Tunneling-assisted emission, 151, 177
Tunneling through insulating layer, 247
Tunneling through Schottky barrier, 200
Tunneling to deep levels, 181
Turning point, 47
Two-photon absorption, 268

Uniaxial crystals, 338
Uniaxial strain, 26
Urbach's rule, 43

Vacancy, 8
Valence band, 2
Valleys of conduction band, 5

van Roosbroeck–Shockley relation, 108
Verdet constant, 346
Vibrational absorption of impurities, 80
Vibrational mode, 77
Visibility, 95
Voigt effect, 347

Wave guiding, 225
Wavelength modulation, 401
Wave vector, 3
Weierstrass sphere, 105
Work function, 288

Zeeman effect, 32
Zener breakdown, 203

A CATALOGUE OF SELECTED DOVER BOOKS IN ALL FIELDS OF INTEREST

AMERICA'S OLD MASTERS, James T. Flexner. Four men emerged unexpectedly from provincial 18th century America to leadership in European art: Benjamin West, J. S. Copley, C. R. Peale, Gilbert Stuart. Brilliant coverage of lives and contributions. Revised, 1967 edition. 69 plates. 365pp. of text.

21806-6 Paperbound \$3.00

FIRST FLOWERS OF OUR WILDERNESS: AMERICAN PAINTING, THE COLONIAL PERIOD, James T. Flexner. Painters, and regional painting traditions from earliest Colonial times up to the emergence of Copley, West and Peale Sr., Foster, Gustavus Hesselius, Feke, John Smibert and many anonymous painters in the primitive manner. Engaging presentation, with 162 illustrations. xxii + 368pp.

22180-6 Paperbound \$3.50

THE LIGHT OF DISTANT SKIES: AMERICAN PAINTING, 1760-1835, James T. Flexner. The great generation of early American painters goes to Europe to learn and to teach: West, Copley, Gilbert Stuart and others. Allston, Trumbull, Morse; also contemporary American painters—primitives, derivatives, academics—who remained in America. 102 illustrations. xiii + 306pp. 22179-2 Paperbound \$3.50

A HISTORY OF THE RISE AND PROGRESS OF THE ARTS OF DESIGN IN THE UNITED STATES, William Dunlap. Much the richest mine of information on early American painters, sculptors, architects, engravers, miniaturists, etc. The only source of information for scores of artists, the major primary source for many others. Unabridged reprint of rare original 1834 edition, with new introduction by James T. Flexner, and 394 new illustrations. Edited by Rita Weiss. 6⅝ x 9⅝.

21695-0, 21696-9, 21697-7 Three volumes, Paperbound \$15.00

EPOCHS OF CHINESE AND JAPANESE ART, Ernest F. Fenollosa. From primitive Chinese art to the 20th century, thorough history, explanation of every important art period and form, including Japanese woodcuts; main stress on China and Japan, but Tibet, Korea also included. Still unexcelled for its detailed, rich coverage of cultural background, aesthetic elements, diffusion studies, particularly of the historical period. 2nd, 1913 edition. 242 illustrations. lii + 439pp. of text.

20364-6, 20365-4 Two volumes, Paperbound \$6.00

THE GENTLE ART OF MAKING ENEMIES, James A. M. Whistler. Greatest wit of his day deflates Oscar Wilde, Ruskin, Swinburne; strikes back at inane critics, exhibitions, art journalism; aesthetics of impressionist revolution in most striking form. Highly readable classic by great painter. Reproduction of edition designed by Whistler. Introduction by Alfred Werner. xxxvi + 334pp.

21875-9 Paperbound \$3.00

ALPHABETS AND ORNAMENTS, Ernst Lehner. Well-known pictorial source for decorative alphabets, script examples, cartouches, frames, decorative title pages, calligraphic initials, borders, similar material. 14th to 19th century, mostly European. Useful in almost any graphic arts designing, varied styles. 750 illustrations. 256pp. 7 x 10. 21905-4 Paperbound $4.00

PAINTING: A CREATIVE APPROACH, Norman Colquhoun. For the beginner simple guide provides an instructive approach to painting: major stumbling blocks for beginner; overcoming them, technical points; paints and pigments; oil painting; watercolor and other media and color. New section on "plastic" paints. Glossary. Formerly *Paint Your Own Pictures*. 221pp. 22000-1 Paperbound $1.75

THE ENJOYMENT AND USE OF COLOR, Walter Sargent. Explanation of the relations between colors themselves and between colors in nature and art, including hundreds of little-known facts about color values, intensities, effects of high and low illumination, complementary colors. Many practical hints for painters, references to great masters. 7 color plates, 29 illustrations. x + 274pp.
20944-X Paperbound $3.00

THE NOTEBOOKS OF LEONARDO DA VINCI, compiled and edited by Jean Paul Richter. 1566 extracts from original manuscripts reveal the full range of Leonardo's versatile genius: all his writings on painting, sculpture, architecture, anatomy, astronomy, geography, topography, physiology, mining, music, etc., in both Italian and English, with 186 plates of manuscript pages and more than 500 additional drawings. Includes studies for the Last Supper, the lost Sforza monument, and other works. Total of xlvii + 866pp. 7⅞ x 10¾.
22572-0, 22573-9 Two volumes, Paperbound $12.00

MONTGOMERY WARD CATALOGUE OF 1895. Tea gowns, yards of flannel and pillow-case lace, stereoscopes, books of gospel hymns, the New Improved Singer Sewing Machine, side saddles, milk skimmers, straight-edged razors, high-button shoes, spittoons, and on and on . . . listing some 25,000 items, practically all illustrated. Essential to the shoppers of the 1890's, it is our truest record of the spirit of the period. Unaltered reprint of Issue No. 57, Spring and Summer 1895. Introduction by Boris Emmet. Innumerable illustrations. xiii + 624pp. 8½ x 11⅝.
22377-9 Paperbound $8.50

THE CRYSTAL PALACE EXHIBITION ILLUSTRATED CATALOGUE (LONDON, 1851). One of the wonders of the modern world—the Crystal Palace Exhibition in which all the nations of the civilized world exhibited their achievements in the arts and sciences—presented in an equally important illustrated catalogue. More than 1700 items pictured with accompanying text—ceramics, textiles, cast-iron work, carpets, pianos, sleds, razors, wall-papers, billiard tables, beehives, silverware and hundreds of other artifacts—represent the focal point of Victorian culture in the Western World. Probably the largest collection of Victorian decorative art ever assembled—indispensable for antiquarians and designers. Unabridged republication of the Art-Journal Catalogue of the Great Exhibition of 1851, with all terminal essays. New introduction by John Gloag, F.S.A. xxxiv + 426pp. 9 x 12.
22503-8 Paperbound $5.00

INCIDENTS OF TRAVEL IN YUCATAN, John L. Stephens. Classic (1843) exploration of jungles of Yucatan, looking for evidences of Maya civilization. Stephens found many ruins; comments on travel adventures, Mexican and Indian culture. 127 striking illustrations by F. Catherwood. Total of 669 pp.
20926-1, 20927-X Two volumes, Paperbound $5.50

INCIDENTS OF TRAVEL IN CENTRAL AMERICA, CHIAPAS, AND YUCATAN, John L. Stephens. An exciting travel journal and an important classic of archeology. Narrative relates his almost single-handed discovery of the Mayan culture, and exploration of the ruined cities of Copan, Palenque, Utatlan and others; the monuments they dug from the earth, the temples buried in the jungle, the customs of poverty-stricken Indians living a stone's throw from the ruined palaces. 115 drawings by F. Catherwood. Portrait of Stephens. xii + 812pp.
22404-X, 22405-8 Two volumes, Paperbound $6.00

A NEW VOYAGE ROUND THE WORLD, William Dampier. Late 17-century naturalist joined the pirates of the Spanish Main to gather information; remarkably vivid account of buccaneers, pirates; detailed, accurate account of botany, zoology, ethnography of lands visited. Probably the most important early English voyage, enormous implications for British exploration, trade, colonial policy. Also most interesting reading. Argonaut edition, introduction by Sir Albert Gray. New introduction by Percy Adams. 6 plates, 7 illustrations. xlvii + 376pp. 6½ x 9¼.
21900-3 Paperbound $3.00

INTERNATIONAL AIRLINE PHRASE BOOK IN SIX LANGUAGES, Joseph W. Bátor. Important phrases and sentences in English paralleled with French, German, Portuguese, Italian, Spanish equivalents, covering all possible airport-travel situations; created for airline personnel as well as tourist by Language Chief, Pan American Airlines. xiv + 204pp.
22017-6 Paperbound $2.25

STAGE COACH AND TAVERN DAYS, Alice Morse Earle. Detailed, lively account of the early days of taverns; their uses and importance in the social, political and military life; furnishings and decorations; locations; food and drink; tavern signs, etc. Second half covers every aspect of early travel; the roads, coaches, drivers, etc. Nostalgic, charming, packed with fascinating material. 157 illustrations, mostly photographs. xiv + 449pp.
22518-6 Paperbound $4.00

NORSE DISCOVERIES AND EXPLORATIONS IN NORTH AMERICA, Hjalmar R. Holand. The perplexing Kensington Stone, found in Minnesota at the end of the 19th century. Is it a record of a Scandinavian expedition to North America in the 14th century? Or is it one of the most successful hoaxes in history. A scientific detective investigation. Formerly *Westward from Vinland*. 31 photographs, 17 figures. x + 354pp.
22014-1 Paperbound $2.75

A BOOK OF OLD MAPS, compiled and edited by Emerson D. Fite and Archibald Freeman. 74 old maps offer an unusual survey of the discovery, settlement and growth of America down to the close of the Revolutionary war: maps showing Norse settlements in Greenland, the explorations of Columbus, Verrazano, Cabot, Champlain, Joliet, Drake, Hudson, etc., campaigns of Revolutionary war battles, and much more. Each map is accompanied by a brief historical essay. xvi + 299pp. 11 x 13¾.
22084-2 Paperbound $7.00